AF682746

Adaptation Biology and Medicine

Volume 6: Cell Adaptations and Challenges

SERIES EDITORS: **P.K. Singal and N. Takeda**

Adaptation Biology and Medicine

Volume 6: Cell Adaptations and Challenges

EDITORS

P. Wang • C.-H. Kuo

N. Takeda • P.K. Singal

Narosa Publishing House

New Delhi Chennai Mumbai Kolkata

Adaptation Biology and Medicine

504 pgs | 152 figs | 18 tbs

SERIES EDITORS: **P.K. Singal and N. Takeda**

EDITORS:

P. Wang
Department and Institute of Physiology
School of Medicine
National Yang-Ming University
Shih-Pai, Taipei, TAIWAN (11221), ROC

C.-H. Kuo
Laboratory of Exercise Biochemistry Taipei
Physical Education College 101, Sec 2,
Jhongcheng Rd.,Shihlin, Taipei City, 111,
TAIWAN, ROC

N. Takeda
Department of Internal Medicine
Aoto Hospital, Jikei University School of Medicine
Tokyo, Japan

P.K. Singal
Institute of Cardiovascular Sciences
R3021- 351 Tache Avenue
Winnipeg, MB R2H 2A6 CANADA

Copyright © 2011, Narosa Publishing House Pvt. Ltd.

NAROSA PUBLISHING HOUSE PVT. LTD.

22 Delhi Medical Association Road, Daryaganj, New Delhi 110 002
35-36 Greams Road, Thousand Lights, Chennai 600 006
306 Shiv Centre, Sector 17, Vashi, Navi Mumbai 400 703
2F-2G Shivam Chambers, 53 Syed Amir Ali Avenue, Kolkata 700 019

www.narosa.com

All rights reserved. No part of this publication may be reproduced, stored in a retrieval system, or transmitted in any form or by any means, electronic, mechanical, photocopying, recording or otherwise, without prior written permission of the publisher.

All export rights for this book vest exclusively with Narosa Publishing House Pvt. Ltd.
Unauthorised export is a violation of terms of sale and is subject to legal action.

ISBN 978-81-7319-935-6

Published by N.K. Mehra for Narosa Publishing House Pvt. Ltd.,
22 Delhi Medical Association Road, Daryaganj, New Delhi 110 002

Printed in India

Prof. Nirmal K. Ganguly, M.D., New Delhi, India

Nirmal Ganguly has been Vice President of the International Society for Adaptive Medicine and is now an active Board Member. He was the key organizer of the IV World Congress of ISAM, held in Chandigarh, India in 1995. His active interest and leadership gave a momentum in the early growth phase of the Society. He is now working as Distinguished Biotechnology Professor & Advisor in the Translational Health Science and Technology Institute, India. He is also the President of Asian Institute of Public Health and Technology. He was Director General, Indian Council of Medical Research, New Delhi for ten years ending in 2007. He was the President of National Academy of Medical Sciences, New Delhi. Prof.Ganguly has been very prolific in publishing and has more than 740 papers to his credit and has guided 130 Ph.D theses. His major research areas have been tropical diseases, cardiovascular diseases and diarrhoeal diseases. He has received more than 100 national and international honors. Most recently, Dr. Ganguly was honoured with the prestigious Padma Bhushan award given by the President of India on 26th of January, 2008 in the field of Medicine. Society is honored by his presence on its Board of Directors. Chandigarh conference will be his legacy in the history of the Society.

Preface

It has been known for some time that gradual adaptation to repeat exposures to different "body stresses" such as exercise, environmental factors, disease conditions, work related stress and drug treatments increases the resistance of the organism not only to that particular condition but also to other potentially harmful challenges. Such protective cross effects have stimulated the development of a new scientific era of adaptive medicine. International Society for Adaptive Medicine (ISAM) was established in 1990 in Freiberg to increase awareness as well as understanding of adaptive changes in human body organs. The Society aims at not only the study of basic mechanisms of adaptation but also their applications, for the treatment and prophylaxis of diseases. Another aim is to understand the science of failed adaptations or maladaptations which result in clinical conditions.

The Society also sponsors regional symposia as well as World Congresses around the world. At present, the Society is comprised of members worldwide and its meetings bring together biologists, clinicians and engineers. The 9th World Congress of the ISAM was held in Taipei, Taiwan during August 2-5, 2009. The previous World Congresses of the Society were held in Freiberg, Moscow, Tokyo, Chandigarh, Framingham, Lyon, San Diego and Moscow. The 10th World Congress is planned to be in Romania in 2012. The conference in Taipei brought together basic and clinical scientists who shared a common interest in discussing the current concepts for adaptation of different body systems in physiological and pathophysiological states and application of this knowledge in therapeutic approaches. More than 300 participants from many parts of the world highlighted major and rapidly expanding areas of adaptive medicine in 6 Landmark lectures, more than 80 symposia presentations. In addition, we had 62 posters and of these 31 participated in the Young Investigator's Award competitions. From the general competition, three students, namely Tran Duc Dung, Yu-Min Cho and Chung-Yu Chen, received the best poster awards and Dr. Pawan K. Singal, Winnipeg, was presented with the Lifetime Achievement Award.

The present book is based on papers selected from the invited symposia speakers as well as poster presentations. Although the entire book is one continuous discussion of issues as well as data dealing with the understanding of the biology of the adaptive responses and their potential applications in different therapies, it has been divided into three sections for an easy reading. Section A. **Adaptation in Cardiovascular Protections**, comprised of eleven chapters, highlights the mechanisms of adaptations in different components of the cardiovascular system. These initial papers are followed by eleven chapters in Section B on **Adaptive Programming in Systematic Functions** and in the remaining twelve papers presented in Section C, **Environmental and Lifestyle Adaptations**, the discussion focuses on the enhancement of subcellular adaptations by different approaches including lifestyle changes.

The book would be of great interest to biologists as well as clinicians interested in the overexpression of good adaptive responses and blunting the expression of maladaptive changes in improving the population health.

P. Wang, *Taipei*
C.-H. Kuo, *Taipei*
N. Takeda, *Tokyo*
P.K. Singal, *Winnipeg*

Acknowledgments

Owing to the efforts of the local organizing team as well as its leaders, the IX World Congress in Taipei was a resounding success. It had one of the strongest scientific programs, active participation by new members, and a most generous Taipei hospitality. Ample time was provided for informal exchanges among delegates in an array of carefully planned social events. All this was made possible by the efforts of many volunteers and financial support from different private and government organizations.

Grants from the National Yang-Ming University, Sport Affairs Council-Executive Yuan, National Science Council-Executive Yuan, Ministry of Economic Affairs-Executive Yuan, Department of Health-Taipei City Government, and Taipei Physical Education College are gratefully acknowledged. An active participation by the Institute of Cardiovascular Sciences was pivotal in the delivery of the conference. Equally important was the role played by Austin Wang (Taipei) in garnering the local support, as well as participation by the local scientists. We would also like to thank the following for their time and outstanding efforts in the organization of the various aspects of this Congress as well as publication of these proceedings: Hsiao-Fung Pu and Chien-Chen Lu and Mu-Tsung Chen (Taipei); and Mary Brown, Jennifer Froese and Dr. Thomas P. Thomas (Winnipeg).

P. Wang, *Taipei*
C.-H. Kuo, *Taipei*
N. Takeda, *Tokyo*
P.K. Singal, *Winnipeg*

Contents

Contributing Authors

A. Adameová
Department of Pharmacology and Toxicology
Faculty of Pharmacy
Slovak Academy of Sciences and Centre of Excellence for Cardiovascular Research
Bratislava, SLOVAK REPUBLIC

M. Al-Bader
Department of Physiology
Faculty of Medicine
Kuwait University, KUWAIT

A.G. Al-Banaw
Department of Medical Laboratory Sciences
Faculty of Allied Health Sciences
Kuwait University, KUWAIT

F. Al-Yatama
Department of Medical Laboratory Sciences
Faculty of Allied Health Sciences
Kuwait University, KUWAIT

Z. Ali
Functional Genomics Unit
Institute of Genomics and Integrative Biology
Mall Road, Delhi-110 007, INDIA

M. Anand-Srivastava
Department of Physiology
Faculty of Medicine
Universite de Montreal
C. P. 6128, Succ. Centre-ville
Montreal, Quebec, CANADA H3C 3J7

Y.V. Arkhipenko
Faculty of Fundamental Medicine
M.V. Lomonosov Moscow State University
Lomonosovsky pr., 31, kor. 5
Moscow 119192, RUSSIA

A. Arneja
Department of Internal Medicine
Faculty of Medicine
University of Manitoba
Winnipeg, Manitoba, CANADA

A. Bajaj
Institute of Cardiovascular Sciences
Department of Physiology
University of Manitoba
Winnipeg, Manitoba CANADA

J.E. Baker
Division of Cardiothoracic Surgery
Medical College of Wisconsin
8701 Watertown Plank Road
Milwaukee, WI 53226, USA

I.V. Bedareva
Faculty of Fundamental Medicine
M.V. Lomonosov Moscow State University
Lomonosovsky pr., 31, kor. 5
Moscow 119192, RUSSIA

D.A. Belousova
Department of Adaptive Biology and Medicine
Institute of General Pathology and Pathophysiology
Russian Academy of Medical Sciences
Baltijskaya Street 8
Moscow 125315 RUSSIA

A.V. Bolotova
Faculty of Fundamental Medicine
M.V. Lomonosov Moscow State University
Lomonosovsky pr., 31, kor. 5
Moscow 119192, RUSSIA

O.A. Borodovitsina
Department of Adaptive Biology and Medicine
Institute of General Pathology and Pathophysiology
Russian Academy of Medical Sciences
Baltijskaya Street 8
Moscow 125315 RUSSIA

A. Bouallegue
Laboratory of Cell Signaling
Montreal Diabetes Research Centre
Centre de Recherche de Centre hospitalier de l, Université de Montréal (CRCHUM) and
Department of Medicine
Université de Montréal
Montreal, Quebec, CANADA

I.G. Bryndina
Department of Pathophysiology
Izhevsk State Medical Academy
Kommunarov St. 281
Izhevsk, UR 426034, RUSSIA

M. Burge
Cancer Care Services
Royal Brisbane Hospital
Herston, QLD. 4029, AUSTRALIA

Y.-S. Chan
Division of Sports Medicine
Department of Orthopedic Surgery and Hyperbaric Oxygen Therapy Center
Chang-Gung Memorial Hospital and Chang Gung University College of Medicine
5 Fu-Hsing St. 333, Kweishan
Taoyuan, TAIWAN

D.G. Chang
Physical Medicine and Rehabilitation
Department of Orthopaedic Surgery
University of California, San Diego 350 Dickinson St., 8894
San Diego, CA 92103-8894, USA

T.-C. Chang
Department of Biochemistry
National Defense Medical Center P.O. Box 90048, Nei-Hu 114
Taipei, TAIWAN, ROC

W.-L. Chang
School of Pharmacy
National Defense Medical Center
Taipei, TAIWAN, ROC

S.-M. Chen
Committee for General Education
Shih Hsin University
Taipei, TAIWAN

M. Chia
Department of Physical Education and Sports Science
National Institute of Education
Nanyang Technological University,1 Nanyang walk, SINGAPORE 637616

J. C.-Y. Chung
Division of Cardiovascular Surgery
Toronto General Research Institute,
University Health Network
University of Toronto
Toronto, Ontario, CANADA

M. Contreras
College of Education and Human Development
Department of Health and Kinesiology
The University of Texas at San Antonio One UTSA Circle
San Antonio, Texas 78249 USA

W. Cooke
Laboratory for Applied Autonomic Neurophysiology
The University of Texas at San Antonio
San Antonio, Texas, 78249 USA

P. d'Alessio
University Paris Sud-11 Inserm Unit 935
ESTEAM Stem Cell Core Facility CHU
Paul Brousse 12, ave Paul Vaillant-Couturier 94807 Villejuif, FRANCE

D.K. Das
Cardiovascular Research Center
University of Connecticut School of Medicine Farmington, CT 06030-1110, USA

M. Descobeth
Department of Physiology Faculty of Medicine
Universite de Montreal C. P. 6128, Succ. Centre-ville
Montreal, Quebec, CANADA H3C 3J7

A. Doppalapudi
Division of Cardiothoracic Surgery
Department of Medicine
Medical College of Wisconsin Milwaukee, WI, USA

Victor E. Dosenko
Bogomoletz Institute of Physiology
National Academy of Sciences of Ukraine
Bogomoletz st. 4, Kiev, 01024, UKRAINE

H.F. Downey
University of North Texas Health Science Center Fort Worth, TX, USA

Tetyana I. Drevytska
Bogomoletz Institute of Physiology
National Academy of Sciences of Ukraine
Bogomoletz st. 4, Kiev, 01024, UKRAINE

B. Falcon
College of Education and Human Development
Department of Health and Kinesiology
The University of Texas at San Antonio One UTSA Circle
San Antonio, Texas 78249 USA

M.S. Faussone-Pellegrini
Department of Anatomy, Histology and Forensic Medicine,
Section of Histology University of Florence
Florence, ITALY

B.L. Fish
Division of Radiation Oncology
Medical College of Wisconsin
Milwaukee, WI, USA

D. Fogt
College of Education and Human Development
Department of Health and Kinesiology
The University of Texas at San Antonio One UTSA Circle
San Antonio, Texas 78249 USA

N.C. Frenkel
University College Utrecht
The Netherlands

P.N. Gerasimov
Department of Pathophysiology
Izhevsk State Medical Academy
Izhevsk, 426034 RUSSIA

E.L. Germanova
Institute of General Pathology and Pathophysiology
Russian Academia Medical Sciences
Moscow, RUSSIA

M. Gherghiceanu
'Victor Babes' National Institute of Pathology
Bucharest, ROMANIA

O. Gomes
Sao Francisco de Assis
Truth is Jesus Cardiovascular Foundation
Rua Jose do Patrocinio 522 - 31525160
Belo Horizonte - MG, BRAZIL

A.V. Goryacheva
Institute of General Pathology and Pathophysiology,
Moscow, RUSSIAN FEDERATION

N. Gurusamy
Cardiovascular Research Center
University of Connecticut School of Medicine
Farmington, CT 06030-1110, USA

G. Gutierrez
College of Education and Human Development
Department of Health and Kinesiology
The University of Texas at San Antonio One UTSA Circle
San Antonio, Texas 78249 USA

A.R. Hargens
Department of Orthopaedic Surgery
University of California
San Diego, CA 92103-8894 USA

L. Harmann
Department of Medicine
Medical College of Wisconsin
Milwaukee, WI, USA

S.T. Haworth
Department of Medicine
Medical College of Wisconsin
Milwaukee, WI, USA

J.W. Hopewell
Department of Clinical Oncology
The Churchill Hospital
Oxford, UNITED KINGDOM

S.-F. Huang
Department of Biochemistry
National Defense Medical Center P.O. Box 90048, Nei-Hu 114
Taipei, TAIWAN, ROC

Y.-C. Huang
Graduate Institute of Life Science
National Defense Medical Center
Taipei, TAIWAN, ROC V. Ishchuk State Institute of Gerontology
Academy of Medical Sciences Kiev, UKRAINE

T. Jones
College of Education and Human Development Department of Health and Kinesiology
The University of Texas at San Antonio
One UTSA Circle
San Antonio, Texas 78249 USA

J. Kalns
Hyperion Biotechnology Inc.
Brooks City Base
San Antonio, Texas, USA

G. Kaushik
Department of Biophysics
Post Graduate Institute of Medical Education and Research
Chandigarh-160012, INDIA

T. Kelly
School of Biology
Aristotle University
Thessaloniki, GREECE

K.L. Khanduja
Department of Biophysics
Post Graduate Institute of Medical Education and Research
Chandigarh-160012, INDIA

N. Khaper
Northern Ontario School of Medicine
Lakehead University
Thunder Bay, Ontario, CANADA

Y.I. Kirova
Institute of General Pathology and Pathophysiology
Russian Academia Medical Sciences
Moscow, RUSSIA

R.A. Komorowski
Division of Pathology Medical College of Wisconsin
Milwaukee, WI, USA

S. Kostin
Max-Planck Institute for Heart and Lung Research (W.C. Kerckhoff Institute)
Bad Nauheim, GERMANY

N.V. Kostina
Faculty of Fundamental Medicine
M.V. Lomonosov Moscow State University
Lomonosovsky pr., 31, kor. 5
Moscow 119192, RUSSIA

R.C. Kukreja
Department of Internal Medicine
Division of Cardiology, Box 980204 Virginia Commonwealth University
Richmond VA 23298, USA

C.-H. Kuo
Laboratory of Exercise Biochemistry Taipei Physical Education College 101, Sec 2, Jhongcheng Rd.,Shihlin, Taipei City, 111, TAIWAN, ROC

A. Lazou
School of Biology, Aristotle University Thessaloniki, GREECE

L.-C. Lee
Committee for General Education
Shih Hsin University Taipei, TAIWAN

R.-K. Li
MaRS Centre
Toronto Medical Discovery Tower Room 3-702, 101 College Street
Toronto, Ontario M5G 1L7 CANADA

X.A. Li
Division of Radiation Oncology
Medical College of Wisconsin
Milwaukee, WI, USA

Y. Li
Department of Physiology Faculty of Medicine
Universite de Montreal C. P. 6128, Succ. Centre-ville
Montreal, Quebec, CANADA H3C 3J7

Jheng-Yu Lin
Department of Biochemistry
National Defense Medical Center P.O. Box 90048, Nei-Hu 114
Taipei, TAIWAN, ROC

D. Lincoln
Entity Systems
Independent Research Foundation
Chapel Hill, QLD, 4069 AUSTRALIA

L.D. Lukyanova
Department of Bioenergetics and Hypoxia Problems
Institute of General Pathology andPathophysiology
Russian Academy of Medical Sciences
Baltijskaya Street 8
Moscow 125315, RUSSIA

S.V. Lyamina
Department of Adaptive Biology and Medicine
Institute of General Pathology and Pathophysiology
Russian Academy of Medical Sciences
Baltijskaya Street 8
Moscow 125315 RUSSIA

R.T. Mallet
University of North Texas Health Science Center
Fort Worth, TX, USA

K. Mallikarjuna
Division of Molecular Biology and Exercise Physiology
Department of Zoology
Sri Venkateswara University
Tirupati – 517 502, A. P, INDIA and
Laboratory of Exercise Biochemistry
Taipei Physical Education College
Taipei-11153, TAIWAN, ROC

I.Y. Malyshev
Laboratory of Stress and Adaptation
Department of Adaptive Biology and Medicine
Institute of General Pathology and Pathophysiology
Russian Academy of Medical Sciences
Baltijskaya Street 8
Moscow 125315 RUSSIA

I. Mankovska
Department of Hypoxia
Bogomoletz Institute of Physiology
National Academy of Sciences of Ukraine
Bogomoletz st. 4, Kiev, 01024, UKRAINE
E. B. Manukhina
Laboratory of Regulatory Mechanisms of Adaptation
Institute of General Pathology and Pathophysiology
Russian Academy of Medical Sciences
Baltijskaya Street 8,
Moscow 125315, RUSSIA

J. Matejíková,
Institute for Heart Research
Slovak Academy of Sciences
Dúbravská cesta 9, P.O.B 104
840 05 Bratislava, SLOVAK REPUBLIC

R.Q. Migrino
Division of Cardiovascular Medicine
Medical College of Wisconsin
Milwaukee, WI, USA

A. Mihic
Division of Cardiovascular Surgery
Toronto General Research Institute, University Health Network
University of Toronto
Toronto, Ontario, CANADA

M. Moghadasian
Department of Human Nutritional Sciences
University of Manitoba
Winnipeg, Manitoba CANADA

F. M. A. Mohammed
Department of Medical Laboratory Sciences
Faculty of Allied Health Sciences
Kuwait University, KUWAIT

E. A. Monastyrskaya
Department of Adaptive Biology and Medicine
Institute of General Pathology and Pathophysiology
Russian Academy of Medical Sciences
Baltijskaya Street 8, Moscow 125315 RUSSIA

J. E. Moulder
Division of Radiation Oncology
Medical College of Wisconsin
Milwaukee, WI, USA

I. Nikolsky
Institute of Genetic and Regenerative Medicine
Academy of Medical Sciences
Kiev, UKRAINE

M.A.Q. Pasha
Functional Genomics Unit
Institute of Genomics and Integrative Biology
Mall Road, Delhi-110 007, INDIA

C. Pathak
Department of Biophysics
Post Graduate Institute of Medical Education and Research
Chandigarh-160012, INDIA

O. Pecháòová
Institute of Normal and Pathological Physiology
Slovak Academy of Sciences and Centre of Excellence for Cardiovascular Research
Bratislava, SLOVAK REPUBLIC

L.M. Popescu
Department of Cellular and Molecular Medicine
"Carol Davila" University of Medicine and Pharmacy P.O. Box 35-29
Bucharest 35, ROMANIA

H.-F. Pu
Department of Physiology School of Medicine
National Yang-Ming University
Taipei 11221 TIAWAN

D.D. Ramdath
Guelph Food Research Centre
Agriculture and Agri-Food Canada
Guelph, Ontario, CANADA

T. Ravingerova
Institute for Heart Research
Slovak Academy of Sciences
Dúbravská cesta 9, P.O.B 104
840 05 Bratislava, SLOVAK REPUBLIC

S.K. Reddy
Division of Molecular Biology
Sri Venkateswara University
Tirupati – 517 502, INDIA

S.N. Reddy
Division of Molecular Biology and Exercise Physiology
Department of Zoology
Sri Venkateswara University
Tirupati – 517 502, A. P, INDIA

F.N. Salloum
Department of Internal Medicine
Division of Cardiology, Box 980204
Virginia Commonwealth University
Richmond VA 23298, USA

T.G. Sazontova
Faculty of Fundamental Medicine
M.V. Lomonosov Moscow State University
Lomonosovsky pr., 31, kor. 5
Moscow 119192, RUSSIA

T.V. Serebrovska
Bogomoletz Institute of Physiology
Bogomoletz str., 4
Kiev, 01024, UKRAINE

K.R. Shanmugam
Division of Molecular Biology and Exercise Physiology
Department of Zoology
Sri Venkateswara University
Tirupati – 517 502, A. P, INDIA

P.K. Singal
Institute of Cardiovascular Sciences
R3021- 351 Tache Avenue
Winnipeg, MB R2H 2A6 CANADA

R.K. Singal
Cardiac Sciences
Faculty of Medicine
University of Manitoba
Winnipeg, Manitoba, CANADA

F. Sinowatz
Institute of Veterinary Anatomy
University of Munich, GERMANY

A.K. Srivastava
Laboratory of Cell Signaling
Montreal Diabetes Research Centre
Centre de Recherche de Centre hospitalier de l,Université de Montréal (CRCHUM) and
Department of Medicine
Université de Montréal Research Centre,
CHUM – Angus Campus 2901, Rachel East Montréal, Québec H1W 4A4 CANADA

J.L. Strande
Department of Medicine
Division of Cardiovascular Medicine
Medical College of Wisconsin
Milwaukee, WI, USA

T. Stobdan
Functional Genomics Unit
Institute of Genomics and Integrative Biology
Mall Road, Delhi-110 007, INDIA

J. Su
Division of Cardiothoracic Surgery
Medical College of Wisconsin
Milwaukee, WI, USA

V. Sukhorukov
Head of the Department of General Pathology
Moscow Institute of Pediatrics and ChildrenSurgery
Taldomskaya str., 2
Moscow 127412, RUSSIA

Z. Sumbalová
Pharmacobiochemical Laboratory
Faculty of Medicine
Comenius University
Bratislava, SLOVAK REPUBLIC

A.F. Vanin
Institute of Chemical Physics
Moscow, RUSSIAN FEDERATION

N.N. Vasilieva
Department of Pathophysiology
Izhevsk State Medical Academy
Izhevsk, 426034 RUSSIA

P.S. Wang
Department and Institute of Physiology
School of Medicine
National Yang-Ming University
Shih-Pai, Taipei, TAIWAN (11221), ROC

R.D. Weisel
Division of Cardiovascular Surgery
Toronto General Research Institute,
University Health Network
University of Toronto
Toronto, Ontario, CANADA

F.A.C. Wiegant
Faculty of Science
Department of Biology
Cellular Dynamics
Utrecht University, P.O. Box 80056 3508 TB Utrecht, THE NETHERLANDS

S.-F. Yet
National Health Research Institutes
Institute of Cellular and Systems Medicine 35 Keyan Road, R2-5021
Zhunan Town, Miaoli County 35053
TAIWAN, ROC

P.-L. Yu
Department of Surgery
Taipei City Hospital
Taipei 10431TIAWAN and Department of Physiology
School of Medicine
National Yang-Ming University
Taipei 11221 TIAWAN

A. R. Yurasov
Faculty of Fundamental Medicine
M.V. Lomonosov Moscow State University
Lomonosovsky pr., 31, kor. 5
Moscow 119192, RUSSIA

Q. Zhang
Department of Orthopaedics
Sahlgrenska University Hospital
Per Dubbsgatan 14, 6th Floor
413 46 Göteborg, SWEDEN

Adaptation Biology and Medicine (Volume 6: Cell Adaptations and Challenges)
Editors: P. Wang, C.-H. Kuo, N. Takeda and P.K. Singal
Copyright © 2011 Narosa Publishing House Pvt. Ltd., New Delhi, India

1. Injectable Scaffolds for Cardiac Regeneration

Jennifer Chia-Ying Chung*, **Anton Mihic***, **Richard D. Weisel, Ren-Ke Li**

Division of Cardiovascular Surgery, Toronto General Research Institute, University Health Network, University of Toronto Toronto, Ontario, Canada

Abstract

Tissue engineering involves the development of biologically compatible materials for use in a therapeutic setting of tissue repair. An injectable scaffold for cardiac regeneration applications is an emerging trend within the field. Compared to bio-engineered tissue grafts, biocompatible injectable hydrogels may represent an advantageous delivery vehicle for cells, factors, and drugs, since invasive procedures are not required for the incorporation of these scaffolds into the damaged heart. Reconstruction of the left ventricle after a myocardial infarction requires a biomaterial scaffold that is mechanically compliant but strong enough to maintain the reconstruction, despite high intramyocardial pressures. New biodegradable biomaterials can be formulated to overcome the challenges of cell retention and survival associated with the direct intramyocardial injection of cells. Scaffolds can also promote neovascularization by providing favorable microenvironments for endogenous cell recruitment and may direct cell proliferation and differentiation in situ. In this review, we discuss the promise and challenges associated with cell therapy for cardiac regeneration and consider the current application of biomaterial scaffolds in vivo to help fulfill the potential of cell therapy.

Keywords: myocardial infarction, injectable scaffolds, tissue engineering, cardiac regeneration, hydrogel, biodegradable materials, cell therapy

Introduction

Myocardial infarction (MI) results from occlusion of one or more coronary arteries supplying the heart with oxygen-rich blood. MI is the leading cause of heart failure, which is a significant source of morbidity and mortality throughout the world.[1] Following an MI, a non-contractile and potentially expanding scar results from extensive cardiomyocyte cell death and inflammation, which then exerts a combined pressure and volume overload on the heart.[2] In the subsequent months, a dense, collagenous scar forms due to the proliferation of fibroblasts and endothelial cells in the infarct region. The heart undergoes compensatory ischemic ventricular remodeling in response to the

*Address for Correspondence: Ren-Ke Li, M.D., Ph.D. MaRS CentreToronto Medical Discovery Tower Room #3-702,101 College St. Toronto, Ontario M5G 1L7 Canada Phone: 416-581-7492, Fax: 416-581-7493 Email: renkeli@uhnres.utoronto.ca
*These authors contributed equally to this manuscript.

impaired contractile function.[3] While initially adaptive, increasing ventricular volume may progress to heart failure. Current medical therapies target some aspects of the compensatory response, including ACE inhibition, α-blockade and aldosterone blockade.[4] These treatments slow the progression of ventricular remodeling to a certain extent, but they do not address the irreversible loss of contractile cells that follows an MI, and they are unable to prevent the progressive ventricular dysfunction that occurs in patients with an ejection fraction of less than 40%.[5] Cardiac transplantation is a life-saving intervention for end-stage heart failure, and there are major limitations to this treatment due to donor organ shortages and complications from immunosuppression following heart transplantation.

The heart has a limited capacity for self-repair after injury.[6-8] The mediators of this endogenous repair response are bone marrow-derived stem cells and resident cardiac progenitor cells, which may partially restore heart function after an MI.[9-11] Cardiac progenitor cells were identified using the stem cell markers Lin- and c-kit+, as well as the transcription factors GATA-4 and Nkx2.5.[12] Although bone marrow-derived stem cells help to partially restore injured cardiomyocytes bordering the infarct region,[13] it is evident that resident cardiac progenitor cells are not of sufficient quantity or proliferative capacity to recover normal function. Therefore, novel therapies must be developed to supplement the endogenous regenerative capacity of the damaged heart, and to reverse detrimental ventricular remodeling.

The Promise of Cell Therapy

Cell therapy is a novel therapeutic option for the prevention of ventricular dilation and cardiac dysfunction following an MI. Over the last decade, a variety of somatic cells and stem cells have been used to regenerate the damaged myocardium in infarcted animal hearts.[14-16] Initial studies demonstrated that the implantation of cells could improve cardiac function, enhance regional perfusion and prevent ventricular remodeling. The specific mechanisms underlying the functional improvement remain unclear; however, recent studies propose that implanted cells produce significant paracrine effects.[17,18] Paracrine influences may enhance angiogenesis, reduce deleterious matrix remodeling, and improve endogenous cell recruitment to the infarct region, thereby limiting scar expansion and ventricular remodeling. More critically, the generation of new, functional myocardial tissue remains a substantial hurdle, as demonstrated by the limited clinical benefits of cell therapy.[19,20]

The challenges associated with cell therapy for cardiac regeneration are substantial, and defining the specific factors that prevent endogenous or transplanted cells from achieving myocardial regeneration remains an arduous task. The infarct region and accompanying scar tissue present an extraordinarily hostile microenvironment characterized by inflammation, ischemia, apoptosis, and fibrosis. Inflammation may promote angiogenesis and the signals required to recruit progenitor cells gradually becomes maladaptive. Obstruction of blood supply to the infarct region after an MI produces local ischemia, triggering rapid necrosis of myocardial cells, impairing the vasculature, and disrupting connective tissue. Ischemia further inhibits the regenerative capacity of the limited population of resident cardiac progenitor cells as well as other endogenous or transplanted cells. Although scar formation and fibrosis are initially beneficial because they prevent myocardial rupture, a thick fibrotic tissue may thin and dilate, and may impede the arrival of regenerating cells.

Approximately 90% of implanted cells of multiple types have been demonstrated to die within the first week after delivery to the infarct region of the heart.[21,22] Genetic manipulation of transplanted cells has been used to limit ischemic injury and apoptosis following cell transplantation. The genetic over-expression of anti-apoptotic proteins such as Akt prevents ventricular remodeling and restores cardiac function in a rat model of MI; however, it does not address the irreversible loss of cardiomyocytes.[23]

Tissue engineered cardiac grafts may help to restore contractile function by repopulating the damaged heart with new cardiomyocytes and may therefore be suitable for the treatment of a number of cardiac defects. Our group was the first to treat an infarct scar with fetal rat ventricular cardiomyocytes seeded in a sponge-like biodegradable gelatin scaffold (Gelfoam).[24,25] The beating cardiac grafts were cultured for 7 days and then surgically implanted onto cryoinjured rat myocardial scar tissue. After 5 weeks, we observed the survival of implanted cells, vascularization of the cardiac graft, and the formation of junctions between implanted and recipient heart cells. Alternatively, 3-dimensional porous alginate scaffolds have been used to deliver fetal rat cardiomyocytes to scar tissue after an MI.[26] Polyglycolic acid (PGA) scaffolds also support the growth and proliferation of rat neonatal cardiac cells.[27] Oxygen and nutrient diffusion limits have been identified as the major roadblocks to cell proliferation within the tissue constructs: diffusion of nutrients and oxygen in normal cardiac tissue is approximately 150 μm and 100 μm, respectively. [28-30] Clinically, the initial results of graft implantation in patients have been mixed.[31] Shin'oka and colleagues demonstrated positive clinical results after repairing congenital heart defects by implanting a vascular autograft composed of a composite biodegradable scaffold seeded with autologous bone marrow cells.[32,33] In contrast, a study using porcine heart valves treated with Synergraft decellularizing technology demonstrated limited cellular engraftment, evidence of early calcification, and significant morbidity and mortality.[34]

Injection is a more attractive delivery method than surgical graft insertion because it is less invasive, and it does not disturb the complicated, synchronized geometry of the heart wall.[35] The astonishing regenerative potential of stem cells has recently spurred numerous clinical trials utilizing a variety of cell types, with mostly mixed results (reviewed extensively by Segers and Lee).[36] A major challenge associated with cell transplantation via injection is poor and extremely variable cell retention, with up to ~90% of cells lost to the circulation from the site of injection.[21] A phase I trial of myoblast transplantation via injection through the epicardium found this procedure to be safe; however, clinical improvement was not demonstrated.[37] Injections are plagued with low seeding efficiency (58% at 0-1 hrs) and low cell viability (24% at 24 hrs).[38] To minimize the limitations of cell therapy, injectable scaffolds for in situ tissue engineering are being developed rapidly. Ideally, these injectable biomaterials would retain the cells within the injured heart, deliver growth factors, and provide a hospitable and customized local microenvironment.

Injectable Biomaterials for Cardiac Regeneration

The requirements for injectable biomaterial scaffolds are similar to those of cardiac grafts; however, the delivery methods are different. In vitro tissue engineering (such as scaffold seeding, cell sheets, or organ printing) can produce myocardial tissue prior to implantation. In contrast, injectable scaffolds conform to the geometry of heart and allow tissue formation to take place in situ. This less invasive approach ultimately has the potential to augment action potential conduction and yield synchronous tissue contraction while preventing harmful mechanical and electrical remodeling.

Injectable biomaterials for producing in situ scaffolds can be divided into two subcategories: synthetic and naturally-derived. The advantages of synthetic materials are that their properties are more easily manipulated, and they do not have issues with immunogenicity, which is critical for human clinical applications.[39] Synthetic materials, including polylactic acid (PLA), polyglycolic acid (PGA), and their copolymers can be custom-tailored to suit the specific mechanical properties of the tissue that they will replace. Adjusting the copolymer ratio can alter their degradation kinetics and importantly, these degradation products are non-cytotoxic.[40] Additionaly, the hydrophobic/hydrophilic ratio of synthetic polymer scaffolds can be adjusted to affect cell adhesion properties. One negative aspect of synthetic biomaterials is that their bulk degradation may ultimately produce local inflammation, which in turn may impair scaffold mechanical properties.[40]

In contrast to synthetics, naturally-derived biomaterials do not provoke foreign body inflammatory responses and they intrinsically express proteins and factors that encourage cell adhesion, proliferation and differentiation.[41] Naturally-derived biomaterials have garnered more attention than synthetics because they are themselves derivatives of extracellular matrix proteins. Collagen, fibrin, alginate, and Matrigel are the most-utilized biomaterials for injectable scaffolds in cardiac tissue engineering. These materials are not without their own set of limitations; for example, they are prone to provoking host immune responses and generally exhibit rapid degradation properties when compared with their synthetic counterparts. In vitro and in vivo assessments of several injectable biomaterial scaffolds and their properties for in situ cardiac regeneration are detailed below.

Biomaterial properties

The design of an ideal biomaterial for use as an injectable scaffold in the treatment of MI requires that several basic challenges are overcome. The material must have two phases: it must be fluidic at working conditions such that it is injectable and able to conform to the natural geometry of the myocardium and it must then form a porous gel matrix inside the body. For instance, alternating hydrophobic and hydrophilic peptides self-assemble into a matrix at physiologic pH.[42] Fibrinogen and thrombin are kept apart prior to injection and combine to form fibrin glue inside the body.[43] Synthetic hydrogels such as PLA and polyethylene glycol (PEG) can be engineered to crosslink when exposed to body temperature.[44] Temperature-responsive versions of Matrigel[35], a preparation of animal-derived basement membrane and chitosan[45], a biocompatible and biodegradable polymer composed of oligosaccharides of varying length, have also been characterized in the context of myocardial tissue engineering. In contrast, alginate can maintain low viscosity until injected into tissues where it encounters physiological calcium levels, at which time it cross-links, forming a hydrogel. Hydrogels can also achieve cross-linking via photoinitiators and UV light exposure, as in the case of composite PEG fibrin (PEGylated fibrin) hydrogels.[46,47] For the specific application of hydrogel injection into the infarct region, temperature-sensitive hydrogels likely offer the most utility and flexibility. An important consideration is the optimal gelation time, which must be reasonably short since continuous contractions of the heart are likely to squeeze out low viscosity hydrogels prior to complete gelation.

Injectable biomaterials should degrade at a controlled rate, without toxicity. Fibrin glue accomplishes this by degrading over a period of 7-10 days in vivo.[48] Many synthetic biomaterials have been developed to achieve controlled biodegradability for the purposes of tissue engineering. For example, PGA can be chemically modified to degrade over a timescale ranging from several weeks to months in vivo.[49] Although the ideal degradation rate has yet to be optimized, it should complement the rate at which cells, including extracellular matrix-secreting fibroblasts, replace the biomaterial. Biodegradation that occurs too quickly would result in premature cell and growth factor leakage, while degredation that occurs too slowly may impair the ability for cells to integrate and form a functional tissue construct in situ.

Controlling the mechanical properties of scaffold materials is of great clinical importance. Stiff or rigid biomaterials provide mechanical support and prevent ventricular dilation by functioning much like a left ventricular restraint device while the damaged myocardium is fragile and susceptible to myocyte slippage.[50] In silico modeling has demonstrated that injecting increasingly stiff, non-contractile substances increases scar thickness, reducing wall stress enough to attenuate deleterious remodeling.[51] In contrast, an excessively stiff material could cause diastolic dysfunction and impede the spontaneous contraction of implanted cardiomyocytes. Manipulation of in situ scaffold stiffness has been demonstrated using both composite PEGylated fibrinogen and self-assembling peptide scaffolds.[42,52] Collagen and alginate are mechanically unstable and collagen is difficult to handle.[53] In this context, one of the main advantages of employing synthetic hydrogels is the ability to readily manipulate their physical properties. Generally, the mechanical properties of an integrated hydrogel scaffold must be compatible with the specific application for cardiac regeneration. For example, the left ventricle is characterized by relatively high systolic pressures, necessitating a less compliant, stiffer biomaterial scaffold than may be employed for right ventricular repair.

In Vivo Utilization of Injectable Biomaterials

The injection of biomaterial hydrogels alone has produced promising results in animal models after MI. Fibrin glue was demonstrated to preserve wall thickness and ventricular size in rats.[50] Likewise, Matrigel was injected into injured heart muscle in a mouse model of MI.[54] In this study, an injectable scaffold conformed to the host heart's geometry and improved fractional shortening and wall thickness relative to a control injection at 2-weeks post MI. Landa *et al.* developed an injectable, bioabsorbable alginate hydrogel that undergoes phase transition in the body as it is exposed to increased calcium.[55] Using a rat model of MI, they demonstrated that alginate performed equally well - if not slightly better - compared to the injection of neonatal cardiomyocytes alone. The alginate scaffold retarded morphological and functional deterioration in both acute (one week) and chronic (two month) myocardial scars. Finally, a coronary-ligation model of MI in the rabbit was used to study a novel, synthetic, thermosensitive hydrogel, Dex-PCL-HEMA/PNIPAAm.[56] This material alone was demonstrated to prevent scar expansion and ventricular wall thinning, and was associated with improved heart function compared to a phosphate buffered saline injection control group. These results are encouraging and suggest that the materials alone may provide therapeutic benefits as an "off the shelf" treatment following MI. The injectable scaffolds contributed to myocardial mechanical and physical support, while providing an improved microenvironment, similar to the preexisting extracellular matrix. Nevertheless, the ultimate goal remains to replace scar tissue in the infarct region with viable and synchronously beating myocardium. Therefore, the potential for these injectable scaffolds to augment the delivery of cells to the myocardium may be their most promising attribute.

Cell retention

One of the greatest initial hurdles associated with cell therapy has been retention of cells after intramyocardial injection, since cardiac contractions relentlessly squeeze newly transplanted cells into the vasculature and away from the infarct region.[57] Injectable biomaterials that form scaffolds in situ are designed to trap cells in place after rapid cross-linking. Kofidis *et al.* demonstrated that Matrigel improves the engraftment rate of mouse embryonic stem cells in a mouse MI model.[54] In this study, the ratio of graft size to infarct size was dramatically increased - from 29.1% for cells injected alone, to 45.5% for cells injected with Matrigel. Long-term cell retention is also vastly improved using injectable scaffolds. Fibrin glue improves the retention and survival of skeletal myoblasts as evidenced by higher cellular densities versus cells alone at 5 weeks post-MI.[48] Histological examination revealed that cells delivered with fibrin glue penetrated into the fibrous scar tissue, whereas those treated with bovine serum albumen (control group) remained along the scar border. Similarly, Ryu *et al.* injected bone marrow mononuclear cells (BMMNCs) into ischemic, cryoinjured rat hearts after 21 days, either with or without fibrin matrix.[58] Fluorescently-labeled BMMNCs survived at 8 weeks after injection. Fibrin was associated with greater overall tissue regeneration and served as a cell implantation matrix supporting extensive neovascularization as demonstrated by significantly increased microvessel density. Appropriately-selected in situ scaffolds provide an initial boost to cell retention following injection and also improve long-term cell retention by providing a rich extracellular matrix environment for the attachment, proliferation, and survival of transplanted cells.

Cell type

As is the case in many fields of regenerative medicine, the selection of an optimal cell type for maximum long-term tissue regeneration is controversial. Cardiomyocytes, embryonic stem cells, skeletal myoblasts, autologous endothelial cells, and BMMNCs have all been investigated in the context of injectable scaffolds. The strengths and weaknesses of the various types have been reviewed extensively elsewhere.[59-61] Cell type may turn out to be crucial, as various materials function optimally with specific cell types. This hypothesis is supported by findings from experiments with platelet-derived growth factor BB (PDGF-BB), an important signaling molecule in cardiovascular development that is involved in cardiomyocyte survival pathways, angiogenesis, and the improvement of left ventricular function post-MI. PDGF-BB bound to self-assembling peptide nanofibers protected neonatal cardiomyocytes more effectively than skeletal myoblasts.[62,63] The mechanism is unclear, but the peptides were also unable to support myoblast growth in vitro. Similarly, a combination of skeletal myoblasts and fibrin injected into infarcted rat myocardium failed to significantly outperform fibrin alone or cells alone as evaluated by fractional shortening and ventricular wall thickness.[50] In contrast, Matrigel combined with embryonic stem cells resulted in a significant functional improvement as evidenced by increased fractional shortening.[35] Despite this, improvements in wall thickness and ventricular dilatation were not convincingly superior to the cells-only and Matrigel-only controls. Moreover, the grafts failed to develop a cardiomyocyte phenotype.

As the controversy over cell type is unlikely to be swiftly resolved, future research by proponents of specific cell types, or even combinations thereof, should compare outcomes with a variety of materials. Ideally, a theoretical framework for cell responsiveness to an assortment of materials should be developed. Important considerations include biocompatibility, biodegradability, mechanical properties and biomimicry – reflecting the properties of the extracellular matrix being replaced. Simultaneously, the current limitations of cell therapy must be overcome in order to enhance cell adhesion, survival, proliferation and functional integration (summarized in Fig. 1).

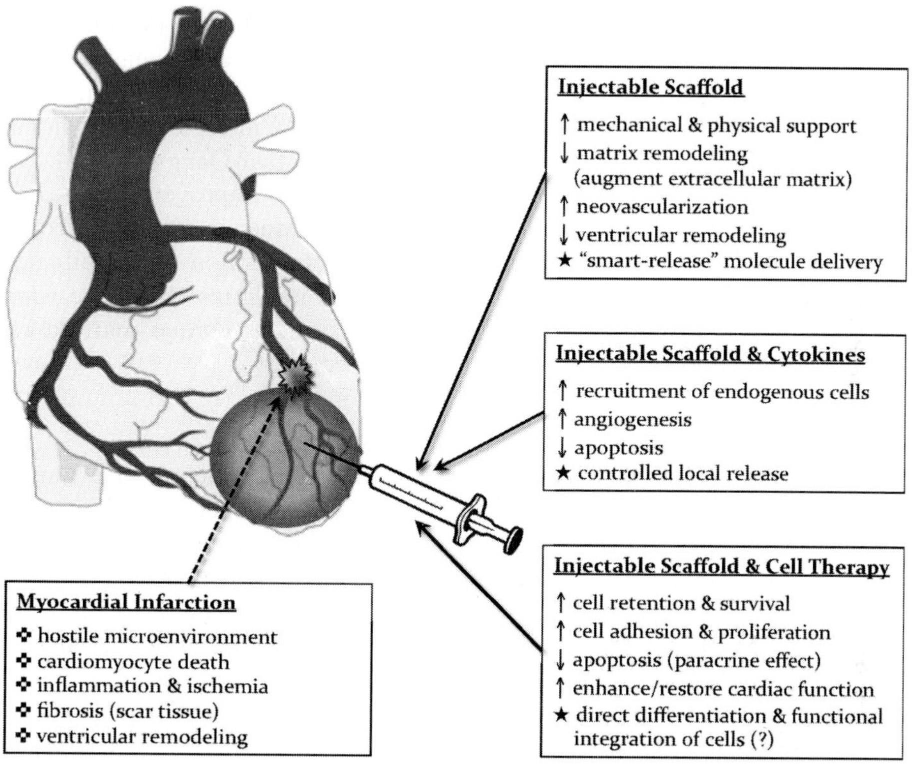

Fig. 1. Therapeutic benefits of injectable scaffolds with or without cytokines or cells following an MI.

Overcoming the Challenges of Cell Therapy with Injectable Scaffolds

Microenvironment

Cardiac tissue engineering provides a unique opportunity to augment the infarcted microenvironment through the use of injectable biomaterials. The biomaterials can serve as scaffolds for the delivery of specific factors into a hostile microenvironment. Laflamme *et al.* used Matrigel with highly purified human embryonic stem cells directed towards cardiomyocyte differentiation.[64] Although Matrigel with cells produced some functional improvement in a rat MI model, the simultaneous administration of a "pro-survival cocktail" produced significantly improved functional recovery compared to control. This cocktail included Matrigel and an assortment of cytokines and drugs, including factors that block mitochondrial pathways, induce ischemic preconditioning, activate the Akt pathway and inhibit caspases. The pro-survival cocktail protected grafted cells along multiple pathways and successfully increased their survival fourfold over cells plus Matrigel alone. Consequently, the combination treatment effectively attenuated ventricular dilation and preserved fractional shortening.

The results of Laflamme *et al.* stress the importance of the microenvironment. In developing the pro-survival cocktail, three primary obstacles causing death of implanted cells were identified: anoikis, ischemia, and inflammation-related factors. Anoikis occurs as a result of cell detachment from the extracellular matrix and induces programmed cell death. Naturally-derived materials such as collagen, fibronectin and laminin contain molecules that allow cell adhesion, such as the integrin adhesion peptide motifs RGD and YIGSR. Self-assembling peptides[65] and synthetic hydrogels[66] can be engineered to incorporate such cell attachment peptides. Peptide modification of alginate in rat MI models has illustrated the differential effects of these molecules. Tsur-Gang *et al.* utilized an alginate scaffold modified by RGD and YIGSR peptides, but observed no improvement in cardiac function over an unmodified alginate control when injected alone.[67] In contrast, Yu *et al.* demonstrated that alginate conjugated to RGD peptide alone improved heart function and facilitated a greater angiogenic response than non-modified alginate.[68] Standard measures should be adopted for the evaluation and comparison of peptides and scaffold materials, as unique scaffold chemical and biophysical traits may affect the therapeutic value of each peptide.

Angiogenesis

Pro-survival cocktails that address ischemia and inflammation-related factors are essential not only to protect implanted cells, but also to salvage cardiomyocytes in the ischemic area, thereby preventing the spread of scar tissue and limiting ventricular dilation. To overcome ischemia, the scarred myocardium must be revascularized. Tissue patches grown in vitro are associated with several challenges in this area.[59] Currently, there is an inadequate ability to design microvascular networks in vitro, thereby limiting the delivery of nutrients for diffusion and restricting the thickness of tissue constructs. If vascularization does not occur prior to transplantation, then the construct cannot maintain cell survival once it is delivered to the hypoxic scar tissue. Tissues grown in situ benefit from endogenous cardiac repair mechanisms such as neovascularization and cell migration into the infarct area. Theoretically, engineering an appropriately porous scaffold microenvironment rich with homing factors may yield a dynamic and interactive relationship between graft and host, thereby enhancing cell survival.

Some materials intrinsically promote angiogenesis. For instance, Matrigel contains a combination of growth factors and the degradation products of fibrin induce angiogenesis.[48] Huang *et al.* compared the angiogenic capacities of collagen I, Matrigel, and fibrin and found that these materials all performed similarly.[69] In particular, at 5 weeks after injection (6 weeks post-MI), all three biomaterials significantly increased the capillary density and functionality of newly formed arterioles, and indicated a trend towards increased myofibroblast infiltration in the infarct area.

Culturing endothelial cells within biomaterial scaffolds also induces angiogenesis. For example, endothelial cells can form functional capillaries within fibrin, as demonstrated by transmission electron microscopy.[70] However, the endothelial cells exhibit high rates of apoptosis within collagen

I, and require a high concentration of angiogenic factors for survival.[71,72] In contrast, endothelial cells in self-assembling peptides can maintain their viability and form capillary-like structures at 14 days in 10% fetal bovine serum, without the addition of vascular endothelial growth factor (VEGF) or basic fibroblast growth factor. Underlying the stark differences in cell survival between the self-assembling peptide and collagen scaffolds are alterations in gene expression. Self-assembling peptide scaffolds possess significantly higher levels of VEGF, anti-apoptotic factors and pro-survival factors.

Inflammation

Yet another barrier to cell survival within biomaterials is inflammation, whether induced as a consequence of the MI or as a reaction to the biomaterial itself. For example, an immunogenic reaction against self-assembling peptide nanofiber scaffolds was evidenced by infiltration of CD-68-expressing macrophages at one month after injection.[63] An exception to this phenomenon is autologous fibrin, which features the advantages of naturally-derived scaffolds that lack the dangers of immune rejection.[58] To suppress inflammation, systemic therapy may be considered. A number of immunosuppressant and anti-inflammatory agents, including dexamethasone, ibuprofen, complement-depleting cobra-venom factor and anti-asialo-GM1, have been utilized in the context of injectable biomaterials for cardiac regeneration.[64] However, translating these types of combined therapy to the clinical setting will likely involve numerous obstacles, as their safety for human patients has yet to be demonstrated.

Delivery of molecules

The most direct way to augment the microenvironment is to supplement the biomaterial with growth factors or drugs. To locally concentrate their delivery, the molecules can be tethered, non-covalently coupled, or engineered for "smart"-release from biomaterials.[73] "Smart" release systems refer to designs that respond to the environment - typically the matrix metalloproteinases present in myocardial remodeling or the mechanical stimuli of contracting myocardium. Insulin-like growth factor1 (IGF-1), a cardiomyocyte growth and survival factor, was tethered to the matrix in a biotinstreptavidin complex, and remained in the infarct area for 28 days.[74] This study demonstrated that the combination both in vitro and in vivo resulted in the activation of survival factor Akt and the decreased activation of apoptotic factor caspase-3. Recently, stromal cell-derived factor-1 (SDF-1), a chemotactic factor involved in the homing of hematopoietic stem cells, was incorporated into a similar biomaterial scaffold.[75] SDF-1 was engineered to be protease-resistant, preventing the formation of a neurotoxic remnant and simultaneously maintaining the protein's bioactivity. A nanofiber matrix sustained peptide release over a 7-day period, resulting in higher capillary density and the presence of cells with stem cell markers. This novel biomaterial was associated with improvements in left ventricular end-systolic volume and ejection fraction when injected immediately after an MI in rats. Recently, the effects of an erythropoietin-gelatin hydrogel drug delivery system were tested in a rabbit model of MI.[76] Infarct size and ventricular remodeling were reduced at 2 weeks and 2 months post-MI, supported by the increased expression of numerous factors that included the erythropoietin receptor (EPO-R), VEGF and matrix metalloproteinase-1 (MMP-1), as well as cardioprotective extracellular-signal-regulated kinase (ERK) and Akt signaling molecules.

There are considerable options for the incorporation of candidate cytokines, growth factors, and drugs into emerging biomaterials for cardiac regenerative therapy. Determining the ideal combination and optimizing the dosages of these molecules for each cell type or combination of cell types is, however, a daunting task. Studies to date, such as those described, are critical in their contribution to proof-of-concept developments in this field. Nevertheless, they are not sufficient for rapid progress. The development of a high throughput in vitro experimental setup would be ideal.

Hao *et al.* contribute an extra dimension of complexity to this picture by proposing the importance of a sequential and regulated release of specific factors.[77] Angiogenesis is a complex, multistep process, and so the delivery of multiple factors at consecutive stages should produce superior results. An alginate hydrogel containing two releasable factors was characterized by an earlier release profile for VEGF than for PDGF-BB, and was delivered via intramyocardial injection in a rat MI model.[77] This combination increased the formation of mature vessels and improved cardiac function as compared to the delivery of single factors.

Further work aimed at optimizing the matrix microenvironment will enhance the general functionality of injected grafts. Once sufficient cell viability is established, the final frontier will involve directing the proliferation and differentiation of implanted cells in a manner that is in syncronization with the surrounding host myocardium. This development would allow new heart tissue to become structurally, mechanically and electrically integrated with the existing myocardium, thereby completely regenerating the damaged heart.

Conclusions

The development of injectable scaffolds for cardiac regeneration is an emerging field. Progress has been made in terms of cell retention and survival within the scaffolds; however, the true potential for these materials to provide a customized microenvironment has yet to be exploited. Because they are injectable, the scaffolds offer a non-invasive clinical option, and in situ tissue engineering allows maximal communication between the graft and the host. Injectable scaffolds may help to fulfill the promise of cell therapy by overcoming hurdles related to tissue inflammation and angiogenesis, as well as cell retention, apoptosis, differentiation and proliferation. New biomaterials hold the potential to achieve the ultimate goal of cardiac tissue engineering: to generate tissue that behaves functionally like myocardial tissue, contracting in electrical and mechanical synchrony with the heart.

Acknowledgements

Our own research was primarily supported by grants to R-KL from the Heart and Stroke Foundation of Ontario and the Canadian Institutes of Health Research. R-KL is a Career Investigator of the Heart and Stroke Foundation of Canada, and holds a Canada Research Chair in cardiac regeneration. JC received a research grant from the Mach-Gaensslen Foundation of Canada. AM is the recipient of a CIHR Canada Graduate Scholarship. We thank Heather McDonald Kinkaid for editorial assistance.

References

1. Lloyd-Jones D, Adams R, Carnethon M, De Simone G, *et al.* Heart disease and stroke statistics-2009 update: a report from the American Heart Association Statistics Committee and Stroke Statistics Subcommittee. Circulation. 2009;119(3): e21-181.

2. Opie LH, Commerford PJ, Gersh BJ, and Pfeffer MA. Controversies in ventricular remodelling. Lancet. 2006;367(9507): 356-67.
3. Bardy GH, Lee KL, Mark DB, Poole JE, *et al.* Amiodarone or an implantable cardioverterdefibrillator for congestive heart failure. N Engl J Med. 2005;352(3): 225-37.
4. McMullen NM and Pasumarthi KB. Donor cell transplantation for myocardial disease: does it complement current pharmacological therapies? Can J Physiol Pharmacol. 2007;85(1): 1-15.
5. McMurray J and Pfeffer MA. New therapeutic options in congestive heart failure: Part I. Circulation. 2002;105(17): 2099-106.
6. Deb A, Wang S, Skelding KA, Miller D, *et al.* Bone marrow-derived cardiomyocytes are present in adult human heart: A study of gender-mismatched bone marrow transplantation patients. Circulation. 2003;107(9): 1247-9.
7. de Weger RA, Verbrugge I, Bruggink AH, van Oosterhout MM, *et al.* Stem cell-derived cardiomyocytes after bone marrow and heart transplantation. Bone Marrow Transplant. 2008;41(6): 563-9.
8. Laflamme MA, Myerson D, Saffitz JE, and Murry CE. Evidence for cardiomyocyte repopulation by extracardiac progenitors in transplanted human hearts. Circ Res. 2002;90(6): 634-40.
9. Orlic D, Kajstura J, Chimenti S, Jakoniuk I, *et al.* Bone marrow cells regenerate infarcted myocardium. Nature. 2001;410(6829): 701-5.
10. Beltrami AP, Barlucchi L, Torella D, Baker M, *et al.* Adult cardiac stem cells are multipotent and support myocardial regeneration. Cell. 2003;114(6): 763-76.
11. Oh H, Bradfute SB, Gallardo TD, Nakamura T, *et al.* Cardiac progenitor cells from adult myocardium: homing, differentiation, and fusion after infarction. Proc Natl Acad Sci U S A. 2003;100(21): 12313-8.
12. Beltrami AP, Urbanek K, Kajstura J, Yan SM, *et al.* Evidence that human cardiac myocytes divide after myocardial infarction. N Engl J Med. 2001;344(23): 1750-7.
13. Hsieh PC, Segers VF, Davis ME, MacGillivray C, *et al.* Evidence from a genetic fate-mapping study that stem cells refresh adult mammalian cardiomyocytes after injury. Nat Med. 2007;13(8): 970-4.
14. Schuster MD, Kocher AA, Seki T, Martens TP, *et al.* Myocardial neovascularization by bone marrow angioblasts results in cardiomyocyte regeneration. Am J Physiol Heart Circ Physiol. 2004;287(2): H525-32.
15. Li RK, Jia ZQ, Weisel RD, Mickle DA, *et al.* Cardiomyocyte transplantation improves heart function. Ann Thorac Surg. 1996;62(3): 654-60; discussion 660-1.
16. Tomita S, Li RK, Weisel RD, Mickle DA, *et al.* Autologous transplantation of bone marrow cells improves damaged heart function. Circulation. 1999;100(19 Suppl): II247-56.
17. Formigli L, Perna AM, Meacci E, Cinci L, *et al.* Paracrine effects of transplanted myoblasts and relaxin on post-infarction heart remodelling. J Cell Mol Med. 2007;11(5): 1087-100.
18. Uemura R, Xu M, Ahmad N, and Ashraf M. Bone marrow stem cells prevent left ventricular remodeling of ischemic heart through paracrine signaling. Circ Res. 2006;98(11): 1414-21.

19. Menasche P. Cell-based therapy for heart disease: a clinically oriented perspective. Mol Ther. 2009;17(5): 758-66.
20. Dimmeler S and Zeiher AM. Cell therapy of acute myocardial infarction: open questions. Cardiology. 2009;113(3): 155-60.
21. Muller-Ehmsen J, Whittaker P, Kloner RA, Dow JS, *et al.* Survival and development of neonatal rat cardiomyocytes transplanted into adult myocardium. J Mol Cell Cardiol. 2002;34(2): 107-16.
22. Zhang M, Methot D, Poppa V, Fujio Y, *et al.* Cardiomyocyte grafting for cardiac repair: graft cell death and anti-death strategies. J Mol Cell Cardiol. 2001;33(5): 907-21.
23. Mangi AA, Noiseux N, Kong D, He H, *et al.* Mesenchymal stem cells modified with Akt prevent remodeling and restore performance of infarcted hearts. Nat Med. 2003;9(9): 1195-201.
24. Li RK, Jia ZQ, Weisel RD, Mickle DA, *et al.* Survival and function of bioengineered cardiac grafts. Circulation. 1999;100(19 Suppl): II63-9.
25. Li RK, Yau TM, Weisel RD, Mickle DA, *et al.* Construction of a bioengineered cardiac graft. J Thorac Cardiovasc Surg. 2000;119(2): 368-75.
26. Leor J, Aboulafia-Etzion S, Dar A, Shapiro L, *et al.* Bioengineered cardiac grafts: A new approach to repair the infarcted myocardium? Circulation. 2000;102(19 Suppl 3): III56-61.
27. Bursac N, Papadaki M, Cohen RJ, Schoen FJ, *et al.* Cardiac muscle tissue engineering: toward an in vitro model for electrophysiological studies. Am J Physiol. 1999;277(2 Pt 2): H433-44.
28. Radisic M, Yang L, Boublik J, Cohen RJ, *et al.* Medium perfusion enables engineering of compact and contractile cardiac tissue. Am J Physiol Heart Circ Physiol. 2004;286(2): H507-16.
29. Carrier RL, Rupnick M, Langer R, Schoen FJ, *et al.* Perfusion improves tissue architecture of engineered cardiac muscle. Tissue Eng. 2002;8(2): 175-88.
30. Zimmermann WH, Schneiderbanger K, Schubert P, Didie M, *et al.* Tissue engineering of a differentiated cardiac muscle construct. Circ Res. 2002;90(2): 223-30.
31. Tang GH, Fazel S, Weisel RD, Van Arsdell GS, *et al.* Cardiovascular tissue engineering therapy: so near, so far? Ann Thorac Surg. 2005;79(6): 1831-3.
32. Shin'oka T, Imai Y, and Ikada Y. Transplantation of a tissue-engineered pulmonary artery. N Engl J Med. 2001;344(7): 532-3.
33. Shin'oka T. [Clinical results of tissue-engineered vascular autografts seeded with autologous bone marrow cells]. Nippon Geka Gakkai Zasshi. 2004;105(8): 459-63.
34. Simon P, Kasimir MT, Seebacher G, Weigel G, *et al.* Early failure of the tissue engineered porcine heart valve SYNERGRAFT in pediatric patients. Eur J Cardiothorac Surg. 2003;23(6): 1002-6; discussion 1006.
35. Kofidis T, de Bruin JL, Hoyt G, Lebl DR, *et al.* Injectable bioartificial myocardial tissue for large-scale intramural cell transfer and functional recovery of injured heart muscle. J Thorac Cardiovasc Surg. 2004;128(4): 571-8.

36. Segers VF and Lee RT. Stem-cell therapy for cardiac disease. Nature. 2008;451(7181): 937-42.
37. Dib N, Michler RE, Pagani FD, Wright S, *et al.* Safety and feasibility of autologous myoblast transplantation in patients with ischemic cardiomyopathy: four-year follow-up. Circulation. 2005;112(12): 1748-55.
38. Reinecke H and Murry CE. Taking the death toll after cardiomyocyte grafting: a reminder of the importance of quantitative biology. J Mol Cell Cardiol. 2002;34(3): 251-3.
39. Chen QZ, Harding SE, Ali NN, Lyon AR, *et al.* Biomaterials in cardiac tissue engineering: Ten years of research survey. Materials Science and Engineering. 2008;R59(
40. Giraud MN, Armbruster C, Carrel T, and Tevaearai HT. Current state of the art in myocardial tissue engineering. Tissue Eng. 2007;13(8): 1825-36.
41. Jawad H, Lyon AR, Harding SE, Ali NN, *et al.* Myocardial tissue engineering. Br Med Bull. 2008;87(31-47.
42. Narmoneva DA, Oni O, Sieminski AL, Zhang S, *et al.* Self-assembling short oligopeptides and the promotion of angiogenesis. Biomaterials. 2005;26(23): 4837-46.
43. Zhang P, Zhang H, Wang H, Wei Y, *et al.* Artificial matrix helps neonatal cardiomyocytes restore injured myocardium in rats. Artif Organs. 2006;30(2): 86-93.
44. Lee J, Bae YH, Sohn YS, and Jeong B. Thermogelling aqueous solutions of alternating multiblock copolymers of poly(L-lactic acid) and poly(ethylene glycol). Biomacromolecules. 2006;7(6): 1729-34.
45. Lu WN, Lu SH, Wang HB, Li DX, *et al.* Functional improvement of infarcted heart by coinjection of embryonic stem cells with temperature-responsive chitosan hydrogel. Tissue Eng Part A. 2009;15(6): 1437-47.
46. Shapira K, Dikovsky D, Habib M, Gepstein L, *et al.* Hydrogels for cardiac tissue regeneration. Biomed Mater Eng. 2008;18(4-5): 309-14.
47. Shapira-Schweitzer K, Habib M, Gepstein L, and Seliktar D. A photopolymerizable hydrogel for 3-D culture of human embryonic stem cell-derived cardiomyocytes and rat neonatal cardiac cells. J Mol Cell Cardiol. 2009;46(2): 213-24.
48. Christman KL, Vardanian AJ, Fang Q, Sievers RE, *et al.* Injectable fibrin scaffold improves cell transplant survival, reduces infarct expansion, and induces neovasculature formation in ischemic myocardium. J Am Coll Cardiol. 2004;44(3): 654-60.
49. Gunatillake PA and Adhikari R. Biodegradable synthetic polymers for tissue engineering. Eur Cell Mater. 2003;5(1-16; discussion 16.
50. Christman KL, Fok HH, Sievers RE, Fang Q, *et al.* Fibrin glue alone and skeletal myoblasts in a fibrin scaffold preserve cardiac function after myocardial infarction. Tissue Eng. 2004;10(3-4): 403-9.
51. Wall ST, Walker JC, Healy KE, Ratcliffe MB, *et al.* Theoretical impact of the injection of material into the myocardium: a finite element model simulation. Circulation. 2006;114(24): 2627-35.

52. Shapira-Schweitzer K and Seliktar D. Matrix stiffness affects spontaneous contraction of cardiomyocytes cultured within a PEGylated fibrinogen biomaterial. Acta Biomater. 2007;3(1): 33-41.
53. Christman KL and Lee RJ. Biomaterials for the treatment of myocardial infarction. J Am Coll Cardiol. 2006;48(5): 907-13.
54. Kofidis T, Lebl DR, Martinez EC, Hoyt G, *et al.* Novel injectable bioartificial tissue facilitates targeted, less invasive, large-scale tissue restoration on the beating heart after myocardial injury. Circulation. 2005;112(9 Suppl): I173-7.
55. Landa N, Miller L, Feinberg MS, Holbova R, *et al.* Effect of injectable alginate implant on cardiac remodeling and function after recent and old infarcts in rat. Circulation. 2008;117(11): 1388-96.
56. Wang T, Wu DQ, Jiang XJ, Zhang XZ, *et al.* Novel thermosensitive hydrogel injection inhibits post-infarct ventricle remodelling. Eur J Heart Fail. 2009;11(1): 14-9.
57. Menasche P. Skeletal myoblasts as a therapeutic agent. Prog Cardiovasc Dis. 2007;50(1): 7-17.
58. Ryu JH, Kim IK, Cho SW, Cho MC, *et al.* Implantation of bone marrow mononuclear cells using injectable fibrin matrix enhances neovascularization in infarcted myocardium. Biomaterials. 2005;26(3): 319-26.
59. Laflamme MA and Murry CE. Regenerating the heart. Nat Biotechnol. 2005;23(7): 845-56.
60. Mazhari R and Hare JM. Advances in cell-based therapy for structural heart disease. Prog Cardiovasc Dis. 2007;49(6): 387-95.
61. Franco D, Moreno N, and Ruiz-Lozano P. Non-resident stem cell populations in regenerative cardiac medicine. Cell Mol Life Sci. 2007;64(6): 683-91.
62. Hsieh PC, Davis ME, Gannon J, MacGillivray C, *et al.* Controlled delivery of PDGF-BB for myocardial protection using injectable self-assembling peptide nanofibers. J Clin Invest. 2006;116(1): 237-48.
63. Dubois G, Segers VF, Bellamy V, Sabbah L, *et al.* Self-assembling peptide nanofibers and skeletal myoblast transplantation in infarcted myocardium. J Biomed Mater Res B Appl Biomater. 2008;87(1): 222-8.
64. Laflamme MA, Chen KY, Naumova AV, Muskheli V, *et al.* Cardiomyocytes derived from human embryonic stem cells in pro-survival factors enhance function of infarcted rat hearts. Nat Biotechnol. 2007;25(9): 1015-24.
65. Zhang S. Fabrication of novel biomaterials through molecular self-assembly. Nat Biotechnol. 2003;21(10): 1171-8.
66. Schmedlen RH, Masters KS, and West JL. Photocrosslinkable polyvinyl alcohol hydrogels that can be modified with cell adhesion peptides for use in tissue engineering. Biomaterials. 2002;23(22): 4325-32.
67. Tsur-Gang O, Ruvinov E, Landa N, Holbova R, *et al.* The effects of peptide-based modification of alginate on left ventricular remodeling and function after myocardial infarction. Biomaterials. 2009;30(2): 189-95.

68. Yu J, Gu Y, Du KT, Mihardja S, *et al.* The effect of injected RGD modified alginate on angiogenesis and left ventricular function in a chronic rat infarct model. Biomaterials. 2009;30(5): 751-6.
69. Huang NF, Yu J, Sievers R, Li S, *et al.* Injectable biopolymers enhance angiogenesis after myocardial infarction. Tissue Eng. 2005;11(11-12): 1860-6.
70. Chekanov V, Akhtar M, Tchekanov G, Dangas G, *et al.* Transplantation of autologous endothelial cells induces angiogenesis. Pacing Clin Electrophysiol. 2003;26(1 Pt 2): 496-9.
71. Satake S, Kuzuya M, Ramos MA, Kanda S, *et al.* Angiogenic stimuli are essential for survival of vascular endothelial cells in three-dimensional collagen lattice. Biochem Biophys Res Commun. 1998;244(3): 642-6.
72. Kuzuya M, Satake S, Ramos MA, Kanda S, *et al.* Induction of apoptotic cell death in vascular endothelial cells cultured in three-dimensional collagen lattice. Exp Cell Res. 1999;248(2): 498-508.
73. Davis ME, Hsieh PC, Grodzinsky AJ, and Lee RT. Custom design of the cardiac microenvironment with biomaterials. Circ Res. 2005;97(1): 8-15.
74. Davis ME, Hsieh PC, Takahashi T, Song Q, *et al.* Local myocardial insulin-like growth factor 1 (IGF-1) delivery with biotinylated peptide nanofibers improves cell therapy for myocardial infarction. Proc Natl Acad Sci U S A. 2006;103(21): 8155-60.
75. Segers VF, Tokunou T, Higgins LJ, MacGillivray C, *et al.* Local delivery of protease-resistant stromal cell derived factor-1 for stem cell recruitment after myocardial infarction. Circulation. 2007;116(15): 1683-92.
76. Kobayashi H, Minatoguchi S, Yasuda S, Bao N, *et al.* Post-infarct treatment with an erythropoietin-gelatin hydrogel drug delivery system for cardiac repair. Cardiovasc Res. 2008;79(4): 611-20.
77. Hao X, Silva EA, Mansson-Broberg A, Grinnemo KH, *et al.* Angiogenic effects of sequential release of VEGF-A165 and PDGF-BB with alginate hydrogels after myocardial infarction. Cardiovasc Res. 2007;75(1): 178-85.

Adaptation Biology and Medicine (Volume 6: Cell Adaptations and Challenges)
Editors: P. Wang, C.-H. Kuo, N. Takeda and P.K. Singal
Copyright © 2011 Narosa Publishing House Pvt. Ltd., New Delhi, India

2. Telocytes and Heart Renewing

L.M. Popescu[1,2], Mihaela Gherghiceanu[2], S. Kostin[3], Maria Simonetta Faussone-Pellegrini[4]

[1]Department of Cellular and Molecular Medicine, 'Carol Davila' University of Medicine and Pharmacy, Bucharest, Romania

[2]'Victor Babes' National Institute of Pathology, Bucharest, Romania

[3]Max-Planck Institute for Heart and Lung Research (W.C. Kerckhoff Institute), Bad Nauheim, Germany

[4]Department of Anatomy, Histology and Forensic Medicine, Section of Histology, University of Florence, Florence, Italy

Abstract

The regenerative mechanisms balance the degenerative changes, which naturally occur in the heart, and it seems reasonable to explore both sides of the process as a whole. Probably, there are rigorous controlling mechanisms that prevent an excessive or anarchic growing of the heart and an argument of this possibility is the low frequency of primitive cardiac tumours. The most conceivable reason for the inefficient regeneration of cardiac tissue after injury is that telocytes change their phenotype in hypoxic condition and their network is broken. Considering all these aspects, a better molecular profiling of telocytes could help us to develop an efficient cellular therapy for heart disease.

Keywords: cardiac renewing, telocytes (TCs), interstitial Cajal-like cells (ICLCs), cardiac stem cell niches (CSCN), TCs as nurse cells, TCs in pathology, towards the future

Introduction

The Rationale for Term "Telocytes" (And "Telopodes")

We previously described (2005) a new type of cell which became known as "interstitial Cajal-like cell" (acronym ICLC). We named these cells ICLC because they apparently seemed to be similar with the well-known gastrointestinal interstitial cells of Cajal (ICC). However, ICLC appeared under the electron microscope and using immunohistochemistry to be completely different from ICC. Therefore, we coined the term TELOCYTE (TC) for these cells and TELOPODES (TP) for their prolongations, to prevent further confusion with other interstitial cell types (e.g. fibroblasts, fibrocytes, fibroblast-like cells, mesenchymal cells and interstitial Cajal cells).

To see the entire "story" visit www.telocytes.com and read: "Telocytes – A case of serendipity: the winding way from Interstitial Cells of Cajal (ICC), via Interstitial Cajal-Like Cells (ICLC) to

*Address for Correspondence: L.M. POPESCU, M.D., Ph.D.,Department of Cellular and Molecular Medicine,"Carol Davila" University of Medicine and Pharmacy,P.O. Box 35-29, Bucharest 35, Romania E-mail: LMP@jcmm.org

*The initial title of this chapter was "Cardiac ICLC (Interstitial Cajal-Like Cells) and Heart Renewing", but this appears now to be obsolete. Actually, this change of nomenclature does not alter the scientific content of this chapter any way.

telocytes" by Popescu LM and Faussone-Pellegrini MS, published in J Cell Mol Med. 2010;14:729-740.

Cardiac Renewing

While spectacular medical advances have been made over the past decades, heart disease remains a leading cause of death worldwide. A therapeutic promise for cardiovascular disease comes from stem cell therapy using the regenerative potential of the organism. Studies in animal models indicate that transplantation of adult or embryonic stem cells [1-9] has the potential to improve the cardiac function after ischemic injury, but long-term engraftment of stem cells is not a solved problem and clear demonstration of *in vivo* cardiomyogenesis from transplantation of stem/progenitor cells does not exist [6-11].

A recent study [12] proved without doubt that human cardiomyocytes renew, about 50 % of cardiomyocytes being replaced during a normal life span. Cardiomyocyte renewal appear as a slow, age-dependent process with rates of about 1% per year at age 25 declining to 0.4% per year at age 75 [12]. It is now accepted that heart regeneration occurs [13] but the basis of the regenerative mechanism remain to be identified.

Mature cardiomyocytes divide after myocardial infarction [13] but mitosis seems to be an uncommon event in adult mammalian heart [14, 15]. Mitosis merely involves karyokinesis without

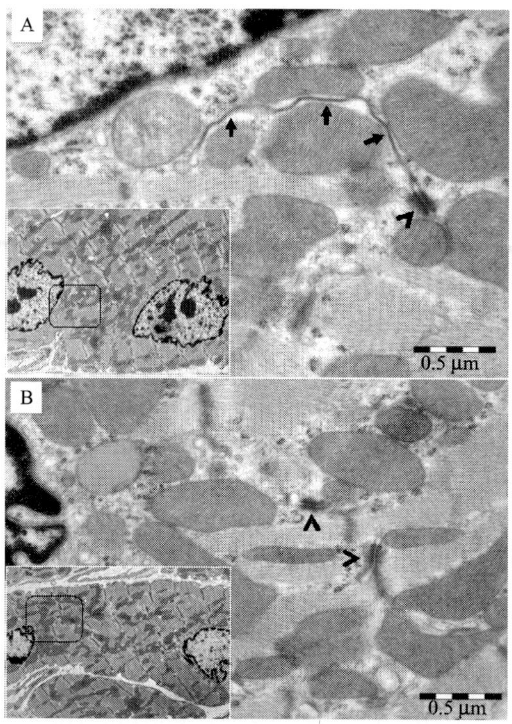

Fig. 1. A and B. Electron micrographs of mouse ventricle. Binucleate cardiomyocytes have intracellularly located gap junctions (arrows) and desmosome-like structures (arrowheads). These junctions located in between nuclei suggest that cytokinesis could be possible but the separation of the two contracting cells is a slow process (insets).

cytokinesis and rare binucleate cardiomyocyte can be found in the adult normal heart. The presence of 'intracellular' gap junctions and desmosome-like structures located in between the two nuclei (Fig. 1) suggests cytokinesis but, because of particular structure of cardiomyocytes, cytokinesis must be a very slow process. Therefore, *mitosis* of adult cardiomyocytes seems to be an *ineffective mechanism* for cardiac repair after injuries. The heart has traditionally been viewed as a postmitotic organ and this might be true considering that cardiomyocytes undergo hypertrophy, rather than hyperplasia, in response to hemodynamic stress.

It is proved now that heart is self-renewing organ [12] but the source of renewing cardiomyocytes is not precisely identified [16]. In the last decade, different types of cardiac stem cells have been isolated from adult heart [16-25] and they seem to be responsible for the baseline turnover of myocardial cells under normal conditions. The distinct phenotypic differences between cardiomyocyte progenitors and their lineage diversity [26], their (ultra)structural profile and their precise localization in the heart remain to be established. We are just beginning to rediscover this highly organised, heterocellular organ – the heart.

Telocytes (TCs)

A distinct type of interstitial cell - **telocyte** - has been recently documented in human and mammalian myocardium and it was initially named interstitial Cajal-like cell (ICLC) [27]. The presence of TCs has been reported first time by Hinescu and Popescu in 2005 in human atrial myocardium [28] and, presently, a few other papers showed that these peculiar cells make a three-dimensional network in the entire heart - in atria [29] ventricles [30], myocardial sleeves [31, 32], and subepicardium [33, 34].

Hitherto, the ultrastructural features (Fig. 2) define the cardiac telocytes: characteristic long (up to 100 μm), moniliform (with thin segments - up to 100nm) cell processes with a dichotomous branching pattern; caveolae and coated vesicles; mitochondria, relatively well developed smooth and rough endoplasmic reticulum; intermediate and thin filaments, microtubules and undetectable thick filaments; no basal lamina; a single cilium; gap junctions; organized in a network - labyrinthic system (overlapping cytoplasmic processes); cell-to-cell contacts with working cardiomyocytes; close vicinity with capillaries and nerve fibres; close contacts with other interstitial cells (e.g. macrophages, mast cells, lymphocytes).

Cardiac TCs are located merely around blood vessels and between fascicles of cardiomyocytes forming an incomplete sheath around them. TCs interconnect with each other through point cell-to-cell contacts or gap junctions thus forming a three-dimensional cellular network into the atria, ventricle, myocardial sleeves and subepicardium. In addition, the TCs are usually in proximity to nerves and make close contact [35] with other interstitial cells (mainly immunocompetent cells as macrophages, mast cell and lymphocytes). Moreover, heterocellular communication occurs between the cardiac TCs and cardiomyocytes through free or budding shed vesicles, exocytotic multi-vesicular bodies and direct contacts [27, 36]. Attachment plaques often connect the TCs to the extracellular matrix suggesting an important role in tensional integrity of the cardiac tissue.

Currently, there is not an established panel of antibodies for TC immunophenotyping. We previously reported that antibodies used against CD117/c-kit resulted in a weakly and inconstantly immunostaining of TCs and most of CD34 and vimentin positive cells might be superposed [29, 30].

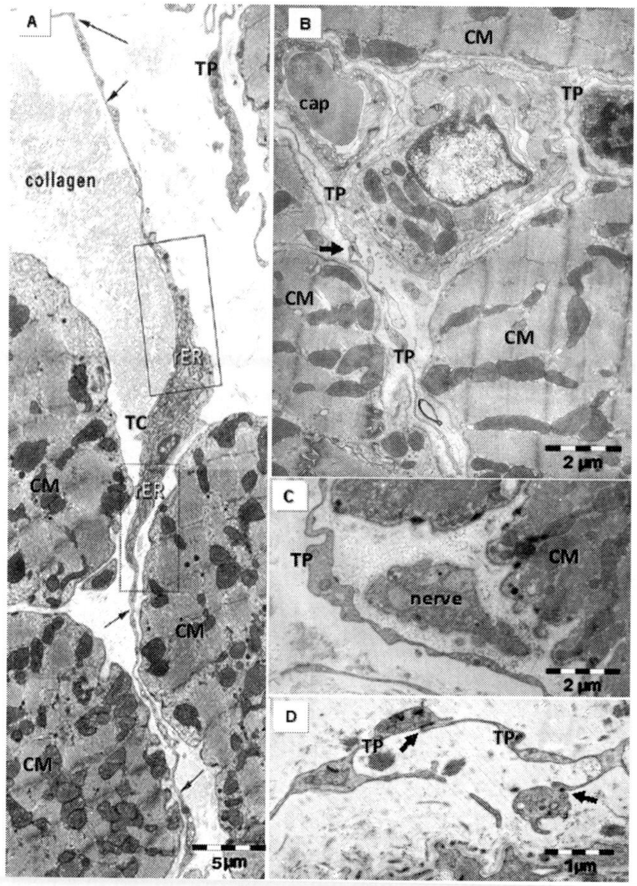

Fig. 2. Transmission electron microscopy (TEM). These images show that telocytes (TC) make a 3D network within the heart. A. TCs in the right atrial interstitium (rat). Note the characteristic aspects of the telopodes (TP, arrows): very long and thin cellular elongations, with uneven calibre (moniliform aspect). Square marks show portions of the TC cell body containing rough endoplasmic reticulum (rER). Commonly, only segments of TP are visible on TEM images. B. TC in the ventricular interstitium (rat). TP form a network in the narrow spaces between cardiomyocytes (CM) and surround blood capillaries (cap). Arrow indicates a gap junction between telopodes. C. TC in close vicinity with a nerve in the right atrial interstitium (rat). D. Small gap junction (arrows) connecting telopodes. Panel A and B - image reprinted with kind permission from Ref. [27].

Immunohistochemistry alone cannot provide precise discrimination among cells in interstitium since technical artefacts could be encountered. For instance, c-kit positive cells with different functions and with distinct ultrastructural features should not be discriminated. They might be stem cells, mast cells [37, 38], or this new type of cells we have described in the mammalian heart – telocytes. All these cells can appear on immunofluorescence as round cell and it is obvious for stem cells and mast cells but this could be likewise for TCs either. The TCs has a thin rim of cytoplasm around nucleus and their long processes are often up to 100 nm thick thus falling below light microscopy resolution. Therefore, thin TC processes - telopodes - can be undetectable under the immunofluorescence microscope. Laser scanning microscopy (LSM) is a more accurate technique

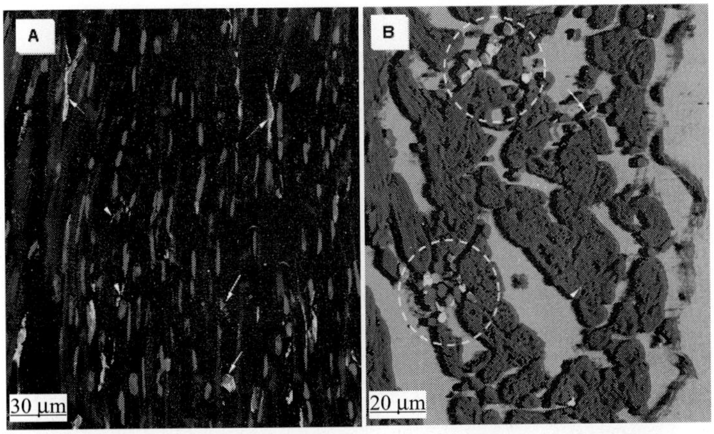

Fig. 3. Confocal microscopy shows c-kit positive cells in human left ventricular myocardium. A. **Maximum projection laser scanning microscopy image of longitudinally oriented fibres. Dashed arrows indicate cellular bodies of c-kit positive cells (green) from where long processes begin and run parallel to the long axis of cardiomyocytes. Straight arrows disclose transversally sectioned c-kit cells at the level of cellular bodies showing a very thin perinuclear rim of the green signal. Arrowheads indicate thin long processes intermingling with cardiomyocytes.** B. **Laser scanning microscopy image of transversally and obliquely oriented fibres. Three-dimensional shadow projection image shows clusters (yellow round marks) or isolated round c-kit positive cells (arrow). Note small green dots (arrowhead) which represent transversally sectioned cellular processes of c-kit positive cells located in close proximity to cardiomyocytes. F-actin is red stained with phalloidin conjugated with TRITC and nuclei are blue stained with DAPI.**

which revealed interstitial c-kit positive cells with long processes have been easily seen intermingling with cardiomyocytes (Fig. 3A) on longitudinally sectioned tissue. Their network-like distribution is consistent with electron microscopy findings. Intriguingly, LSM images in section of transversally and obliquely oriented fibres show clusters or isolated round c-kit positive cells in human left ventricular myocardium (Fig. 3B) as they appear in the cardiac stem cell niches [39, 40]. Isolated c-kit positive cells with long processes were also observed under epicardial mesothelial cells (Fig. 4A, B) and TEM shows that these cells are identifiable as TCs (Fig. 4C).

Although the exact phenotypic identity and function of the TCs are still unknown, they have a peculiar ultrastructural appearance because of their extremely long and moniliform processes with a dichotomous branching pattern that give to the TC a dendritic appearance. The characteristic processes of TC sare able to extend to a long distance (tens of micrometers) and they are connected to each other thereby generating a cellular network in the entire heart. We hypothesized TCs have an integrative function supporting the three-dimensional organization of the cardiomyocytes, their physiology and their renewal [27]. One of the most remarkable aspects of our own data is that TCs are located together with cardiomyocyte progenitors in a particular environment - cardiac stem cell niche [34].

Cardiac Stem Cell Niches (CSCN)

The recent discovery in the adult mammalian myocardium of a small c-kit positive cell population with the phenotype, behaviour and regenerative potential of cardiac stem and progenitor cells has proposed these cells as the most appropriate for source for cell therapy [16]. Clusters of c-kit, MDR1,

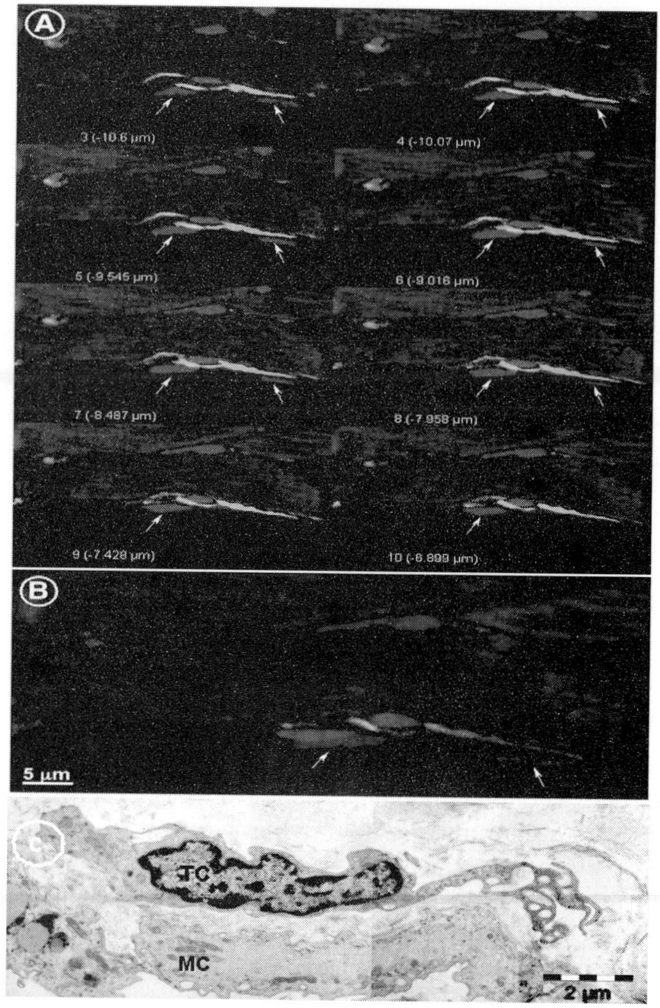

Fig. 4. A and B. Confocal microscopy. C-kit positive cells in human subepicardial area. A. **Gallery of LSM images obtained at 0.529 µm steps showing the spatial distribution of c-kit positive cells with long processes under mesothelial cells (arrows) in human myocardium.** B. **Maximum projection of LSM images shown in panel** A. **Mesothelial cells are indicated with arrows.** C. **TEM. One TC located under the mesothelial cells (MC) of the human epicardium.**

sca-1 positive cells have been described in selected areas of the heart, mostly in atria and apex, but also in ventricles [37, 39, 40]. These clusters seem to form functional units, which are supposed to be cardiac stem cell niches in murine [39] and human [40] hearts.

Using electron microscopy, we have found cardiomyocyte progenitors (CMPs) in adult mouse heart present only in the subepicardium surrounding coronary artery at the emergence point from aorta (Fig. 5A). In this area, the connective tissue is loose and particularly rich in capillaries and interstitial cells (Fig. 5B, C). Mast cells, adipocytes, macrophages, lymphocytes, monocytes, fibroblasts and nerve cells reside here along with numerous TCs and cardiomyocyte progenitors.

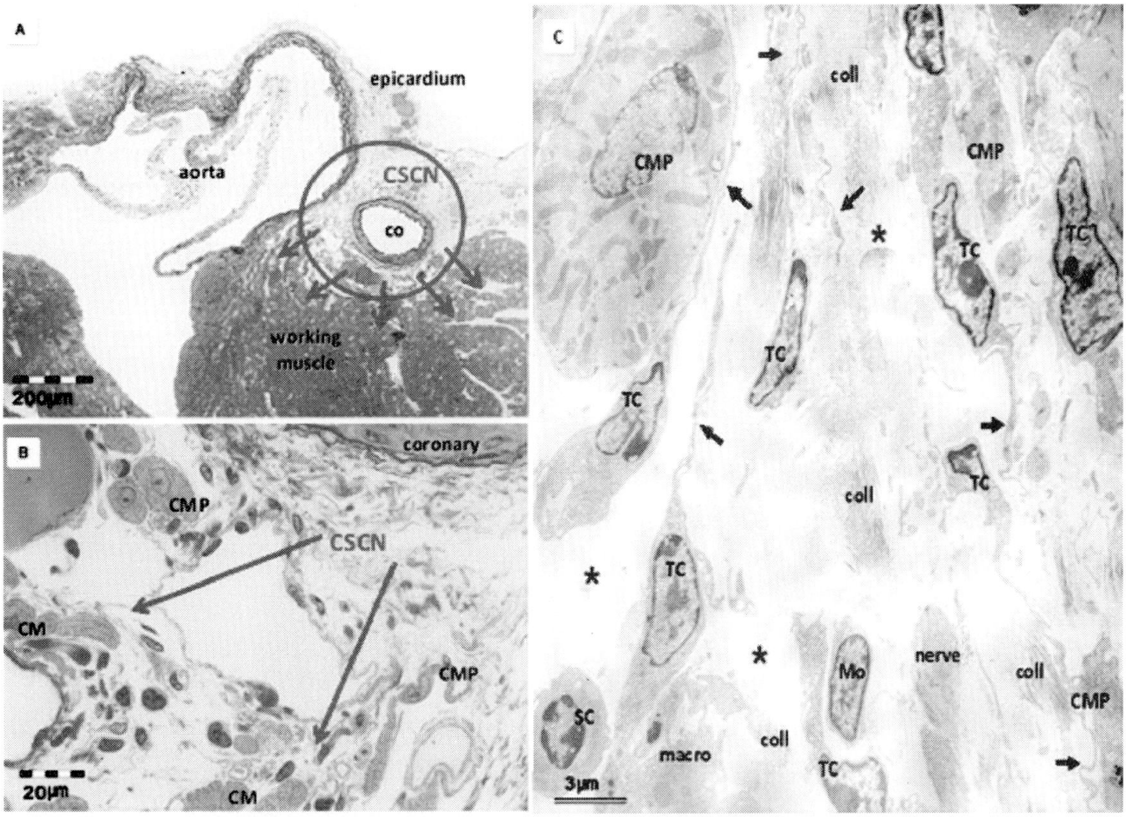

Fig. 5. Cardiac stem cell niche in mouse heart. A and B. **Light microscopy of toluidine-blue stained semithin sections of epicardium from 16 weeks old mice.** A. **Cardiac stem cell niche (CSCN) located in the subepicardial area, around the coronary artery and next to the emergence point from aorta (red circle). Image reprinted with kind permission from Ref. [34]** B. **Details from the lower side of the encircled area in A. CSCN is located at the boundary of working cardiomyocytes (CM). Cardiac renewing could be a vectorial process (red arrows), newly formed cardiomyocytes (CMP) integration taking place in the periphery of the cardiac muscle (CM).** C. **Transmission electron microscopy image of the area depicted in B shows the diversity of cellular populations in CSCN. Telocytes (TC) nerves, macrophages (macro), monocytes (Mo), putative stem cells (SC) and cardiomyocyte progenitors (CMP). Telopodes (arrows) spread into the extracellular matrix making a 3D network. The extracellular matrix contains bundles of collagen fibres (coll) and large areas of electron-lucent ground substance (asterisks).**

The identity of putative resident cardiac stem cells is still unknown and their ultrastructural phenotype should be of particular interest for their identification in situ. Their precise localization in complex architecture of the myocardium must be equally important for our understanding of regenerative physiology. Significant evidences in this direction come from one of a few studies using electron microscopy to characterize stem cells. This study has clearly shown that ultrastructural differences exist among human mesenchymal stem cells with similar immunophenotype and this is probably related with their differentiation potential [41].

We have identified two types of **putative cardiac stem cells** (**CSC**) using exclusion criteria. **Type I** of SC are small oval cells (around 10 micrometers) with large nuclei with scattered chromatin, very few mitochondria and smooth endoplasmic reticulum cisternae, free ribosomes and a few short cellular processes (Fig. 6A). **Type II** of SC are also oval, small cells (10-15 micrometers) but with large and indented nuclei with less scattered chromatin and prominent nucleoli, smooth and rough endoplasmic reticulum cisternae, very few mitochondria, clusters of small and clear vacuoles generating a multilocular appearance (similar to bone marrow mesenchymal stem cells [41]), a few dense granules and lysosomes and filopodia (Fig. 6B). These CSCs were found in the loose connective tissue surrounding coronary in the epicardial area.

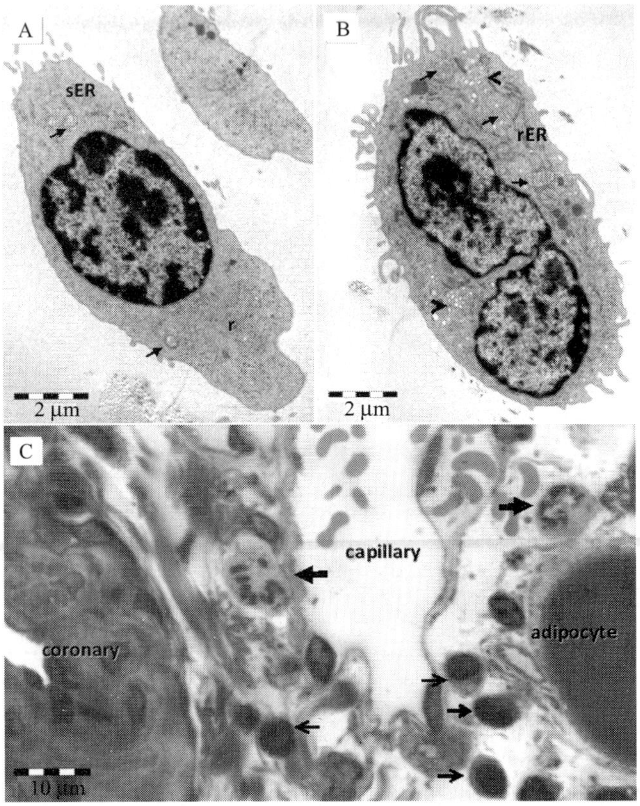

Fig. 6. Putative cardiac stem cells (SC) in the adult mouse epicardial stem cell niche. A and B. **Transmission electron microscopy.** A. **Type I of SC: small oval cells (around 10 µm) with large nucleus with scattered chromatin, very few mitochondria (arrows), smooth endoplasmic reticulum cisternae (sER) and free ribosomes (r).** B. **Type II of SC: cells with an indented nucleus with less scattered chromatin and prominent nucleoli, smooth and rough endoplasmic reticulum (rER), mitochondria (arrows), clusters of small and clear vacuoles generating a multilocular appearance (arrowheads), few dense granules, lysosomes and numerous processes.** C. **Light microscopy of toluidine-blue stained semithin section in subepicardial area. Mitosis (arrows) of small round cells (thin arrows) in a cardiac stem cell niche next to a branch of the coronary artery.**

In contrast with stem cells, the **cardiomyocyte progenitors (CMPs)** are recognizable without difficulty. These cells displayed typical ultrastructural features of immature cardiomyocytes [34] including: high nucleo-cytoplasmic ratio, unorganized bundles of filaments, lipid droplets, intracytoplasmic dense bodies similar to primordial Z lines, intracytoplasmic desmosome-like structures (primordial intercalated discs) and cortical leptomeric fibrils. Moreover, these cells have large mitochondria, numerous caveolae and a continuous basal lamina.

The CMPs in the niche showed to be in different stages of development (Fig. 7) and some CMPs showed autophagic and oncotic changes suggesting that not all of these cells are fully developed and integrated into the cardiac muscle. All these suggest that there is a pool of cells in a continuous process of commitment and differentiation and that **cardiac renewing is a dynamic process.**

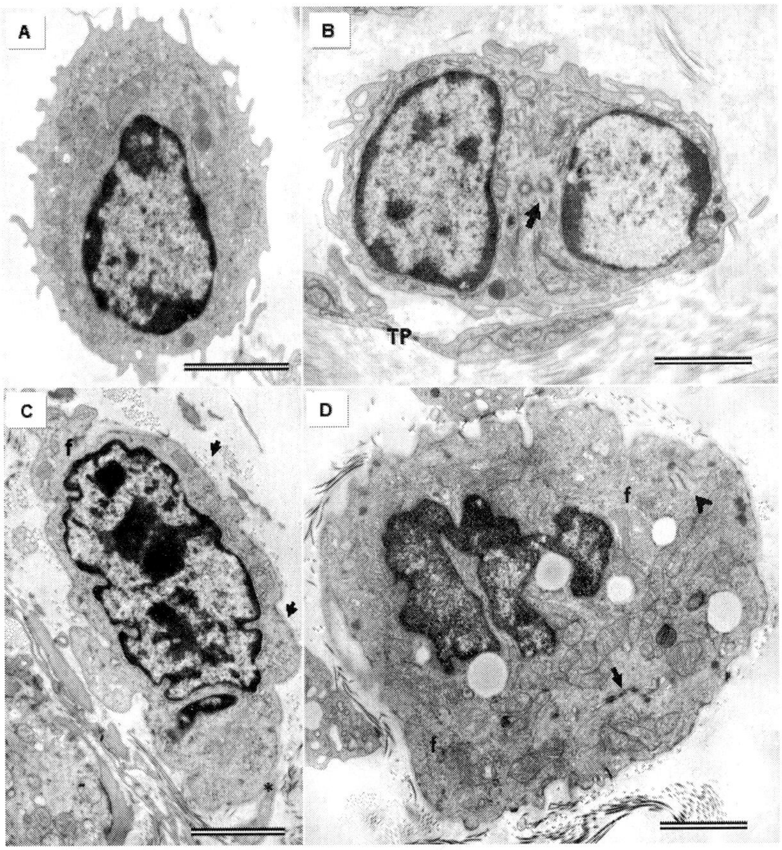

Fig. 7. **Transmission electron microscopy. Cardiomyocyte progenitors (CMP) in the adult mouse epicardial stem cell niche.** A. **Cardiac stem cell type II.** B. **Cardiac stem binucleate cell with more elaborate endoplasmic reticulum and centrosomes, probably in late telophase in mitosis.** C. **Cardiomyocyte progenitor with immature features: a high nucleo-cytoplasmatic ratio, an incomplete basal lamina (arrows); numerous mitochondria and ribosomes; few endoplasmic reticulum cisternae, small patches of filaments (f) and numerous caveolae (asterisks).** D. **More mature cardiomyocyte progenitor with large mitochondria, lipid droplets, bundles of unorganized filaments (f) and dense bodies, leptomeric fibrils (arrowhead), and desmosome-like intracytoplasmatic structures. Scale bar 2 μm.**

The CMPs discovered around coronary reside here in a particular environment with all characteristic attributes of the stem-cell niche. The concept of a stem-cell niche was proposed by the first time by Schofield in 1978 [42] who hypothesized the existence of a specialized place for localization and regulation of hematopoietic stem cells in the bone marrow. Since then, much progress has been achieved [43] and it is now generally accepted that adult stem cells reside in specialized small areas named stem-cell niches that specifically coordinate adult tissues renewal in several organs [44-53].

A stem cell niche (SCN) as a coherent and dynamic structure is considered essential for stem cells survival, commitment and differentiation. Irrespective of location, a SCN capable to house stem cells entails few constitutive elements with regulatory properties [45, 46, 50]: stromal supporting cells (also named: nurse cells, niche cells or supporting cells) that interact directly with the stem cells and with each other; extracellular matrix proteins that provide mechanical signals to the niche; blood vessels that transmit systemic signals and bring circulating stem cells if needed; neural inputs that might communicate distant physiological cues to the stem cell microenvironment.

The cardiac stem cell niche we have reported in the adult mouse heart has the complete set of elements (Fig. 8): cardiomyocyte progenitors, stromal interconnected cells, blood vessels, nerve cells and extracellular matrix. Stromal cell of the niche nurse CMPs and fulfil all the criteria for TCs.

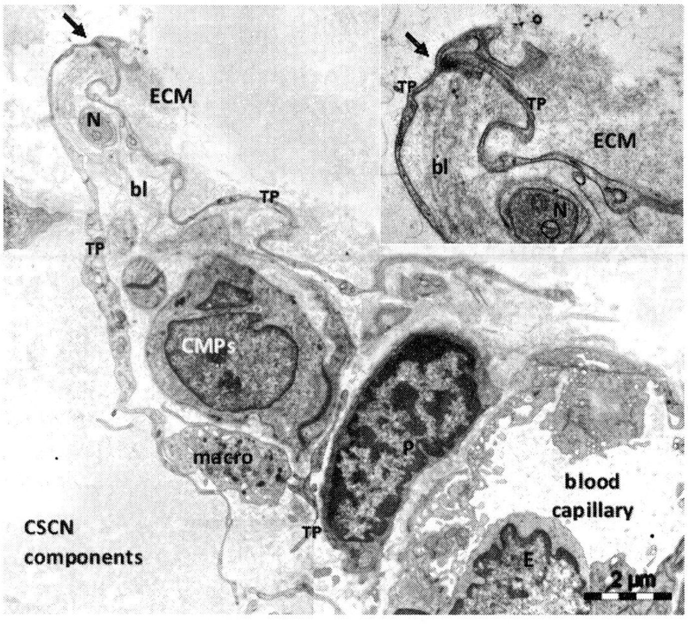

Fig. 8. Transmission electron microscopy. Adult mouse epicardial stem cell niche. Cardiac stem cell niche components: the cardiomyocyte progenitor (CMP) with extending basal lamina (bl); nerve ending (N) blood capillary and stromal supporting cells, which are TCs processes. In this micrograph the telopodes (TP) are interconnected (arrow) and surround all elements of the niche. In the inset, a gap junction between telopodes processes (arrow). Extracellular matrix (ECM) is composed by amorphous electron-dense material. End – endothelial cell, P-pericyte, macro – macrophage, ECM – extracellular matrix.

The extracellular matrix (ECM) is loose and contains ground substance and spare bundles of collagen fibres and few elastic fibres. Extracellular matrix produced by resident cells in CSCN seems to be different in this site since they contain large amounts of electron-lucent or amorphous ground substance beside laminins and non-fibrillar collagen (in basal lamina of the CMPs). These data are important since ECM components and their adhesion with cells within the niche are essential for the establishment and maintenance of niche architecture and control stem cell self-renewal and differentiation [51, 53].

A crucial element of the niche is represented by the stromal supporting cells and electron microscopy showed that TCs are a real candidate for this role offering physical and informational support for CMPs [34].

Telocytes as Nurse Cells

The TCs in the CSCN interconnect with each other through point cell-to-cell contacts or gap junctions, thus forming a three-dimensional network. Attachment plaques connect the TCs to the extracellular matrix and the perivascular elastic material or to the basal lamina of the CMPs. Moreover, close contacts exist between the subepicardial TCs and the myocardial TCs. The subepicardial TCs accompany all the blood vessels forming an incomplete sheath around them and the telopodes surround the cross-sectioned coronary artery wall. In addition, the TCs have close vicinity with nerve cells, blood capillaries and interstitial immunocompetent cells [34].

The TCs accompany all the CMPs independently of their degree of differentiation. Sometimes, CMPs are grouped and the TCs border these supra-cellular structures (Fig. 9) and seem to direct their growing and fusion with mature cardiomyocytes in the periphery of the working muscle (Fig. 10).

Often, the telopodes are directly anchored to CMPs basal lamina in correspondence with plasmalemmal areas reinforced by an electron-dense material. Surprisingly, the CMPs basal lamina, which is often thick, goes over the cells and is closely surrounded by the telopodes (Fig 9). Another type of heterocellular communication occurs through shed microvesicles or exocytotic multi-vesicular bodies that are frequently seen in between TC and CMP [34]. We have reported that these types of intercellular communication regularly occur between TCs and adult cardiomyocytes [36].

Electron microscopy showed no cellular connections of the CMPs with other cell types except other CMPs and TCs. Conversely, TCs establish cell-to-cell contacts with all types of CSCN cells: putative stem cells, monocytes, macrophages and mast cells. All the components of the CSCN are necessarily connected by the TCs in an elusive anatomical structure.

We consider it necessary to draw attention to the presence of cells with role of immune surveillance in CSCN because their influence on the cardiac renewing could be extremely important. TCs in the cardiac niche establish close contacts [35] with all these cells proving another evidence for an **integrative role** for this type of cells.

The diversity of the structural components of the CSCN (Figs. 5, 8) makes us to believe that TCs integrate complex molecular networks, which control the fate of cardiac stem cells in the niche and their incorporation in the myocardium.

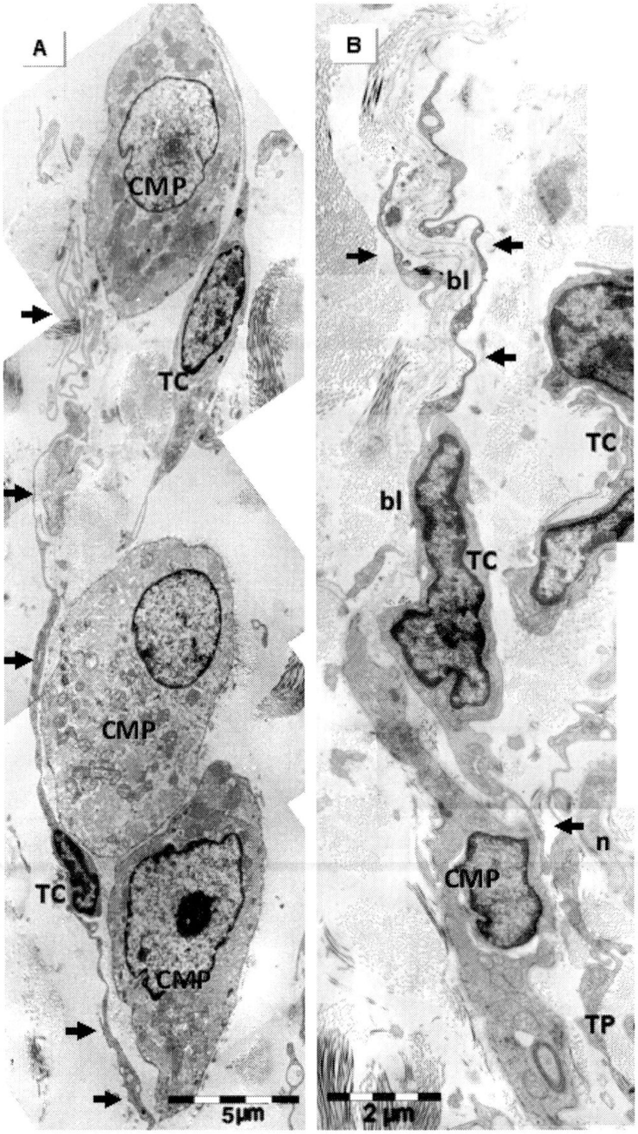

Fig. 9. A and B. Telocytes (TC) as niche cells. These electron micrographs illustrate the relationships of the TCs with cardiomyocyte progenitors (CMP). The telopodes (TP) run parallel (arrows) with the main axis of the CMP and seem to establish their direction of development. In panel B, the TCs sustain a direction (arrows) for the basal lamina (bl) which goes beyond the CMP body. Images reprinted with kind permission from Ref. [34].

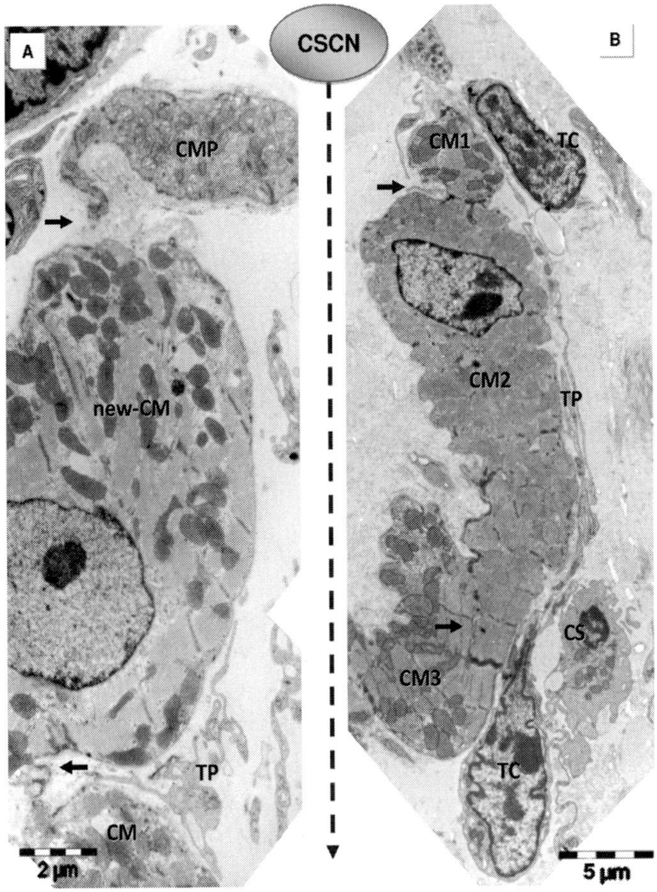

Fig. 10. A and B. Electron micrographs of telocytes (TC) as niche cells. Telopodes (TP) seem to direct the cardiomyocytes progenitor (CMP) and the newly formed cardiomyocytes (new-CM) connection with the adult cardiomyocytes (CM). Cardiomyocyte progenitors fuse (arrows) with adult cardiomyocytes in the periphery of ventricular muscle (A) and intercalated discs (arrows in B) have developed between cardiomyocytes (CM1-3). The cardiac muscle grows from the periphery in a centripetal process. Panel A reproduced with kind permission from Ref. [34].

Even though tens of atrial and ventricular areas have been processed for electron microscopy, we could not find any CMPs in another location except at the peri-coronary epicardium. The last point where we have found CMPs was the entrance of the coronary artery in the cardiac muscle. One difference in cardiac structure of man and mouse must be mentioned – in the mouse, the coronary arteries are embedded within the myocardium after a short epicardial trajectory. This could be an important aspect in evaluation of the CSCN on human heart. In our opinion, cardiac renewing could be a **spatially restricted process** - coronary emergence from aorta. This site is located in the periphery of the myocardium and newly formed CM could be easily integrated without perturbing contraction of the working CM and disarray them. Cardiac renewing might be a **vectorial process** – cardiac growing with a **centripetal direction** could be necessary to preserve the complex architecture of cardiac muscle.

Encouraging evidence comes from the newts. The newt heart is a model for adult heart regeneration as newts can functionally regenerate their heart after amputation of the apex of the ventricle. In this regenerative process, a network of stromal supporting cells is primarily developed [54, 55] and, intriguingly, these cells fulfil all ultrastructural criteria for TCs (Fig. 11).

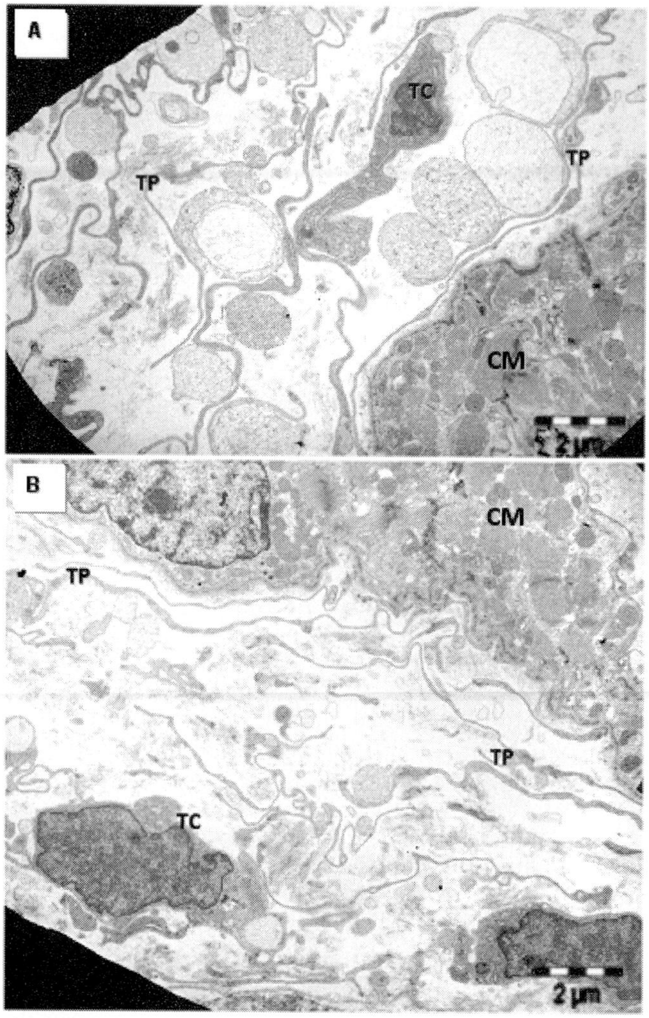

Fig. 11. Transmission electron microscopy. TCs in the injured heart of the newt „*Notophthalmus viridescens*". A. Numerous telopodes (TP)are present in a healed myocardial region 48 hours after amputation of the ventricular apex. TCs seem to originate from the intact myocardium and to extend to the injured site where are located in close vicinity to vacuolar cellular debris. B. TCs 4 days after injury appear more compactly arranged and tend to form a cellular network.

Understanding the physiology of the environment in which the amplified autologous stem cells will be delivered is equally important for an effective therapeutic strategy. A major objective is to identify molecular targets for TCs physiology modulation as supporting stromal cells.

Telocytes in Pathology

The heart muscle regenerates poorly after injury. It can be envisaged that the major reason for the poor regenerative response of myocardium must therefore lie in the peculiarity of tissular architecture, which entails long-living cells. The heart has a unique architecture responsible for the contractile properties and, in this circumstance, even cell death shows particular features [56, 57]. The slow replacement of cardiomyocytes by stem cell differentiation requires an intracellular renewing process involving merely few organelles for the maintenance of heart function. Autophagy is a distinct process involving the degradation of a cardiomyocyte's own components in normal and pathological conditions [58-61]. For instance, in failing human heart, a loss of cells occurs due to a concurrent incidence of oncosis, apoptosis and excessive autophagy [56]. Autophagic cell death is one of several types of programmed cell death (Type II) [58] and could be the major type of death in the myocardium.

Often dying cardiomyocytes still preserve their connections with neighbouring cells and appear as "ghost" cells – a bag of basal lamina filled with autophagosomes. This feature could be also encountered in CSCN where autophagosomes could be seen in the dying CMPs that are not integrated in working muscle.

The ischemic injury could promote interstitial remodelling, TCs phenotypic changes and fibrosis instead of regeneration. Little is known about interstitial remodelling in cardiac pathologies [64-69] but as we have seen in atrial fibrillation for instance (unpublished data), interstitial cells die, fibroblasts increase in number, interstitial fibrosis occurs, and isolated atrial amyloid deposition increases in the space between TCs and cardiomyocytes (Fig. 12). All these changes support our hypothesis that TCs are important cellular components of the cardiac tissue architecture implying that a thorough and detailed phenotypic characterization of the TCs will lead to a better understanding of the delicate balance between regeneration and cardiac cell death.

Towards the Future

The precise function(s) of telocytes are still obscure. However, several potential roles in the cardiac physiology and pathology are proposed.

First, TCs are involved in **intercellular signalling at distance** since they have a strategic position, very close to cardiomyocytes, blood capillaries and nerve endings and in addition, TCs form a 3D-network. The TC cardiac network suggests the existence of a local regulatory compartment involving long-distance communication.

At least two mechanisms should be considered: (i) a paracrine and/or juxtacrine secretion of small signal molecules and (ii) shedding microvesicles that play unique roles in the 'horizontal' transfer of important macromolecules (e.g. proteins or RNAs) among neighbouring cells, necessary for the rapid phenotype adjustments in a variety of conditions [70-73].

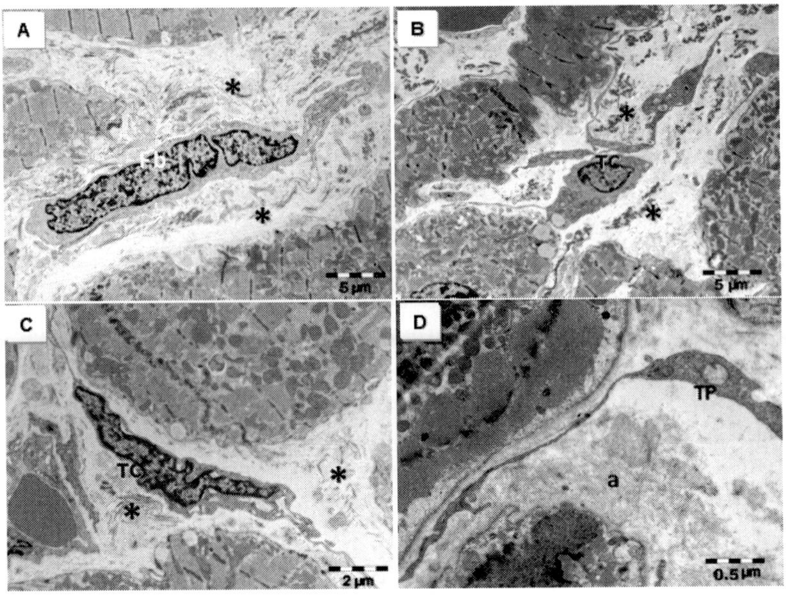

Fig. 12. Transmission electron microscopy. Human atrial myocardium in atrial fibrillation. Fibroblast (Fb) in panel A and TCs in panels B and C. **Extracellular matrix containing many bundles of collagen fibres (asterisks).** D. **Amyloid deposits (a) in isolated atrial amyloidosis increase TC- cardiomyocyte distance.**

Second, TCs may have a role in mechanotransduction and **tensional integration** (tensegrity) [74-78] of the heart, in view of their characteristic ultrastructure (extremely long and convoluted processe (telopodes) with intermediate filaments and microtubules parallel to the long axis of the cell, attachment plaques connecting it to the extracellular matrix). This might be particularly important in atrial fibrillation. Moreover, TCs may participate in pacemaking and/or arrhythmogenesis. For instance, a significant population of TCs and of c-kit+ cells have been recently described in the myocardial sleeves of the pulmonary veins [31, 32]. Although, the mechanisms of atrial fibrillation are still subjects of controversies [79], many authors favour 'the pulmonary vein wave' hypothesis [80-85].

Third, TCs are key players in **cardiac renewal.** Based on c-kit positivity, it was previously supposed that TCs might be considered as adult stromal mesenchymal stem cells or an unexplored population of uncommitted resident cells [27, 86]. Our recent study [34] makes this hypothesis unlikely since we have found very few TCs wrapped in a basal lamina [34]. More likely, the main role of TCs in cardiac renewal is to nurse cardiomyocyte progenitors in CSCN. Briefly, our new hypothesis is that TCs, by their network, have the competence to offer tissue-wide signals essential for stem cells development in the CSCN and also to provide signals for the recruitment of ancillary cells (stem or not) from the circulating blood into the niche. We propose a schema (Fig. 13) to emphasize the role of the TCs in cardiac renewing and their relationships with the components of the CSCN [34] and the working myocardium [36].

Fourth, TCs are likely involved in **cardiac repair**. In the CSCN, TCs accompany the immature cardiac cells and assist their fusion with adult cardiomyocytes. At the same time, the TCs of the niche are connected with the TCs of the working myocardium. Using this network, TCs might offer support

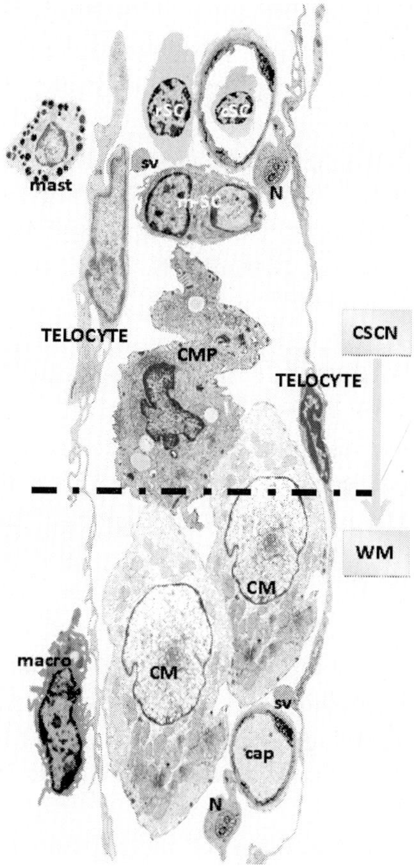

Fig. 13. This schematic representation shows that the telocyte's network (blue) is a prerequisite for myocardial cellular homeostasis shaping the scaffold required to activate exogenous (cSC) and/or resident stem cells (rSC). TCs guide cardiomyocyte progenitors' (CMP, brown) migration from cardiac stem cell niche (CSCN) and fusion with adult cardiomyocytes (CM, light brown) in the working muscle (WM). Myocardial TCs communicate at distance with nurse TCs from epicardial niche and deliver information 'packages' via shed vesicles (sv). Blood vessels (cap - capillary), nerves (green), macrophages and mast cells assist the TCs to monitor the cardiac renewing. m-SC – mitotic stem cell. Note: working myocardium contains all the elements of a stem niche except stem cells. Modified schema with kind permission from Ref. [34].

for CMPs migration at the site of an injury [87]. Another possibility is that the cardiac TCs outside the niche control the integration of the circulating bone-marrow derived cells that express early myocardial markers [88] into the injured myocardium. As we figured in the proposed schema, all the compulsory components defining the niche exist in the working myocardium. TCs might form a transitory niche for the recruitment of circulating stem cells to the site of injury. It is difficult to understand why the cardiac muscle does not properly regenerate after injury if all these multipotent cardiac stem cells exist.

Discovering the molecular mechanisms involved in telocytes-cardiac stem cell communication could be a chance to improve cell therapy for cardiac diseases. Anyway, some perspectives do exist [89].

References

1. Kumar D, Kamp TJ and LeWinter MM. Embryonic stem cells: differentiation into cardiomyocytes and potential for heart repair and regeneration. *Coron Artery Dis.* 2005; 16: 111-116.
2. Marín-García J and Goldenthal MJ. Application of stem cells in cardiology: where we are and where we are going. *Curr Stem Cell Res Ther.* 2006; 1: 1-11.
3. Formigli L, Perna AM, Meacci E, Cinci L, Margheri M, Nistri S, Tani A, Silvertown J, Orlandini G, Porciani C, Zecchi-Orlandini S, Medin J and Bani D. Paracrine effects of transplanted myoblasts and relaxin on post-infarction heart remodeling. *J Cell Mol Med.* 2007; 11: 1087-1100.
4. Roccio M, Goumans MJ, Sluijter JP and Doevendans PA. Stem cell sources for cardiac regeneration. *Panminerva Med.* 2008; 50: 19-30.
5. Liu J, Sluijter JP, Goumans MJ, Smits AM, van der Spoel T, Nathoe H and Doevendans PA. Cell therapy for myocardial regeneration. *Curr Mol Med.* 2009; 9:287-298.
6. Gersh BJ, Simari RD, Behfar A, Terzic CM and Terzic A. Cardiac cell repair therapy: a clinical perspective. *Mayo Clin Proc.* 2009; 84: 876-892.
7. Menasché P. Stem cell therapy for heart failure: are arrhythmias a real safety concern? *Circulation.* 2009;119: 2735-2740.
8. Zhang F and Pasumarthi KB. Embryonic stem cell transplantation: promise and progress in the treatment of heart disease. *BioDrugs.* 2008; 22: 361-74.
9. Perin EC and Silva GV. Autologous cell-based therapy for ischemic heart disease: clinical evidence, proposed mechanisms of action, and current limitations. *Catheter Cardiovasc Interv.* 2009; 73: 281-288.
10. Joggerst SJ and Hatzopoulos AK. Stem cell therapy for cardiac repair: benefits and barriers. *Expert Rev Mol Med.* 2009; 11: e20.
11. Boudoulas KD and Hatzopoulos AK. Cardiac repair and regeneration: the Rubik's cube of cell therapy for heart disease. *Dis Model Mech.* 2009;2: 344-358.
12. Bergmann O, Bhardwaj RD, Bernard S, Zdunek S, Barnabé-Heider F, Walsh S, Zupicich J, Alkass K, Buchholz·BA, Druid H, Jovinge S and Frisén J. Evidence for cardiomyocyte renewal in humans. *Science.* 2009; 324: 98-102.
13. Beltrami AP, Urbanek K, Kajstura J, Yan SM, Finato N, Bussani R, Nadal-Ginard B, Silvestri F, Leri A, Beltrami CA and Anversa P. Evidence that human cardiac myocytes divide after myocardial infarction. *N. Engl. J. Med.* 2001; 344: 1750–1757.
14. van Amerongen MJ and Engel FB. Features of cardiomyocyte proliferation and its potential for cardiac regeneration. *J Cell Mol Med.* 2008;12: 2233-2244.
15. Stephen MJ, Poindexter BJ, Moolman JA, Sheikh-Hamad D and Bick RJ. Do binucleate cardiomyocytes have a role in myocardial repair? Insights using isolated rodent myocytes and cell culture. *Open Cardiovasc Med J.* 2009;3: 1-7.
16. Parmacek MS and Epstein JA. Cardiomyocyte renewal. *N Engl J Med.* 2009; 361: 86-88.

17. Ellison GM, Torella D, Karakikes I and Nadal-Ginard B. Myocyte death and renewal: modern concepts of cardiac cellular homeostasis. *Nat Clin Pract Cardiovasc Med.* 2007;4 Suppl 1: S52-59.
18. Barile L, Messina E, Giacomello A and Marbán E. Endogenous cardiac stem cells. *Prog Cardiovasc Dis.* 2007;50: 31-48.
19. Lyngbaek S, Schneider M, Hansen JL and Sheikh SP. Cardiac regeneration by resident stem and progenitor cells in the adult heart. *Basic Res Cardiol.* 2007;102: 101-114.
20. Ott HC, Matthiesen TS, Brechtken J, Grindle S, Goh SK, Nelson W and Taylor DA. The adult human heart as a source for stem cells: repair strategies with embryonic-like progenitor cells. *Nat Clin Pract Cardiovasc Med.* 2007;4 Suppl 1: S27-39.
21. Zhou B and Pu WT. More than a cover: epicardium as a novel source of cardiac progenitor cells. *Regen Med.* 2008;3: 633-635.
22. Kajstura J, Urbanek K, Rota M, Bearzi C, Hosoda T, Bolli R, Anversa P and Leri A. Cardiac stem cells and myocardial disease. *J Mol Cell Cardiol.* 2008;45: 505-513.
23. Stamm C, Choi YH, Nasseri B and Hetzer R. A heart full of stem cells: the spectrum of myocardial progenitor cells in the postnatal heart. *Ther Adv Cardiovasc Dis.* 2009;3:215-229.
24. Stastna M, Abraham MR and Van Eyk JE. Cardiac stem/progenitor cells, secreted proteins, and proteomics. *FEBS Lett.* 2009;583: 1800-1807.
25. Chamuleau SA, Vrijsen KR, Rokosh DG, Tang XL, Piek JJ, Bolli R. Cell therapy for ischaemic heart disease: focus on the role of resident cardiac stem cells. *Neth Heart J.* 2009;17: 199-207.
26. Martin-Puig S, Wang Z, Chien KR. Lives of a heart cell: tracing the origins of cardiac progenitors. *Cell Stem Cell.* 2008;2: 320-31.
27. Kostin S and Popescu LM. A distinct type of cell in myocardium: interstitial Cajal-like cells (ICLCs). *J Cell Mol Med.* 2009;13: 295-308.
28. Hinescu ME and Popescu LM. Interstitial Cajal-like cells (ICLC) in human atrial myocardium. *J Cell Mol Med.* 2005;9: 972-975.
29. Hinescu ME, Gherghiceanu M, Mandache E, Ciontea SM and Popescu LM. Interstitial Cajal-like cells (ICLC) in atrial myocardium: ultrastructural and immunohistochemical characterization. *J Cell Mol Med.* 2006;10: 243-257.
30. Popescu LM, Gherghiceanu M, Hinescu ME, Cretoiu D, Ceafalan L, Regalia T, Popescu AC, Ardeleanu C and Mandache E. Insights into the interstitium of ventricular myocardium: interstitial Cajal-like cells (ICLC). *J Cell Mol Med.* 200;10: 429-458.
31. Gherghiceanu M, Hinescu ME, Andrei F, Mandache E, Macarie CE, Faussone-Pellegrini MS and Popescu LM. Interstitial Cajal-like cells (ICLC) in myocardial sleeves of human pulmonary veins. *J Cell Mol Med.* 2008;12: 1777-1781.
32. Morel E, Meyronet D, Thivolet-Bejuy F and Chevalier P. Identification and distribution of interstitial Cajal cells in human pulmonary veins. Heart Rhythm. 2008; 5:1063–7.
33. Suciu L, Popescu LM, Regalia T, Ardelean A and Manole CG. Epicardium: interstitial Cajal-like cells (ICLC) highlighted by immunofluorescence. *J Cell Mol Med.* 2009;13: 771-777.

34. Popescu LM, Gherghiceanu M, Manole CG and Faussone-Pellegrini MS. Cardiac renewing: interstitial Cajal-like cells nurse cardiomyocyte progenitors in epicardial stem cell niches. *J Cell Mol Med.* 2009;13: 866-886.
35. Popescu LM, Gherghiceanu M, Cretoiu D and Radu E. The connective connection: interstitial cells of Cajal (ICC) and ICC-like cells establish synapses with immunoreactive cells. Electron microscope study in situ. *J Cell Mol Med.* 2005;9: 714-730.
36. Mandache E, Popescu LM and Gherghiceanu M. Myocardial interstitial Cajal-like cells (ICLC) and their nanostructural relationships with intercalated discs: shed vesicles as intermediates. *J Cell Mol Med.* 2007;11: 1175-1184.
37. Anversa P, Leri A, Rota M, Hosoda T, Bearzi C, Urbanek K, Kajstura J and Bolli R. Concise review: stem cells, myocardial regeneration, and methodological artifacts. *Stem Cells.* 2007;25: 589-601.
38. Pouly J, Bruneval P, Mandet C, Proksch S, Peyrard S, Amrein C, Bousseaux V, Guillemain R, Deloche A, Fabiani JN and Menasché P. Cardiac stem cells in the real world. *J Thorac Cardiovasc Surg.* 2008;135:673-678.
39. Urbanek K, Cesselli D, Rota M, Nascimbene A, De Angelis A, Hosoda T, Bearzi C, Boni A, Bolli R, Kajstura J, Anversa P and Leri A. Stem cell niches in the adult mouse heart. *Proc Natl Acad Sci U S A.* 2006;103: 9226-9231.
40. Bearzi C, Rota M, Hosoda T, Tillmanns J, Nascimbene A, De Angelis A, Yasuzawa-Amano S, Trofimova I, Siggins RW, Lecapitaine N, Cascapera S, Beltrami AP, D'Alessandro DA, Zias E, Quaini F, Urbanek K, Michler RE, Bolli R, Kajstura J, Leri A and Anversa P. Human cardiac stem cells. *Proc Natl Acad Sci U S A.* 2007;104:14068-14073.
41. Pasquinelli G, Tazzari P, Ricci F, Vaselli C, Buzzi M, Conte R, Orrico C, Foroni L, Stella A, Alviano F, Bagnara GP and Lucarelli E. Ultrastructural characteristics of human mesenchymal stromal (stem) cells derived from bone marrow and term placenta. *Ultrastruct Pathol.* 2007;31: 23-31.
42. Schofield R. The relationship between the spleen colony-forming cell and the haemopoietic stem cell. *Blood Cells.* 1978;4: 7-25
43. Papayannopoulou T and Scadden DT. Stem-cell ecology and stem cells in motion. *Blood.* 2008;111: 3923-3930.
44. Fuchs E, Tumbar T and Guasch G. Socializing with the neighbors: stem cells and their niche. *Cell.* 2004;116:769-778.
45. Li L and Xie T. Stem cell niche: structure and function. *Annu Rev Cell Dev Biol.* 2005;21: 605-631.
46. Scadden DT. The stem-cell niche as an entity of action. *Nature.* 2006;441: 1075-1079.
47. Quesenberry PJ and Aliotta JM. The paradoxical dynamism of marrow stem cells: considerations of stem cells, niches, and microvesicles. *Stem Cell Rev.* 2008;4: 137-147.
48. Drummond-Barbosa D. Stem cells, their niches and the systemic environment: an aging network. *Genetics.* 2008;180: 1787-1797.

49. Dellatore SM, Garcia AS and Miller WM. Mimicking stem cell niches to increase stem cell expansion. *Curr Opin Biotechnol.* 2008;19: 534-540.
50. Jones DL and Wagers AJ. No place like home: anatomy and function of the stem cell niche. *Nat Rev Mol Cell Biol.* 2008;9: 11-21.
51. Daley WP, Peters SB and Larsen M. Extracellular matrix dynamics in development and regenerative medicine. *J Cell Sci.* 2008;121(Pt 3): 255-264.
52. Walker MR, Patel KK and Stappenbeck TS. The stem cell niche. *J Pathol.* 2009;217: 169-180.
53. Raymond K, Deugnier MA, Faraldo MM, Glukhova MA. Adhesion within the stem cell niches. *Curr Opin Cell Biol.* 2009;21:623-629.
54. Borchardt T and Kostin S. Plastizität von Herzmusklezellen des Molches. *Max-Planck-Gesellschaft, Entwicklungs-& Evolutionsbiologie,* Tätigkeitsbericht. 2007; S:237-243
55. Kruse J, Borchardt T, Kostin S and Braun T. Analysis of heart regeneration in the newt Notophthalmus viridescens. Max-Planck- Institute for Heart and Lung Research, Bad Nauheim, Germany. EMBO ConferenceSeries, 3rd European Conference on Regeneration, *"Cellular and Molecular Basis of Regeneration and Tissue Repair"*. Ascona Switzerland?, 10-15 September 2006.
56. Kostin S, Pool L, Elsässer A, Hein S, Drexler HC, Arnon E, Hayakawa Y, Zimmermann R, Bauer E, Klövekorn WP, Schaper J. Myocytes die by multiple mechanisms in failing human hearts. *Circ Res.* 2003;92: 715-724.
57. Kostin S. Pathways of myocyte death: implications for development of clinical laboratory biomarkers. *Adv Clin Chem.* 2005;40: 37-98.
58. Terman A and Brunk UT. Autophagy in cardiac myocyte homeostasis, aging, and pathology. *Cardiovasc Res.* 2005;68: 355-365.
59. Goswami SK and Das DK. *Exp Clin Cardiol.* 2006;11:183-188.
60. Das M, Gherghiceanu M, Lekli I, Mukherjee S, Popescu LM and Das DK. Essential role of lipid raft in ischemic preconditioning. *Cell Physiol Biochem.* 2008;21: 325-334.
61. Gurusamy N and Das DK. Is autophagy a double-edged sword for the heart? *Acta Physiol Hung.* 2009;96:267-276.
62. Kostin S, Hein S, Arnon E, Scholz D and Schaper J. The cytoskeleton and related proteins in the human failing heart. *Heart Fail Rev.* 2000;5: 271-280.
63. Hein S, Elsässer A, Kostin S, Zimmermann R and Schaper J.Functional disturbances due to structural remodeling in the failing human heart. *Arch Mal Coeur Vaiss.* 2002;95:815-820.
64. Everett TH 4th, Olgin JE. Atrial fibrosis and the mechanisms of atrial fibrillation. *Heart Rhythm.* 2007;4: S24-27.
65. Corradi D, Callegari S, Maestri R, Benussi S and Alfieri O. Structural remodeling in atrial fibrillation. *Nat Clin Pract Cardiovasc Med.* 2008;5:782-796.
66. Burstein B and Nattel S. Atrial structural remodeling as an antiarrhythmic target. *J Cardiovasc Pharmacol.* 2008;52: 4-10.
67. Nguyen BL, Fishbein MC, Chen LS, Chen PS and Masroor S. Histopathological substrate for chronic atrial fibrillation in humans. *Heart Rhythm.* 2009;6: 454-460.

68. Pellman J, Lyon RC and Sheikh F. Extracellular matrix remodeling in atrial fibrosis: Mechanisms and implications in atrial fibrillation. *J Mol Cell Cardiol.* 2009. doi:10.1016/j.yjmcc.2009.09.001

69. Slezak J, Tribulova N, Okruhlicova L, Dhingra R, Bajaj A, Freed D and Singal P. Hibernating myocardium: pathophysiology, diagnosis, and treatment. *Can J Physiol Pharmacol.* 2009;87:252-265.

70. Cocucci E, Racchetti G, Meldolesi J. Shedding microvesicles: artefacts no more. *Trends Cell Biol.* 2009;19: 43-51.

71. Yuan A, Farber EL, Rapoport AL, Tejada D, Deniskin R, Akhmedov NB and Farber DB. Transfer of microRNAs by embryonic stem cell microvesicles. *PLoS ONE.* 2009; 4:e4722.

72. Gillette JM, Larochelle A, Dunbar CE and Lippincott-Schwartz J. Intercellular transfer to signalling endosomes regulates an ex vivo bone marrow niche. *Nat Cell Biol.* 2009;11: 303-311.

73. Gnecchi M, Zhang Z, Ni A and Dzau VJ. Paracrine mechanisms in adult stem cell signaling and therapy. *Circ Res.* 2008;103: 1204-1219.

74. Ingber DE. Tensegrity and mechanotransduction. J Bodyw Mov Ther. 2008 Jul;12(3):198-200.

75. Ingberg DE. Cellular mechanotransduction: putting all the pieces together again. *FASEB J.* 2006;20: 811–827

76. Berbari NF, O'Connor AK, Haycraft CJ and Yoder BK.The primary cilium as a complex signaling center. *Curr Biol.* 2009;19: R526-535.

77. Chiquet M, Gelman L, Lutz R and Maier S. From mechanotransduction to extracellular matrix gene expression in fibroblasts. *Biochim Biophys Acta.* 2009;1793: 911-920.

78. Bergmann MW and De Windt LJ. Linking cardiac mechanosensing at the sarcomere M-band, nuclear factor kappaB signaling, and cardiac remodeling. *Hypertension.* 2007;49: 1225-1227.

79. Nattel S and Opie LH. Controversies in atrial fibrillation. *Lancet.* 2006; 367: 262–272.

80. Haissaguerre M, Sanders P, Hocini M, Jais P and Clémenty J. Pulmonary veins in the substrate for atrial fibrillation: the "venous wave" hypothesis. *J Am Coll Cardiol.* 2004;43: 2290–2292.

81. Huang CX, Hu CL and Li YB. Atrial fibrillation may be a vascular disease: the role of the pulmonary vein. *Med Hypotheses.* 2007;68: 629–634.

82. Noheria A, Kumar A, Wylie JV Jr and Josephson ME. Catheter ablation vs. antiarrhythmic drug therapy for atrial fibrillation: a systematic review. *Arch Intern Med.* 2008;168: 581–586.

83. Rostock T, Steven D, Lutomsky B, Servatius H, Drewitz I, Klemm H, Müllerleile K, Ventura R, Meinertz T and Willems S. Atrial fibrillation begets atrial fibrillation in the pulmonary veins on the impact of atrial fibrillation on the electrophysiological properties of the pulmonary veins in humans. *J Am Coll Cardiol.* 2008;51: 2153–2160.

84. Seol CA, Kim WT, Ha JM, Choe H, Jang YJ, Youm JB, Earm YE and Leem CH. Stretch activated currents in cardiomyocytes isolated from rabbit pulmonary veins. Prog Biophys Mol Biol. 2008; 97: 217–31.

85. Anderson RH. The anatomic substrates for abnormal automaticity in the pulmonary veins. *Heart Rhythm.* 2008;5: 1068–1069.

86. Nesselmann C, Ma N, Bieback K, Wagner W, Ho A, Konttinen YT, Zhang H, Hinescu ME and Steinhoff G. Mesenchymal stem cells and cardiac repair. *J Cell Mol Med.* 2008;12: 1795–1810.

87. Gonzalez A, Rota M, Nurzynska D, Misao Y, Tillmanns J, Ojaimi C, Padin-Iruegas ME, Müller P, Esposito G, Bearzi C, Vitale S, Dawn B, Sanganalmath SK, Baker M, Hintze TH, Bolli R, Urbanek K, Hosoda T, Anversa P, Kajstura J and Leri A. Activation of cardiac progenitor cells reverses the failing heart senescent phenotype and prolongs lifespan. *Circ Res.* 2008;102: 597-606.

88. Kucia M, Dawn B, Hunt G, Guo Y, Wysoczynski M, Majka M, Ratajczak J, Rezzoug F, Ildstad ST, Bolli R and Ratajczak MZ. Cells expressing early cardiac markers reside in the bone marrow and are mobilized into the peripheral blood after myocardial infarction. *Circ Res.* 2004;95: 1191-1199.

Adaptation Biology and Medicine (Volume 6: Cell Adaptations and Challenges)
Editors: P. Wang, C.-H. Kuo, N. Takeda and P.K. Singal
Copyright © 2011 Narosa Publishing House Pvt. Ltd., New Delhi, India

3. Stem Cells and Cardioprotection

Shaw-Fang Yet, PhD*

Institute of Cellular and System Medicine, National Health Research Institutes, Zhunan, Miaoli County 35053, Taiwan

Abstract

Ischemic heart disease and heart failure are leading causes of morbidity and mortality. Regenerating and repairing the damaged heart after loss of myocardium to ischemic injury seemed was impossible until recently, because cardiomyocytes had long been considered terminally differentiated, incapable of proliferating again. However, with the advent of stem cells this old concept has been challenged. Pluripotent embryonic stem cells (ESCs) have the potential of differentiating into cell types of all germ layers, including components in the heart—cardiomyocytes and blood vessels (endothelial and vascular smooth muscle cells). In addition to ESCs, multipotent adult stem/progenitor cells from different organs (e.g., bone marrow and heart) also have been isolated. These embryonic and adult stem/progenitor cells could be used to mend the damaged heart. Animal studies and clinical trials have shown a great promise of stem cell therapy for cardioprotection and cardiac repair, although some issues still need to be resolved. Transplantation of stem cells-derived cardiomyocytes into infarcted myocardium showed short-term improvement in heart function, although engraftment of cardiomyocytes was limited. In contrast, direct transplantation of stem cells has a long-term risk of tumorigenesis in the heart. Thus, therapeutic approaches using multipotent stem or progenitor cells rather than pluripotent ESCs are perhaps better alternatives. Furthermore, increasing evidence indicates that paracrine factors induced by transplanted stem/progenitor cells may have significantly contributed to cardioprtotection and restoring cardiac performance after myocardial infarction. Paracrine factors can activate cardiomyocyte survival pathways, promote resident and transplanted stem cells differentiation into cardiomyocytes, reduce inflammatory cells infiltration, and enhance angiogenesis for better perfusion. Given the great potential of stem cell therapy for the heart, further research investigating the underlying mechanisms of stem cells in cardioprotection will provide additional insights and better therapeutic approaches.

Keywords: Ischemic heart disease, cardiomyocytes, embryonic stem cells

Introduction

Heart disease is the leading cause of death in the United States for both men and women across all ages, more than cancer and chronic lower respiratory disease (1). In addition to high mortality rate, heart disease also has a high socioeconomic price; it is projected to cost more than $300 billion in 2009, including health care services, medications, and lost productivity (2). As such, it is imperative to improve therapeutic options for heart related diseases.

*Address for Correspondence: Shaw-Fang Yet, Ph.D. Institute of Cellular and System Medicine, National Health Research Institutes, 35 Keyan Road, Zhunan, Miaoli County 35053, Taiwan. E-mail: syet@nhri.org.tw

Heart is one of the least regenerative organs in the body; the old dogma is that cardiomyocytes are terminally differentiated and incapable of regeneration. However, with the advent of stem cell biology, this concept is being challenged (3, 4). In response to myocardial injury, the heart attempts to repair the damage by formation of noncontractile scar tissues. This results in loss of myocardium and often leads to heart failure. One option to repair and protect a failing heart is to replace the whole heart by heart transplantation. However, this is limited by the number of donor organs available and allograft rejection. Another promising option emerging is stem cell transplantation therapy. Stem cells are multi- and pluripotent cells, capable of self-renewal and can differentiate into different cell types, including cardiomyocytes.

Cell Therapy for the Damaged Heart

There are two major types of stem cells, adult and embryonic. Adult stem cells are multipotent, can be identified and isolated from adult tissues. For example, skeletal myoblasts (satellite cells) are adult skeletal muscle stem cells. Bone marrow cells are from adult bone marrow and cardiac stem cells from adult heart. Pluripotent embryonic stem cells are isolated from the inner cell mass of blastocysts. All these stem cells have been investigated for cell therapy to repair damaged hearts.

Skeletal Myoblasts

Satellite cells are undifferentiated skeletal myoblasts responsible for skeletal muscle formation during development. In normal conditions, satellite cells are in a quiescent state. However, in response to injury, satellite cells proliferate and differentiate into muscle cells to regenerate damaged skeletal muscle. In contrast, damaged myocardium is replaced with scar tissues by endogenous repairing mechanisms. Because of the regenerating property of skeletal myoblasts, they are one of the early candidate cells for cardiac repair (5-7). There are several advantages of them: 1) autologous origin, can be isolated from patients themselves; 2) relatively easy to isolate; 3) great growth potential and can be scaled up in vitro; and 4) importantly, they are committed to myogenic lineage and thus lack of tumorigenicity.

Early studies showed that satellite cells could differentiate into cardiac-like cells (5) and improve myocardial performance up to 6 weeks post implantation (6). To assess the long-term efficacy of myoblast transplantation, a sheep model was used (8). Experimental myocardial infarction was induced in sheep and 14 days later either culture medium control or myoblasts were injected into the myocardium across and around infarcted area. Global and regional left ventricular (LV) functions were evaluated before, 4, and 12 months after in-scar injections of autologous skeletal myoblasts or culture medium. In the control group, LV ejection fraction (LVEF) deteriorated at 4 months; in contrast, myoblast transplantation limited LVEF deterioration (8). Wall motion score increased in the control group at 4 months, indicating a decrease in contractility; but the score was decreased in the myoblast group, suggesting improvement of contractility. Histological examination of the control hearts revealed that infarct scars were transmural with fibrous tissue intermingled with fat tissue. In contrast, myoblast transplanted hearts had large areas of grafted cells, with features of well-differentiated skeletal muscle cells and a tendency to organize in parallel. At 1 year, histological analysis revealed that viable myotubes and myoblasts were present within the scar. These results suggested that myoblast transplantation might improve systolic scar function through colonization of fibrosis by skeletal muscle cells with the ability to withstand a cardiac-type workload (8).

However, the mechanisms responsible for the cardiac functional improvement by myoblast transplantation remained unclear. To this end, EGFP-labeled skeletal myobalsts were transplanted into rat infarcted myocardium (1 week) and analyzed 4 weeks later with intracelluar recordings and fluorescence microscopy (9). Interestingly, grafted myoblasts differentiated into peculiar hyper-excitable myotubes with a contractile activity fully independent of neighboring cardiomyocytes, suggesting mechanisms other than electromechanical coupling between grafted and host cells were involved in the improvement of cardiac function. Subsequently, a hypothesis of cell-mediated paracrine mechanisms was proposed (10). Analysis of gene expression and growth factors released by undifferentiated myoblasts and myotubes revealed that myoblast-secreted factors may contribute to the beneficial effects of myogenic cell transplantation in infarcted myocardium (10).

Although the underlying mechanisms remained unclear, improvement of cardiac performance by myoblasts transplantation in several animal models (5, 6, 8) prompted a first clinical study on a patient with severe ischemic heart failure (7). Autologous skeletal myoblasts were implanted into the postinfarction scar during coronary artery bypass grafting of remote myocardial areas. Five months later, there was evidence of contraction and viability in the grafted scar on echocardiography and positron emission tomography (7). Although this result was encouraging, it required validation by additional studies. In the first phase I cohort of grafted patients in a single center, there were 9 patients with severe heart failure (61.8±11.6 yrs; EF =35%) in the study group (11). Bypass surgery was performed on these patients and autologous myoblasts were transplanted at the same time. These patients were followed for ~4 years. In this cohort, myoblast transplantation improved both clinical status and EF with low incidence of hospitalizations for heart failure. Although there was an increased risk of ventricular arrhythmias, it could be controlled by medical therapy and/or automatic cardiac defibrillator implantation.

Subsequently, 97 patients with severe LV dysfunction (ejection fraction ≤35%) scheduled for bypass surgery were included in the MAGIC trial—the first randomized, placebo-controlled, double-blind, multicenter study (12). Each patient received either cells grown from a skeletal muscle biopsy or a placebo solution injected in and around the scar. All patients received an implantable cardioverter-defibrillator. Global and regional LV functions were assessed by echocardiography 6 months after transplantation. The results from this MAGIC trial indicated that although myoblast injections combined with coronary surgery in patients with depressed LV function failed to improve echocardiographic heart function, the high dose myoblast transplantation was associated with a significant anti-remodeling effect (12).

Myoblast therapy has shown promise for cardiac repair, however, there are still hurdles to overcome before myoblast transplantation becomes a potentially effective adjunct to current heart failure therapies (13).

Bone Marrow Cells

One other candidate cells for cardiac cell therapy to repair injured heart are bone marrow cells (BMCs). For cell therapy, BMCs have the advantages of being autologous origin, containing various lineage cells, and can be expanded in vitro. In an early study, Lineage negative c-kit positive ($Lin^{-}c\text{-}kit^{POS}$) or

Lin^{-}c-kit^{NEG} BMCs were injected (3-5 h after infarction) into mouse myocardium bordering the infarct (14). When hearts were examined 9 days later, transplanted Lin^{-}c-kit^{POS} but not Lin^{-}c-kit^{NEG} cells regenerated infarcted myocardium and improved ventricular functions. Intriguingly, in addition to proliferating myocytes, the developing tissue also contained vascular structures (14). Bone marrow-derived mesenchymal stem cells (MSCs) are a small percentage (~0.001 to 0.01%) of cells from bone marrow aspirate with multilineage potential (15). Purified human MSCs from adult bone marrow have been shown to engraft in the mouse myocardium and appeared to differentiate into cardiomyocytes (16). Thus, the persistence of the engrafted human MSCs and their in situ differentiation in the heart may represent the basis for using these adult stem cells for cellular cardiomyoplasty (16). The ability of MSCs to differentiate into a variety of cells, including cardiomyocytes and vascular endothelial cells, makes them an attractive therapeutic tool for heart failure (17).

Experimental studies showed that BMC transplantation contributed to improvement of cardiac repair/function of infarcted myocardium and inducing myogenesis and angiogenesis (14, 16, 18-20). In patients with ischemic heart disease, autologous BMC transfer concomitant with coronary artery bypass grafting showed specific improvement in coronary perfusion in 3 of the 5 patients (21). In another clinical study, 20 patients underwent standard therapy and 10 of them received additional intracoronary BMC transplantation (22). The results revealed that transplanted autologous BMCs might lead to repair of infarcted myocardium (myocardial regeneration and neovascularization) when applied during the immediate postinfarction period (22).

To assess whether intracoronary transfer of autologous BMCs could improve global LVEF, a randomized, controlled BOOST clinical trial was initiated (23). Total of 60 patients (30 in the control group and 30 in the BMC group) were included in the study. Autologous BMCs were infused into infarct-related coronary artery during the early postinfarction period and patients were evaluated after 6 months. The results indicated that transplantation of autologous BMCs significantly improved LV systolic function as evidenced by increased global LVEF (23). However, a follow-up study at 18 months revealed that although there was an acceleration of LVEF recovery, a single dose of intracoronary BMCs did not provide long-term benefit on LV systolic function (24).

In the REPAIR-AMI trial (25), 204 patients were randomly assigned for the trial. Three to 7 days after successful reperfusion therapy, 103 patients received medium in placebo group and 101 patients received BMCs. At 4 months, BMC group showed greater improvement of LVEF (25) and at 1 year, BMC group has a significant reduction of death, recurrence of myocardial infarction (MI), and revascularization procedures (26). The results indicated that BMC cell therapy reduced adverse cardiovascular events after acute MI (AMI). Thus, large-scale studies are warranted to confirm the effects of BMC administration on mortality and morbidity in patients with AMIs (26). Interestingly, in a substudy of REPAIR-AMI trial (27), BMC therapy restored microcirculation function of the infarct-related artery and improved vascular conductance capacity. This study provides a clinical proof of concept that progenitor cell transplantation promotes vascular repair.

Matsumoto *et al.* (28) hypothesized that because of the angiogenic activity of VEGF, grafting of VEGF-expressing MSCs might provide advanced benefits on treatment for MI over MSCs alone. Indeed, their data suggested that transplantation of VEGF-expressing MSCs could effectively treat AMI by providing enhanced cardioprotection, followed by angiogenic effects in salvaging ischemic myocardium (28). To test whether paracrine factors released from BMCs play a role in angiogenesis and cardioprotection, BMCs and peripheral blood leucocytes from 15 patients with AMI were cultured in serum-free medium to obtain conditioned supernatant (27). These conditioned supernatants were found to stimulate human coronary artery endothelial cell proliferation, migration, tube formation, and also induced cell sprouting. Moreover, BMC supernatants protected rat cardiomyocytes from ischemia or ischemia/reperfusion-induced cell death (27). Furthermore, ProteinChip and GeneChip array analyses revealed that BMCs expressed distinct patterns of pro-angiogenic and cytoprotective secreted factors, supporting the hypothesis that paracrine factors contributed to the therapeutic effects (29).

Consistent with the concept that BMCs secrete protective paracrine factors in cardiac protection, bone marrow mononuclear cells (BM-MNCs) were found to secrete significant amount of the pleiotropic cytokine IL-10 (30). To test the effect of IL-10, wild type or IL-10 knockout mouse BM-MNCs were intramyocardially injected into the infarct region of mouse hearts immediately after left anterior descending (LAD) coronary artery ligation. At 14 days post infarction, LV end-diastolic pressure (LVEDP) was substantially higher in diluent group when compared with sham control; transplantation of wild type BM-MNCs significantly reduced LVEDP while IL-10 knockout BM-MNCs had little effect. LV end-systolic volume also had a similar trend, suggesting IL-10 mediated the improvement of LV function. Consistent with functional data, IL-10 knockout cells did not improve LV remodeling as wild type cells did (30).

Although the underlying mechanisms are still not well defined, experimental studies and clinical trials have shown therapeutic benefits of BMCs in the treatment of heart disease, particularly AMI. Further investigation and clinical trials would provide insights into the long-term efficacy of BMC cell therapy for the heart.

Cardiac-derived Stem Cells

Several studies demonstrated that although the adult heart is mainly composed of terminally differentiated cells, it contains multipotent stem cells for cardiac regeneration (31-33). Cell transplantation therapy using cardiac stem cells derived from patient's own biopsies would avoid the issue of rejection. Smith *et al.* (34) found that cardiosphere-derived cells (CDCs) expanded from percutaneous endomyocardial biosy specimens exhibit regenerative potential. In a mouse study, following injection into the border zone of myocardial infarcts, human CDCs engrafted and migrated into the infarct zone. After 20 days, the percentage of viable myocardium within the infarct zone was greater in the CDC-treated group than in the fibroblast-treated control group. Furthermore, LVEF was higher in the CDC-treated group, suggesting improvement of ventricular function (34).

The advantages of CDCs for cardiac repair are 1) CDCs can be isolated from minimally invasive percutaneous endomyocardial biopsies and grown to large numbers in a few weeks and 2) CDCs can be preprogrammed to differentiate into cardiac and vascular lineages for the heart. A subsequent study to assess the safety and efficacy in large animals was conducted with the goal of clinical translation in the future (35). Intracoronary delivery of CDCs in a preclinical porcine model of postinfarct left ventricular dysfunction resulted in formation of new cardiac tissue, reduced relative infarct size, attenuated adverse remodeling, and improved hemodynamics. The results indicate that CDCs are a novel, safe, and potentially effective therapy for ischemic cardiomyopathy (35). With the therapeutic potential of CDCs, the National Institutes of Health has sponsored a trial to study intracoronay delivery of autologous CDCs in patients with ischemic cardiomyopathy (35).

Embryonic Stem Cells

Pluripotent embryonic stem cells (ESCs) are derived from the inner cell mass of blastocyst embryos. ESCs have the ability to proliferate extensively and remain in an undifferentiated state. Given the pluripotency of ESCs, they have the potential of differentiating into cell types of all three germ layers. Thus, ESCs hold a great promise for cell therapy; ESCs might serve as a critical source for cell transplantation in many devastating diseases including heart disease, spinal cord injury, Parkinson's disease, and diabetes, just to name a few.

To test the therapeutic potential of ESCs, undifferentiated mouse ESCs were injected into the myocardium 5 min after LAD ligation (36). Surprisingly, ESC-transplanted hearts developed tumor 4 weeks after transplantation. Consistent with these findings, a 2007 study also demonstrated that transplantation of undifferentiated ESCs into bordering zone of infarct resulted in teratoma formation 3 weeks later (37). These studies suggested that cardiac repair strategies involving ESCs would need to control cardiomyocyte differentiation and avoid undifferentiated ESCs.

To isolate ESC-derived cardiomyocytes for transplantation, Kolossov *et al.* (38) generated transgenic ESCs expressing puromycin resistant gene and EGFP under the control of cardiac specific α-MHC promoter. Selection with puromycin resulted in highly purified cardiomyocyte population. These ESC-derived cardiomyocytes engrafted into the infarcted myocardium as evidenced by green fluorescence. The ESC-derived EGFP-positive cells displayed elongated cell shapes and also expressed α-actinin, characteristics of cardiomyocytes. Furthermore, connexin 43 staining revealed gap juction formation between transplanted cardiomyocytes. ESC-derived cardiomyocytes transplantation improved heart function by increasing LVEF with a very low incidence of tumor formation. The authors concluded that ESC-based cell therapy was a promising approach for the treatment of impaired myocardial function and provided better results than BM-derived cells (38).

In another study, activin A and BMP4 were used to direct differentiation of human ESCs (hESCs) to cardiomyocytes (39). These cardiomyocytes were delivered into infarcted rat hearts in a cocktail of pro-survival factors that limited cardiomyocyte death after transplantation. When the hearts were analyzed a week later, both β-MHC and human centromere positive cells indicated grafting of transplanted myocytes. Cells delivered in prosurvival cocktail improved graft survival in infarcted hearts compared with control groups. Moreover, hESC-derived cardiomyocyte transplantation delivered in prosurvival cocktail improved postinfarct ventricular function as assessed by decreased LVEDD and LVESD, and increases in fractional shortening.

Consistent with the potential of hESC-derived cardiomyocytes for myocardial cell therapy strategies, Caspi *et al.* (40) showed that transplantation of hESC-derived cardiomyocytes into chronically infarcted rat hearts resulted in the formation of stable cardiomyocyte grafts and formed gap junctions with host cardiac tissue. In addition to attenuation of the remodeling process, transplantation of hESC-derived cardiomyocytes also improved myocardial performance (40). In contrast, transplantation of undifferentiated hESCs resulted in the formation of teratoma-like structures (40).

Induced Pluripotent Stem Cells

In a landmark study in 2006 (41), Yamanaka's group used retroviruses to introduce various transcription factors into mouse embryonic fibroblasts in which Fbx15 locus had been knocked in with a βgeo cassette. After G418 selection, fibroblasts that had been converted to embryonic stem cell-like cells were selected. They found that 4 transcription factors — Oct3/4, Sox2, Klf4, and c-Myc — are sufficient to reprogram fibroblasts into ESC-like cells, designated induced pluripotent stem (iPS) cells. The iPS cells exhibit the morphology and growth properties of ESCs and express ESC marker genes. Subcutaneous transplantation of iPS cells into nude mice resulted in tumors containing a variety of tissues from all three germ layers. These data demonstrate that pluripotent stem cells can be directly generated from fibroblast cultures by the addition of only 4 transcription factors (41). The possibility of reprogramming somatic cells into iPS cells may eventually allow the creation of pluripotent cells directly from somatic cells of patients for therapeutic purposes.

With the advent of iPS technology, directional differentiation of iPS cells into different cell types for patient-specific cell therapy would have great clinical applications. Narazaki *et al.*(42) found that various cardiovascular cells can be systematically induced from iPS cells. The differentiation properties of iPS cells are almost completely identical to those of ESCs, raising the possibility of developing novel cardiovascular regenerative medicine with iPS cells (42). To investigate the cardiac differentiation potential of murine iPS cells in comparison to a well-established murine ESC line (43), a well-established standard differentiation protocol was used. Analyses on molecular, structural, and functional levels demonstrated that iPS cells differentiated into functional cardiomyocytes with typical features of ESC-derived cardiomyocytes (43).

A recent paper reported that iPS fate is determined by host competency (44). Subcutaneous injection of iPS cells resulted in teratoma formation in immunodeficient nude mice, but not in immunocompetent mice. Similarly, transplantation of iPS cells into infarcted heart led to gross tumor formation in nude mice. Furthermore, compared with fibroblasts, iPS therapy restored myocardial performance as evidenced by increased ejection fraction, fractional shortening, and systolic wall thickness (44). These results established iPS as a promising new cell source in the treatment of heart disease.

Conclusion

Stem cell transplantation to regenerate damaged myocardium has emerged as an attractive therapeutic strategy with great promise. Although much had been done but there are still many issues to be resolved regarding the efficacy and safety of stem cell therapy for treating heart disease. Several sources of stem cells are potential candidates for cardiac regeneration; however, it appears that to match particular target clinical setting the right stem cell type might be selected. For example, autologous skeletal myoblast therapy is promising for heart failure while BMCs are promising for acute myocardial infarction. With the advancement of iPS technology, individual patient-based therapy might be possible. Taken together, mending a broken heart may not be just a dream anymore.

Sources of Funding

This work was supported in part by National Science Council (Taiwan) Grant 97-3111-B-400-005 and National Health Research Institutes (Taiwan) Grant CS-098-PP-06.

References

1. Kung HC, Hoyert DL, Xu J and Murphy SL. Deaths: final data for 2005. *Natl Vital Stat Rep.* 2008;56: 1-120.
2. American Heart Association. Heart disease and stroke statistics – 2009 update. Dallas, Texas: *American Heart Association*; 2009.
3. Pasumarthi KB and Field LJ. Cardiomyocyte cell cycle regulation. *Circ Res.* 2002;90: 1044-1054.
4. Rubart M and Field LJ. Cardiac regeneration: repopulating the heart. *Annu Rev Physiol.* 2006;68: 29-49.
5. Chiu RC, Zibaitis A and Kao RL. Cellular cardiomyoplasty: myocardial regeneration with satellite cell implantation. *Ann Thorac Surg.* 1995;60: 12-18.
6. Taylor DA, Atkins BZ, Hungspreugs P, Jones TR, Reedy MC, Hutcheson KA, Glower DD and Kraus WE. Regenerating functional myocardium: improved performance after skeletal myoblast transplantation. *Nat Med.* 1998;4: 929-933.
7. Menasche P, Hagege AA, Scorsin M, Pouzet B, Desnos M, Duboc D, Schwartz K, Vilquin JT and Marolleau JP. Myoblast transplantation for heart failure. *Lancet.* 2001;357: 279-280.
8. Ghostine S, Carrion C, Souza LC, Richard P, Bruneval P, Vilquin JT, Pouzet B, Schwartz K, Menasche P and Hagege AA. Long-term efficacy of myoblast transplantation on regional structure and function after myocardial infarction. *Circulation.* 2002;106: I131-136.
9. Leobon B, Garcin I, Menasche P, Vilquin JT, Audinat E and Charpak S. Myoblasts transplanted into rat infarcted myocardium are functionally isolated from their host. *Proc Natl Acad Sci U S A.* 2003;100: 7808-7811.
10. Perez-Ilzarbe M, Agbulut O, Pelacho B, Ciorba C, San Jose-Eneriz E, Desnos M, Hagege AA, Aranda P, Andreu EJ, Menasche P and Prosper F. Characterization of the paracrine effects of human skeletal myoblasts transplanted in infarcted myocardium. *Eur J Heart Fail.* 2008;10: 1065-1072.
11. Hagege AA, Marolleau JP, Vilquin JT, Alheritiere A, Peyrard S, Duboc D, Abergel E, Messas E, Mousseaux E, Schwartz K, Desnos M and Menasche P. Skeletal myoblast transplantation in ischemic heart failure: long-term follow-up of the first phase I cohort of patients. *Circulation.* 2006;114: I108-113.
12. Menasche P, Alfieri O, Janssens S, McKenna W, Reichenspurner H, Trinquart L, Vilquin JT, Marolleau JP, Seymour B, Larghero J, Lake S, Chatellier G, Solomon S, Desnos M and Hagege AA. The Myoblast Autologous Grafting in Ischemic Cardiomyopathy (MAGIC) trial: first randomized placebo-controlled study of myoblast transplantation. *Circulation.* 2008;117: 1189-1200.
13. Menasche P. Skeletal myoblasts and cardiac repair. *J Mol Cell Cardiol.* 2008;45: 545-553.
14. Orlic D, Kajstura J, Chimenti S, Jakoniuk I, Anderson SM, Li B, Pickel J, McKay R, Nadal-Ginard B, Bodine DM, Leri A and Anversa P. Bone marrow cells regenerate infarcted myocardium. *Nature.* 2001;410: 701-705.

15. Pittenger MF, Mackay AM, Beck SC, Jaiswal RK, Douglas R, Mosca JD, Moorman MA, Simonetti DW, Craig S and Marshak DR. Multilineage potential of adult human mesenchymal stem cells. *Science*. 1999;284: 143-147.
16. Toma C, Pittenger MF, Cahill KS, Byrne BJ and Kessler PD. Human mesenchymal stem cells differentiate to a cardiomyocyte phenotype in the adult murine heart. *Circulation*. 2002;105: 93-98.
17. Ohnishi S, Ohgushi H, Kitamura S and Nagaya N. Mesenchymal stem cells for the treatment of heart failure. *Int J Hematol*. 2007;86: 17-21.
18. Orlic D, Kajstura J, Chimenti S, Limana F, Jakoniuk I, Quaini F, Nadal-Ginard B, Bodine DM, Leri A and Anversa P. Mobilized bone marrow cells repair the infarcted heart, improving function and survival. *Proc Natl Acad Sci U S A*. 2001;98: 10344-10349.
19. Kocher AA, Schuster MD, Szabolcs MJ, Takuma S, Burkhoff D, Wang J, Homma S, Edwards NM and Itescu S. Neovascularization of ischemic myocardium by human bone-marrow-derived angioblasts prevents cardiomyocyte apoptosis, reduces remodeling and improves cardiac function. *Nat Med*. 2001;7: 430-436.
20. Tomita S, Mickle DA, Weisel RD, Jia ZQ, Tumiati LC, Allidina Y, Liu P and Li RK. Improved heart function with myogenesis and angiogenesis after autologous porcine bone marrow stromal cell transplantation. *J Thorac Cardiovasc Surg*. 2002;123: 1132-1140.
21. Hamano K, Nishida M, Hirata K, Mikamo A, Li TS, Harada M, Miura T, Matsuzaki M and Esato K. Local implantation of autologous bone marrow cells for therapeutic angiogenesis in patients with ischemic heart disease: clinical trial and preliminary results. *Jpn Circ J*. 2001;65: 845-847.
22. Strauer BE, Brehm M, Zeus T, Kostering M, Hernandez A, Sorg RV, Kogler G and Wernet P. Repair of infarcted myocardium by autologous intracoronary mononuclear bone marrow cell transplantation in humans. *Circulation*. 2002;106: 1913-1918.
23. Wollert KC, Meyer GP, Lotz J, Ringes-Lichtenberg S, Lippolt P, Breidenbach C, Fichtner S, Korte T, Hornig B, Messinger D, Arseniev L, Hertenstein B, Ganser A and Drexler H. Intracoronary autologous bone-marrow cell transfer after myocardial infarction: the BOOST randomised controlled clinical trial. *Lancet*. 2004;364: 141-148.
24. Meyer GP, Wollert KC, Lotz J, Steffens J, Lippolt P, Fichtner S, Hecker H, Schaefer A, Arseniev L, Hertenstein B, Ganser A and Drexler H. Intracoronary bone marrow cell transfer after myocardial infarction: eighteen months' follow-up data from the randomized, controlled BOOST (BOne marrOw transfer to enhance ST-elevation infarct regeneration) trial. *Circulation*. 2006;113: 1287-1294.
25. Schachinger V, Erbs S, Elsasser A, Haberbosch W, Hambrecht R, Holschermann H, Yu J, Corti R, Mathey DG, Hamm CW, Suselbeck T, Assmus B, Tonn T, Dimmeler S and Zeiher AM. Intracoronary bone marrow-derived progenitor cells in acute myocardial infarction. *N Engl J Med*. 2006;355: 1210-1221.
26. Schachinger V, Erbs S, Elsasser A, Haberbosch W, Hambrecht R, Holschermann H, Yu J, Corti R, Mathey DG, Hamm CW, Suselbeck T, Werner N, Haase J, Neuzner J, Germing A, Mark B, Assmus B, Tonn T, Dimmeler S and Zeiher AM. Improved clinical outcome after intracoronary administration of bone-marrow-derived progenitor cells in acute myocardial infarction: final 1-year results of the REPAIR-AMI trial. *Eur Heart J*. 2006;27: 2775-2783.

27. Erbs S, Linke A, Schachinger V, Assmus B, Thiele H, Diederich KW, Hoffmann C, Dimmeler S, Tonn T, Hambrecht R, Zeiher AM and Schuler G. Restoration of microvascular function in the infarct-related artery by intracoronary transplantation of bone marrow progenitor cells in patients with acute myocardial infarction: the Doppler Substudy of the Reinfusion of Enriched Progenitor Cells and Infarct Remodeling in Acute Myocardial Infarction (REPAIR-AMI) trial. *Circulation*. 2007;116: 366-374.
28. Matsumoto R, Omura T, Yoshiyama M, Hayashi T, Inamoto S, Koh KR, Ohta K, Izumi Y, Nakamura Y, Akioka K, Kitaura Y, Takeuchi K and Yoshikawa J. Vascular endothelial growth factor-expressing mesenchymal stem cell transplantation for the treatment of acute myocardial infarction. *Arterioscler Thromb Vasc Biol*. 2005;25: 1168-1173.
29. Korf-Klingebiel M, Kempf T, Sauer T, Brinkmann E, Fischer P, Meyer GP, Ganser A, Drexler H and Wollert KC. Bone marrow cells are a rich source of growth factors and cytokines: implications for cell therapy trials after myocardial infarction. *Eur Heart J*. 2008;29: 2851-2858.
30. Burchfield JS, Iwasaki M, Koyanagi M, Urbich C, Rosenthal N, Zeiher AM and Dimmeler S. Interleukin-10 from transplanted bone marrow mononuclear cells contributes to cardiac protection after myocardial infarction. *Circ Res*. 2008;103: 203-211.
31. Beltrami AP, Barlucchi L, Torella D, Baker M, Limana F, Chimenti S, Kasahara H, Rota M, Musso E, Urbanek K, Leri A, Kajstura J, Nadal-Ginard B and Anversa P. Adult cardiac stem cells are multipotent and support myocardial regeneration. *Cell*. 2003;114: 763-776.
32. Oh H, Bradfute SB, Gallardo TD, Nakamura T, Gaussin V, Mishina Y, Pocius J, Michael LH, Behringer RR, Garry DJ, Entman ML and Schneider MD. Cardiac progenitor cells from adult myocardium: homing, differentiation, and fusion after infarction. *Proc Natl Acad Sci U S A*. 2003;100: 12313-12318.
33. Urbanek K, Torella D, Sheikh F, De Angelis A, Nurzynska D, Silvestri F, Beltrami CA, Bussani R, Beltrami AP, Quaini F, Bolli R, Leri A, Kajstura J and Anversa P. Myocardial regeneration by activation of multipotent cardiac stem cells in ischemic heart failure. *Proc Natl Acad Sci U S A*. 2005;102: 8692-8697.
34. Smith RR, Barile L, Cho HC, Leppo MK, Hare JM, Messina E, Giacomello A, Abraham MR and Marban E. Regenerative potential of cardiosphere-derived cells expanded from percutaneous endomyocardial biopsy specimens. *Circulation*. 2007;115: 896-908.
35. Johnston PV, Sasano T, Mills K, Evers R, Lee ST, Smith RR, Lardo AC, Lai S, Steenbergen C, Gerstenblith G, Lange R and Marban E. Engraftment, differentiation, and functional benefits of autologous cardiosphere-derived cells in porcine ischemic cardiomyopathy. *Circulation*. 2009;120: 1075-1083.
36. Swijnenburg RJ, Tanaka M, Vogel H, Baker J, Kofidis T, Gunawan F, Lebl DR, Caffarelli AD, de Bruin JL, Fedoseyeva EV and Robbins RC. Embryonic stem cell immunogenicity increases upon differentiation after transplantation into ischemic myocardium. *Circulation*. 2005;112: I166-172.
37. Nussbaum J, Minami E, Laflamme MA, Virag JA, Ware CB, Masino A, Muskheli V, Pabon L, Reinecke H and Murry CE. Transplantation of undifferentiated murine embryonic stem cells in the heart: teratoma formation and immune response. *FASEB J*. 2007;21: 1345-1357.

38. Kolossov E, Bostani T, Roell W, Breitbach M, Pillekamp F, Nygren JM, Sasse P, Rubenchik O, Fries JW, Wenzel D, Geisen C, Xia Y, Lu Z, Duan Y, Kettenhofen R, Jovinge S, Bloch W, Bohlen H, Welz A, Hescheler J, Jacobsen SE and Fleischmann BK. Engraftment of engineered ES cell-derived cardiomyocytes but not BM cells restores contractile function to the infarcted myocardium. *J Exp Med.* 2006;203: 2315-2327.
39. Laflamme MA, Chen KY, Naumova AV, Muskheli V, Fugate JA, Dupras SK, Reinecke H, Xu C, Hassanipour M, Police S, O'Sullivan C, Collins L, Chen Y, Minami E, Gill EA, Ueno S, Yuan C, Gold J and Murry CE. Cardiomyocytes derived from human embryonic stem cells in pro-survival factors enhance function of infarcted rat hearts. *Nat Biotechnol.* 2007;25: 1015-1024.
40. Caspi O, Huber I, Kehat I, Habib M, Arbel G, Gepstein A, Yankelson L, Aronson D, Beyar R and Gepstein L. Transplantation of human embryonic stem cell-derived cardiomyocytes improves myocardial performance in infarcted rat hearts. *J Am Coll Cardiol.* 2007;50: 1884-1893.
41. Takahashi K and Yamanaka S. Induction of pluripotent stem cells from mouse embryonic and adult fibroblast cultures by defined factors. *Cell.* 2006;126: 663-676.
42. Narazaki G, Uosaki H, Teranishi M, Okita K, Kim B, Matsuoka S, Yamanaka S and Yamashita JK. Directed and systematic differentiation of cardiovascular cells from mouse induced pluripotent stem cells. *Circulation.* 2008;118: 498-506.
43. Mauritz C, Schwanke K, Reppel M, Neef S, Katsirntaki K, Maier LS, Nguemo F, Menke S, Haustein M, Hescheler J, Hasenfuss G and Martin U. Generation of functional murine cardiac myocytes from induced pluripotent stem cells. *Circulation.* 2008;118: 507-517.
44. Nelson TJ, Martinez-Fernandez A, Yamada S, Perez-Terzic C, Ikeda Y and Terzic A. Repair of acute myocardial infarction by human stemness factors induced pluripotent stem cells. *Circulation.* 2009;120: 408-416.

Adaptation Biology and Medicine (Volume 6: Cell Adaptations and Challenges)
Editors: P. Wang, C.-H. Kuo, N. Takeda and P.K. Singal
Copyright © 2011 Narosa Publishing House Pvt. Ltd., New Delhi, India

4. Synergistic Effect of Magnesium and Diltiazem on Reversal of Myocardial Cell Membrane Stress and Dyskinesia

Otoni Moreira Gomes[1*], Neelam Khaper[2] and Pawan K. Singal[3]

[1]São Francisco de Assis Truth is Jesus Cardiovascular Foundation, Belo Horizonte, Brazil; [2]Northern Ontario School of Medicine, Lakehead University, Thunder Bay, Ontario, Canada; [3]Institute of Cardiovascular Sciences, University of Manitoba, Winnipeg, Canada

Abstract

Administration of the calcium antagonist, diltiazem, abrogates the exercise stress test-induced occurrence of the silent ST segment depression in many patients without symptoms of coronary artery disease, suggesting that myocardial dyskinesia plays an important role in this phenomenon. In order to avoid daily diltiazem dosages higher than 270mg, the possibility of co-treatment with magnesium, was investigated. Analysis of the effect of co-treatment of diltiazem-magnesium in the reversal of ST segment depression during treadmill electrocardiography exercise testing was undertaken in 10 patients, who did not exhibit clinical symptoms of coronary artery disease, but still presented ST segment depression during exercise stress testing. Seven of these patients were males, between 22 and 63 years old, (mean age was 42.1 years]. These patients were already receiving atenolol 25mg (MID) and diltiazem 90mg (TID). Following baseline testing, magnesium (150 mg BID) was also initiated and the treadmill electrocardiography exercise testing was repeated under the standard Bruce protocol five to six days later. The magnesium therapy corrected the STj and STy deviations without significant alteration in the heart rate in the control (157.1± 23.2 bpm) versus the treatment (131.1 ± 11.2 bpm ($p > 0.05$). STj depression changed from mean values of 2.7 ± 0.3 mm in the control to 0.9 ± 0.25 mm after treatment ($p < 0.001$); STy depression changed from mean values of 2.7 ± 0.5 mm in the control to mean values of 0.7 ± 0.06 mm after treatment ($p < 0.001$). We found that Mg corrected the STj and STy deviations without a significant alteration of the heart rate. These results indicate that magnesium, improved diltiazem's effect in treating myocardial dyskinesia.

Keywords: myocardial dyskinesia, silent myocardial ischemia, coronary artery disease, exercise testing

Introduction

The horizontal depression to the ST segment has often been interpreted as representing a physiological response of tachycardia and considered to be dependent on the heart rate [1]. Indeed, ST/FC index, as a diagnostic element has been evaluated in exercise stress testing [2,3]. However, this interpretation has

*Address for Correspondence: Otoni Moreira Gomes São Francisco de Assis Truth is Jesus Cardiovascular Foundation, Rua José do Patrocínio, 522 - 31525160 Belo Horizonte - MG - Brazil E-mail: gomes@ servcor.com

been challenged by others, arguing that it does not have a greater diagnostic significance than the simple ST variations [4-7]. In patients, without previous myocardial infarction, the isolated ST depression features a sensitivity rate of 72% and a specificity rate of 62.5% [8]. Increases in the R-wave sensitivity (58.5%) and specificity (67.5%) as well as in the QRS complex sensitivity (78%) and specificity (57.5%) were also reported [8]. These were comparable with the classical signs of pain (sensitivity 63% and specificity 75%) thereby suggesting that the reliability of exercise stress testing can be increased when the above mentioned parameters are taken into consideration. Furthermore, the positivity of only one signal corresponds to normal coronary circulation in most cases (8). This complexity in using the electrocardiographic diagnosis to guide patient care has often been highlighted in recent studies [9-11].

A significant increase in cardiovascular death rate (3.5 times) and death due to brain strokes (2.2 times) have been reported in asymptomatic patients with a horizontal depression of the ST segment during exercise stress testing [12]. With the exception of patients with normal basal ECG, this feature seemed to indicate the need for complementary angiographic, angiokinetic or scintigraphic studies with cineangiography being the most reliable for a definitive diagnosis. Studies of patients without coronary artery obstructions presenting downslope alterations of the ST segment during exercise stress test and with adequate long-term monitoring and follow-up remain rare and/or relatively incomplete. However, there is already sufficient evidence to justify preventive treatment in these cases, considering the inherent high risk of evolution toward myocardium infarction [12].

Exercise stress testing induces the occurrence of the silent ST segment depression in many patients without symptoms of coronary artery disease. Administration of the calcium inhibitor, diltiazem, abrogates this electrocardiographic feature, suggesting that myocardial cellular membrane ionic stress dyskinesia is an important part of this phenomenon [13-15]. In patients with acute myocardial infarction, diltiazem treatment attenuated myocardial dyskinesia which correlated with reduced infarct size [16]. Magnesium is known to block some of the physiological actions of calcium [17]. Magnesium treatment in patients with coronary artery disease has also been reported to significantly increase exercise duration time, lessen exercise induced chest pain and improve endothelial function [18, 19]. The main objective of the present investigation, therefore, was to evaluate the effect of diltiazem-Mg in the reversal of ST depression during treadmill exercise stress testing.

Methods

With the approval of the Bio-ethics Commission of the Institution, ten consecutive patients (seven of these patients were males, between 22 and 63 years old (mean age was 42.1 years) were included in the study. These patients without clinical symptoms and without scintillographic evidence of coronary artery disease, but still presenting ST segment depression during exercise stress testing were studied. The patients were already receiving atenolol 25mg (MID) and diltiazem 90mg (TID). A positive exercise stress test was qualified by the occurrence of ST segment horizontal depression without any evidence of precordial pain.

Following baseline testing, 150mg (BID) Mg was initiated and the treadmill electrocardiography exercise testing was repeated under the standard Bruce protocol after five or six days, The examinations were performed using a computerized treadmill (model Digistress/ MG), following the standard Bruce protocol. The J (STj) and Y (STy) point variations of the ST segment, the maximum

VO_2 reached (mL/kg/min), heart rate (bpm), the double product (DP), the METs reached and the functional class (AHA) were analyzed and graded, using the following criteria: 1- excellent, 2- good, 3- regular (medium), 4- weak and 5- very weak. The Student's t-test was employed to evaluate the statistical significance of differences between two parallel samples, and a p value less than 0.05 was considered statistically significant. The World Health Organization's EPIINFO computer program (version 6.04) was used.

Results

The Mg added therapy corrected the STj and STy deviations without significant alteration of the heart rate in the control (157.1± 23.2 bpm) compared with the treatment (131.1 ± 11.2 bpm) ($p > 0.05$). STj depression changed from mean values of 2.7± 0.3 mm, in the control to 0.9 ± 0.25 mm after treatment ($p < 0.001$); STy depression changed from mean values of 2.7± 0.5 mm in the control and mean values of 0.7 mm ± 0.06 mm after treatment ($p < 0.001$) [Table 1 and Fig.1].

Table 1. Effects of magnesium and diltiazem on the reversal of ST segment.

AGE (y)	METs mL/kg /min	VO_2 bpmXmmHg	DP	HR (bpm)	AHA	ST_J (mm)	ST_Y (mm)	PAIN	SAP (mmHg)	DAP (mmHg)
Atenolol (25 mg, MID) + Diltiazem (90-120 mg, TID)										
44	10.4	36.3	23409	157	2.6	–2.7	–2.7	–	155	69
Atenolol (25 mg, MID) + Diltiazem (90-120 mg, TID) + Magnesium (150 mg, BID)										
44	11.7	36.3	21381	131	2.4	–0.9*	-0.7*	–	149	66

* $p < 0.001$

Discussion

The myocardial disease responsible for the appearance of effort-induced depression of the ST segment, even without obstructive coronary lesions is an important pathology, causing high morbidity and mortality rates compared to other heart diseases [20-22]. In some studies, the incidence is similar to that of hypertension, with possible significant regional variations [23]. The positive exercise stress test in asymptomatic patients was registered in 0.6 to 15% of the studied population and in a maximum of 0.06 to 1.6% cases, it is associated with significant arterial disease (narrowing of 50% or more of the epicardial coronary artery). By comparison, the myocardial stress induced dyskinesia occurs in 0.54 to 13.4% of the asymptomatic population [24]. A significant increase in cardiovascular death (3.5 times) and of stroke (2.2 times) greater in asymptomatic individuals with horizontal depression of the ST segment evidenced during exercise stress testing has been demonstrated [12].

Our previous finding demonstrated the efficacy of using 90-120 mg diltiazem (TID) to treat the complications of myocardial ionic stress dyskinesia. The use of diltiazem increases patient safety during the initial period of rehabilitation, leading to effective myocardial conditioning. This leads to a faster improvement of the functional performance, as well as providing psychological and functional benefits in terms of motivation and adherence [13-15]. Improvement in myocardial dyskinesia with diltiazem has also been reported in patients with acute myocardial infarction [16].

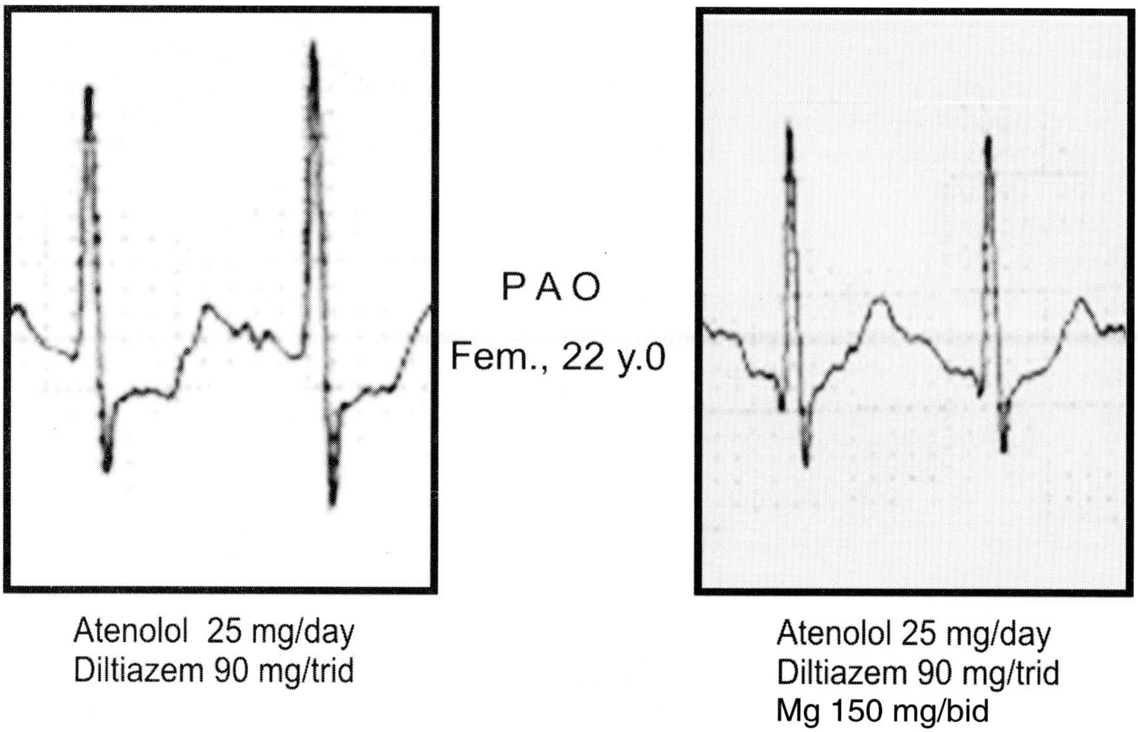

Fig.1. ST downslope correction by Mg interactive effect

It is known that certain patients require diltiazem dosages higher than 270 mg daily (90 mg TID). Present study shows the possibility of a positive synergistic effect of Mg co-treatment. We found that 150mg Mg (BID) corrected the STj and STy deviations without significant alteration of the heart rate. In previous studies, comparing the efficacy of Mg and diltiazem, it was found that magnesium sulfate had favorable effects on ventricular rate control and restoration of sinus rhythm in patients with atrial fibrillation [25]. Oral magnesium therapy in patients with coronary artery disease was also associated with significant improvement in branchial artery endothelial function, exercise tolerance and improvement in functional capacity [18, 19].

Improvement in ejection fraction and lower incidence of heart failure with oral magnesium therapy has also been reported in patients with acute myocardial infarction [26]. Pretreatment with magnesium-sulfate in patients with acute myocardial infarction was also protective against reperfusion injury by attenuating the levels of inflammatory mediators such as interleukin-6 and matrix metalloproteinase -1, thereby preventing the increase in myocardial lesion by calcium overload [27]. Although, the mechanism/s of the beneficial effects of magnesium in these pathophysiological conditions are still unclear, some studies suggest its role in reducing calcium overload, and improving the vascular tone by reducing afterload and attenuating cardiac arrhythmia [17-19, 26-28]. The results from the present study indicate that Mg improved the efficacy of diltiazem in treating myocardial cell membrane ionic stress dyskinesia.

Acknowledgements

Dr. Pawan Singal is the holder of the Naranjan Dhalla Chair in Cardiovascular Research supported by the St. Boniface Hospital and Research Foundation.

References

1. Okin PM, Kligfield P. Computer-based implementation of the ST-segment/heart rate slope. Am J Cardiol 1989;64:926-30.
2. Detrano R, Salcedo E, Passalacqua M, Friis R. Exercise electrocardiographic variables: a critical appraisal. J Am Coll Cardiol 1986;8:836-47.
3. Lachterman B, Lehmann KG, Detrano R, Neutel J, Froelicher VF. Comparison of ST segment/ heart rate index to standard ST criteria for analysis of exercise electrocardiogram. Circulation 1990;82:44-50.
4. Froelicher VF, Lehmann KG, Thomas R, Goldman R, Morrison D, Edson R et al. The electrocardiographic exercise test in a population with reduced workup bias: diagnostic performance, computerized interpretation, and multivariable prediction. Ann Intern Med 1998;128:965-74.
5. Okin PM, Grandits G, Rautaharju PM, Prineas RP, Cohen JD, Crow RS et al. Prognostic value of heart rate adjustment of exercise-induced ST segment depression in the multiple risk factor intervention trial. J Am Coll Cardiol 1996;27:1437-43.
6. Okin PM, Anderson KM, Levy D, Kligfield P. Heart rate adjustment of exercise-induced ST segment depression: improved risk stratification in the Framingham Offspring Study. Circulation 1991;83: 866-74.
7. Gibbons RJ, Balady GJ, Bricker JT, Chaitman BR, Fletcher GF, Froelicher VF, et al. ACC/AHA 2002 Guideline Update for Exercise Testing: Summary Article: Circulation 2002;106:1883-92.
8. Pic A, Broustet JP. Diagnostic value of amplitude variations of the QRS complex during in computerized exercise testing. Arch Mal Coeur Vaiss 1984;77:54-63.
9. Severi S, Orsini E, Marracini P, Michelassi C, L'Abbate A. The basal electrocardiogram and the exercise stress test in assessing prognosis in patients with unstable angina. Eur Heart J 1988;9:441-6.
10. Fruegaard P, Launbjerg J, Jacobsen HL, Madsen JK. Seven year prognostic value of the electrocardiogram at rest and an exercise test in patients admitted for, but without, confirmed myocardial infarction. Eur Heart J 1993;14:499-504.
11. Grundy D, Gibler WB, Bassan R. O eletrocardiograma e o monitor de tendência do segmento ST na valiação diagnóstica e no prognóstico da dor torácica. In: Bassan R, editor. Síndrome coronariana aguda nas unidades de dor torácica. Rio de Janeiro:Atheneu;2000. p.71-95.
12. Kurl S, Laukkanen JA, Tuomainen TP, Rauramaa R, Lakka TA, Salonen R et al. Association of exercise-induced, silent ST-segment depression with the risk of stroke and cardiovascular disease in men. Stroke 2003;34:1760-5.
13. Gomes OM, Gomes ES, Faraj M. Myocardium stress diskinesia disease Braz J Cardiovasc Surg 2004; 19(4): 378-85

14. Gomes OM, Gomes ES. Myocardial cell membrane stress ionic dyskinesia reversal by diltiazem. Exp and Clin Cardiology 2006; 11(3); 239-42.
15. Gomes O.M, Gomes E.S. Discinesia iônica de estresse da membrana celular miocárdica:novo paradigma para diagnóstico e avaliação terapêutica pelo teste ergométrico. Revista do DERC 2007;39:18-20.
16. Pizzetti G, Mailhac A, Li Volsi L, Di Marco F, Lu C, Margonato A, et al. Beneficial effects of diltiazem during myocardial reperfusion: a randomized trial in acute myocardial infarction. Ital Heart J 2001; 2: 757-65.
17. Iseri LT, French JH. Magnesium: nature's physiologic calcium blocker. Am Heart J 1984; 108: 188-93.
18. Shecter M, Bairey Merz CN, Stuehlinger HG, Slany J, Pachinger O, Rabinowitz B. Effects of oral magenisium therapy on exercise tolerance, exercise-induced chest pain and quality of life in patients with coronary artery disease. Am J Cardiol 2003;91:517-21.
19. Shechter M, Sharir M, Labrador MP, Forrester J, Silver B, Merz B. Oral magnesium therapy improves endothelial function in patients with coronary artery disease. Circulation 2000;102:2353-58.
20. Gomes OM, Crizola R, Kazzaz NM, Anjos MLM, Caetano MC, Pitchon M et al. Censo pressórico: estudo Belo Horizonte. Coração 1989;1:20-5.
21. Gomes OM, Caetano MC, Pitchon M, Teixeira RMB, Barros MVL, Gomes ES et al. Censo pressórico: estudo Belo Horizonte II/Seguimento tardio. Coração 1990;2:13-6.
22. Hedbald B, Juul-Moller S, Svemsson K, Hanson B, Isacasson SO, Janson L, et al. Increased mortality in men with ST segment depression during 24 h ambulatory long-term ECG recording. Results from prospective population study 'Men born in 1914', from Malmo Sweden. European Heart Journal 1989;10:149-58
23. Hedblad B, Janzon L. Hypertension and ST segment depression during ambulatory electrocardiographic recording. Results from the prospective population study 'men born in 1914' from Malmo, Sweden. Hypertension 1992;20:32-7
24. Fowler-Brown A, Pignone M, Pletcher M, Tice JA, Sutton SF, Lohr KN. Exercise tolerance testing to screen for coronary heart disease: a systematic review for the technical support for the US Preventive Services Task Force. Ann Inter Med 2004;140:W9-24.
25. Chiladakis JA, Stathopoulos C, Davlouros P, Manolis AS. Intravenous magnesium sulphate versus diltiazem in paroxysmal atrial fibrillation. Int J Cardiology 2001:79:287-91.
26. Shechter M, Hod H, Rabinowitz B, Boyko V, Chouraqui P. Long term outcome of intravenous magnesium therapy in thrombolysis-ineligible acute MI patients. Cardiology 2003 99;205-10.
27. Shibata M, Ueshima K, Harada M, Nakamura M, Hiramori K, Endo S. et al. Effect of magnesium sulfate pretreatment and significànce of matrix metalloproteinase-1 and interleukin-6 levels in coronary reperfusion therapy for patients with acute myocardial infarction. Angiology 1999 50; 573-582.
28. Kuguyama K, Yasue H, Okumura K, Goto K, Munoda K, Muyagi H. et al. Suppression of exercise induced angina by magnesium sulfate in patients with variant angina. J Am Coll Cardiol. 1988. 12:1177-1183.

Adaptation Biology and Medicine (Volume 6: Cell Adaptations and Challenges)
Editors: P. Wang, C.-H. Kuo, N. Takeda and P.K. Singal
Copyright © 2011 Narosa Publishing House Pvt. Ltd., New Delhi, India

5. Adaptive Mechanisms in the Diabetic Myocardium: An Alternative form of Endogenous Cardioprotection and its Loss

T. Ravingerová[*], J. Matejíková, A. Adameová[1], M. Nemceková, O. Pechánová[2], T. Kelly[3], Z. Sumbalová[4], A. Lazou[3]

Institute for Heart Research, [2]Institute of Normal and Pathological Physiology, Slovak Academy of Sciences and Centre of Excellence for Cardiovascular Research SAS; [1]Department of Pharmacology and Toxicology, Faculty of Pharmacy; [4]Pharmacobiochemical Laboratory, Faculty of Medicine, Comenius University, Bratislava, Slovakia; [3]School of Biology, Aristotle University, Thessaloniki, Greece

Summary

High glucose (HG) is one of the factors determining the outcome of myocardial ischemic insult. However, it is still not clear whether it is casually related to a decreased ischemic tolerance in diabetic patients. In contrast to clinical and epidemiological studies demonstrating a higher risk of cardiovascular disorders in the diabetics, experimental data are not unequivocal and suggest that besides higher myocardial vulnerability, diabetes mellitus may be associated with induction of adaptive processes leading to paradoxically lower susceptibility to ischemia. It has been proposed that this phenomenon might share some molecular pathways with short-term endogenous protection (preconditioning) against ischemia/reperfusion injury in non-diseased heart. This chapter reviews some controversial findings of enhanced resistance to ischemia in the diabetic heart that stem from the experimental studies in different models of myocardial ischemia/reperfusion injury. Specifically, it addresses the issue of potential mechanisms of increased resistance to ischemia in the experimental model of streptozotocin-induced diabetes. A particular focus has been drawn to the role of reactive oxygen species and hyperglycemia as some of the stressful factors, as well as to the cell signalling mechanisms mediated by "pro-survival" cascades of protein kinases, in relation to the mechanisms of classical ischemic preconditioning. Finally, mechanisms involved in suppression of protection in the diabetic myocardium including the effect of concomitant pathology, such as hypercholesterolemia, are also discussed.

Keywords: Hyperglycemia, oxidative stress, myocardial ischemia, adaptation, cell signaling

[*]Address for Correspondence: Dr. T. Ravingerová, Institute for Heart Research, Slovak Academy of Sciences, POB 104, Dúbravská cesta 9 840 05 Bratislava, Slovak Republic, E-mail: usrdravi@savba.sk, Phone: 421 2 5477 4405
Fax: 421 2 5477 6637

Introduction

Susceptibility to Ischemia/Reperfusion Injury in the Diabetic Heart: Controversies of the Experimental Studies

Diabetes mellitus is associated with a higher risk of congestive heart failure and ischemic heart disease including myocardial infarction that has been described in various clinical and epidemiological studies (Kannel and McGee 1979; Dubrey *et al.* 1994). Deterioration of heart function and rhythm disorders that can develop in diabetic patients even in the absence of coronary artery disease are common manifestations of diabetic cardiomyopathy associated with an impaired homeostasis of Na^+, Ca^{2+} and K^+ in the cardiomyocytes. Development of diabetes leads to oxidative stress (Singal *et al.* 2001; Maritim *et al.* 2003) and defects in the cell sarcolemmal and sarcoplasmic reticular membranes, as well as to alterations in the function of ion transport systems (Na^+/K^+- and Ca^{2+}-ATPase, Na^+/H^+ and Na^+/Ca^{2+} exchangers and Ca^{2+} channels) resulting in the abnormal Na^+ and Ca^{2+} handling in the diabetic myocardium that might compromise its tolerance to ischemia (Dhalla *et al.* 1998; Khandoudi *et al.* 1990; Lee *et al.* 1992; Pi-erce *et al.* 1990; Anzawa *et al.* 2006).

However, in contrast to human studies, experimental data are not unequivocal suggesting that diabetic hearts are not uniform in their susceptibility to ischemic injury that may be not only increased, but also unchanged or even decreased (Feuvray and Lopaschuk 1997; Galinanes and Fowler 2004). One of the main reasons for the differences between the clinical and experimental studies may be the contribution of other pathologies in the diabetic patients, such as hypertension, hypercholesterolemia, microvascular dysfunction, neuropathy and nephropathy, as well as the treatment with antidiabetic drugs, in particular, with sulfonylureas that block the preconditioning effect during myocardial ischemia by preventing opening of the ATP-sensitive mitochondrial [K(ATP] channels (Brady and Terzic 1998; Galinanes and Fowler 2004; Loubani *et al.* 2005; Fisman *et al.* 2008). These circumstances may determine the overall outcome of myocardial response to ischemia in the diabetics.

First paradoxical results showing better recovery of contractile function in the diabetic rat hearts subjected to ischemia/reperfusion (I/R) were reported by Tani and Neely (1988). Later on, Liu *et al.* (1993) have also shown that acute streptozotocin (STZ)-induced diabetes reduced the size of infarction measured by tetrazolium staining in a rat model of regional I/R by nearly 70%. Several mechanisms have been proposed to explain the discrepancies in the experimental studies. It was suggested that differences between the duration, severity and type of the diabetes, different protocols of the studies testing the effect of ischemia alone or ischemia and subsequent reperfusion, as well as between different end-points may account for the conflicting results. Further explanations, such as severity of ischemic protocol (global zero-flow versus low-flow ischemia), the presence of the metabolic substrate in the perfusion medium and the differences related to the experimental models utilized in the studies (open-chest animals or isolated hearts) can also be taken into consideration. Last but not least, differences between the species and the distribution of coronary collaterals (and myocardial perfusion) may contribute to the diversity of the results (Paulson 1997; Feuvray and Lopaschuk 1997).

Thus, studies of infarct size indicate that diabetic dogs develop larger infarcts than normal ones (Forrat *et al.* 1993), whereas in the diabetic rabbits and rats, infarct size tends to be smaller (Hadour *et al.* 1998; Liu *et al.* 1993). Similarly, susceptibility to arrhythmias in the diabetic hearts has been found both enhanced and reduced (Hekimian *et al.* 1985; Kusama *et al.* 1992). Functional recovery upon reperfusion in the diabetic rats has been reported to be improved in the early stage and impaired in the later phase of the disease (Tosaki *et al.* 1996).

Significant alterations of the heart function occur in the chronic (9 weeks) phase of STZ-induced diabetes (increased ratio of heart to body weight, decreased heart rate and coronary flow, as well as lower rates of contraction and relaxation, along with reduced myocardial levels of high energy phosphates and morphological deteriorations in the endothelial capillary cells, as a manifestation of developing diabetic cardiomyopathy. Surprisingly, these diabetic hearts were more resistant to Ca^{2+} oveload after transient depletion of Ca^{2+} followed by its repletion, when 85% of the diabetic hearts survived Ca^{2+} paradox-induced injury in contrast to 0% in non-diabetics (Ravingerova *et al.* 1996). In these surviving hearts, not only almost complete restoration of electrical activity and coronary flow, but an improved recovery of contractile function that reached 75% of the baseline values, was observed. Moreover, recovery of systolic function was accompanied by a remarkable attenuation of irreversible contracture and diastolic dysfunction that was associated with a tremendous rise in left ventricular enddiastolic pressure in the non-diabetic controls as a consequence of Ca^{2+} overload in cardiomyocytes. Importantly, contents of ATP and total adenine nucleotides in the diabetic hearts that survived Ca^{2+} paradox injury were significantly higher than in the non-diabetic controls (Ziegelhoffer *et al.* 1997). The latter was accompanied by a better preservation of the cell ultrastructure manifested by the absence of hypercontracted and ruptured myofibrils and by the maintenance of sarcolemmal and mitochondrial integrity (Tribulova *et al.* 1996).

Our in *vivo* studies in the open-chest anesthetized rats with acute (1 week) STZ-induced diabetes clearly demonstrated the differences in the susceptibility of the diabetic hearts to ischemia-induced ventricular arrhythmias and irreversible myocardial damage: the size of myocardial infarction (IS; as a percentage of the area at risk; AR) in the diabetic rats reached only 67% of that in the non-diabetic controls, whereas the incidence and severity of ventricular atrrhythmias were not influenced by the diabetic state (Ravingerova *et al.* 2003). The latter was also confirmed in our recent *in vivo* studies showing that the incidence of severe ventricular arrhythmias during LAD coronary artery occlusion and reperfusion, such as ventricular tachycardia (VT) and fibrillation (VF), was similar in the acutely diabetic hearts and in the non-diabetic controls (Adameova *et al.* 2007a,b).

On the other hand, the studies performed in the Langendorff-perfused hearts of rats in the same model of acute STZ-induced diabetes revealed both, smaller infarctions (IS/AR reduced by 60% in the diabetic hearts) (Fig. 1A) and a lower vulnerability to ischemia-induced severe ventricular arrhythmias (Fig. 1B,C). A pattern of distribution of ectopic activity during 30 min of ischemia was not modified in the diabetics (Fig. 1B), however, the total number of premature ventricular complexes (PVC) and total duration of the episodes of VT (Fig. 1C) was significantly reduced in the diabetic hearts (Matejikova *et al.* 2008). In addition, postischemic recovery of left ventricular function after global I/R was also significantly improved in the diabetic hearts (Fig. 1D).

Our studies in this model also demonstrated that despite the development of cardiomyopathy already in the early stage of the disease (documented by bradycardia and reduced rates of contraction and relaxation, as well as by decreased pressure-rate product, an index of cardiac contractile performance), only 42% of these hearts exhibited VT and 16% of them had short episodes of VF and no sustained VF (SVF), different to from VT 100%, VF 70% and SVF 36% in the non-diabetics ($P<0.05$) (Ravingerova *et al.* 2000). Moreover, this antiarrhythmic protection was maintained even in

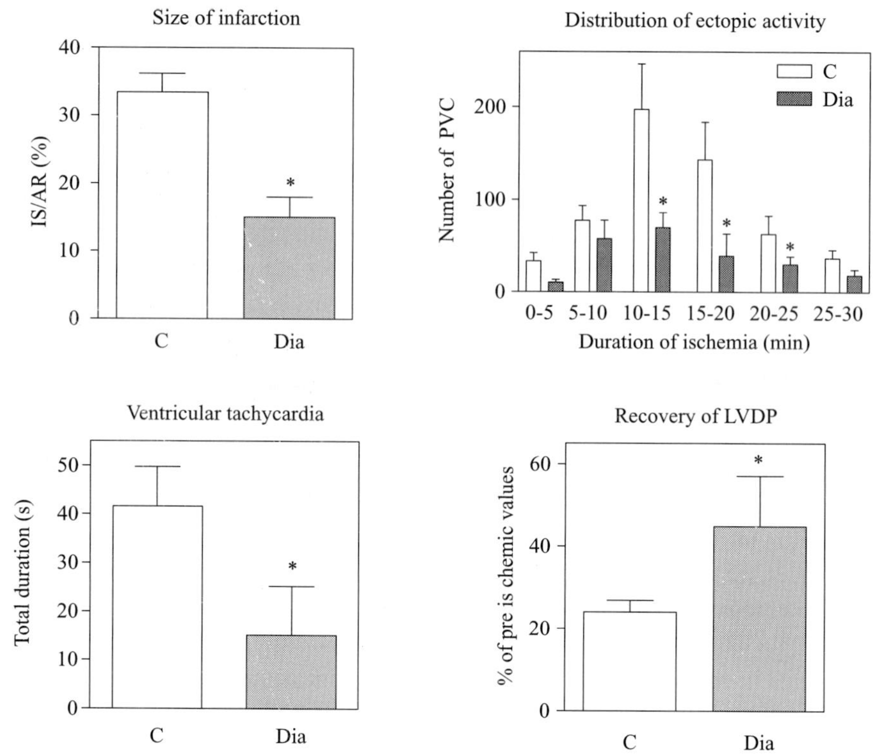

Fig. 1. Effect of 1-week streptozotocin-induced diabetes on the size of myocardial infarction (A), ischemia-induced ventricular arrhythmias (B,C) and postischemic recovery of left ventricular developed pressure (LVDP) in the Langendorff-perfused rat hearts. Infarct size (IS) area is expressed in % of the area at risk (AR) area. Values are means ± S.E.M. from 10-12 hearts per group. * P<0.05 vs. non-diabetic control (C) group.

the chronic (8 weeks) phase of the disease, and although VT occurred in almost all of the diabetic hearts, the total duration of both VT and VF was substantially shorter than in the respective controls (Ravingerova *et al.* 2001).

Furthermore, in chronically diabetic rats, antiarrhythmic protection has been observed not only in the isolated heart preparation, but in the diabetic animals *in vivo*, in which the ectopic activity was also lower, the incidence of VT was reduced by more than 2-fold as compared with the respective controls, and lethal SVF was completely abolished. In contrast, antiinfarct protection in these animals was attenuated as compared with the acute phase of the disease indicating a dichotomy in the susceptibility of the diabetic hearts to ischemia-induced ventricular arrhythmias and irreversible injury in relationship with the duration of the diabetic state (Ravingerova *et al.* 2003).

Taken together, these results support a view of rather lower sensitivity of the diabetic rat heart to ischemia-reperfusion injury in both stages of the disease, albeit with different manifestations of an enhanced ischemic tolerance.

Mechanisms of Lower Sensitivity to I/R Injury in the Diabetic Heart

Metabolic alterations in the diabetic heart involve impaired metabolism of carbohydrates including glucose transport and utilization (Rodrigues *et al.* 1995). The role of glycogen metabolism during ischemia is still a matter of discussion since on the one hand, anaerobic glycolysis provides a major source of ATP production, but on the other hand, it also results in the increased accumulation of acid glycolytic metabolites that decrease intracellular pH and contribute to the severity of injury. It is suggested that processes related to the alterations in glucose metabolism and to the regulation of the intracellular pH might be responsible for the reduced sensitivity to ischemia under certain experimental conditions (Khandoudi *et al.* 1990; Pierce *et al.* 1990; Feuvray and Lopaschuk 1997; Galinanes and Fowler 2004). E.g., in contrast to less severe models utilizing low-flow or moderate ischemia (with lower degree of acidosis due to the partial washout of lactate), in a setting of severe zero-flow ischemia, decreased glycolytic flux might be actually beneficial and attenuate production of glycolytic products and H^+ load. Under conditions of global I/R associated with Na^+ and Ca^{2+} gain, a lower rate of glycolysis in the myocardium due to inhibition of phosphofructokinase activity, the rate-limiting step in glycolysis, by enhanced levels of fatty acids (Opie 1996) and decreased activity of Na^+/H^+- exchanger altogether or separately may result in a lower gain of Na^+ during ischemia and thus, reduced influx of Ca^{2+} via Na^+/Ca^{2+} exchange during early reperfusion (Khandoudi *et al.* 1990; Pierce *et al.* 1990; Feuvray and Lopaschuk 1997).

Furthermore, it was found that in the acutely diabetic rat heart, the mitochondrial mechanisms of energy transfer can adapt to the conditions of increased energy demands due to augmented calcium transients and that the latter may also account for its reduced susceptibility to injury (Ziegelhoffer *et al.* 2002).

Antiarrhythmic Mechanisms in the Diabetic Myocardium

Concerning the mechanisms of arrhythmogenesis, it is known that the outward potassium currents are reduced to a different extent in the epi- and endocardial layers of the diabetic myocardium (Shimoni *et al.* 1995) thus attenuating dispersion of refractoriness as a substrate for reentry arrhythmias occurring during ischemia. Moreover, K(ATP) channels in the diabetic cardiomyocytes have been found to be much more sensitive and open at higher levels of ATP (Smith and Wahler 1996).

Activation of K(ATP) channels has been considered as one of the mechanisms of cardioprotection in general (Noma *et al.* 1983) including protection against arrhythmias related to triggered activity due to

Table 1. Potential mechanisms of preconditioning-like protection in the diabetic myocardium. Relevance to ischemic preconditioning.

Alterations in the metabolism of glucose (Goto *et al.* 1995), *high glucose* acting as *a PC mimetic* in the non-diabetic myocytes.
Abnormal calcium handling – *calcium PC* (Meldrum *et al.* 1996).
Increased production of ROS (Kakkar *et al.* 1996; Maritim *et al.* 2003); higher levels of antioxidants (Chen *et al.* 2006; Matejikova *et al.* 2008).
↑ Activation of PKCε (Ooie *et al.* 2003; Malhotra *et al.* 2005)
↑ MAPK activity – enhanced ERK1/2 activation (Strniskova *et al.* 2003; Xu *et al.* 2004)
↑ PI3K/Akt activity, eNOS activity (Ma *et al.* 2006)

enhanced Ca^{2+} influx (Spinelli *et al.* 1991), and sarcolemmal K(ATP) channels appeared to be involved in the antiarrhythmic protection in the rabbit (Tan *et al.* 1993) and canine (Gross and Auchampach 1992) myocardium. In addition, resistance to arrhythmias caused by high calcium in the diabetic rat hearts can be explained by the alterations in the properties of L-type Ca^{2+} channels and reduced Ca^{2+} influx (Lee *et al.* 1992).

Moreover, accummulation of lactate during ischemia is substantially lower in the diabetic myocardium and does not reach its level in the non-diabetic hearts (Ravingerova *et al.* 2000, 2001). Thus, a decreased production of acid metabolites and subsequently reduced local acidosis as a source of alterations in the cell membrane currents and electrophysiological changes promoting arrhythmogenesis (Cascio *et al.* 1995) might contribute to suppression of arrhythmias occurring during ischemia in the diabetic hearts. In addition, lower susceptibility to reperfusion-induced arrhythmias in the diabetic rat hearts is supported by the observation of an increased resistance to calcium overload in the diabetic myocardium (Ziegelhoffer *et al.* 1999).

Diabetes as an Alternative Form of Adaptive Preconditioning-like Phenomena

Potential mechanisms of preconditioning-like protection in the diabetic myocardium are summarized in Table 1. Altered metabolism of glucose has been proposed to play a role in cardioprotective mechanisms in the diabetic myocardium based upon the findings that inhibition of glucose uptake by the cells can mimic ischemic preconditioning (I-PC) and salvage normal myocardium (Goto *et al.* 1995), whereas restoration of glycogen stores depleted by I-PC leads to a loss of myocardial protection (Wolfe *et al.* 1993). Moreover, high glucose (25 mM) treatment has been shown to render normal cardiomyocytes resistant to chronic hypoxia-induced apoptosis and necrosis by preventing the accumulation of Ca^{2+} during hypoxia (Schaffer *et al.* 2000). Furthermore, it has been demonstrated that an enhanced glucose represents a stressful stimulus able to trigger an adaptive response in the diabetic heart (Chen *et al.* 2006).

In addition, increased resistance to ischemia in the experimental models of diabetes can also be related to the alterations in intracellular calcium signaling since it has been demonstrated that increased calcium can induce preconditioning-like protection in the normal heart (Meldrum *et al.* 1996).

Thus, greater tolerance to ischemic injury observed in the diabetic heart can be considered as an alternative form of intrinsic cardioprotection analogous to that induced by I-PC in the normal heart, or by adaptation to chronic myocardial hypoxia, when numerous metabolic stimuli, in particular those related to oxidative stress and increased intracellular calcium signaling can trigger protection against acute I/R.

The Role of Free Radicals and Pro/Antioxidative State in the Diabetic Heart

Kakkar *et al.* (1996) have demonstrated that although increased generation of reactive oxygen species (ROS) can be detected in the very early stage of STZ-induced diabetes, the latter was also accompanied by an increased level of antioxidative enzymes (catalase, superoxide dismutase and glutathione peroxidase). Enhanced levels of endogenous antioxidants may be thus considered as a manifestation of the adaptive response in the rat diabetic myocardium (Kucharska *et al.* 2001) induced by increased ROS similar to that demonstrated in the non-diabetic hearts exposed to I-PC (Das *et al.* 1992).

Increased generation of ROS coupled with higher activities of antioxidant enzymes and concentration of antioxidants in the models of experimental diabetes have been reported by several authors (Kakkar *et al.* 1996; Chen *et al.* 2006). In our recent study we found that one week after the induction of diabetes in a STZ rat model, a decreased susceptibility to ventricular arrhythmias manifested by a lower total number of PVC and shorter duration of VT episodes during 30 min of ischemia was associated with significantly elevated levels of endogenous antioxidants, such as CoQ10 and, in particular, CoQ_9 in the left ventricular myocardium of the diabetic animals. In addition, the levels of a-tocopherol were increased by 22.5% in the diabetic hearts (Matejikova *et al.* 2008). In the study of Chen *et al.* (2006), higher resistance of the diabetic hearts to I/R injury was associated with an increased antioxidant capacity of the myocardium already in the acute phase of STZ-induced diabetes with severe hyperglycemia (blood glucose > 20 mM), or when the hearts were perfused with a similarly high concentration of glucose.

The results of our study have also pointed out to an important aspect related to the antioxidant treatment of the diabetics. While treatment with antioxidant N-acetylcysteine (NAC) and NOS inhibitor L-NAME conferred an effective protection against ischemia-induced arrhythmias, both interventions were not effective in the diabetic rat heart and did not afford any additional myocardial protection during I/R (Matejikova *et al.* 2008). Thus, it is possible that changes in the pro/antioxidant state might modify a response of the diabetic heart to antioxidant treatment. The latter might result in a reduced effectiveness of NAC to further suppress arrhythmias in the diabetic myocardium. Similarly, pretreatment with L-NAME that was found cardioprotective and reduced the extent of I/R injury in the normal non-preconditioned hearts (Andelova *et al.* 2005), was antiarrhythmic only in the non-diabetic hearts (Matejikova *et al.* 2008). On the other hand, since in the acutely preconditioned non-diabetic myocardium L-NAME suppressed an increased ischemic tolerance (Andelova *et al.* 2005), its antiarrhythmic effect might also be attenuated under conditions of the diabetic state.

It is thus conceivable that both, cardioprotection conferred by I-PC in the healthy heart, and increased resistance to ischemia in the diabetic myocardium might share similar molecular protective pathways, such as ROS and NO signaling. These data are supported by the study of Chen *et al.* (2006) who demonstrated that in the presence of L-NAME, reperfusion arrhythmias were increased in the diabetic heart, suggesting an involvement of NO in the mechanisms of protection in the diabetic myocardium, and, more importantly, that the efficacy of antiradical interventions may differ in the non-adapted and adapted myocardium.

The role of ROS-mediated signaling in the mechanisms of higher ischemic tolerance of the diabetic myocardium has been demonstrated in the experiments performed in acutely diabetic rat hearts, which showed an increased resistance to calcium overload despite markedly impaired biophysical properties (decreased fluidity) of the sarcolemmal membrane due to the processes of ROS formation and non-enzymatic glycation of the membrane proteins (Ziegelhoffer at al. 1997). Moreover, reversal of these deleterious effects of ROS on protein glycation by the treatement with resorcylidene aminoguanidine attenuated not only the changes in the membrane properties and its enhanced rigidity, but abolished calcium tolerance of the diabetic hearts indicating that a certain amount of ROS formation is required for the triggering of adaptive processes in the diabetic myocardium (Ziegelhoffer at al. 1999). In addition, moderately increased myocardial production of ROS during the preconditioning procedure (Matejíková *et al.* 2009) is likely to play a role in myocardial preservation through the attenuation in

ROS levels during the subsequent phase of sustained ischemia (Halestrap *et al.* 2007). Similarly, in the diabetic hearts, cardioprotective effects during ischemia were associated with lower production of ROS (reduced TBARS levels) than in the non-diabetic rat myocardium (Fig. 2).

It appears thus that radicals may play a dual role in the heart: apart from their damaging role in the non-adapted myocardium, they may act as signaling molecules involved in protective mechanisms of both, short-term cardioprotection and long-lasting cardiac adaptation (Andelova *et al.* 2005; Kolar *et al.* 2007). These findings further address the issue of the effectiveness of antiradical interventions in the intact non-adapted (unstressed) myocardium and in the myocardium that has shifted to a defensive phenotype in response to stress.

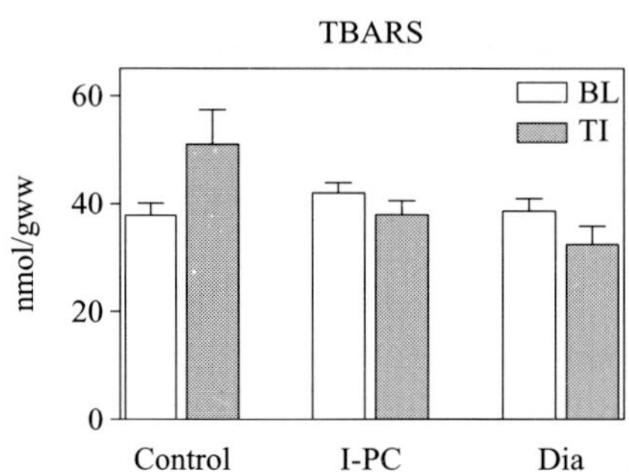

Fig. 2. Myocardial concentration of thiobarbituric acid reactive substances (TBARS) at baseline (BL) and after test ischemia (TI) in the non-preconditioned controls, hearts preconditioned with one episode of ischemia and reperfusion (5 min each; I-PC) and hearts from 1-week diabetic rats (Dia). Values are means ± S.E.M. from 6-8 hearts per group. * P<0.05 vs. BL; # P<0.05 vs. non-diabetic controls.

The Role of Protein Kinases in Cardioprotection in the Diabetic Myocardium

It has been found that calcium-induced protection against I/R is associated with the translocation of Protein Kinase C (PKC) (Meldrum *et al.* 1996). Moreover, hyperglycemia is a powerful stimulus for the PKC activation that is known to occur in the diabetic myocardium even in the early phase of the disease (Malhotra *et al.* 1997). In addition, its translocation has been shown to mediate cardioprotection in the STZ-induced diabetic rat heart (Moon *et al.* 1999), whereas PKC inhibition abolished an increased resistance to I/R injury in the diabetic hearts (Ooie *et al.* 2003). Furthermore, it was shown that it is the activation of cardiac PKC in genetically engineered mice that mediates antiapoptotic effects and leads to the preservation of left ventricular pump function, which promotes the survival phenotype and inhibits negative inotropic properties of chronic hyperglycemia (Malhotra *et al.* 2005).

In line with the studies demonstrating an important role of protein kinase signaling in cardioprotection in the diabetic myocardium (Ooie *et al.* 2003; Ma *et al.* 2006), we have detected a nearly 3-fold increase in the levels of phosphorylated ERK1/2 (extracellular signal-regulated kinase) in the diabetic hearts in the acute (1-week) phase that correlated with a reduced size of infarction (Strniskova *et al.* 2003) suggesting a potential positive role of this cascade in the response of the diabetic heart to an acute ischemic challenge. In concert, the study of Xu *et al.* (2004) clearly demonstrated that short- and long-term duration of hyperglycemia exert opposite influences on the intracellular signaling mediated by ERK1/2 pathway that correlated with the size of myocardial infarction and the intensity of apoptotic processes. While in the acute phase of STZ-induced diabetes, an increased ERK1/2 activity, reduced infarct size and smaller number of dead myocytes were observed, and these protective effects were suppressed with the duration of the disease.

Since phosphatidylinositol 3-kinase and its effector Akt (PI3K/Akt) cascade plays an important role in cell survival mechanisms, as well as in the regulation of glucose metabolism in the normal heart (Murphy and Steenbergen 2007), it could be expected to be involved in the changes in ischemic tolerance in the diabetics. In accordance, it has been reported by Ma *et al.* (2006) that PI3K/Akt and eNOS become transiently activated in the early stage of diabetes with a subsequent increase in NO formation, and that this may protect the heart from I/R injury.

To elucidate the involvement of PI3K/Akt in the PC-like effect of high glucose (22 mM) in the normal heart (glucose preconditioning, GPC), as well as in response to I/R injury in the diabetic heart (blood glucose >20 mM), PI3K/Akt inhibitor wortmannin was chosen to clarify the role of this cascade in lethal injury and given in the concentration (100 nM) that suppressed PC-induced infarct size limitation in the normal hearts (Ravingerová *et al.* 2009a). While in the control groups the size of infarction was not affected by wortmannin, in both GPC (Fig. 3A) and diabetic (Fig. 3B) groups, cardioprotection was abolished by treatment with this compound indicating that activation of PI3K/Akt cascade is involved in the mechanisms of antiinfarct protection triggerred by high glucose.

Therefore, it seems that in the diabetic hearts, numerous molecular mechanisms can be responsible for the activation of the endogenous cardioprotection, at least in the early phase of the disease.

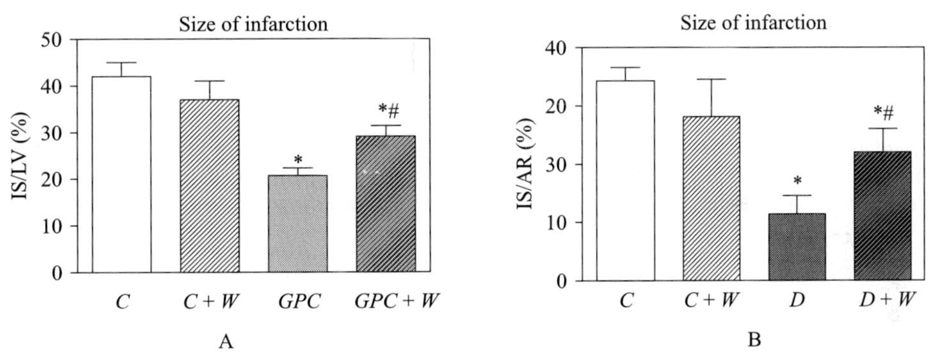

Fig. 3. Effect of PI3K inhibitor wortmannin (W; 100 nM) on the size of myocardial infarction in the Langendorff-perfused hearts preconditioned with high glucose (GPC; A) and in the hearts of 1-week diabetic (D) rats (B). Values are means ± S.E.M. from 8-10 hearts per group. * P<0.05 vs. controls (C); # P<0.05 vs. GPC or D.

Effect of Preconditioning in the Diabetic Heart

Whether or not the diseased myocardium can be preconditioned similarly as a normal one, is still a matter of discussion. Some authors consider preconditioning as a „healthy heart" phenomenon, and it is well documented that an enhanced resistance to I/R in the normal preconditioned myocardium may be abolished by concurrent pathological conditions, such as hypercholesterolemia (Giricz *et al.* 2006).

In this respect, of major concern may be hypoglycemic therapy with oral antidiabetic drugs, in particular, sulphonylurea drugs (e.g., glibenclamide) in humans, since K(ATP) inhibition as a

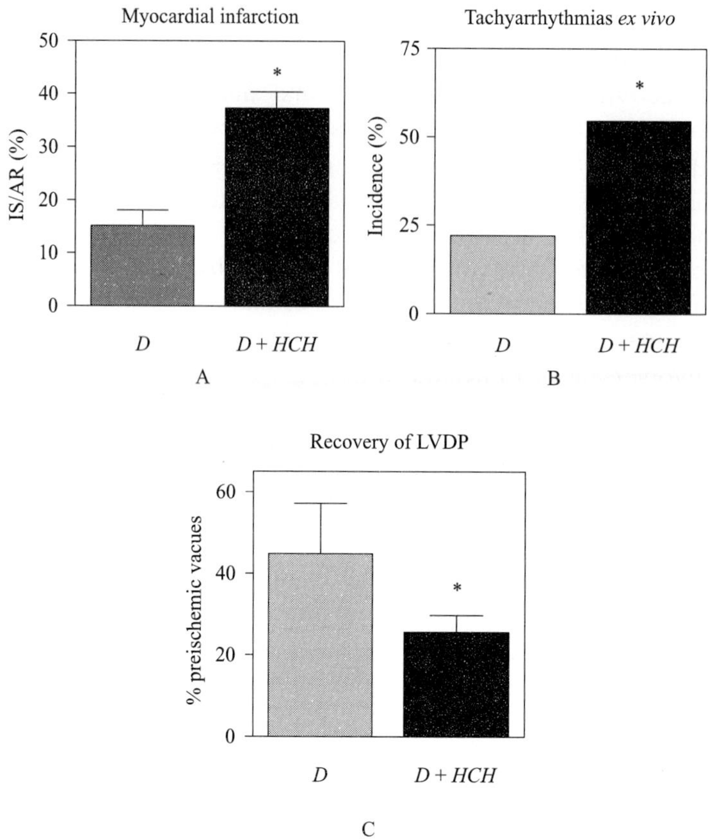

Fig. 4. Effect of hypercholesterolemia (HCH) induced by fat-cholesterol diet in acutely diabetic (D) rat hearts on the size of myocardial infarction (A), ventricular tachyarrhythmias (B) and recovery of LVDP (C) in the isolated hearts. Values are means ± S.E.M. from 10-12 hearts per group. * P<0.05 vs. diabetic hearts.

consequence of their action (and suppression of clinical manifestations of preconditioning under conditions of myocardial ischemia) may be one of the causes of higher mortality in diabetic patients (Tomai *et al.* 1994; Brady and Terzic 1998; Fisman *et al.* 2008). In this regard, Loubani *et al.* (2005) explain a failure to precondition human cardiomyocytes by the treatment of patients with oral hypoglycemics. In addition, Hassouna *et al.* (2006) reported that preconditioning with ischemia, phenylephrine, adenosine, or diazoxide failed to protect diabetic myocardium. However, activation of PKC or p38-MAPK was still protective. These authors suggest that it is the mitochondrial dysfunction in the diabetic myocardium, possibly dysfunctional mitoK(ATP) channels, that leads to the inability to respond to preconditioning.

On the other hand, several authors have demonstrated in other experimental models that under certain conditions, the diabetic myocardium can be preconditioned, albeit with a stronger preconditioning stimulus. In other words, increased threshold for the cardioprotection in the diabetic myocardium requires higher intensity of preconditioning. Thus, Tsang *et al.* (2005) reported that only three cycles of I-PC significantly reduced the size of infarction in the diabetic rats as compared with the efficiency of one cycle in the non-diabetic animals. Cardioprotective effects correlated with significant elevation in Akt phosphorylation only after three cycles of I-PC in diabetics, whereas similar changes in Akt activity were detected in the non-diabetic hearts already after one cycle of I-PC. The latter indicates that the diabetic heart still maintains a potential to be preconditioned, however, to achieve preconditioning protection of the diabetic myocardium, it is necessary to increase the I-PC stimulus to reach the threshold for cardioprotection and a critical level of Akt phosphorylation to mediate protective effect.

In accord, it has been shown in our study in the isolated rat heart that one cycle of I-PC did not confer any additional antiarrhythmic protection and any further reduction in the accumulation of lactate during ischemia of acutely diabetic hearts as compared with the non-preconditioned ones (Ravingerova *et al.* 2000). It seems that since the diabetic myocardium appears to be already "preconditioned", the latter might mask the effect of any further protective interventions.

In agreement with our results, Galagudza *et al.* (2007) have recently reported that rats with alloxan-induced insulin-dependent diabetes mellitus also exhibited smaller infarct size and lower susceptibility to ventricular tachyarrhythmias. However, the same procedure of one cycle of I-PC was ineffective in the diabetic animals, different from the healthy ones supporting the existence of an endogenous cardioprotective phenotype (metabolic preconditioning) in experimental diabetes.

It is also possible that due to induction of the adaptive mechanisms, the diabetic state increases the threshold for cardioprotection and thus, a stronger preconditioning stimulus is required to achive this threshold than in the non-adapted myocardium.

Loss of Endogenous Cardioprotection in the Diabetic Heart

Duration of the diabetic state markedly determines myocardial response to I/R. It was shown in our studies that an increased resistance to ischemia-induced arrhythmias in the acutely diabetic myocardium can be attenuated in the later phase of the disease with more severe manifestations of the diabetic state and development of diabetic cardiomyopathy (Ravingerova *et al.* 2000; 2001). Thus, in the chronic (8-week) phase of the disease, the severity of arrhythmias in the isolated rat hearts was higher than in the acute phase, since the incidence of severe ventricular arrhythmias, such as VT and VF, was increased as compared with 1-week diabetes. Moreover, antiinfarct protection in these animals was blunted as compared with the acute phase of the disease (Ravingerova *et al.* 2003). Xu *et al.* (2004) have also demonstrated that in the chronic phase of STZ-induced diabetes, antiinfarct protection was lost, and this effect was associated with a decrease in ERK1/2 phosphorylation as compared with time-matched non-diabetic controls indicating that long-term duration of hyperglycemia exerts a negative influence on ischemic myocardial injury that may depend on an ERK1/2-mediated intracellular signaling pathway.

Similarly to the effect of I-PC in the non-diabetic heart that was abrogated by an additional pathological condition (Giricz *et al.* 2006), hypercholesterolemia (HCH) induced by high fat-cholesterol diet appeared to be one of the reasons for the loss of the antiinfarct protection in the acutely diabetic Langendorff-perfused rat hearts in our study (Fig. 4A). In addition, in this "double disease" model, we observed that HCH exacerbated severe ventricular arrhythmias in the open-chest *in vivo* animals subjected to LAD occlusion and in the *ex vivo* model (Fig. 4B) and impaired postischemic recovery of left ventricular function in the diabetic hearts subjected to global ischemia and reperfusion (Fig. 4C) indicating that combination of diabetes and other pathologies might blunt endogenous cardioprotection in the diabetic heart as well (Adameova *et al.* 2007a,b).

Although peroxisome proliferator-activated receptors (PPAR) as key transcriptional regulators of lipid metabolism, energy production (Huss and Kelly 2004) and inflammation (Smeets *et al.* 2007) have been suggested to play an important role in susceptibility to myocardial I/R injury (Ravingerová *et al.* 2009b), their role in cardioprotection, in particular, in pathological models, such as diabetes and HCH, is not fully elucidated. To clarify this issue, we explored a potential link between a response to I/R in the hearts of both, diabetic and diabetic-hypercholesterolemic rats and changes in cardiac PPAR gene expression. Enhanced resistance to I/R injury in the diabetic hearts was associated with preservation of baseline mRNA levels of PPARγ upon I/R in contrast to its marked downregulation in the non-diabetic controls. Concurrent HCH attenuated both, cardioprotection and changes in PPAR levels in diabetics (Fig. 5). These findings indicate that PPAR activation might be involved in the adaptive cardioprotective mechanisms activated in the diabetic myocardium in the acute phase to counteract metabolic disorders and, on the other hand, the loss of cardioprotection might be potentially related to the downregulation of PPAR promoting pro-inflammatory effects. Thus, PPAR might represent an important therapeutic target in the management of ischemic heart disease in the diabetics.

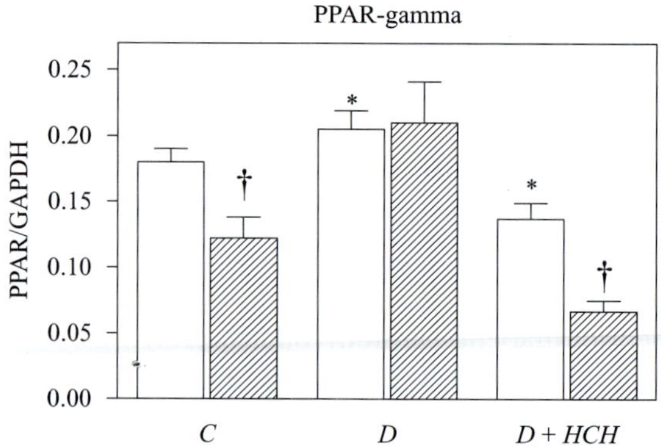

Fig. 5. Expression of PPAR gamma mRNA in the hearts of control (C), diabetic (D) and diabetic-hypercholesterolemic (D+HCH) rats at baseline (empty bars) and after 30 min global ischemia and 2-h reperfusion (filled bars). Values are means ± S.E.M. for at least four hearts. † P<0.05 vs. baseline; * P<0.05 vs. baseline in controls.

In conclusion, experimental diabetes mellitus might induce adaptive processes in the myocardium at least in the early stage of the disease that are analogous to short-term cardioprotection in the non-diseased preconditioned myocardium. Response to I/R in the diabetic heart may be attenuated with the duration of the diabetic state and by concomitant pathology, such as HCH, and in both conditions is associated with the changes in gene expression of PPAR.

Acknowledgement

Supported by grants VEGA SR 2/0173/08, 2/0054/11, 1/0620/10, APVV 0538-07, APVV-LPP-0393-09 and GSRT 5190/2005-759.

References

1. Adameová A, Kuzelová M, Andelová E, Faberová V, Pancza D, Švec P, Ziegelhöffer A, Ravingerová T. Hypercholesterolemia abrogates an increased resistance of diabetic rat hearts to ischemia-reperfusion injury. *Mol Cell Biochem.* 2007a; 295: 129-136.
2. Adameová A, Ravingerová T, Švec P, Fáberová V, Kuzelová M. The myocardial infarct size-limiting and antiarrhythmic effects of acyl-CoA: Cholesterol acyltransferase inhibitor VULM 1457 protect the hearts of diabetic-hypercholesterolaemic rats against ischaemia/reperfusion injury both in vitro and in vivo. *Eur J Pharmacol.* 2007b; 576: 114-121.
3. Andelová E, Barteková M, Pancza D, Styk J, Ravingerová T. The role of NO in ischemia/ reperfusion injury in isolated rat heart. *Gen Physiol Biophys.* 2005; 24, 411-26.
4. Anzawa R, Bernard M, Tamareille S, Baetz D, Confort-Gouny S, Gascard JP, Cozzone P, Feuvray D. Intracellular sodium increase and susceptibility to ischaemia in hearts from type 2 diabetic db/db mice. *Diabetologia* 2006 ; 49: 598-606.
5. Brady PA, Terzic A. The sulfonylurea controversy: more questions from the heart. *J Am Coll Cardiol.* 1998; 31: 950-956.
6. Cascio WE, Johnson TA, Gettes LS. Electrophysiologic changes in ischemic ventricular myocardium: I. Influence of ionic, metabolic and energetic changes. *J Cardiovasc Electrophysiol.* 1995; 11: 1039-1062.
7. Chen H, Shen WL, Wang XH, Chen HZ, Gu JZ, Fu J, Ni YF, Gao PJ, Zhu DL, Higashino H. Paradoxically enhanced heart tolerance to ischaemia in type 1 diabetes and role of increased osmolarity. *Clin Exp Pharmacol Physiol.* 2006 ; 33: 910-916.
8. Das DK, Prasad MR, Lu D, Jones RM. Preconditioning of heart by repeated stunning. Adaptive modification of antioxidative defense system. *Cell Mol Bio.* 1992; 38: 739-49.
9. Dhalla NS, Liu X, Panagia V and Takeda A. Subcellular remodeling and heart dysfunction in chronic diabetes. *Cardiovasc Res.* 1998;40: 239-247.
10. Dubrey SW, Reaveley DR, Seed M, Lane DA, Ireland H, O'Donnell M, O'Connor B, Noble MI, David R, Leslie G. Risk factors for cardiovascular disease in IDDM. A study of identical twins. *Diabetes* 1994; 43, 831-835.
11. Feuvray D and Lopaschuk GD. Controversies on the sensitivity of the diabetic heart to ischemic injury: the sensitivity of the diabetic heart to ischemic injury is decreased. *Cardiovasc Res.* 1997; 34, 113-120.
12. Fisman EZ, Motro M, Tenenbaum A. Non-insulin antidiabetic therapy in cardiac patients: current problems and future prospects. *Adv Cardiol.* 2008; 45: 154-170.
13. Forrat R, Sebbag L, Wiensperger A, Guidollet J, Delaye J, De Lorgeril M. Acute myocardial infarction in diabetic dogs with experimental diabetes. *Cardiovasc Res.* 1993; 27: 1908-1912.
14. Galagudza MM, Nekrasova MK, Syrenskii AV, Nifontov EM. Resistance of the myocardium to ischemia and the efficacy of ischemic preconditioning in experimental diabetes mellitus. *Neurosci. Behav Physiol.* 2007; 37: 489-493.

15. Galinanes M, Fowler AG. Role of clinical pathologies in myocardial injury following ischaemia and reperfusion. *Cardiovasc Res.* 2004; 61: 512-521.
16. Giricz Z, Lalu MM, Csonka C, Bencsik P, Schulz R, Ferdinandy P. Hyperlipidemia attenuates the infarct size-limiting effect of ischemic preconditioning: role of matrix metalloproteinase-2 inhibition. *J Pharmacol Exp Ther.* 2006; 316: 154-161.
17. Goto M, Tsuchida A, Liu Y, Cohen MV, Downey JM. Transient inhibition of glucose uptake mimics ischemic preconditioning by salvaging ischemic myocardium in the rabbit heart. *J Mol Cell Cardiol.* 1995; 27: 1883-1894.
18. Gross GJ, Auchampach JA. Blockade of ATP-sensitive potassium channel prevents myocardial preconditioning in dogs. *Circ Res.* 1992; 70: 223-233.
19. Hadour G, Ferrera R, Sebbag L, Forrat R, Delaye J, De Lorgeril M. Improved myocardial tolerance to ischemia in the diabetic rabbit. *J Mol Cell Cardiol.* 1998; 30:1869-1875.
20. Halestrap AP, Clarke SJ. Khaliulin I. The role of mitochondria in protection of the heart by preconditioning *Biochim Biophys Acta.* 2007; 1767: 1007-1031.
21. Hassouna A, Loubani M, Matata BM, Fowler A, Standen NB, Galiñanes M. Mitochondrial dysfunction as the cause of the failure to precondition the diabetic human myocardium. *Cardiovasc Res.* 2006; 69: 450-458.
22. Hekimian G, Khandoudi N, Feuvray D, Beigelman PM. Abnormal cardiac rhythm in diabetic rats. *Life Sci.* 1985; 37: 547-551.
23. Huss JM and Kelly DP. Nuclear receptor signaling and cardiac energetics. *Circ Res.* 2004; 95: 568-578.
24. Kakkar R, Mantha SV, Kalra J, Prasad K. Time course study of oxidative stress in aorta and heart of diabetic rat. *Clin Sci. (Lond).* 1996; 91: 441-448.
25. Kannel WB and McGee DL. Diabetes and cardiovascular risk factors. The Framingham Study. *Circulation* 1979; 59: 8-13.
26. Khandoudi A, Bernard M, Cozzone P, Feuvray D. Intracellular pH and role of Na^+/H^+ exchange during ischemia and reperfusion of normal and diabetic rat hearts. *Cardiovasc Res.* 1990;24: 873-878.
27. Kolár F, Jezková J, Balková P, Breh J, Neckár J, Novák F, Nováková O, Tomásová H, Srbová M, Ost'ádal B, Wilhelm J, Herget J. Role of oxidative stress in PKC-delta upregulation and cardioprotection induced by chronic intermittent hypoxia. Am J *Physiol Heart Circ Physiol.* 2007; 292(1):H224-30.
28. Kucharská J, Gvozdjaková A, Štefek M, Sotníková R, Sumbalova Z. Adaptive changes of antioxidant status in development of experimental diabetes. *Bratisl Lek Listy* 2001; 102: 515-519.
29. Kusama Y, Hearse DJ, Avkiran M. Diabetes and susceptibility to reperfusion-induced ventricular arrhythmias. *J Mol Cell Cardiol.* 1992; 24: 411-421.
30. Lee SL, Ostadalova I, Kolar F, Dhalla NS. Alterations in Ca^{2+}-channels during the development of diabetic cardiomyopathy. *Mol Cell Biochem.* 1992; 109: 173-179.

31. Liu Y, Thornton JD, Cohen MV, Downey JM, Schaffer SW. Streptozotocin-induced non-insulin dependent diabetes protects the heart from infarction. *Circulation* 1993; 88: 1273-1278.
32. Loubani M, Fowler A, Standen NB, Galinanes M. The effect of gliclazide and glibenclamide on preconditioning of the human myocardium. *Eur J Pharmacol.* 2005; 515: 142-9.
33. Ma G, Al-Shabrawey M, Johnson JA, Datar R, Tawfik HE, Guo D, Caldwell RB, Caldwell RW. Protection against myocardial ischemia/reperfusion injury by short-term diabetes: enhancement of VEGF formation, capillary density, and activation of cell survival signaling. *Naunyn Schmiedebergs Arch Pharmacol.* 2006; 373: 415-27.
34. Malhotra A, Reich D, Reich D, Nakouzi A, Sanghi V, Geenen DL, Buttrick PM. Experimental diabetes is associated with functional activation of protein kinase C? and phosphorylation of troponin I in the heart, which are prevented by angiotensin II receptor blockade. *Circ Res.* 1997; 81: 1027-1033.
35. Malhotra A, Begley R, Kang BP, Rana I, Liu J, Yang G, Mochly-Rosen D, Meggs LG. PKC-{epsilon}-dependent survival signals in diabetic hearts. *Am J Physiol Heart Circ Physiol.* 2005; 289: H1343-50.
36. Maritim AC, Sanders RA and Watkins JB. Diabetes, oxidative stress, and antioxidants: a review. *J Biol Mol Toxicol.* 2003;17:24-38.
37. Matejikova J, Kucharska J, Pancza D, Ravingerova T. The effect of antioxidant treatment and NOS inhibition on the incidence of ischaemia-induced arrhythmias in the diabetic rat heart. *Physiol Res*, 2008; 57: S55-60.
38. Matejíková J, Kucharská J, Pintérová M, Pancza D, Ravingerová T. Protection against ischemia-induced ventricular arrhythmias and myocardial dysfunction conferred by preconditioning in the rat heart: involvement of mitochondrial K(ATP) channels and reactive oxygen species. *Physiol Res.* 2009; 58: 9-19.
39. Meldrum DR, Cleveland JC Jr, Sheridan BC, Rowland RT, Banerjee A, Harken AH. Cardiac preconditioning with calcium: clinically accessible myocardial protection. *J Thorac Cardiovasc Surg.* 1996; 112: 778-786.
40. Moon CH, Jung JS, Lee SH, Baik EJ. Protein kinase C inhibitors abolish the increased resistance of diabetic rat heart to ischemia-reperfusion injury. *Jpn J Physiol.* 1999; 49: 409-415.
41. Murphy E and Steenbergen C. Preconditioning: the mitochondrial connection. *Annu Rev Physiol.* 2007; 69: 51-67.
42. Noma A. ATP-regulated K^+ channels in cardiac muscle. *Nature* 1983; 305: 147-148.
43. Ooie T, Takahashi N, Nawata T, Arikawa M, Yamanaka K, Kajimoto M, Shinobara T, Shigematsu S, Hara M, Yoshimatsu H, Saikawa T. Ischemia-induced translocation of protein kinase C-epsilon mediates cardioprotection in the streptozotocin-induced diabetic rat. *Circ J.* 2003; 67: 955-961.
44. Opie LH. Myocardial ischemia, reperfusion and cytoprotection. *Rev Port Cardiol.* 1996; 15: 703-708.
45. Paulson D. The diabetic heart is more sensitive to ischemic injury. *Cardiovasc Res.* 1997; 34: 104-112.

46. Pierce GN, Ramjiawan B, Dhalla NS, Ferrari R. Na^+/H^+ exchange in cardiac sarcolemmal vesicles isolated from diabetic rats. *Am J Physiol.* 1990; 258: H255-E261.

47. Ravingerová T, Styk J, Pancza D, Tribulová N, Šeboková J, Volkovová K, Ziegelhöffer A, Slezák J. Diabetic cardiomyopathy in rats: alleviation of myocardial dysfunction caused by Ca^{2+} overload. *Diabetes Res Clin Pract.* 1996; 31 Suppl: S105-S112.

48. Ravingerova T, Stetka R, Volkovova K, Pancza D, Dzurba A, Ziegelhöffer A, Styk J. Acute diabetes modulates response to ischemia in isolated rat heart. *Mol Cell Biochem.* 2000; 210: 143-151.

49. Ravingerova T, Neckar J, Kolar F, Stetka R, Volkovova K, Ziegelhöffer A, Styk J. Ventricular arrhythmias following coronary artery occlusion in rats: is the diabetic heart less or more sensitive to ischaemia? *Basic Res Cardiol.* 2001; 96:160-168.

50. Ravingerová T, Neckár J, Kolár F. Ischemic tolerance of rat hearts in acute and chronic phases of experimental diabetes. *Mol Cell Biochem* 2003; 249: 167-174.

51. Ravingerová T, Matejíková J, Pancza D, Kolár F. Reduced susceptibility to ischemia-induced arrhythmias in the preconditioned rat heart is independent of PI3-kinase/Akt. *Physiol Res.* 2009a; 58: 443-447.

52. Ravingerová T, Adameová A, Kelly T, Antonopoulou E, Pancza D, Ondrejèáková M, Khandelwal VMK, Carnická S, Lazou A. Changes in PPAR gene expression and myocardial tolerance to ischaemia: Relevance to pleiotropic effects of statins. *Can J Physiol Pharmacol.* 2009b, in press.

53. Rodrigues B., Cam M.C., McNeill J.H. Myocardial substrate metabolism: implications for diabetic cardiomyopathy. *J Mol Cell Cardiol.* 1995; 27:169-179.

54. Schaffer SW, Croft CB, Solodushko V. Cardioprotective effect of chronic hyperglycemia: effect on hypoxia-induced apoptosis and necrosis. Am J Physiol Heart Circ Physiol. 2000; 278: H1948- H1954.

55. Shimoni Y, Severson D, Giles W. Thyroid status and diabetes modulate regional differences in potassium currents in rat ventricle. *J Physiol (Lond)* 1995; 488: 673-688.

56. Singal PK, Bello-Klein A, Farahmand F, Sandhawalia V. Oxidative stress and functional deficit in diabetic cardiomyopathy. *Adv Exp Med Biol.* 2001; 498: 213-220.

57. Smeets PJH, Planavila A, van der Vusse GJ, van Bilsen M. Peroxisome proliferator-activated receptors and inflammation: take it to heart. *Acta Physiol.* 2007; 191: 171-188.

58. Smith JM, Wahler GM. ATP-sensitive potassium channels are altered in ventricular myocytes from diabetic rats. *Mol Cell Biochem.* 1996; 158: 43-51.

59. Spinelli W, Sorota S, Siegal M, Hoffman BF. Antiarrhythmic actions of the ATP-regulated K^+ current activated by pinacidil. *Circ Res.* 1991; 68: 1127-1137.

60. Strnisková M, Barancik M, Neckar J, Ravingerova T. Mitogen-activated protein kinases in the acute diabetic myocardium. *Mol Cell Biochem.* 2003; 249: 59-65.

61. Tan HL, Mazon P, Verberne HJ, Sleeswijk ME, Coronel R, Opthof T, Janse M. Ischemic preconditioning delays ischemia induced cellular electrical uncoupling in rabbit myocardium by activation of ATP sensitive potassium channels. *Cardiovasc Res.* 1993; 27: 644-651.

62. Tani M, Neely JR. Hearts from diabetic rats are more resistant to in vitro ischemia: Possible role of altered Ca2+ metabolism. *Circ Res.* 1988; 62: 931-940.

63. Tomai F, Crea F, Gaspardone A, Versaci F, De Paulis R, Penta de Peppo A, Chiariello L, Gioffre PA. Ischemic preconditioning during coronary angioplasty is prevented by glibenclamide, a selective ATP-sensitive K+ channel blocker. *Circulation* 1994; 90: 700-705.

64. Tosaki A, Engelman DT, Engelman RM, Das DK. The evolution of diabetic response to ischemia/reperfusion and preconditioning in isolated working rat hearts. *Cardiovasc Res.* 1996; 31: 526-536.

65. Tribulová N, Ravingerová T, Volkovová K, Ziegelhoffer A, Okruhlicová L, Ziegelhoffer B, Styk J, Slezák J. Resistance of diabetic rat hearts to Ca overload-related injury. Histochemical and ultrastructural study. *Diabetes Res Clin Pract.* 1996; 31: S113-22.

66. Tsang A, Hausenloy DJMocanu MM, Carr RD, Yellon DM. Preconditioning the diabetic heart: the importance of Akt phosphorylation. *Diabetes* 2005; 54: 2360-2364.

67. Xu G, Takashi E, Kudo M, Ishiwata T, Naito Z. Contradictory effects of short- and long-term hyperglycemias on ischemic injury of myocardium via intracellular signaling pathway. *Exp Mol Pathol.* 2004; 76: 57-65.

68. Wolfe CL, Sievers RE, Visseren FLJ, Donnelly TJ. Loss of myocardial protection after preconditioning correlates with the time course of glycogen recovery within the preconditioned segment. *Circulation* 1993; 87: 881-892.

69. Ziegelhoffer A, Ravingerova T, Styk J, Sebokova J, Waczulikova I, Breier A, Dzurba A, Volkovova K, Carsky J, Turecky L. Mechanisms that may be involved in calcium tolerance of the diabetic heart. *Mol Cell Biochem* 1997; 176: 191-198.

70. Ziegelhoffer A, Styk J, Ravingerová T, Šeboková J, Volkovová K, Waczuliková I, Carský J, Dzurba A, Docolomanský P. Prevention of processes coupled with free radical formation prevents also the development of calcium resistance in the diabetic heart. *Life Sci.* 1999; 18/19: 1999-2003.

71. Ziegelhöffer A, Ravingerová T, Waczuliková I, Cársky J, Neckár J, Ziegelhöffer-Mihalovicová B, Styk J. Energy transfer in acute diabetic rat hearts: adaptation to increased energy demands due to augmented calcium transients. *Ann NY Acad Sci.* 2002; 967: 63-468.

Adaptation Biology and Medicine (Volume 6: Cell Adaptations and Challenges)
Editors: P. Wang, C.-H. Kuo, N. Takeda and P.K. Singal
Copyright © 2011 Narosa Publishing House Pvt. Ltd., New Delhi, India

6. Quantitative Alterations in Mitochondria: Adaptation Contra Violation

Vladimir Sukhorukov*

Moscow Research Institute of Pediatrics and Child's Surgery Taldomskaya, 2, Moscow

Abstract

We postulate that mitochondrial proliferation may be an important adaptation mechanism in various pathological conditions which was demonstrated in different tissues. One of the examples used was our discovery of higher mitochondria accumulation in muscles of the patients with congenital central core myopathy. For clinical precision of this mechanism significance we compared central core disease patients with and without mitochondrial accumulation. We have revealed the correlation between muscle mitochondria amount and the central core myopathy clinical severity. Our data allow us to conclude that at least in certain forms of congenital myopathy, mitochondrial proliferation may affect the clinical picture of the disease.

Keywords: Mitochondrion, mitochondrial fission, congenital myopathy, central core disease, ragged-red fibers

Introduction

It has been more than 40 years since investigators in the field of clinical medicine have paid attention to the study of mitochondria and the role they have in various pathological processes. The primary etiological factor of the disease caused by mitochondrial changes in structure and function forms a new group of nosological conditions called "mitochondrial disease".

Mitochondrial diseases are a heterogeneous group of diseases caused by genetic structural and functional defects of mitochondria. The precise classification distinctions and concepts of the aetiopathogenic conditions are not yet worked out as well as the treatment modalities are also not known.

Rolf Luft (1962) was the first to point out the possible direct connection between mitochondria and the determination of pathological conditions. He has given the description of a patient with no thyroid hyper metabolism in mitochondria where oxidative phosphorylation was impaired (18). In 1963, when mitochondrial DNA (mtDNA) was discovered for the first time, Engel and Cunningham described the

*Address for Correspondence: Vladimir Sukhorukov Moscow Research Institute of Pediatrics and Child's Surgery Taldomskaya 2, Moscow 125412 Russia E-mail: vsukhorukov@inbox.ru

so called "ragged-red fibers" (RRF) which were alterations of Gomory stained muscle fibers (9). It has been shown that this development was due to the abnormal proliferation of mitochondria and this phenomenon was observed in various mitochondrial diseases. During the following years, a large number of diseases due to mitochondrial genome defect were discovered and this gave rise to a different classification of mitochondrial diseases into an independent group.

Mitochondrion is a universal cytoplasmic organelle in all tissues. It seems that clinical result of energetic system lesion cannot considerably depend on molecular defect localization. The main outcome is the oxidative phosphorylation and ATP synthesis disorder. In contrast there is a large amount of mitochondrial pathology forms. There are a number of mitochondrial changes which can manifest differently in different tissues, at different times and conditions. This may be due to heterogeneous tissue dependence on energy impairment. Various tissues and organs depend on mitochondrial energy differently (20, 23, 35, 36). The nervous system leads the first followed by cardiac muscle, skeletal muscle, kidney, endocrine glands and liver. The amount of mitochondria increases and their structural defects are noticeable in endothelial cells, smooth muscle cells, myocytes and pericytes in the blood vessels (21).

It was proposed that phenotypic displays of some mitochondrial defects can undergo gradual evolution depending on a complex of many factors, so the nosologically outlined forms are only stages (or branchings) in the common pathologic process [1]. In the mentioned work it concerns syndromes MERRF (myoclonus epilepsy and ragged-red fibers), MELAS (mitochondrial encephalomyopathy, lactic acidosis, stroke-like episodes) and Leigh disease. Similar reasoning can be related to the full set of mitochondrial pathology.

Though mitochondria are considered as the most independent cytoplasmic organelle, they are only a part of cellular homeostasis. Mitochondrial disorders consist of "primary" and "secondary" where primary changes are connected with the disturbance of the genetic control of mitochondrial proteins. Secondary changes are connected with many infringements of cellular and organ systems like in patients with peroxisomal diseases (3), pulmonary disorders (17) or various forms of myositis (25). We have made several descriptions of the secondary mitochondrial infringements in different tissues at various diseases as already reported in previous ISAM congress [31, 32]. Probable "secondary" mitochondrial changes complicate the diagnostics of energy infringements and the revealing of "primary" changes of these organelles connected with a disturbance of the genetic control of mitochondrial proteins.

There is no doubt that mitochondrial disorders studying at the above mentioned diseases is actual not only from the point of view of the analysis of specific pathogenesis in this or that illness, but also for revealing the common laws of mitochondria changes in a sick and healthy organism. Though mitochondrial diseases are an adequate natural model for research of many functional features of these organelles, in the present article we shall stop only on some aspects connected to increase of mitochondria amount in cells.

Mitochondrial Amount Increase: Violation Aspects

It was historically developed that the main morphological criterion of mitochondrial diseases - RRF presence in skeletal muscles - for the first time was investigated of full value in syndromes that are aetiopathogenetically connected with mtDNA mutations (so-called "primary" mitochondrial diseases).

Abbreviation "RRF" was even included in the name of the nosological form in one of them - syndrome MERRF. All this has led to the widespread point of view that the presence of RRF is an attribute of mtDNA mutation, or an attribute of a primary (in aetiopathogenetic sense) mitochondrial pathology [6, 8, 27, 29].

According to Di Mauro and Moraes (7), the homogeneous enough muscle tissue is an ideal model for pathogenetic consequences of mtDNA mutations studying. Mitochondrial distress signals are sent out by means of their proliferation and RRF development. De Vivo (5) expresses the RRF's strategic point of mutations involving tRNA genes as well as interfering with the intra mitochondrial protein synthesis. Though there is a correlation between RRF expressiveness and the mutant label accumulation degree in RRF in molecular genetically proved primary mitochondrial myopathies [14], it is not clear whether RRF develops due to the presence of mutant genes or if it is due to a compensatory effect. Anyhow, it was generally accepted that RRF represent a morphological substratum of mitochondrial insufficiency, and (when the certain quantitative threshold is achieved) the standard parameter of mtDNA damage.

About 15 years ago, Di Mauro and Moraes (7) postulated the following representations which became basic in many respects:

1. It is possible to regard RRF as a visual correlation of mtDNA mutations threshold effect (because of presence of plenty of mtDNA mutations in RRF and often cytochrome oxidase activity reduction in RRF); **2.** Protein synthesis in RRF is reduced; **3.** The trigger mechanisms of mitochondrial proliferation and RRF formation is not known yet; **4.** The mitochondrial genome damage with this development of RRF phenomenon can be both congenital and acquired. These postulates still remain even after the accumulation of new evidences in the area of triggering the mechanism of mitochondrial proliferation.

The mitochondrial number increases in cells as an attribute of genetic infringements have a negative influence. The mitochondrial hyperplasia interferes with the oxidizing process and many other functions in which mitochondria play a key role. An increase in reactive oxygen species (ROS), an overabundance of mitochondrial proteins of a thermal shock, an oxidizing stress and protection against calcium reduction (37) are some of the examples. As a result of these changes, various biomolecules are damaged including nucleic acids and components of membranes which cause cell death eventually.

One more negative consequence of mitochondrial amount increase is the resorption mechanisms of autoimmunity. Here we are dealing with sarcoplasmic masses which are homogenous congestions of sarcoplasm in the subsarcolemmal areas of muscle fibers. This feature is typical for patients with mitochondrial myopathies. In our data, the presence of developed sarcoplasmic masses, is the most typical syndrome of MERRF. These masses are rather free from myofibrils and may surround these muscle fibers completely. Histological examination shows their homogeneous nature and fibrillar appearance which differ from adjoining darker intact myofibrils. This histochemical studies show the hyperactivity of sarcoplasmic masses of the oxidizing and glycolytic enzymes. The arrangements of rare bunches of myofilaments are not in order. Singular tubular structures, lysosomes and lipofuscin congestions have also been observed. Though sarcoplasmic masses are frequently found in myotonic dystrophy, they have also been found in muscular dystrophies (like the limb-girdle) which is of our special interest, in hypothyroid myopathy (3,8). Energy imbalance may be responsible for the pathogenesis of these structures in hypothyroid and mitochondrial myopathies.

The frequency of sarcoplasmic masses in MERRF syndrome (primary mitochondrial myopathy), location of these masses (place of probable previous subsarcolemmal congestion of mitochondria), lysis of the content of the masses and the presence in an organism of insufficiently studied antimitochondrial antibodies system [26] - all this as a whole allows to assume about the presence of an autoimmune component in the pathogenesis of some mitochondrial myopathies.

The sequence of events hypothesized are as follows:

1. mtDNA mutations result in energy insufficiency; **2.** Growth (probably compensatory) of mitochondrial pools in energy dependent tissues like the skeletal muscles and the development of mitochondrial congestions in the subsarcolemmal regions and RRF as well; **3.** An increase in the amount and size of mitochondria can lead to mutant mitochondrial units which accumulate in the subsarcolemmal areas; **4.** There is autoimmune influence in the subsarcolemmal congestions of mutant mitochondria by antimitochondrial antibodies.

The elimination of sarcoplasm by the autoimmune system could explain the presence of small regions of necroses in muscle fibers of patients with mitochondrial myopathy. Lytic damages of some subsarcolemmal regions can probably result subsequently to secondary nonspecific (concerning such autoimmune process) focal coagulate necroses of adjacent areas of sarcoplasm involving separate leukocytes, plasmorrhexis, and the subsequent reparation processes. At the same time it is impossible to exclude that mitochondrial accumulation in separate sarcoplasm regions can start processes of autophagy [see 4] due to which sarcoplasmic masses may be formed. And in that case it can have positive for a cell effect of clarification from defective structures.

Adaptive Value of Mitochondrial Amount Increase

However, the natural assumption that the mitochondria accumulate to raise an energy potential of a tissue for a long time expresses in different works at a speculative level, and is sometimes presented as the fact which stands to reason. At the same time the strict proofs of it are not so numerous. Wiesner et al (38) have counted the mitochondria with the help of an electron microscope in rat cardiomyocytes during creatine depletion. What they found was an increase of mitochondrial density by 75% during the reduction of energy supply in cardiomyocytes. Levels of representative mRNA, encoded on mtDNA or on nuclear chromosomes were elevated 1.5 to 2 fold, while mtDNA content was not changed. The mRNA of the nuclear encoded mitochondrial transcription factor A (mtTFA) was increased after GPA feeding. So the authors concluded that an impairment of mitochondrial energy supply may cause stimulation of gene expression resulting in mitochondrial proliferation as a compensatory mechanism. The activation of mtTFA gene has confirmed the important function of this protein in the nuclear-mitochondrial communication.

Based on our data from "ragged-red fibers" (RRF) and other morphological attributes, there are mitochondrial dysfunctions in diseases like tuberous sclerosis, Lennox-Gastaut syndrome, Coffin-Lowry syndrome and a syndrome of a ring chromosome 14, where primary mitochondrial genesis is very unlikely. In our opinion, RRF is not an unequivocal certificate on an mtDNA mutation but only morphological display of mitochondrial functions insufficiency. So the RRF presence in itself is not a specific attribute in relation to the genesis of diseases. A suitable illustration of this position can be indications that the contractile activity in skeletal muscles (10) induces mitochondrial biogenesis. The next example is the large mitochondrial congestions in the diaphragm during chronic obstructive pulmonary disease (17). In the latissimus dorsi muscle, such changes do not happen and in the intercostals muscles, such changes are much more moderate. It is very probable that when the functional loading on a muscle increases, the secondary mitochondrial changes also increase proportionately. Even the mtDNA defects can be secondary in relation to defects of nuclear genome as for example in myotomic dystrophy (24).

All aforesaid and also representation that RRF are found out in sufficient for the diagnosis (more than 5%) quantity not in all cases of true mitochondrial insufficiency (see for example 2, 35), testifies to necessity of cautious treatment of the data on presence or absence RRF in muscle bioptates. We have postulated that mitochondrial proliferation in many respects is determined by a reparation potential of a mitochondrial pool and may be an important adaptation mechanism in various pathological conditions in different tissues (30-32).

- Clinical debut more frequently in neonates but may be later
- Non- severe muscle hypotonia in infancy
- Motor delay
- Stable generalised muscle weakness
- Possibility of cardiac myopathy
- Skeletal deformations
- Variable phenotype (from slight to moderate severe; may be asymptomatic)

Fig. 1. Main clinical characteristics of central core disese

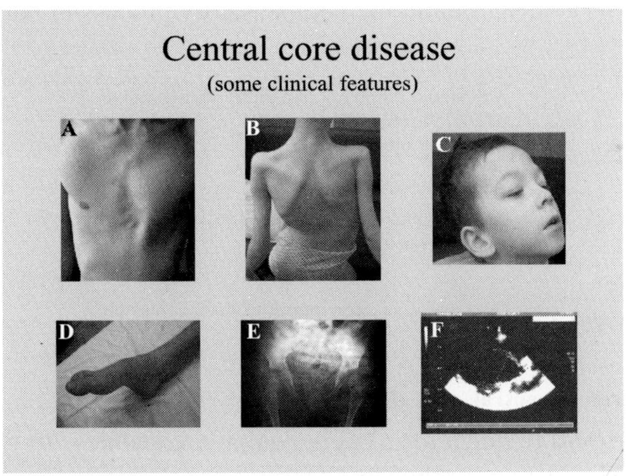

Fig. 2. Some clinical features of patients with central core disease.
A - thoracic deformity; B - scoliosis; C - ptosis; D - pes cavus; E - hip dislocation; F - cardiomyopathy

At the last ISAM Congress in Moscow (32), we pointed out the following examples:

1. Families that had one or more patients with primary mitochondrial diseases and their relatives who did not, or only slightly, show any clinical manifestations of mitochondrial deficiency, would test for considerable amount of RRF, well exceeding the amount in patients themselves.
2. RRF is very often found in patients with non mitochondrial diseases among whom there are hereditary diseases with well known primary mutation of the nucleus genome (for example, tuberose sclerosis, Marfan and Ehlers-Danlos syndromes). The availability of abnormal mitochondrial proliferation in these diseases allows considering this phenomenon not so much as a marker of mitochondrial mutations, but as a marker of some functional mitochondrial reaction.
3. Another observation we consider as one more proof is that several patients with congenital myopathy, showed for the first time, myopathic symptomatic complex where the patients were over 40 years old. Their biopsies showed considerable amount of abnormal subsarclemmal masses of mitochondria. We believe that the structural and functional defects in these cases are efficiently compensated by mitochondrial proliferation and as a result, by elevated energy supply of the remaining contractile muscle elements.
4. In addition to the proof observed in skeletal muscle tissue, we have demonstrated some homologous examples in other tissue elements as well: smooth muscle tissue of urinary channels, kidney epithelium, basal layer of epidermis and leukocytes.

One of the examples used was our discovery of higher mitochondrial accumulation in the muscles of patients with congenital "central core" myopathy - a non mitochondrial disease.

According to our preliminary data, the presence of mitochondrial masses correlates with more compensated clinical course of myopathy. For clinical precision of this significant mechanism, we tried to find the proof of its contribution to the compensatory process in the central core disease.

- Autosomal dominant hereditary with variable q12 allelic to 19-Locus-p13.1. Copenetration. malignant hyperthermia.
- Enzyme (including mitochondrial) inactivity in central areas of muscle fibers.
- Disorder of calcium output from sarcoplasmic reticulum

Fig. 3. Pathogenetic features of central core disease

Central core disease is the most frequent form of congenital myopathies. Its clinical characteristics are in Figs. 1-2 and some etiological and pathogenic features are shown in Figs. 3 and 4.

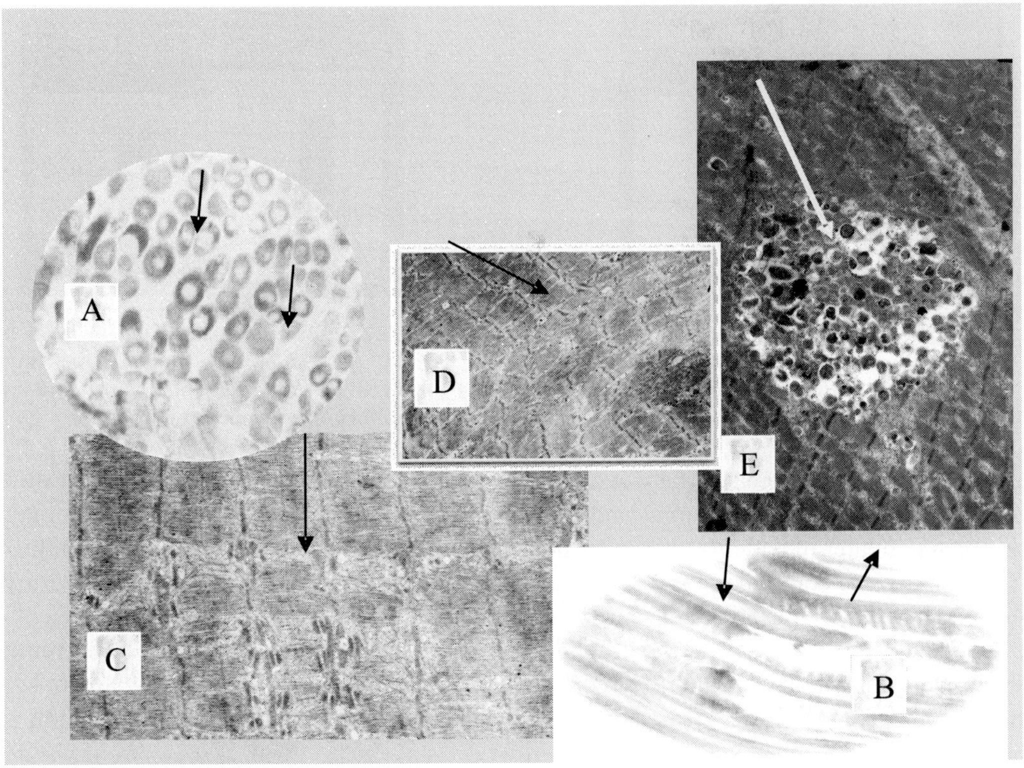

Fig. 4. Morphological characteristics of skeletal muscle in patients with central core disease.
A, B - Light microscopy of histochemical activity of succinic dehydrogenase; there are central zones without enzyme activity (arrows) in each muscle fiber; A - cross section from the 6 months old patient, B - longitudinal section from the 12 years old patient.
C, D, E - electronic microscopy of muscle fibers of different central core disease patients: there are central sarcomere zones (arrows) with various degree of damage of ultra structures.

In biopsy material from patients with this disease, we could often find abnormal subsarcolemmal masses of mitochondria (RRF). The aim of our investigation was to compare clinical and laboratory features in central core disease in patients with and without subsarcolemmal masses of mitochondria.

We have examined 36 patients (5 adults and 31 children of different ages) with morphologically proved central core disease. Twenty-three patients had subsarcolemmal masses of mitochondria and 13 did not. The degree of functional decompensation was estimated by use of several muscle activity evaluation scales. We evaluated clinical and various laboratory parameters: biochemical (pyruvate and lactate concentrations in blood and others), cytochemical (mitochondrial enzyme activity in lymphocytes), histochemical

(mitochondrial enzyme activity in muscle, subsarcolemmal mitochondria accumulations morphometry) and ultrastructural (electron microscopy evaluation of mitochondrial quality and quantity changes).

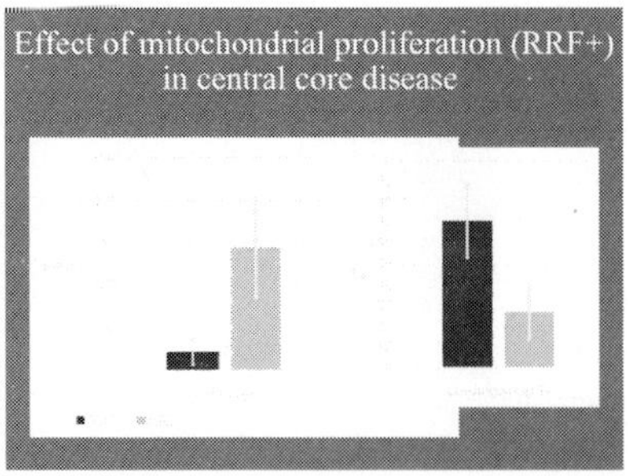

Fig. 5. The difference in some clinical features in the group of central core myopathy between the patients withand without mitochondrial accumulation

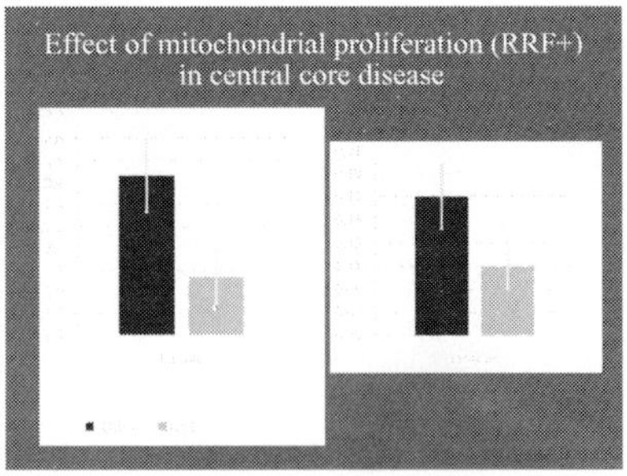

Fig. 6. The difference in some biochemical features in the group of central core myopathy between the patients with and without mitochondrial accumulation

We have seen clear differences between the two groups. First of all, there was a great difference between the times of clinical symptoms and their manifestation. The debut (and so the severity) of the disease was much later in patients with mitochondrial accumulation (Fig. 5). Cardiomyopathy, which is the severest disease complication, was not common in these patients. So this is a trustworthy proof of our idea. There was a great difference between lactate and pyruvate levels in blood of patients of the two groups analysed. Their levels three hours after glucose loading are in Fig. 6. The remarkable differences were shown in the biochemical parameters of calcium (Fig. 7). The total calcium level was less and the ionizated calcium level was higher in the group with mitochondrial accumulation.

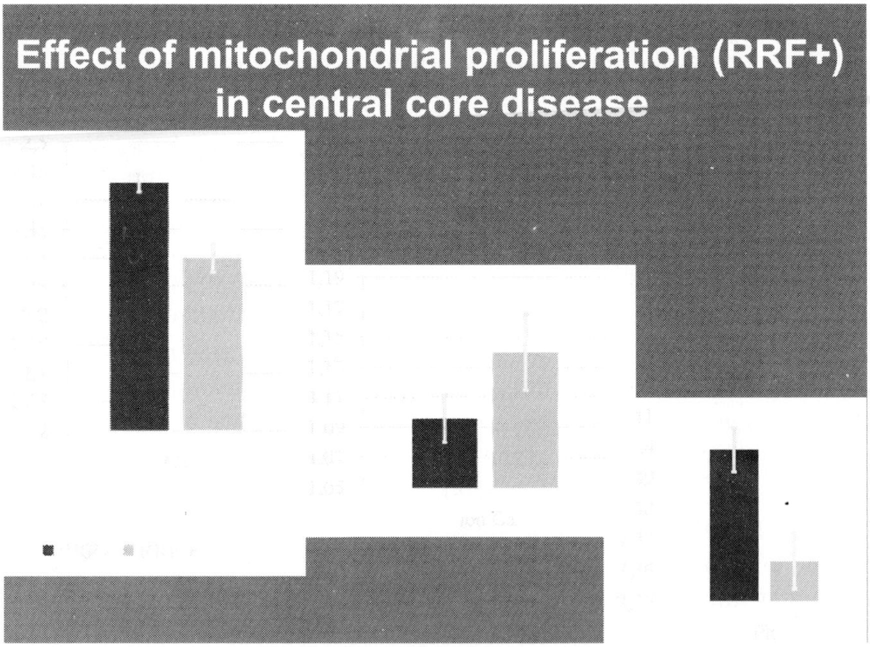

Fig. 7. The difference in calcium and phosphor parameters in the group of central core myopathy between the patients with and without mitochondrial accumulation

We are sure that at least in certain forms of congenital myopathy, mitochondrial dysfunctions may affect the clinical picture of the disease. Our supplementary data about the effectiveness of enegotropic medicines (complex of L-carnitine, coenzyme Q10 and some others) in our patients are the additional proof of this (Fig. 8).

Our main purpose was to evaluate if the mitochondrial conglomerates are of any importance or not. Now we can say with confidence that an increase of these organells in muscle may have a compensatory effect. This conclusion confirms our earlier assumptions and is proved by the facts we have shown about the correlation between the amount of muscle mitochondria and the central core myopathy in clinical severity.

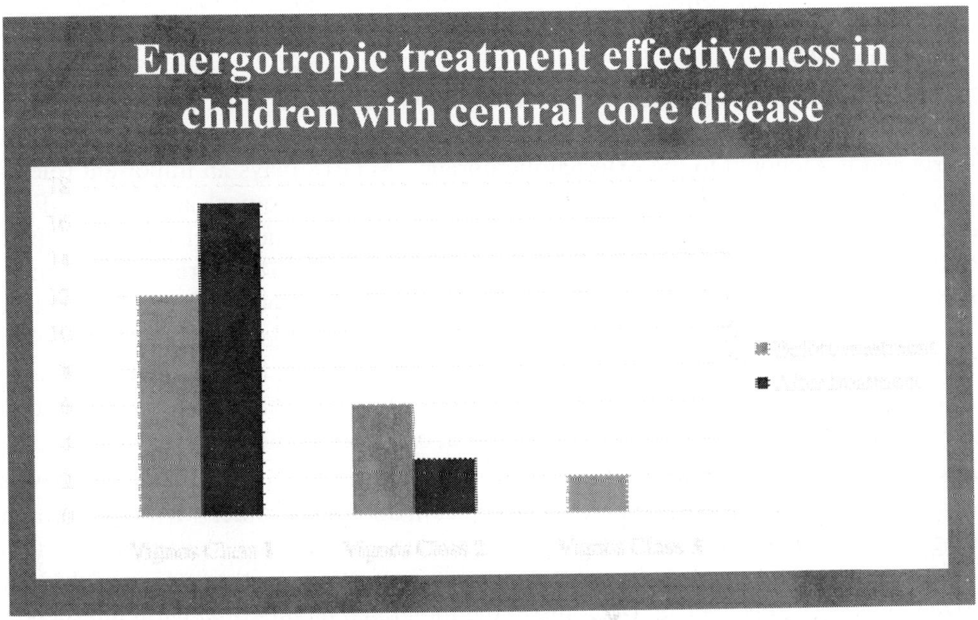

Fig. 8. The clinical effectiveness of energotropic treatment in patient with central core disease. The distribution of our patients before and after treatment in Vignos scale - test for muscle function evaluation (the more Vignos class number, the worse muscle activity).

What can be the reason for the increased level of mitochondrial proliferation? As it was indicated above, in 1993 patriarchs of mitochondriology postulated the thesis that the trigger of mitochondrial proliferation is unknown (7). However, various works on this theme soon began to appear. The data that free fatty acid levels can have influence on the mitochondrial amount have appeared in 1995 (16). A candidate for the role of a mitochondrial proliferation regulator, the nuclear respiratory factor 1 (NRF-1) has been named (39).

Previous years were marked by significant promotion in the study of mitochondrial proliferation mechanisms. It has been shown that the system of molecular complexes regulates mitochondrial fusion and fission. Intramitochondrial mitofusin GTPases play an important role in the former of these processes (12, 22). The group of the proteins connected with the external mitochondrial membrane is engaged apparently in fission of mitochondria (review in 34). Cytoplasmic GTPase dynamin 1-like protein (DNM1L) is recruited to the mitochondrial outer membrane to exert fission of mitochondria. Human Fis1 (hFis1) is a highly conserved 17-kDa integral mitochondrial protein of the outer membrane. It is involved in mediating mitochondrial fission, presumably by setting up the fission complex on the cytoplasmic face of the outer membrane through (direct or indirect) recruitment of DNM1L.

In addition, hFis1 has been implicated in the initial step of apoptosis. Short term overexpression of this protein leads to excessive fission followed by the exit of cytochrome C in cytosol and apoptosis starts. In addition to the interaction between DNM1L and hFis1, there is interaction of DNM1L in the external mitochondrial membrane with the ubiquitin-conjugating enzyme 9 (Ubc9) and the small ubiquitin-like modifier 1 (Sumo 1). In mammalian cells, the E3 ubiquitin ligase controls the mitochondrial fission (15).

Daniel Tondera and his co-authors (33, 34) have shown one more participant of fission machinery - the human mitochondrial protein called MTP18. This protein is part of the internal membrane of mitochondria and is anchored to the external membrane. MTP18 plays an important role during the final stages of mitochondrial fission. It is important in the physiological activity of hFis1 and the division of mitochondria. These authors (34) have hypothesized the precise activity of these two proteins; if overexpression of hFis1 takes place (but not MTP18) it results in mitochondrial fragmentation, an output of cytochrome C and apoptosis. On the other hand, if overexpression of MTP18 takes place, fission stops forming a high grade mitochondria without exit cytochrome C in cytosol and programmed cell death. Recently some researchers have focused their attention to a new protein of the external mitochondrial membrane known as the Mitochondrial Fission Factor (MFF), which together with hFis1, supervises the proliferation of these organelles (11).

It stands to reason that processes of mitochondria biogenesis should be strictly coordinated on the one hand with the energy metabolism activity in the cell, and on the other - with the basic processes determining existence of the cell during this or that moment of time, that is characteristic for different stages of the cell cycle. Taking into account, that passage of those or other stages of the cell cycle is characterized by energy dependence, we come to the integrated picture of general interdependence of the main cellular events. For example insufficient provision of energy results in activation of AMP-activated protein kinase which is successfully named a metabolic sensor of energy ensuring in eukaryotic cells [19].

We believe that the mitochondrial proliferation by its nature is an adaptation process. The increase in the absolute number of these organelles helps the organism to cope with the poor quality of cellular energy metabolism. The adaptation potential of proliferating mitochondria under functional tissue deficiency is remarkable. We can very well, understand the importance of having a working model for stimulating the mitochondrial proliferation features to solve the problems of increasing adaptive abilities of the organism.

References

1. Bercovic SF, Carpenter S, Evans A, Karpati G, Shoubridge EA, Andermann F, Meyer E, Tyler JL, Diksic M, Arnold D, Wolfe LS, Andermann E and Hakim AM. Myoclonus epilepsy and ragged-red fibers (MERRF). 1. A clinical, pathological, biochemical, magnetic resonance spectrographic and positron emission tomographic study. Brain 1989; 112: 1231-1260.

2. Blok RB, Thorburn DR, Danks DM and Dahl HH. mtDNA deletion in a patient with symptoms of mitochondrial cytopathy but without ragged-red fibers. Biochem Mol Med 1995; 56:1: 26-30.
3. Crespo-Armas A, Azavache V, Torres SH, Anchustegui B and Cordero Z. Changes produced by experimental hypothyroidism in fiber type composition and mitochondrial properties of rat slow and fast twitch muscles. Acta Cient Venez 1994; 451: 42-44.
4. Das DK. Role of autophagy in myocardial adaptation to oxidative stress. IX Workd ISAM Congress, Taiwan, 2009 August 2-5.
5. De Vivo D. The expanding spectrum of mitochondrial diseases. Brain and Development 1993; 15: 1-22.
6. DiMauro S and Bonilla E, Zeciani M, Nakagawa M, DeVivo DC. Mitochondrial myopathies. Ann Neurol 1985; 17: 521-538.
7. DiMauro S and Moraes C. Mitochondrial encephalomyopathies. Arch Neurol 1993; 50: 1197-1208.
8. Engel AG and Banker BQ. Myology. McGraw-Hill Book Company, N-Y e.a., 1986
9. Engel WK and Cunninham GG. Rapid examination of muscle tissue: an improved trichrome stain method for fresh frozen biopsy sections. Neurology 1963; 13: 919-926.
10. Essig DA. Contractile activity-induced mitochondrial biogenesis in skeletal muscle. Exerc Sport Sci Rev 1996; 24: 289-319.
11. Gandre-Babbe S and van der Bliek AM. The novel tail-anchored membrane protein Mff controls mitochondrial and peroxisomal fission in mammalian cells. Mol Biol Cell 2008; 19: 2402-2412.
12. Hales KG, Fuller MT. Developmentally regulated mitochondrial fusion mediated by a conserved, novel predicted GTPase. Cell 1997; 90: 121-129.
13. Holmes RD, Moore KJ, Ofenstein JP, Tsatsos P and Klechle FL. Lactic acidosis and mitochondrial dysfunction in two children with peroxisomal disorders. J Inherit Metab Disease 1993; 16: 368-380.
14. Houshmand M, Larsson NG, Oldfors A, Tulinius M and Holme E. Fatal mitochondrial myopathy, lactiv acidosis and complex I deficiency associated with a heteroplasmic A -- > G mutation at position 3251 in the mitochondrial tRNALeu (UUR) gene. Hum Genet 1996; 97: 3: 269-273.
15. Karbowski M, Neutzner A and Youle RJ. The mitochondrial E3 ubiquitin ligase MARCH5 is required for Drp1 dependent mitochondrial division. J Cell Biol 2007; 178: 71-84.
16. Levak-Frank S, Radner H, Walsh A, Stolberger R, Knipping G, Hoefler G, Sattler W, Weinstock PH, Breslow JL and Zechner K. Muscle-specific overexpression of lipoprotein lipase causes a severe myopathy characterized by proliferation of mitochondria and peroxisomes in transgenic mice. J Clin Invest 1995; 96; 976-986.
17. Lloreta J, Orozco M, Gea J, Corominas JM and Serrano S. Selective diaphragmatic mitochondrial abnormalities in a patient with marked air flow obstruction. Ultrastruct Pathol 1996; 20:1: 67-71.

18. Luft R, Ikkos D, Palmieri G, Ernster L and Afzelius BA. A case if severe hypermetabolism of non-thyroid origin with a defect in the maintenance of mitochondrial repiratory control: a correlated clinical, biochemical and morphological study. J Clin Invest 1962; 41: 1776-1804.
19. Martinez-Diez M, Santamaria G, Ortega AD and Cuezva JM. Biogenesis and dynamics of mitochondria during the cell cycle: significance of 39UTRs. PLoS ONE 2006; 1:1: e107.
20. Mierzewska H. Mitochondrial diseases. Neurol Neurochir Pol 1996; 30:2: 265-292.
21. Molnar M, Neudecker S and Schroder JM. Increase of mitochondria in vasa nervorum of cases with mitochondrial myopathy, Kearns-Sayre syndrome, progressive external ophthalmoplegia and MELAS. Neuropathol Appl Neurobiol 1995; 21:5: 432-439.
22. Mozdy AD and Shaw JM. A fuzzy mitochondrial fusion apparatus comes into focus. Nat Rev Mol Cell Biol 2003; 4: 468-478.
23. Munnich A, Rustin P, Rötig A, Chretlen D, Bonnefont J-P, Nuttin C, Cormler V, Vassault A, Parvy P, Bardet J, Charpentler C, Rabler D and Saudurbay J-M. Clinical aspects of mitochondrial disorders. J Inherit Metab Disease 1992; 15: 448-455.
24. Poulton J, Harley HG, Dasmahapatra J, Brown GK, Poatter CG and Sykes B. Mitochondrial DNA does not appear to influence the congenital onset type of myotonic dystrophy. J Med Genet 1995; 32:9: 732-735.
25. Ricci E. Servidei S, Broccolini A, Mirabella M, Manfredi G, Bertini E, Rana M, Papacci M, Di Trapani G and Tonali P. Mitochondrial abnormalities in inflammatory myopathies. Clin Neuropathol 1993; 12(Suppl): 60.
26. Schifter T, Zahavi I and Moroz C. Antimitochondrial antibodies after acute myocardial infarction. Cardiology 1996; 87:1: 67-70.
27. Sengers RC, Stadhouders AM and Trijbels JM. Mitochondrial myopathies. Clinical morphological and biochemical aspects. Eur J Pediatr 1984; 141: 192-207.
28. Sengers RC and Stadhouders AM. Secondary mitochondrial pathology. J Inherit Metab Dis 1987; 10(Suppl): 98-104.
29. Stadhouders AM and Sengers RCA. Morphological observations in skeletal muscle from patients with a mitochondrial myopathy. J Inher Metab Dis 1987; 10(Suppl 1): 62-80.
30. Sukhorukov VS. Mitochondrial insufficiency and individuality: introducing to a problem. Mitochondrion 2002; 1:6: 528.
31. Sukhorukov VS. Energy deficient diathesis as energy metabolism disadaptation in children. Mat of VII World Congress of Int. Soc. For Adaptive Medicine, Moscow 2006; II-1.19: 54.
32. Sukhorukov VS. Mitochondrial proliferation as adaptation mechanism in various diseases. In: Lukyanova L, Takeda N, Singal PK, eds., Adaptation Biology and Medicine: Health Potentials. Narosa Publishers, New Delhi, 2008: Vol. 5, 295-305.
33. Tondera D, Santel A, Schwarzer R, Dames S, Giese K, Klippel A and Kaufmann J. Knockdown of MTP18, a novel phosphatidylinositol 3-kinase dependent protein, affects mitochondrial morphology and induces apoptosis. J Biol Chem 2004; 279: 31544-31555.

34. Tondera D, Czauderna F, Paulick K, Schwarzer R, Kaufmann J and Santel A. The mitochondrial protein MTP18 contributes to mitochondrial fission in mammalian cells. Journal of Cell Science 2005; 118: 3049-3059.
35. Uemura O, Goto Y, Iwasa M, Ando T, Sato K, Tominaga Y, Uchida K, Ichiki T and Sugiyama N. Secondary carnitine palmitoyltransferase deficiency in chronic renal failure and secondary hyperparathyroidism. Tohoku J Exp Med 1996; 178:3: 307-314.
36. Wallace DC, Lott MT, Shoffner JM and Ballinger S. Mitochondrial DNA mutations in epilepsy and neurological disease. Epilepsia 1994; 35(Suppl 1): S43-S50.
37. Wallace DC and Melov S. Radicals raging. Nat Genet 1998; 19: 105-106.
38. Wiesner RJ, Hornung TV, Garman JD, Clayton DA, O'Gorman E and Wallimann t. Stimulation of mitochondrial gene expression and proliferation of mitochondria following impairment of cellular energy transfer by inhibition of the phosphocreatine circuit in rat hearts. Journal of Bioenergetics and Biomembranes 1999; 31:6: 559-567.
39. Xia Y, Buja LM, Scarpulla RC and McMillin JB. Electrical stimulation of neonatal cardiomyocytes results in the sequential activation of nuclear genes governing mitochondrial proliferation and differentiation. Proc Natl Acad Sci USA 1997; 94:21: 11399-11404.

Adaptation Biology and Medicine (Volume 6: Cell Adaptations and Challenges)
Editors: P. Wang, C.-H. Kuo, N. Takeda and P.K. Singal
Copyright © 2011 Narosa Publishing House Pvt. Ltd., New Delhi, India

7. Role of Sildenafil (Viagra™) in Cardioprotection and Treatment of Heart Failure

Fadi N. Salloum and Rakesh C. Kukreja*

Division of Cardiology, Department of Internal Medicine, *Virginia Commonwealth University, Richmond, VA 23298-0204, USA*

Summary

Based on the data reviewed, we suggest that phosphodiesterase-5 (PDE-5) inhibitors hold an immense promise for further development as novel drug therapies for myocardial infarction, cardiac arrhythmias, cardiac hypertrophy, and heart failure. Moreover, with the advancement in management of patients with cardiovascular disease and improvement in survival following cardiovascular events, the incidence of heart failure, especially in patients of age 65 and older, is increasing. These drugs may not only delay or reduce the pathological damage or defects in various vital organs, but also improve the overall well-being and quality of life in patients.

Keywords: Sildenafil, angina, caver nosum, phosphodiesterase, Langendorff mode, nitric oxide synthase

1. Introduction

PDE-5 Inhibitors and Nitric Oxide-driven cGMP Signaling

Over the past decade, the potential application of phosphodiesterase-5 (PDE-5) inhibitors has evolved from what these drugs were originally designed for – angina – to erectile dysfunction (1,2), by serendipity, to pulmonary arterial hypertension (PAH) after rigorous research (3). Prior to their FDA approval to treat PAH, PDE-5 inhibitors were especially attractive for their capacity of restoring nitric oxide (NO)-driven cGMP signaling. PDE-5 inhibitors are among a large family of 11 PDEs that have been identified in mammalian tissues (4,5). PDEs are found in all tissues, but their distribution and type vary among different tissues (6). For example, PDE-5 is abundant in the corpus cavernosum and is the predominant PDE in this tissue (7). Although other PDEs are present in the corpus cavernosum, they do not appear to significantly modulate changes in cGMP levels associated with the ability to achieve

*Address for Correspondence: Rakesh C. Kukreja, Ph.D. Professor of Medicine, Physiology, Biochemistry and Emergency Medicine Eric Lipman Professor of Cardiology Division of Cardiology, Box 980204 Virginia Commonwealth University 1101 East Marshall Street, Room 7-020D Richmond, VA 23298, USA Telephone: (804) 628-5521 Fax: (804) 828-8700 E-mail: rakesh@vcu.edu Fax: 421 2 5477 6637

penile erection. Immunohistochemical studies have demonstrated the presence of PDE-5 in vascular and bronchial smooth muscle and in platelets. Moreover, recent studies have provided evidence for PDE-5A expression in canine (8) and mouse heart (9). The PDEs vary in their substrate specificity for cyclic adenosine monophosphate (cAMP) and cGMP: PDE-5, PDE-6, and PDE-9 are specific for cGMP; PDE-4, PDE-7, and PDE-8 are specific for cAMP; and PDE-1, PDE-2, PDE-3, PDE-10, and PDE-11 have mixed specificity for cAMP/ cGMP (10). Although PDE-3 has similar affinity for cAMP and cGMP, it demonstrates lower enzymatic activity against cGMP. Sildenafil is structurally similar to cGMP and therefore competes with cGMP for binding to PDE-5 at the catalytic site (11). Interestingly, PDE expression also changes in pathologic conditions, in addition to a decline in endothelial function and disruption of NO homeostasis. For instance, in patients with cardiovascular disease or diabetes, NO levels are suboptimal due to endothelial dysfunction (damaged NOS), and recently myocardial PDE-5 expression has been shown to increase in patients with heart failure (12). In this regard, targeting PDE-5 is also necessary therapeutically rather than relying solely on conventional attempts to improve NOS activity and NO production in diseased states.

Sildenafil citrate (Viagra™) is the first PDE-5 inhibitor approved for treatment of erectile dysfunction. The discovery of this drug in 1989 was the result of extensive research on chemical agents that might potentially be useful in the treatment of coronary heart disease. Initial clinical studies on sildenafil in the early 1990s were not promising with respect to its anti-anginal potential. However, a remarkable side effect was reported by a number of volunteers participating in these investigations; sildenafil seemed to enhance penile erections, which soon thereafter became the main focus of further studies. More than 10 million men worldwide have been treated with sildenafil since its market debut in 1998. Sildenafil is highly specific for PDE-5 inhibition with relatively minor cross-reactivity with PDE-6 (13). It has a chemical structure similar to cGMP (**Fig. 1.**) and inhibits PDE-5 by binding to the cGMP-catalytic sites (4) thereby allowing the accumulation of cGMP in the erectile tissue. Two additional agents in this class (tadalafil [Cialis™], vardenafil [Levitra™]) have also been developed and approved by the FDA for treatment of erectile dysfunction.

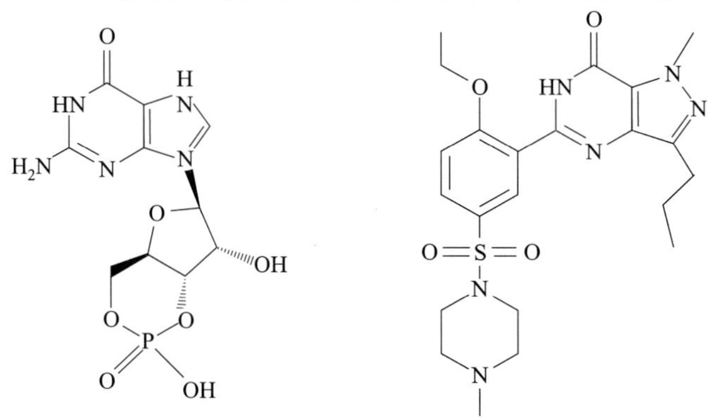

cGMP **Sildenafil**

Fig. 1. Structural resemblance of cGMP and sildenafil

2. Sildenafil and Myocardial Protection against Ischemia/Reperfusion Injury

Our landmark study published in 2002 demonstrated the protective effect of sildenafil against experimental myocardial ischemia/reperfusion in rabbits (14). In this study, rabbits were treated with sildenafil (0.7 mg/kg, *i.v.*) or volume-matched saline either 30 min (acute phase) or 24 h (delayed phase) prior to 30 min of regional ischemia by coronary artery occlusion followed by 3 h of reperfusion. Since sildenafil was administered intravenously, the dose of 0.7 mg/kg, *i.v.* was chosen to simulate the serum level of sildenafil for a patient with 70 kg body weight after orally taking a 100 mg tablet of sildenafil (*i.e.* 100 mg/70 kg × 49% = 0.70 mg/kg, considering approximately 49% bio-availability through oral administration according to Pfizer, Inc.). The infarct size (% of risk area) was reduced from 33.8±1.7% in the saline treated control group to 10.8±0.9% in the acute phase (68% reduction) and 19.9±2.0% in the delayed phase (41% reduction) in the sildenafil-treated rabbits. To gain a better understanding of the mechanism of action through which sildenafil exerted its cardioprotective effects, the role of mitochondrial ATP-sensitive potassium (K_{ATP}) channel was examined using the channel blocker, 5-hydroxydecanoate (5-HD, 5 mg/kg, *i.v.*), which was given 10 min before ischemia/reperfusion. The results showed that the infarct-limiting effect of sildenafil was abolished in animals treated with 5-HD as shown by significant increase in infarct size to 35.6±0.4% in the acute phase and 36.8±1.6% in the delayed phase. Control animals treated with 5-HD had an infarct size of 33.5±1.9%, which was not different from the control animals. Shortly after these studies, Das *et al.* (15) confirmed that sildenafil at 0.05 mg/kg provided significant protection in the isolated perfused rat heart following global ischemia and reperfusion in Langendorff mode. These investigators observed improved post-ischemic recovery of ventricular function, reduced incidence of ventricular fibrillation and decreased myocardial infarction. However, at higher doses, sildenafil caused a significant increase in the incidence of ventricular fibrillation while at very low doses it had no effect on cardiac function. In 2003, Salloum *et al.* further demonstrated a powerful delayed preconditioning-like effect, with a clinically relevant dose of 0.7 mg/kg of sildenafil, against global ischemia/reperfusion in the isolated-perfused mouse heart (16). This protection was mediated by inducible nitric oxide synthase (iNOS), and sildenafil was shown to up-regulate iNOS expression in the heart at 24 h after treatment (**Fig. 2**). Moreover, the iNOS inhibitor, 1400W, blocked the protective effect of sildenafil in this model. At this point, the interest in investigating the effect of this relatively novel PDE-5 inhibitor on the cardiovascular system and attenuation of myocardial infarction following ischemia was increasing. In 2004, Krolikowski *et al.* also reported that sildenafil induced an infarct-sparing effect in a dog model of infarction (17), suggesting a translational promise of this drug for clinical use against ischemia/reperfusion. Although initial studies by Reffelmann and Kloner were not supportive of the cardioprtotective effects of PDE-5 inhibitors (18), later studies by the same group provided increasing proof that PDE-5 inhibition could in fact be a new therapeutic tool to combat cardiovascular disease (19).

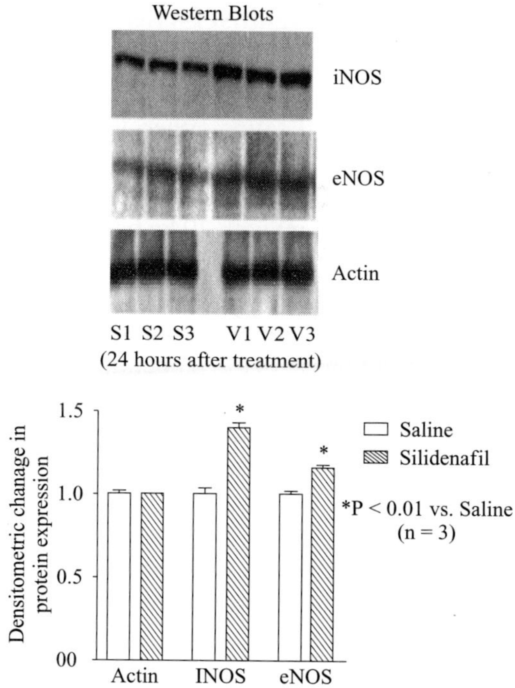

Fig. 2. Sildenafil up-regulates eNOS and iNOS 24 h after treatment

In order to evaluate the safety margin of sildenafil administration on systemic hemodynamics, we measured systolic, diastolic and mean arterial pressures prior to, during, and following sildenafil administration. The intravenous administration of sildenafil caused a rapid decrease in hemodynamics as indicated by 24.5%, 47.3% and 38.8% decline in systolic, diastolic and mean arterial pressures, respectively, within 2 minutes. Interestingly, the systemic hemodynamics nearly returned to baseline levels by 5 minutes after treatment with sildenafil (14). However, no significant changes in heart rate were observed following treatment with sildenafil.

To examine whether the cardioprotective effect of sildenafil was independent of the vasculature and systemic hemodynamics, we isolated adult murine cardiomyocytes. These studies provided the first evidence that adult mouse cardiomyocytes express PDE-5, and that sildenafil protected these cells against simulated ischemia/re-oxygenation *in vitro* by attenuating necrosis as well as apoptosis (9). These cells were incubated under 37°C and 5% CO_2, for 1 h with or without 1 μM or 10 mM sildenafil and then subjected to simulated ischemia for 40 minutes by replacing the cell medium with an "ischemia buffer" which contained (in mM): 118 NaCl, 24 $NaHCO_3$, 1.0 NaH_2PO_4, 2.5 $CaCl_2$-$2H_2O$, 1.2 $MgCl_2$, 20 sodium lactate, 16 KCl, 10 2-deoxyglucose (pH adjusted to 6.2) similar to those previously published (20). After 40 minutes of SI, the percentage of viable cardiomyocytes decreased with respect to non-ischemic control cardiomyocytes. Pretreatment with 1 μM and 10 mM sildenafil significantly improved viability of cardiomyocytes. In addition, the number of TUNEL positive cells (apoptotic nuclei) after 40 min of SI and 18 hours of RO were significantly reduced in cardiomyocytes treated with sildenafil. Further studies in adult mouse cardiomyocytes characterized protein kinase G

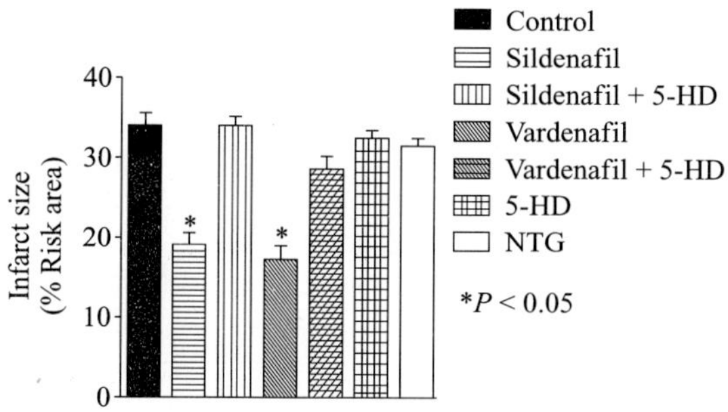

Fig. 3. Sildenafil an vardenafil protect the heart against ischemia/reperfusion when administered at the time of reperfusion

(PKG), ERK, iNOS and GSK-3ß as key players promoting protection of these cells with sildenafil against simulated ischemia (21,22).

3. Sildenafil and Attenuation of Arrhythmias

Following the study by Das *et al.* (15) that showed reduced incidence of ventricular fibrillation paralleled with decreased myocardial infarction with sildenafil treatment in the isolated-perfused rat heart, Nagy *et al.* (23) examined the effects of sildenafil on the severity of life-threatening ventricular arrhythmias secondary to acute myocardial infarction. In these studies, sildenafil (2 mg/kg) was administered orally in dogs 24 h prior to ischemia/reperfusion. There were fewer and less serious ventricular arrhythmias during ischemia in the sildenafil-treated group (ventricular fibrillation 17% vs. 60%; ventricular premature beats 140±52 vs. 437±127% and episodes of ventricular tachycardia 4.0±3.2 vs 19.3±7.7%). These studies suggested that sildenafil also leads to a delayed preconditioning against ventricular arrhythmias. Some recent studies do not support the anti-arrhythmic effect of sildenafil as they demonstrate that supratherapeutic levels of sildenafil significantly decreased the ventricular fibrillation threshold and increased the upper limit of vulnerability, resulting in an expansion of the ventricular fibrillation induction window during vulnerable periods in a swine model (24). This observation, however, has not been reported with small therapeutic doses of the drug.

4. Sildenafil and Cardioprotection Against Ischemia when Administered During Reperfusion

The majority of the studies involving sildenafil and cardioprotection focused on the drug's ability to exert a preconditioning-like effect against ischemia. In order to address yet another clinical challenge, the protective capacity of sildenafil was examined when the drug was given at the time of reperfusion in a rabbit model of ischemia/reperfusion (25). In these studies, adult rabbits underwent 30 minutes of ischemia followed by 3 h of reperfusion. At 5 minutes prior to reperfusion, sildenafil was administered at a dose of 0.7 mg/kg; *iv* in a total volume of 55 ml saline infused at a constant rate of 50 ml/h for 65 minutes. Another PDE-5 inhibitor, vardenafil (0.014 mg/kg, *iv*) was used in this study to demonstrate that cardioprotection with PDE-5 inhibitors is a class effect rather than an off-target effect of sildenafil. The infarct size (% of risk area, mean ± SE, $P < 0.05$) decreased from 33.8 ± 1.7% in the

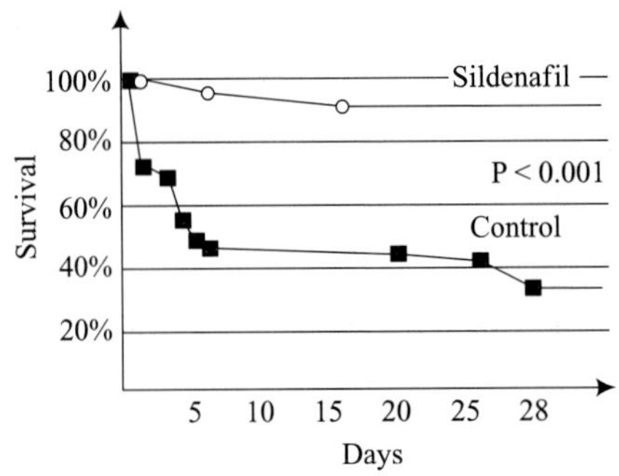

Fig. 4. Kaplan-Meier survival curve showing improved survival with sildenafil versus saline up to 28 days following permanent coronary artery occlusion

saline-treated control group to 19.2 ± 1.3% and 17.0 ± 2.0% in the sildenafil and vardenafil-treated rabbits, respectively (**Fig. 3**). Since there is ample evidence which suggests that exogenous NO that results from administration of a NO donor is cardioprotective (26,27,28), the effect of nitroglycerin (NTG, at a clinically relevant dose of 2μg/kg/min) was tested when infused at the time of reperfusion. Surprisingly, NTG failed to mimic the protective effect of PDE-5 inhibitors as indicated by an infarct size of 31.5 ± 0.8%. Similar to preconditioning, cardioprotection with sildenafil and vardenafil at reperfusion was also mediated by opening of mitoK_{ATP} channel because 5-HD blocked this protection as indicated by a significant increase in infarct size. Even though PDE-5 inhibitors or NTG are known to cause a decline in blood pressure when administered prior to I/R, a much milder effect on hemodynamics was observed when these drugs were administered in a slow infusion protocol with cumulative doses similar to the pre-treatment protocol. This method of administration was chosen to avoid a detrimental decline in systemic blood pressure in the ischemic hearts. One interesting finding in this study is that NTG failed to reduce infarct size at reperfusion. The differences in these results may be attributed to the timing of infusion and the kinetics of cGMP accumulation following treatment with NTG or PDE-5 inhibitors. PDE-5 inhibition with potent drugs including sildenafil and vardenafil causes a prompt accumulation of cGMP (by preventing its degradation) in the tissue (29). The higher availability of NO with NTG infusion would also increase cGMP by activating soluble guanylate cyclase in the presence of GTP, but probably not as swiftly as would selective inhibition of PDE-5. As a result, there may be insufficient availability of cGMP to prevent ischemic injury with NTG at the time of reperfusion. Another possible explanation could be the surge of superoxide radical that occurs at the onset of reperfusion (30,31) which may interact with NO derived from NTG to form the peroxynitrite anion ($ONOO^-$). $ONOO^-$ formation during the early period of reperfusion results in protein nitration and myocardial injury (32). Thus, it appears that rapid up-regulation of cGMP by PDE-5 inhibition, rather than increasing its synthesis through NO, is an effective approach to prevent reperfusion injury in the heart.

5. Sildenafil and Attenuation of Ischemic Cardiomyopathy and Left Ventricular Dysfunction

Due to increasing evidence of cardioprotection with sildenafil owing to its anti-ischemic effects in the setting of ischemia/reperfusion injury, scientists were compelled to investigate the effect of this drug in a more severe model of ischemia-induced heart failure. In a murine model of post-MI remodeling by permanent ligation of the left coronary artery, the effect of chronic treatment with sildenafil was studied on myocardial necrosis at 24 h post myocardial infarction (33). In addition, survival, cardiomyocyte apoptosis, cardiac hypertrophy, pulmonary edema and left ventricular function were examined at 7 and 28 days post myocardial infarction. In this study, 28 out of 30 mice survived in the sildenafil treated group (93%) as compared to 14 out of 39 in the saline control group (36%, P<0.001). The survival rate was 100 % in sham operated mice (**Fig. 4**). Moreover, the infarct size (% of risk area) was reduced from 69.6±4.1% in the saline-treated group to 40.0±4.6% in the sildenafil-treated mice 24 h after infarction (P<0.05). The infarct-sparing effect of sildenafil was abolished with the non-selective NOS inhibitor L-NAME as shown by an increase in infarct size to 60.2±1.6% (P<0.05). Control animals treated with L-NAME alone had an infarct size of 66.7±2.3%, which was not different from the saline group (P>0.05). The risk areas (% LV) were not statistically different between the groups. Since this study was chronic, it involved more time points for evaluation of the various parameters. Apoptosis was assessed using TUNEL assay at 7 and 28 days. The TUNEL positive apoptotic cells in saline-treated hearts were 2.4 ± 0.3% cells as compared to 1.2 ± 0.1% in the sildenafil group on day 7 post infarction in the peri-infarct area (P<0.05). Similarly the TUNEL positive cells decreased from 2.0 ± 0.2% in the saline group to 1.2 ± 0.1% in the sildenafil group on day 28 post infarction (P<0.05). As for left ventricular function, M-mode echocardiography was assessed utilizing the Vevo770^{TM} imaging system (VisualSonics Inc., Toronto, Canada). The hearts from sham and sildenafil treated mice exhibited a smaller left ventricular cavity and thicker infarct wall compared with the saline-treated mice. Increase in left ventricular end diastolic diameter (LVEDD), end systolic diameter (LVESD) and a decrease in anterior wall diastolic thickness (AWDT), systolic thickness (AWST) and fractional shortening (FS) in saline- and sildenafil-treated mice (vs. baseline and sham) were observed on day 7 and 28 (**Fig. 5**). Sildenafil-treated mice had smaller LVEDD, LVESD, greater FS, and lower Tei index (indicating better myocardial performance) on day 7 and 28 as compared to saline-treated group (P<0.05). Sildenafil-treated animals also had a shorter isovolumetric relaxation time values (reflective of lower LV end-diastolic pressure) 28 days after acute MI when compared to saline-treated animals (11±3 vs. 27±7 ms, respectively, P=0.030), which was not different from sham operated animals (10±3, P=NS). AWDT and AWST were also greater in sildenafil-treated animals (vs. saline-treated animals, P<0.05) on day 7 and 28 post MI showing a protective effect in the peri-infarct region, while no differences in posterior wall diastolic and systolic thickness were seen. Aneurysmatic dilatation of the anterior wall and apex was observed on day 28 in 90% of saline-treated mice and 62% of sildenafil treated animals (P>0.05). Moreover, the number of aneurysmatic segments [based on a 16-segment map (34)] was 2.9 in saline-treated animals vs. 1.1 in sildenafil-treated animals (P<0.05).

To further support their results, we carried out biochemical studies assessing pro- and anti-apoptotic proteins, eNOS and iNOS expression. Bcl-2 (anti-apoptotic) and Bax (pro-apoptotic) expression was measured in myocardial samples 2 h after coronary artery occlusion in saline and sildenafil treated mice. Protein was extracted and Bcl-2, Bax and ß-actin were identified by Western blotting as described previously. The expression of Bcl-2 and Bax was normalized with ß-actin. Moreover, the ratio of Bcl-2 to Bax was calculated as an index of apoptotic signaling. As shown in the representative blots and quantitative bar graphs in **Fig. 6**, myocardial levels of Bcl-2 were increased in the sildenafil-treated mice as compared to saline controls (P<0.05, n=3), whereas Bax levels did not change significantly (P>0.05). Bcl-2/Bax ratio was also increased in the sildenafil-treated hearts as compared to the saline-treated hearts (P<0.05). eNOS and iNOS expression was measured in myocardial samples 24 h after coronary occlusion in saline and sildenafil treated mice. Protein was extracted and eNOS, iNOS and ß-actin were identified by Western blotting. The ratio of eNOS and iNOS to ß-actin was used to express the results. Myocardial levels of both eNOS and iNOS were increased in the sildenafil-treated mice as compared to saline controls (P<0.05), providing more evidence that NO mediates cardioprotection with sildenafil.

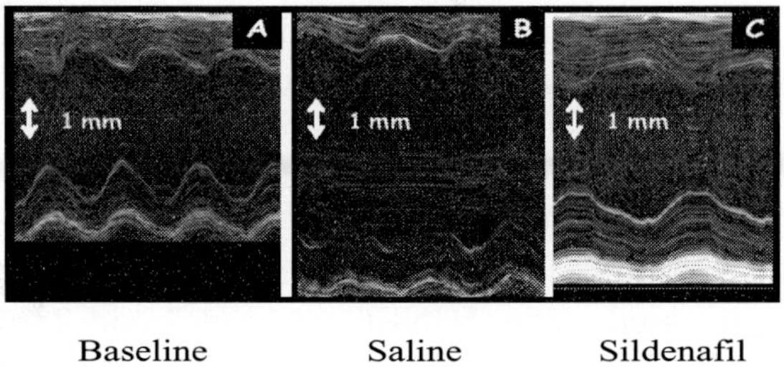

Fig. 5. M-mode echocardiography illustrating less left ventricular dilatation and more contractility in sildenafil-treated hearts compared to saline at 28 days following permanent coronary artery occlusion.

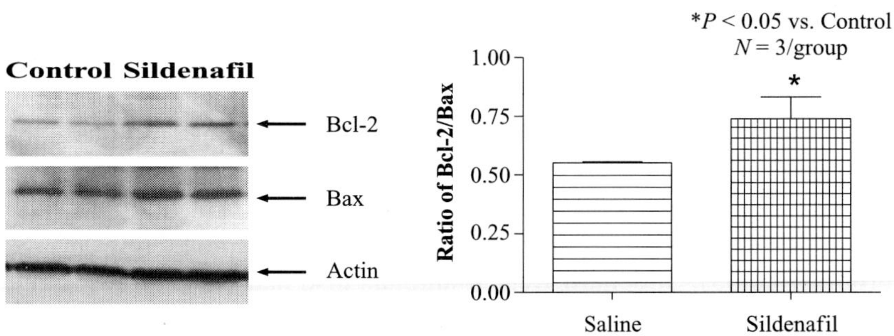

Fig. 6. LeftL: Western bolts illustrating an increase in Bcl-2 expression following sildenafil treatment; Right: Desitometric analysis showing an increase in Bcl-2/ ratio with sidenfil versus saline.

5.1 PDE-5 and Heart Failure

PDE-5A gene is present in the human heart (35) although protein expression and enzyme activity were questioned by Ito *et al.* (36) and Wallis et al (37). A recent study by Pokreisz *et al.* showed increased cardiac left ventricular PDE-5A expression in patients with heart failure (12). In this study, the authors also generated a PDE-5A transgenic mouse in which adverse ventricular remodeling and dysfunction were exacerbated following experimental myocardial infarction.

Moreover, we recently demonstrated that PDE-5-induced cytosolic cGMP hydrolytic activity increases with heart failure in mice when compared to control non-failing hearts (38). Since NO donors remain essential in the management of patients with heart failure, it is plausible that exogenous NO is required due to impairment/inefficiency of NOS and endogenous production of NO, or due to increased activity and/or expression of PDE-5 with heart failure. More studies are needed to evaluate the expression of PDE-5 in the human heart as well as its behavior under diseased states such as myocardial infarction or heart failure. Such studies should address whether the increase in PDE-5 in heart failure patients reported by Pokreisz *et al.* is in fact a cause or an effect of the disease.

6. Sildenafil and Attenuation of Cardiac Hypertrophy

In 2005, Takimoto *et al.* demonstrated that chronic administration of sildenafil not only prevents, but also reverses cardiac hypertrophy induced by transverse aortic constriction (39). In this study, the authors show that blocking the intrinsic catabolism of cGMP with oral sildenafil administration suppresses chamber function and myocyte hypertrophy, and improves in vivo heart function in mice exposed to chronic pressure overload. Sildenafil also reversed pre-established hypertrophy induced by pressure load while restoring chamber function to normal. The authors also reported that cGMP catabolism by PDE-5A increases in pressure-loaded hearts, and therefore inhibition of PDE-5A leads to restoration of cGMP signaling and activation of cGMP-dependent protein kinase (PKG). In this respect, PDE-5A inhibition may provide a new treatment strategy for cardiac hypertrophy and remodeling.

Recently, we also observed that chronic treatment with sildenafil attenuates cardiac hypertrophy and pulmonary edema secondary to ischemic cardiomyopathy in mice (33). These findings collectively support the notion of cardioprotection with sildenafil against chronic pressure overload and ischemia, both of which lead to adverse cardiac remodeling and failure.

7. Sildenafil in Left Ventricular Assist Device (LVAD) Therapy and Heart Transplantation

It is well known that pulmonary hypertension secondary to left heart failure portends a poor prognosis and is a relative contraindication to heart transplantation. A recent study tested the hypothesis that sildenafil would decrease pulmonary hypertension when it persists after adequate left ventricle unloading via LVAD therapy (40). In this study, the authors performed an open-label clinical trial using control patients not receiving therapy. Between 1999 and 2007, 138 consecutive patients undergoing cardiac transplantation evaluation with advanced left ventricular dysfunction, an elevated pulmonary capillary wedge pressure, and pulmonary hypertension (defined by a pulmonary vascular resistance (PVR) >3 Woods Units), were treated with LVAD therapy. The pulmonary capillary wedge pressure was reduced in 58 of these patients to a value <15 mm Hg (11.8±2.0 mm Hg from baseline 23.2±6.2 mm Hg) 1 to 2 weeks after LVAD implantation, but despite this improvement, the PVR of these patients was minimally affected (5.65±3.00 to 5.39±1.78 Wood Units). Twenty-six consecutive patients from this group with persistently elevated PVR received oral sildenafil titrated to an average of dose of 51.9 mg by mouth 3 times per day. The average PVR in the sildenafil-treated group decreased from 5.87+/-1.93 to 2.96±0.92 Wood Units (P<0.001) and the mean pulmonary artery pressure declined from 36.5±8.6 to 24.3±3.6 mm Hg (P<0.0001) and was significantly lower when compared with the 32 LVAD recipients not receiving sildenafil at weeks 12 to 15 after the initial post-LVAD hemodynamic measurements. In addition, hemodynamic measurement of right ventricular function in sildenafil-treated patients was also improved compared with patients not receiving sildenafil. These results implicate PDE-5A inhibition with sildenafil as a treatment modality to decrease PVR in patients with persistent pulmonary hypertension after recent LVAD placement.

8. Future Perspectives

Clinical studies of sildenafil on heart failure patients have reported improved exercise capacity, coupled with reduced pulmonary vascular resistance and better endothelial function (41,42). Based on such findings, the National Institutes of Health has initiated a multi-center clinical trial of sildenafil for treatment of heart failure with a preserved ejection fraction. Moreover, the recent studies showing that sildenafil decreases pulmonary vascular resistance in patients with persistent pulmonary hypertension after recent LVAD placement argue for a need to consider sildenafil as a mainstay in these patients awaiting transplantation. Several other studies also indicated that PDE-5 inhibition with sildenafil has a therapeutic promise for stroke, neurodegenerative diseases and potentially other circulatory disorders (43).

References

1. Boolell, M., Allen, M.J., Ballard, S.A., Gepi-Attee, S., Muirhead, G.J., Naylor, A.M., Osterloh, I.H., Gingell, C. Sildenafil: an orally active type 5 cyclic GMP-specific phosphodiesterase inhibitor for the treatment of penile erectile dysfunction. *Int. J. Impot. Res.* 1996;8: 47-52.
2. Porst, H., Rosen, R., Padma-Nathan, H., Goldstein, I., Giuliano, F., Ulbrich, E., Bandel T. The efficacy and tolerability of vardenafil, a new, oral, selective phosphodiesterase type 5 inhibitor, in patients with erectile dysfunction: the first at-home clinical trial. *Int. J. Impot. Res.* 2001;13: 192–199.
3. Kane LB., Klings ES. Present and future treatment strategies for pulmonary arterial hypertension: focus on phosphodiesterase-5 inhibitors. *Treat Respir Med.* 2006;5: 271-282.
4. Corbin, J.D., Francis, S.H.. Pharmacology of phosphodiesterase-5 inhibitors. *Int. J. Clin. Pract.* 2002;56: 453-459.
5. Corbin, J.D., Francis, S.H., Webb, D.J. Phosphodiesterase type 5 as a pharmacologic target in erectile dysfunction. *Urology*. 2002;60 (suppl. 2): 4-11.
6. Wallis, R.M., Corbin, J.D., Francis, S.H., Ellis, P. Tissue distribution of phosphodiesterase families and the effects of sildenafil on tissue cyclic nucleotides, platelet function, and the contractile responses of trabeculae carneae and aortic rings in vitro. *Am. J. Cardiol.* 1999;83 (suppl.): 3C–12C.
7. Gopal, V.K., Francis, S.H., Corbin, J.D. Allosteric sites of phosphodiesterase-5 (PDE5). Apotential role in negative feedback regulation of cGMP signaling in corpus cavernosum. *Eur. J. Biochem.* 2001;268: 3304–3312.
8. Senzaki, H., Smith, C.J., Juang, G.J., Isoda, T., Mayer, S.P., Ohler, A., Paolocci, N., Tomaselli, G.F., Hare, J.M., Kass, D.A. Cardiac phosphodiesterase 5 (cGMP-specific) modulates beta-adrenergic signaling in vivo and is down-regulated in heart failure. *FASEB J.* 2001;15: 1718–1726.
9. Das, A., Ockaili, R., Salloum, F., Kukreja, R.C.. Protein kinase C plays an essential role in sildenafil-induced cardioprotection in rabbits. *Am. J. Physiol. Heart Circ. Physiol.* 2004;286: H1455-H1460.
10. Rotella, D. P. Phosphodiesterase 5 inhibitors: current status and potential applications. *Nature Rev. Drug Discovery.* 2002;1: 674-682.

11. Kukreja RC, Salloum F, Das A, Ockaili R, Yin C, Bremer YA, Fisher PW, Wittkamp M, Hawkins J, Chou E, Kukreja AK, Wang X, Marwaha V, and Xi L. Pharmacological Preconditioning with Sildenafil: Basic Mechanisms and Clinical Implications. *Vascul Pharmacol.* 2005; 42: 219-232.

12. Pokreisz P, Vandenwijngaert S, Bito V, Van den Bergh A, Lenaerts I, Busch C, Marsboom G, Gheysens O, Vermeersch P, Biesmans L, Liu X, Gillijns H, Pellens M, Van Lommel A, Buys E, Schoonjans L, Vanhaecke J, Verbeken E, Sipido K, Herijgers P, Bloch KD, Janssens SP. Ventricular phosphodiesterase-5 expression is increased in patients with advanced heart failure and contributes to adverse ventricular remodeling after myocardial infarction in mice. *Circulation.* 2009;119: 408-416.

13. Laties, A.M., Fraunfelder, F.T. Ocular safety of Viagra (sildenafil citrate). *Trans Am. Ophthalmol. Soc.* 1999;97: 115–128.

14. Ockaili, R., Salloum, F., Hawkins, J., Kukreja, R.C. Sildenafil (Viagra) induces powerful cardioprotective effect via opening of mitochondrial K_{ATP} channels in rabbits. *Am. J. Physiol. Heart Circ. Physiol.* 2002;283: H1263-H1269.

15. Das, S., Maulik, N., Das, D.K., Kadowitz, P.J., Bivalacqua, T.J. Cardioprotection with sildenafil, a selective inhibitor of cyclic 3',5'-monophosphate-specific phosphodiesterase 5. *Drugs Exp. Clin. Res.* 2002;28: 213-219.

16. Salloum, F., Yin, C., Xi, L., Kukreja, R.C. Sildenafil induces delayed preconditioning through inducible nitric oxide synthase-dependent pathway in mouse heart. *Circ. Res.* 2003;92: 595-597.

17. Krolikowski, J.G., Pagel, P.S., Warltier, D.C., Kersten, J.R. Diabetes abolishes sildenafil-induced preconditioning in dogs. *FASEB J.* 2004;18: A262, (Abstract).

18. Reffelmann, T., Kloner, R.A. Effects of sildenafil on myocardial infarct size, microvascular function, and acute ischemic left ventricular dilation. *Cardiovasc. Res.* 2003;59: 441-449.

19. Sesti C, Florio V, Johnson EG, Kloner RA. The phosphodiesterase-5 inhibitor tadalafil reduces myocardial infarct size. *Int J Impot Res.* 2007;19: 55-61.

20. Rakhit, R.D., Edwards, R.J., Mockridge, J.W., Baydoun, A.R., Wyatt, A.W., Mann, G.E., Marber, M.S. Nitric oxide-induced cardioprotection in cultured rat ventricular myocytes. *Am. J. Physiol. Heart Circ. Physiol.* 2000;278: H1211-H1217.

21. Das A, Xi L, Kukreja RC. Protein kinase G-dependent cardioprotective mechanism of phosphodiesterase-5 inhibition involves phosphorylation of ERK and GSK3beta. *J Biol Chem.* 2008;283: 29572-29585.

22. Das A, Salloum FN, Xi L, Rao YJ, Kukreja RC. Role of ERK1/2 Phosphorylation in Sildenafil-Induced Myocardial Protection against Ischemia-Reperfusion Injury in Mice. *Am J Physiol Heart Circ Physiol.* 2009;296: H1236-H1243.

23. Nagy, O., Hajnal, A., Parratt, J.R., Vegh, A. Sildenafil (Viagra) reduces arrhythmia severity during ischaemia 24 h after oral administration in dogs. *Br. J. Pharmacol.* 2004;141: 549-551.

24. Kanlop N, Shinlapawittayatorn K, Sungnoon R, Chattipakorn S, Lailerd N, Chattipakorn N. Sildenafil citrate on the inducibility of ventricular fibrillation and upper limit of vulnerability in swine. *Med Sci Monit.* 2008;14: BR205-BR209.

25. Salloum FN, Takenoshita Y, Ockaili RA, Daoud VP, Chou E, Yoshida K, and Kukreja R.C. Sildenafil and vardenafil but not nitroglycerin limit myocardial infarction through opening of mitochondrial K_{ATP} channels when administered at reperfusion following ischemia in rabbits. *J Mol Cell Cardiol.* 2007;42: 453-458.
26. Lefer DJ, Nakanishi K, Johnston WE, and Vinten-Johansen J. Antineutrophil and myocardial protecting actions of a novel nitric oxide donor after acute myocardial ischemia and reperfusion in dogs. *Circulation.* 1993;88: 2337–2350.
27. Lochner A, Marais E, Genade S, and Moolman JA. Nitric oxide: a trigger for classic preconditioning? *Am J Physiol Heart Circ Physiol.* 2000;279**:** H2752–H2765.
28. Johnson G 3rd, Tsao PS, Lefer AM. Cardioprotective effects of authentic nitric oxide in myocardial ischemia with reperfusion. *Crit Care Med.* 1991;9: 244-252.
29. Francis SH, Corbin JD. Phosphodiesterase-5 inhibition: the molecular biology of erectile function and dysfunction. *Urol Clin North Am.* 2005;32: 419-429.
30. Zweier JL. Measurement of superoxide-derived free radicals in the reperfused heart. Evidence for a free radical mechanism of reperfusion injury. *J Biol Chem.* 1988;263: 1353-1357.
31. Kukreja, R.C. and Hess, M.L. The oxygen free radical system: from equations through membrane protein interaction to myocardial injury and protection. *Cardiovasc Res.* 1992;26**:** 641-655.
32. Zweier JL, Fertmann J, Wei G. Nitric oxide and peroxynitrite in postischemic myocardium. *Antioxid Redox Signal.* 2001;3: 11-22.
33. Salloum FN, Abbate A, Das A, Houser J, Mudrick CA, Qureshi IZ, Hoke NN, Roy SK, Brown WR, Prabhakar S, and Kukreja RC. Sildenafil (Viagra) Attenuates Ischemic Cardiomyopathy and Improves Left Ventricular Function in Mice. *Am J Physiol Heart Circ Physiol.* 2008;294: H1398-H1406.
34. Schiller NB, Shah PM, Crawford M, DeMaria A, Devereux R, Feigenbaum H, Gutgesell H, Reichek N, Sahn D, Schnittger I. Recommendations for quantitation of the left ventricle by two-dimensional echocardiography. American Society of Echocardiography Committee on Standards, Subcommittee on Quantitation of Two-Dimensional Echocardiograms. *J Am Soc Echocardiogr* 1989;2: 358-367.
35. Sebkhi A, Strange JW, Phillips SC, Wharton J, Wilkins MR. Phosphodiesterase type 5 as a target for the treatment of hypoxia-induced pulmonary hypertension. *Circulation.* 2003;107: 3230–3235.
36. Ito M, Nishikawa M, Fujioka M, Miyahara M, Isaka N, Shiku H, Nakano T. Characterization of the isoenzymes of cyclic nucleotide phosphodiesterase in human platelets and the effects of E4021. *Cell Signal.* 1996;8: 575-81.
37. Wallis RM, Corbin JD, Francis SH, Ellis P. Tissue distribution of phosphodiesterase families and the effects of sildenafil on tissue cyclic nucleotides, platelet function, and the contractile responses of trabeculae carneae and aortic rings in vitro. *Am J Cardiol.* 1999; 83:3C-12C.
38. Vandeput F, Krall J, Ockaili R, Salloum FN, Florio V, Corbin JD, Francis SH, Kukreja RC and Movsesian MA. cGMP-hydrolytic activity and its inhibition by sildenafil in normal and failing human and mouse myocardium. *J Pharmacol Exp Ther.* 2009;330: 884-891.

39. Takimoto E, Champion HC, Li M, Belardi D, Ren S, Rodriguez ER, Bedja D, Gabrielson KL, Wang Y, Kass DA. Chronic inhibition of cyclic GMP phosphodiesterase 5A prevents and reverses cardiac hypertrophy. *Nat Med.* 2005;11: 214-222.
40. Tedford RJ, Hemnes AR, Russell SD, Wittstein IS, Mahmud M, Zaiman AL, Mathai SC, Thiemann DR, Hassoun PM, Girgis RE, Orens JB, Shah AS, Yuh D, Conte JV, Champion HC. PDE5A inhibitor treatment of persistent pulmonary hypertension after mechanical circulatory support. *Circ Heart Fail.* 2008;1: 213-9.
41. Lewis GD, Lachmann J, camuso J, Lepore JJ, Shin J, Martinovic ME, Systrom DM, Bloch KD, Semigran MJ. Sildenafil improves exercise hemodynamics and oxygen uptake in patients with systolic heart failure. *Circulation.* 2007;115: 59-66.
42. Guazzi M, Samaja M, Arena R, Vicenzi M, Guazzi MD. Long-term use of sildenafil in the therapeutic management of heart failure. *J Am Coll Cardiol.* 2007;50: 2145-2147.
43. Kukreja RC, Salloum FN, and Xi L. Nonurologic Applications of Phosphodiesterase Type 5 Inhibitors. *Current Sexual Health Reports.* 2007;4: 64-70.

Adaptation Biology and Medicine (Volume 6: Cell Adaptations and Challenges)
Editors: P. Wang, C.-H. Kuo, N. Takeda and P.K. Singal
Copyright © 2011 Narosa Publishing House Pvt. Ltd., New Delhi, India

8. Diabetes/Hyperglycemia differentially Regulate Giα and Gqα Protein Expression and Associated Signalling in Vascular Smooth Muscle Cells

Yuan Li, Magda Descobeth and Madhu B. Anand-Srivastava*

Department of Physiology, Université de Montréal, Quebec, Canada

Abstract

The impaired vascular contractility and increased cell proliferation are the most common vascular complications associated with diabetes and the chronic hyperglycemia appears to be an important contributing factor in this process. Various signalling pathways are implicated in diabetes/hyperglycemia-induced impaired vascular functions. Increased oxidative stress, non-enzymatic glycation, enhanced production of diacylglycerol, increased activity of protein kinase C, MAP Kinases and other signalling pathways have been proposed to explain the adverse effects of hyperglycemia on vascular smooth muscle cells (VSMC). Hyperglycemia-induced stimulation of L-type Ca^{2+} channel via G-protein-coupled adenylyl cyclase/cAMP and phospholipase C/PKC pathways has also been shown. In addition, hyperglycemia has been reported to decrease the availability of nitric oxide and increase the formation of peroxynitrite which may contribute to all the hemodynamic and physiological changes occurring in diabetes. Furthermore, G-protein-adenylyl cyclase signaling and phosphatidyl inositol (PI) turnover that play an important role in the regulation of cardiovascular functions has also been reported to be impaired in diabetes and under hyperglycemic conditions. In this review, we have highlighted some key signalling pathways, such as G-protein/adenylyl cyclase and Gq protein/PI turnover, and have discussed their differential regulation in diabetes and in response to hyperglycemia as well as their contributions in the development of vascular complications.

Keywords: Hyperglycemia, vascular smooth muscle cells, peroxynitrite, adenylyl cyclase, heterotrimeric proteins, streptozotocin.

G-protein-adenylyl Cyclase Signal Transduction System

The adenylyl cyclase/cAMP is one of the signal transduction systems implicated in the regulation of cardiovascular functions including arterial tone, reactivity and cell proliferation. The hormone-sensitive

*Address for Correspondence: Dr Madhu B. Anand-Srivastava Department of Physiology, Faculty of Medicine Université de Montréal C.P. 6128 Succ. Centre-Ville Montréal (Québec) H2X 1H1 Canada Tel: +1 514-343-2091 Fax: +1 514-343-2111 E-mail: madhu.anand-srivastava@umontreal.ca

adenylyl cyclase system is composed of three components; receptor, catalytic subunit and G-proteins; stimulatory (Gs) and inhibitory (Gi) mediate the stimulatory and inhibitory responses of hormones on adenylyl cyclase respectively (Gilman, 1984; Stryer & Bourne, 1986; Spiegel, 1987). G-proteins are heterotrimeric proteins composed of α, β and γ subunits, and the specificity of G-proteins is attributed to α-subunits (Stryer & Bourne, 1986). Four different forms of Gsα and three distinct forms of Giα; Giα-1, Giα-2 and Giα-3, have been identified. All three forms of Giα ($Gi_{\alpha 1\text{-}3}$) are implicated in adenylyl cyclase inhibition (Wong *et al.* 1992). Both the Gα and Gβγ mediate G-protein signalling. Five different β subunit of 35-36 KDa and seven γ subunit of 8-10 KDa have been identified by molecular cloning (Strathmann *et al.* 1989; Cali *et al.* 1992).The effectors regulated by Gβγ include K^+ channels, phospholipase C-β and adenylyl cyclase (Strathmann *et al.* 1989; Tang & Gilman, 1991; Wickman *et al.* 1994). Of the nine types of adenylyl cyclase that have been cloned and expressed (Ishikawa & Homcy, 1997), only two types, types V and VI have been identified in heart, aorta and brain (Katsushika *et al.* 1992; Premont *et al.* 1992). Adenylyl cyclase type II and IV are activated by Gβγ in the presence of Gsα, type I is inhibited by Gβγ and type III, V and VI do not appear to be directly regulated by Gβγ (Toro *et al.* 1987; Taussig *et al.* 1993).

G-proteins and Adenylyl Cyclase Signaling in Diabetes

Several abnormalities in G-proteins expression and adenylyl cyclase activity have been shown in various pathophysiological conditions including diabetes (Lynch *et al.* 1989). The decreased expression of Giα proteins has been reported in hepatocytes from human diabetics and STZ-diabetic rats (Bushfield *et al.* 1991; Caro *et al.* 1994) whereas an increase in the levels and functions of Giα was shown in diabetic adipocytes from a genetic model of diabetes (Strassheim *et al.* 1991; Livingstone *et al.* 1991). These have also shown a decreased expression of Giα proteins in platelets from diabetic subjects as compared to non diabetic subjects. In addition, diabetic retina has been shown to exhibit decreased levels and functions of Giα (Hadjiconstantinou *et al.* 1988). Hattori *et al.*. (2000) have reported a similar decrease in Gi protein in aorta from long-term diabetic rats, however, these investigators did not examine adenylyl cyclase-Gi-protein signaling in their studies. On the other hand, Weber and McLeod (1997) were unable to observe any changes in the levels of Giα proteins in aorta or caudal artery from 12-14 week-STZ-diabetic rats as compared to control rats. Further support and involvement of Giα-2 protein in pathogenesis of diabetes has been provided by the studies showing that the overexpression of constitutively activated Giα-2 ameliorates STZ-induced diabetes in rats (Zhang *et al.* 2000). In addition, a complete knockout of the Giα-2 gene that has been reported to produce a metabolic state resembling type II diabetes suggests the relationship between the decreased levels of Giα protein and diabetes (Moxham & Malbon, 1996). However, we show the aorta as well as aortic VSMC from STZ-diabetic rats exhibit a decreased expression of Giα-2 and Giα-3 but not of Gsα proteins (Hashim *et al.* 2002) (Fig. 1.) An unaltered expression of Gsα in hearts from STZ-induced diabetic rat has also been reported (Griffiths *et al.* 1990; Hashim *et al.* 2002). A relationship between the development of diabetes and Giα protein expression was also demonstrated (Hashim *et al.* 2002). Rats treated with STZ showed enhanced blood glucose levels within 2 days after injection with a concurrent decrease in the levels of Giα proteins, suggesting that the decrease in the levels of Giα proteins is associated with the development of diabetes. Subsequent increase in the levels of blood glucose through day 5 resulted in further decrease in the levels of Giα proteins suggesting a close

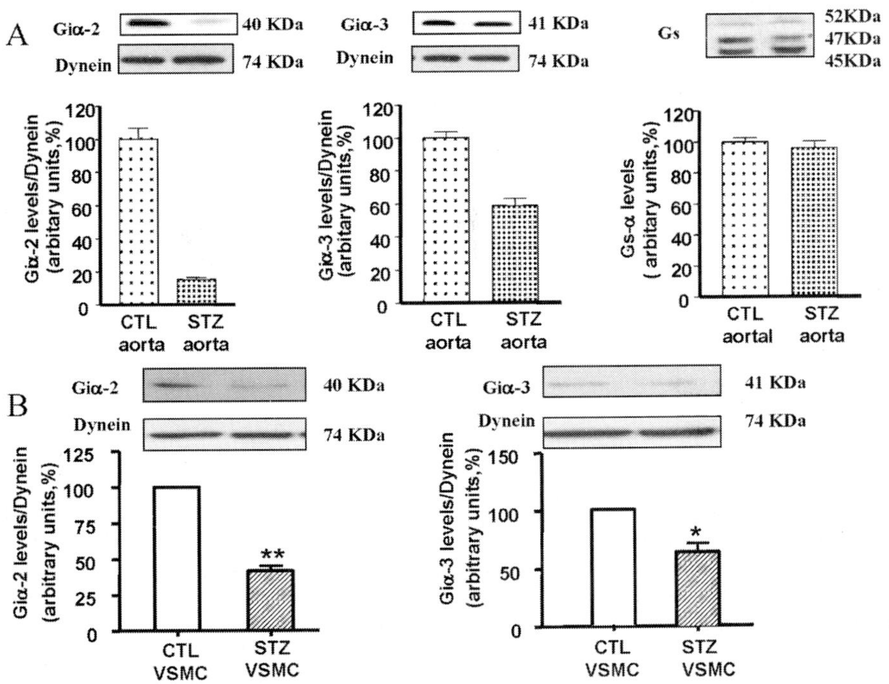

Fig. 1. Quantification of Giα-2, Giα-3 and Gsa proteins by immunoblotting in aorta (A) and aortic vascular smooth muscle cells (VSMC, B) from control (CTL) and STZ-induced diabetic rats (STZ). The membrane proteins (30ug) from control and STZ were separated on SDS-PAGE and transferred to nitrocellulose membrane, which was then immunoblotted using AS/7, EC/2 and RM/7 antibodies for Giα-2, Giα-3 and Gsα proteins respectively as described previously (Hashim et al. 2002). The results are expressed as percentage of controls taken as 100%. Values are means ±S.E.M. of 4 separate experiments utilizing 4 separate aorta preparations from control and STZ-rats. Adapted from (Hashim *et al.* 2002).

relationship between decreased levels of Giα proteins and severity of diabetes (Hashim *et al.* 2002). The decreased levels of Giα proteins were reflected in decreased Gi functions, i.e. the inhibitory effect of GTPγS on FSK-stimulated adenylyl cyclase activity was attenuated (Hashim *et al.* 2002) (Fig. 2). In addition, the inhibitory effect of Ang II, oxotremorine and C-ANP$_{4\text{-}23}$ on adenylyl cyclase activity was completely abolished (Hashim *et al.* 2002). On the other hand, the Gsα–mediated stimulatory effects of hormones as well as of FSK and NaF on adenylyl cyclase were augmented in STZ-aorta as compared to control aorta resulting in an enhanced levels of cAMP, whereas the basal cAMP levels were reduced in diabetic aorta (Hashim *et al.* 2002).

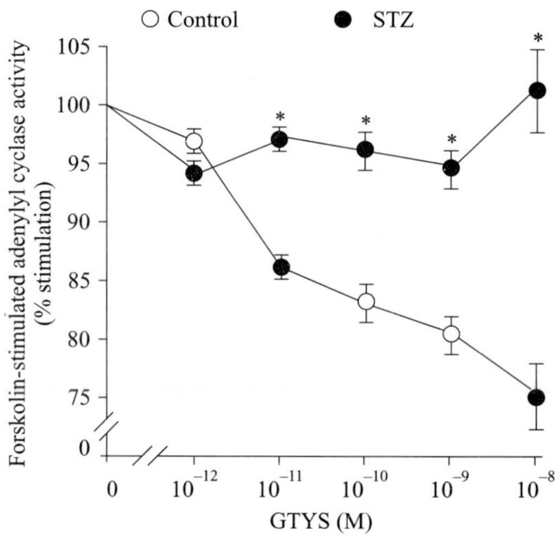

Fig. 2. **Effect of GTPγS on Forskolin-stimulated adenylyl cyclase activity in aorta from control and STZ-induced diabetic rats. Adenylyl cyclase activity was determined in the presence or absence of 100 μM FSK alone or in combination with various concentrations of GTPγS in aorta from control and STZ-induced diabetic rats as described previously (Hashim *et al.* 2002). Values are mean ± SEM of 3 separate experiments. Six animals were utilised for each experiment. Adenylyl cyclase activity in the presence of 100 μM FSK in aorta from control (CTRL) and STZ rats were 603 ± 25 and 381 ± 27 pmol of cAMP/10 min per mg of protein respectively. * P<0.05. Adapted from (Hashim *et al.* 2002)**

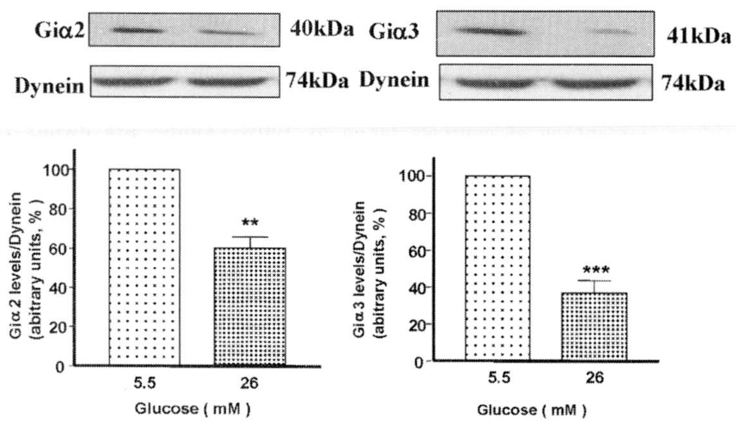

Fig. 3. **Effect of high glucose on the expression of Giα proteins in A10 vascular smooth muscle cells (A10 VSMC). A10 VSMC were exposed to different concentrations of glucose (5.5 mM, control) (CTL) or 26 mM glucose (hyperglycemia) for 72 hrs. The membrane proteins (30 μg) were resolved by SDS/PAGE and transferred to nictrocellulose, which was then immunoblotted with antibody AS/7 for Giα-2 (A) and EC/2 for Giα-3 (B) and detected by using chemiluminescence (ECL) Western blotting technique as described previously (Hashim *et al.* 2002). The immunoblots are representative of three separate experiments (upper panel). Quantification of G proteins was performed by densitometric scanning using an enhanced laser densitometer (LKB) (lower panel). The results are expressed as a percentage of control taken as 100%. Values are means ± SEM of three separate experiments. **P < 0.01, ***P < 0.001. Adapted from (Hashim *et al.* 2002)**

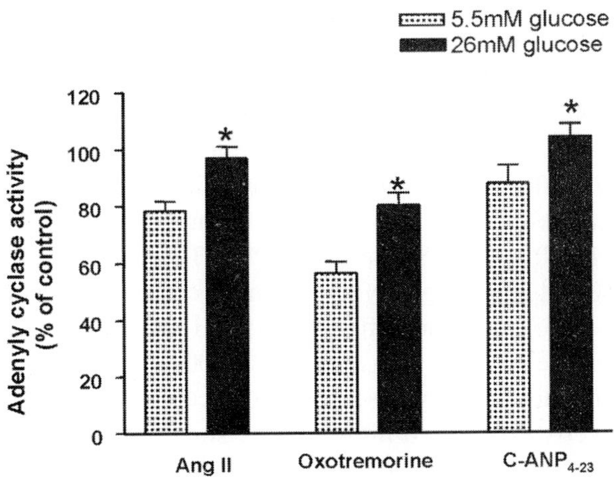

Fig. 4. Effect of high glucose on hormonal inhibition of adenylyl cyclase activity, in A10 vascular smooth muscle cells (VSMC). Adenylyl cyclase activity was determined in the presence of 10 μM GTPγS alone or in combination with 10 μM Angiotensin II (Ang II), 50 μM oxotremorine (OXO) or 0.1 μM C-ANP$_{4-23}$ in VSMC exposed to 5.5 or 26 mM glucose for 72 h as described previously (Hashim *et al.* 2004). Adenylyl cyclase activities in the presence of 10 μM GTPγS in VSMC exposed to 5.5 or 26 mM glucose were 250.0 ± 12.3 and 204.5 ± 5.3 pmol cAMP (mg protein 10 min)^1 respectively. Values are means ± SEM of three separate experiments. *P <0.05.

G-proteins and Adenylyl Cyclase Signaling in Hyperglycemia

VSMC exposed to high glucose (26 mM) that simulate diabetic state also exhibited decreased levels of Giα-2 and Giα-3 proteins, whereas the levels of Gsα were not altered (Hashim *et al.* 2004)(Fig. 3). The decreased expression of Giα proteins was shown to be concentration and time-dependent (Hashim *et al.* 2004). A significant decrease was observed at 20 mM glucose and below that concentration, the levels of Giα proteins were not altered. However, VSMC exposed to higher concentration of glucose (52 mM) or treated for longer period of time (96 hrs) resulted in further decrease in the levels of Giα proteins, whereas the levels of Gsα were not altered under these conditions (Hashim *et al.* 2004). These data indicate a correlation between the levels of glucose (in vivo and in vitro) and decreased expression of Giα proteins and suggest that hyperglycemia may be a contributing factor in diabetes-induced decreased expression of Giα proteins. Hyperglycemia was also shown to stimulate adenylyl cyclase activity in bovine aortic endothelial cells which causes an inhibition of glucose-6-phosphate dehydrogenase and thereby results in decreased levels of NADPH that may be responsible for hyperglycemia-induced apoptosis (Zhang *et al.* 2000). In addition, an increased stimulation of cAMP levels by OP-1206, alpha CD, an analog of prostaglandin E1 (PGE1), was reported in the sciatic nerve from STZ-diabetic rats that were shown to increase Na^+/K^+ ATPase activity (Yasuda *et al.* 1994). In support of these studies, Hashim *et al.* (2004) have recently reported an increased stimulation of adenylyl cyclase activity by isoproterenol and glucagon in A10 vascular smooth muscle cells exposed to high glucose (26 mM) as compared to the cells exposed to normal glucose (5.5 mM). These enhanced stimulations were shown to be attributed to the decreased expression of Giα-2 and Giα-3 levels and not to increased levels of Gsα proteins, because the levels of

Gsα proteins were not altered in hyperglycemia cells. Taken together, it may be suggested that hyperglycemia may be a contributing factor in diabetes-induced decreased expression of Giα proteins. However, (Mancusi *et al.* 1996) were unable to show any changes in Gi protein expression in human umbilical vein endothelial cells (HUVEC) exposed to high glucose for 15 days.

High glucose-induced decreased expression of Giα proteins was shown to be reflected in decreased Gi functions (Hashim *et al.* 2006). For example, Angiotensin II (Ang II), oxotremorine (Oxo) and C-$ANP_{4\text{-}23}$ (a ring deleted peptide of ANP) which inhibit adenylyl cyclase activity through Gi proteins (Anand-Srivastava *et al.* 1987; Anand-Srivastava, 1989; Tucek *et al.* 2001) inhibited the enzyme activity by about 20%, 40% and 25% respectively in control cells. However, the exposure of these cells to high glucose eliminated the inhibitory effect of Ang II and C-$ANP_{4\text{-}23}$, whereas oxo-mediated inhibition was only diminished by 50% (Hashim *et al.* 2004)(Fig.4). In addition, GTPγS-mediated inhibition of FSK-stimulated adenylyl cyclase activity (receptor-independent functions of Giα) was shown to be abolished in cells exposed to high glucose suggesting a correlation between decreased levels of Giα proteins and functions. Thus, it appears that about 50-60% decrease in Giα-2 and Giα-3 proteins by high glucose may be sufficient to inhibit Gi functions and to uncouple the hormone receptors from adenylyl cyclase system, or alternatively, some other mechanisms at the receptor level, such as receptor down-regulation, may also be responsible for a complete attenuation of inhibitory responses on adenylyl cyclase. In this context, acute hyperglycemia induced by STZ or alloxan has been shown to decrease the levels of vascular ANP-C, AT1 and arginine-vasopressin (AVP) receptors (Williams *et al.* 1992; Williams *et al.* 1997; Kook *et al.* 2002). Hyperglycemia has also been shown to impair voltage gated K^+ channel current in rat small coronary VSMC (Liu *et al.* 2001). Since Giα proteins are implicated in the activation of K^+ channels, it may be possible that the impairment of K^+ channel activity may be attributed to the decreased levels of Giα protein induced by high glucose. On the other hand, basal adenylyl cyclase activity was significantly decreased in A10 VSMC or aorta exposed to high glucose (Hashim *et al.* 2004). Since decreased cAMP levels have been shown to augment cell proliferation (Hayashi *et al.* 2000), it may be possible that the decreased basal adenylyl cyclase activity and thereby decreased cAMP levels induced by high glucose may be a contributing factor in increased cell proliferation observed under hyperglycaemic conditions and diabetes (Fujita *et al.* 2002).

Oxidative Stress and G-protein/Adenylyl Cyclase Signaling in Hyperglycemia

Hyperglycemia-induced enhanced oxidative stress has also been reported in cultured VSMC and different tissues from STZ-diabetic rats (Baynes, 1991; Baynes & Thorpe, 1999; Cai & Harrison, 2000). In addition, the contribution of enhanced production of superoxide anion (O_2^-) in the decreased expression of Giα proteins has recently been reported in aortic VSMC from STZ-diabetic rats and A10 cells exposed to high glucose (Li *et al.* 2008). Antioxidants such as α-tocopherol, NAC; scavengers of O_2^- and DPI, an inhibitor of NADPH oxidase that were shown to restore the enhanced levels of O_2^- to control levels (Li *et al.* 2008) were also able to restore the hyperglycemia-induced decreased expression of Giα-2 and Giα-3 to control levels (Fig.5). These studies suggest the implication of NADPH oxidase/ O_2^- in hyperglycemia-evoked decreased expression of Giα proteins. The restoration of decreased expression of Giα proteins to control levels by antioxidants also resulted in the restoration of the decreased Gi-mediated functions (receptor-dependent and–independent), as demonstarated by the restoration of decreased inhibition of adenylyl cyclase by Ang II, C-ANP$_{4-23}$ and oxotremorine to control levels (Li *et al.* 2008). In addition, GTPγS-mediated decreased inhibition of forskolin-stimulated adenylyl cyclase activity, (receptor-independent functions of Giα proteins) as well as enhanced stimulation of adenylyl cyclase by GTPγS and stimulatory hormones such as isoproterenol and glucagon in hyperglycemic cells were also restored to control levels by DPI (Li *et al.* 2008).

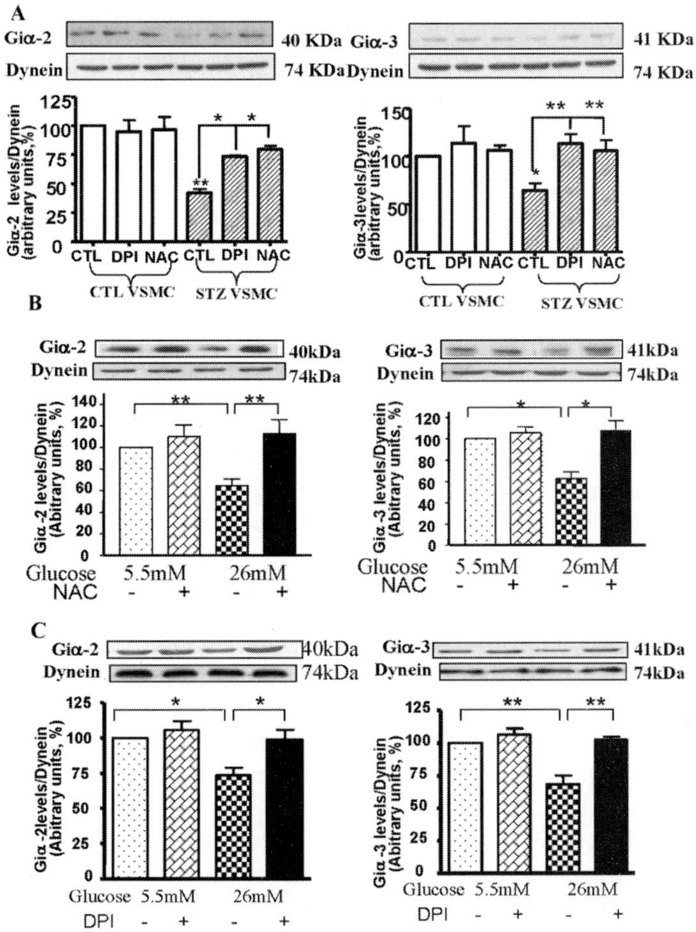

Fig. 5. Effect of antioxidants on the expression of Gi proteins in aortic vascular smooth muscle cells (VSMC) from control (CTL) and STZ-induced diabetic rats (STZ) (A) and A10 vascular smooth muscle cells (A10 VSMC)(B,C). VSMC and A10 cells were incubated in the presence of 5.5 mM (control) or 26 mM glucose (hyperglycemia) alone or in combination with 10 mM N-Acetyl-L-cysteine (NAC) or 10 μM diphenyleneiodonium (DPI) for 72 hours. The cell lysates were prepared and subjected to Western blotting using specific antibodies for Giα-2 (left, upper panel) and Giα-3 (right, upper panel) as described previously (Li *et al.* 2008). The proteins were quantified by densitometric scanning (lower panel). The results are expressed as percentage of control, taken as 100%. Values are means ± SEM of three separate experiments. *P<0.05, ** P<0.01. Adapted from (Descorbeth & Anand-Srivastava, 2008; Li *et al.* 2008).

Hyperglycemia-induced decreased expression of Giα proteins was shown to be attributed to the increased levels of peroxynitrite because scavengers of peroxynitrite, uric acid and MnTBAP were able to restore the hyperglycemia-induced decreased expression of Giα proteins to control levels (Li *et al.* 2008).The implication of peroxynitrite in NO-induced decreased expression of Giα proteins in aortic and A10 VSMC has recently been shown (Bassil & Anand-Srivastava, 2006). In addition, the treatment of VSMC with peroxynitrite has also been reported to result in decreased expression of Giα proteins (Bassil & Anand-Srivastava, 2006). There is accumulating evidence that supports the hypothesis that diabetes is associated with increased nitrosative stress and peroxynitrite formation in several tissues both in experimental animals and humans (Pacher *et al.* 2005). The increased levels of nitrotyrosine, a relatively specific marker of peroxynitrite formation has been shown in different tissues from STZ-diabetic rats and diabetic subjects (Pacher & Szabo, 2006). For example, an increased nitrotyrosine plasma levels were shown in type 2 diabetic patients (Ceriello *et al.* 2001) and iNOs-dependent peroxynitrite production was shown to be increased in platelets from diabetic individuals (Tannous *et al.* 1999). In addition, hyperglycemia has also been reported to induce increased nitrotyrosine formation in the artery wall of monkeys (Pennathur *et al.* 2001). Thus, it may be possible that hyperglycemia-induced enhanced levels of peroxynitrite, formed by the interaction of NO and O_2^-, may contribute to hyperglycemia-induced decreased expression of Giα proteins in VSMC. Taken together, diabetes/hyperglycemia was associated with a decreased expression of Giα proteins and associated adenylyl cyclase signaling which may be attributed to the augmented levels of vasoactive peptides such as Ang II and ET-1 that enhance the oxidative stress by increasing the levels of O_2^- and $ONOO^-$ The ability of antioxidants to reverse the hyperglycemia-induced decreased expression of Giα proteins and adenylyl cyclase signaling to control levels, suggests a key role of ROS generation in this process.

G-protein/Phosphatidyl Inositol Turnover System in Diabetes/Hyperglycemia

Gq family of G-proteins regulate PI turnover signaling. The activation of Gqα by a G-protein coupled receptor (GPCR) stimulates PLCβ which hydrolyzes inositol biphosphates (PIP_2) to produce the intracellular second messengers, inositol triphosphates (IP_3) and diacylglycerol (DAG) (Berridge, 1987; Smrcka *et al.* 1991). The IP_3 stimulates the release of calcium (Ca^{2+}) from the intracellular stores and DAG activates the protein kinase C (PKC). The release of intracellular calcium activates the calcium channels localized in the cell membrane and allows the uptake of extracellular calcium inside the cell (Neves *et al.* 2002). The Gqα subgroup consists of G11α, G14α, and G16α which are encoded by four distinct genes. Gqα and G11α are 88 % identical and are expressed ubiquitously (Strathmann & Simon, 1990). Alterations in the levels of Gqα and associated signaling components result in the impaired cellular functions resulting in various pathological states including diabetes, hypertrophy etc (D'Angelo *et al.* 1997). A genetic ablation of Gqα in mice has been shown to result in a cardiac malformation and craniofacial defects (Offermanns *et al.* 1998), whereas, an overstimulation of Gq pathway in mice was shown in the development of hypertrophic cardiomyopathy (D'Angelo *et al.* 1997). An enhanced expression of Gqα, PKC and DAG was reported in different tissus from STZ-induced diabetic rats as well as in Bio-Breeding (BB) rats (Inoguchi *et al.* 1992; Shiba *et al.* 1993; Goraya *et al.* 1995; Yang *et al.* 1999; Tappia *et al.* 2004). In addition, high glucose treatment of cultured aortic endothelial and vascular smooth muscle cells (VSMC) has also been reported to enhance the activity of PKC and DAG (Inoguchi *et al.* 1992; Williams *et al.* 1997; Ramana *et al.* 2005). However, the studies on the levels of Gqα, an important upstream signaling molecule of PI turnover in diabetic aorta or in VSMC under hyperglycemic conditions are lacking. We demonstrate for the first time that treatment of A10 VSMC with high glucose significantly increased the expression of Gq/11α (Descorbeth & Anand-Srivastava, 2008) (Fig.6), which was not due to increased osmolarity because mannitol (26 mM) did not have any effect on the expression of Gq/11α proteins(Descorbeth & Anand-Srivastava, 2008). In addition, we also report that aortic VSMC from STZ-diabetic rats also exhibit increased expression of Gqα proteins, suggesting that hyperglycemia could mimic the diabetic state. An increased expression of Gqα in sciatic nerves (Goraya *et al.* 1995) and hearts (Yang *et al.* 1999, Tappia *et al.* 2004) from STZ-diabetic rats has also been reported. However, Ceccarelli *et al.* (2001) showed a decreased expression of Gqα in bovine retinal pericites exposed to 25 mmol/l of glucose. In addition, an unaltered or a decreased expression of Gq/11α protein was also shown in gastric VSMC from 10 weeks STZ-diabetic and a genetic model of non-insulin-dependent diabetes (11-12 months) WBN/Kob diabetic rats (Lin *et al.* 2000) respectively. Furthermore, the expression of PLCβ-1 and PLCβ-2, but not PLCβ-3 proteins was also shown to be augmented in VSMC exposed to high glucose (Descorbeth & Anand-Srivastava, 2008) (Fig.7).

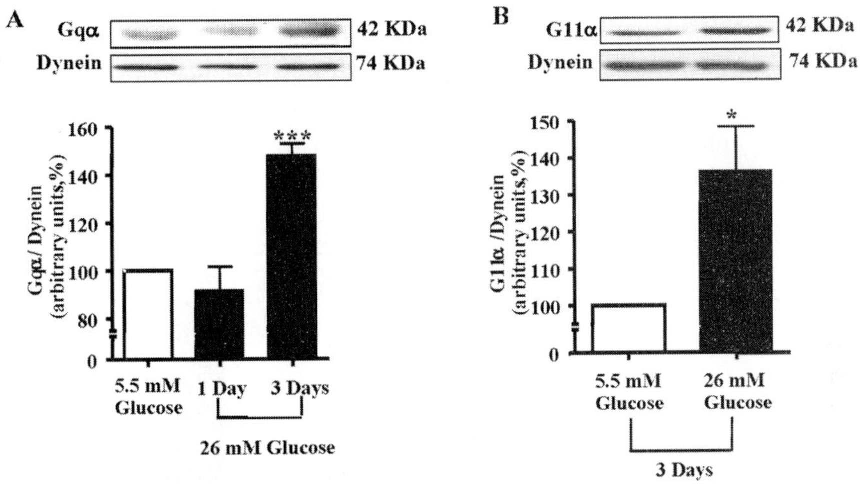

Fig. 6. Effect of high glucose on the expression of Gq/11α proteins in A10 vascular smooth muscle cells (VSMC). A10 VSMC were incubated in the presence of 5.5 mM (control) or 26 mM glucose for 1 and 3 days at 37°C. Membranes were prepared and subjected to Western blotting using specific antibodies against Gqα (A), G11α (B) (upper panel) as described previously (Descorbeth & Anand-Srivastava, 2008). The proteins were quantified by densitometric scanning (lower panel). The results are expressed as a percentage of control taken as 100%. Values are means ± SEM of three separate experiments.*P<0.05,* P<0.001 vs control. Adapted from (Descorbeth & Anand-Srivastava, 2008)**

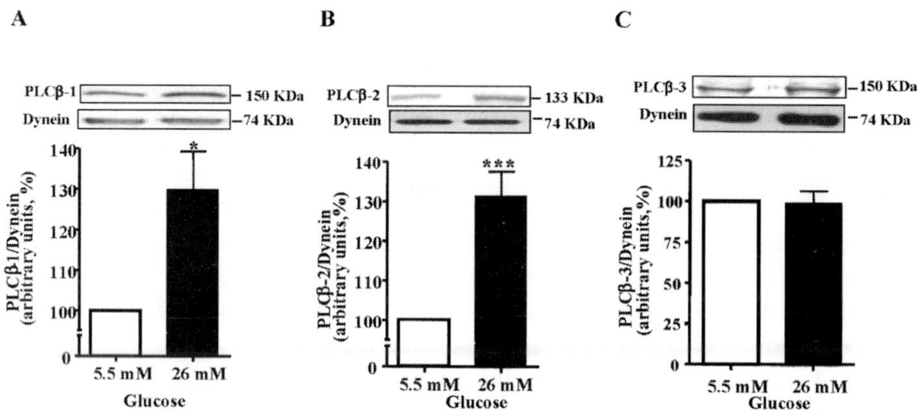

Fig. 7. Effect of high glucose on the expression of PLCβ isoforms in A10 vascular smooth muscle cells (VSMC). A10 VSMC were incubated in the presence of 5.5 mM (control) or 26 mM glucose for 3 days at 37°C. Membranes were prepared and subjected to Western blotting using specific antibodies against PLCβ-1 (A), PLCβ-2 (B) and PLCβ-3 (C) (upper panel) as described previously (Descorbeth & Anand-Srivastava, 2008). The proteins were quantified by densitometric scanning (lower panel). The results are expressed as a percentage of control taken as 100%. Values are means ± SEM of three separate experiments. *P<0.05' *** P<0.001 vs control. Adapted from (Descorbeth & Anand-Srivastava, 2008)

In addition, aortic VSMC from STZ-diabetic rats were also shown to increased expression of PLCβ-1 proteins (Descorbeth & Anand-Srivastava, 2008). The increased expression of Gq and PLC proteins resulted in enhanced formation of IP_3 by ET-1 (Descorbeth & Anand-Srivastava, 2008)(Fig.8). Abebe and MacLeod (1991) using aorta

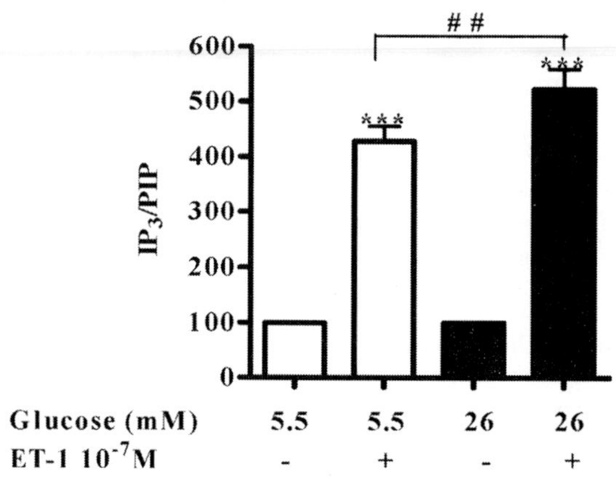

Fig. 8. Effect of high glucose on ET-1-induced IP_3 formation in A10 vascular smooth muscle cells (VSMC). A10 VSMC were incubated in the presence of 5.5 mM (control) or 26 mM glucose for 3 days at 37°C. [myo-2-^{3}H]inositol (5μCi/ml) was added on the 3rd day for the last 24 h of the treatment as described previously (Descorbeth & Anand-Srivastava, 2008). The cells were further incubated in the absence or presence of ET-1 (10^{-7} M) for 2 h, and IP_3 formation was determined as described previously (Descorbeth & Anand-Srivastava, 2008). The results are expressed as a percentage of control taken as 100%.Values are the means ± S.E.M. of 3-4 separate experiments performed in triplicate. *** P< 0.001 vs control, $^{\#\#}$ P < 0.01. Adapted from (Descorbeth & Anand-Srivastava, 2008)

from STZ-diabetic rats have also reported that norepinephrine-induced, but not basal formation of IP_3 was greater as compared to control rats during contraction. In addition, the arteries from STZ-induced diabetic rats were shown to be more responsive to the contractile effects of norepineprine than the arteries from control animals (Abebe & MacLeod, 1990) which may also be due to the increased formation of IP_3 induced by norepinephrine. The increased expression of Gq/11α and PLCβ results in enhanced production in IP_3 which by increasing the intracellular levels of Ca^{2+} may contribute to the vascular complications observed in diabetes (Fig. 9).

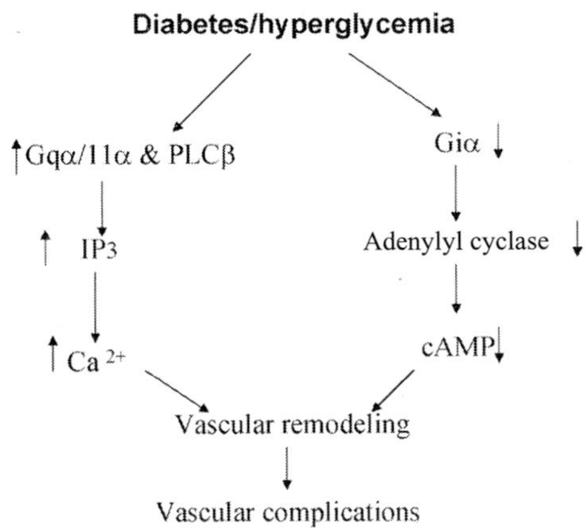

Fig. 9. Schematic diagram depicting the differential regulation of Giα and Gqα expression in Diabetes / hyperglycemia. Diabetes/Hyerglycemia decreases the levels of Giα proteins and augments the levels of Gqα/11α and PLCβ in VSMC. The alterations in these levels result in impaired formation of second messengers such as cAMP and Ca^{2+} respectively and thus may contribute to the vascular complications associated with diabetes.

Conclusion

Diabetes is associated with alterations in the activation of several signalling events which play key role in mediating cell growth, hypertrophy, migration, proliferation and contractility. Many of these signalling components that include G proteins/PLC/PKC and adenylyl cyclases are activated in a differential manner in response to hyperglycemia. The levels of Gi proteins, important component of adenylyl cyclase system are decreased, whereas the levels of Gq/PLC, the signaling molecules of PI turnover are enhanced. The dysregulation of these pathways may contribute to the cardiovascular complications associated with diabetes and could serve as potential targets in pharmacotherapy of the disease.

References

1. Abebe, W., & MacLeod, K. M. (1990). Protein kinase C-mediated contractile responses of arteries from diabetic rats. *Br J Pharmacol, 101*(2), 465-471.
2. Abebe, W., & MacLeod, K. M. (1991). Influence of diabetes on norepinephrine-induced inositol 1,4,5-trisphosphate levels in rat aorta. *Life Sci, 49*(13), PL85-90.

3. Anand-Srivastava, M. B. (1989). Angiotensin II receptors negatively coupled to adenylate cyclase in rat myocardial sarcolemma. Involvement of inhibitory guanine nucleotide regulatory protein. *Biochem Pharmacol, 38*(3), 489-496.
4. Anand-Srivastava, M. B., Srivastava, A. K., & Cantin, M. (1987). Pertussis toxin attenuates atrial natriuretic factor-mediated inhibition of adenylate cyclase. Involvement of inhibitory guanine nucleotide regulatory protein. *J Biol Chem, 262*(11), 4931-4934.
5. Bassil, M., & Anand-Srivastava, M. B. (2006). Nitric oxide modulates Gi-protein expression and adenylyl cyclase signaling in vascular smooth muscle cells. *Free Radic Biol Med, 41*(7), 1162-1173.
6. Baynes, J. W. (1991). Role of oxidative stress in development of complications in diabetes. *Diabetes, 40*(4), 405-412.
7. Baynes, J. W., & Thorpe, S. R. (1999). Role of oxidative stress in diabetic complications: a new perspective on an old paradigm. *Diabetes, 48*(1), 1-9.
8. Berridge, M. J. (1987). Inositol trisphosphate as a second messenger in signal transduction. *Ann N Y Acad Sci, 494*, 39-51.
9. Bushfield, M., Lavan, B. E., & Houslay, M. D. (1991). Okadaic acid identifies a phosphorylation/dephosphorylation cycle controlling the inhibitory guanine-nucleotide-binding regulatory protein Gi2. *Biochem J, 274 (Pt 2)*, 317-321.
10. Cai, H., & Harrison, D. G. (2000). Endothelial dysfunction in cardiovascular diseases: the role of oxidant stress. *Circ Res, 87*(10), 840-844.
11. Cali, J. J., Balcueva, E. A., Rybalkin, I., & Robishaw, J. D. (1992). Selective tissue distribution of G protein gamma subunits, including a new form of the gamma subunits identified by cDNA cloning. *J Biol Chem, 267*(33), 24023-24027.
12. Caro, J. F., Raju, M. S., Caro, M., Lynch, C. J., Poulos, J., Exton, J. H., *et al.*. (1994). Guanine nucleotide binding regulatory proteins in liver from obese humans with and without type II diabetes: evidence for altered «cross-talk» between the insulin receptor and Gi-proteins. *J Cell Biochem, 54*(3), 309-319.
13. Ceccarelli, F., Rosa Mazzoni, M., & Chakravarthy, U. (2001). Binding characteristics of ET receptors in retinal pericytes and effects of high glucose incubation. *Curr Eye Res, 23*(4), 263-270.
14. Ceriello, A., Mercuri, F., Quagliaro, L., Assaloni, R., Motz, E., Tonutti, L., *et al.*. (2001). Detection of nitrotyrosine in the diabetic plasma: evidence of oxidative stress. *Diabetologia, 44*(7), 834-838.
15. D'Angelo, D. D., Sakata, Y., Lorenz, J. N., Boivin, G. P., Walsh, R. A., Liggett, S. B., *et al.* (1997). Transgenic Galphaq overexpression induces cardiac contractile failure in mice. *Proc Natl Acad Sci U S A, 94*(15), 8121-8126.
16. Descorbeth, M., & Anand-Srivastava, M. B. (2008). High glucose increases the expression of Gq/11alpha and PLC-beta proteins and associated signaling in vascular smooth muscle cells. *Am J Physiol Heart Circ Physiol, 295*(5), H2135-2142.

17. Fujita, N., Furukawa, Y., Du, J., Itabashi, N., Fujisawa, G., Okada, K., *et al.*. (2002). Hyperglycemia enhances VSMC proliferation with NF-kappaB activation by angiotensin II and E2F-1 augmentation by growth factors. *Mol Cell Endocrinol, 192*(1-2), 75-84.
18. Gilman, A. G. (1984). G proteins and dual control of adenylate cyclase. *Cell, 36*(3), 577-579.
19. Goraya, T. Y., Wilkins, P., Douglas, J. G., Zhou, J., & Berti-Mattera, L. N. (1995). Signal transduction alterations in peripheral nerves from streptozotocin-induced diabetic rats. *J Neurosci Res, 41*(4), 518-525.
20. Griffiths, S. L., Knowler, J. T., & Houslay, M. D. (1990). Diabetes-induced changes in guanine-nucleotide-regulatory-protein mRNA detected using synthetic oligonucleotide probes. *Eur J Biochem, 193*(2), 367-374.
21. Hashim, S., Li, Y., & Anand-Srivastava, M. B. (2006). G protein-linked cell signaling and cardiovascular functions in diabetes/hyperglycemia. *Cell Biochem Biophys, 44*(1), 51-64.
22. Hashim, S., Li, Y., Nagakura, A., Takeo, S., & Anand-Srivastava, M. B. (2004). Modulation of G-protein expression and adenylyl cyclase signaling by high glucose in vascular smooth muscle. *Cardiovasc Res, 63*(4), 709-718.
23. Hashim, S., Liu, Y. Y., Wang, R., & Anand-Srivastava, M. B. (2002). Streptozotocin-induced diabetes impairs G-protein linked signal transduction in vascular smooth muscle. *Mol Cell Biochem, 240*(1-2), 57-65.
24. Hattori, Y., Matsuda, N., Sato, A., Watanuki, S., Tomioka, H., Kawasaki, H., *et al.*. (2000). Predominant contribution of the G protein-mediated mechanism to NaF-induced vascular contractions in diabetic rats: association with an increased level of G(qalpha) expression. *J Pharmacol Exp Ther, 292*(2), 761-768.
25. Hayashi, S., Morishita, R., Matsushita, H., Nakagami, H., Taniyama, Y., Nakamura, T., *et al.*. (2000). Cyclic AMP inhibited proliferation of human aortic vascular smooth muscle cells, accompanied by induction of p53 and p21. *Hypertension, 35*(1 Pt 2), 237-243.
26. Inoguchi, T., Battan, R., Handler, E., Sportsman, J. R., Heath, W., & King, G. L. (1992). Preferential elevation of protein kinase C isoform beta II and diacylglycerol levels in the aorta and heart of diabetic rats: differential reversibility to glycemic control by islet cell transplantation. *Proc Natl Acad Sci U S A, 89*(22), 11059-11063.
27. Ishikawa, Y., & Homcy, C. J. (1997). The adenylyl cyclases as integrators of transmembrane signal transduction. *Circ Res, 80*(3), 297-304.
28. Katsushika, S., Chen, L., Kawabe, J., Nilakantan, R., Halnon, N. J., Homcy, C. J., *et al.*. (1992). Cloning and characterization of a sixth adenylyl cyclase isoform: types V and VI constitute a subgroup within the mammalian adenylyl cyclase family. *Proc Natl Acad Sci U S A, 89*(18), 8774-8778.
29. Kook, H., Lee, J., Kim, S. W., & Baik, Y. H. (2002). Augmented natriuretic peptide-induced guanylyl cyclase activity and vasodilation in experimental hyperglycemic rats. *Jpn J Pharmacol, 88*(2), 167-173.

30. Li, Y., Descorbeth, M., & Anand-Srivastava, M. B. (2008). Role of oxidative stress in high glucose-induced decreased expression of Gialpha proteins and adenylyl cyclase signaling in vascular smooth muscle cells. *Am J Physiol Heart Circ Physiol, 294*(6), H2845-2854.

31. Lin, S., Kajimura, M., Takeuchi, K., Kodaira, M., Hanai, H., Nishimura, M., *et al.*. (2000). Alterations of GTP-binding proteins (Gsalpha and Gq/11alpha) in gastric smooth muscle cells from streptozotocin-induced and WBN/Kob diabetic rats. *Dig Dis Sci, 45*(8), 1517-1524.

32 .Liu, Y., Terata, K., Rusch, N. J., & Gutterman, D. D. (2001). High glucose impairs voltage-gated K(+) channel current in rat small coronary arteries. *Circ Res, 89*(2), 146-152.

33. Livingstone, C., McLellan, A. R., McGregor, M. A., Wilson, A., Connell, J. M., Small, M., *et al.*. (1991). Altered G-protein expression and adenylate cyclase activity in platelets of non-insulin-dependent diabetic (NIDDM) male subjects. *Biochim Biophys Acta, 1096*(2), 127-133.

34. Lynch, C. J., Blackmore, P. F., Johnson, E. H., Wange, R. L., Krone, P. K., & Exton, J. H. (1989). Guanine nucleotide binding regulatory proteins and adenylate cyclase in livers of streptozotocin- and BB/Wor-diabetic rats. Immunodetection of Gs and Gi with antisera prepared against synthetic peptides. *J Clin Invest, 83*(6), 2050-2062.

35. Mancusi, G., Hutter, C., Baumgartner-Parzer, S., Schmidt, K., Schutz, W., & Sexl, V. (1996). High-glucose incubation of human umbilical-vein endothelial cells does not alter expression and function either of G-protein alpha-subunits or of endothelial NO synthase. *Biochem J, 315 (Pt 1)*, 281-287.

36. Moxham, C. M., & Malbon, C. C. (1996). Insulin action impaired by deficiency of the G-protein subunit G ialpha2. *Nature, 379*(6568), 840-844.

37. Neves, S. R., Ram, P. T., & Iyengar, R. (2002). G protein pathways. *Science, 296*(5573), 1636-1639.

38. Offermanns, S., Zhao, L. P., Gohla, A., Sarosi, I., Simon, M. I., & Wilkie, T. M. (1998). Embryonic cardiomyocyte hypoplasia and craniofacial defects in G alpha q/G alpha 11-mutant mice. *EMBO J, 17*(15), 4304-4312.

39. Pacher, P., Obrosova, I. G., Mabley, J. G., & Szabo, C. (2005). Role of nitrosative stress and peroxynitrite in the pathogenesis of diabetic complications. Emerging new therapeutical strategies. *Curr Med Chem, 12*(3), 267-275.

40. Pacher, P., & Szabo, C. (2006). Role of peroxynitrite in the pathogenesis of cardiovascular complications of diabetes. *Curr Opin Pharmacol, 6*(2), 136-141.

41. Pennathur, S., Wagner, J. D., Leeuwenburgh, C., Litwak, K. N., & Heinecke, J. W. (2001). A hydroxyl radical-like species oxidizes cynomolgus monkey artery wall proteins in early diabetic vascular disease. *J Clin Invest, 107*(7), 853-860.

42. Premont, R. T., Chen, J., Ma, H. W., Ponnapalli, M., & Iyengar, R. (1992). Two members of a widely expressed subfamily of hormone-stimulated adenylyl cyclases. *Proc Natl Acad Sci U S A, 89*(20), 9809-9813.

43. Ramana, K. V., Friedrich, B., Tammali, R., West, M. B., Bhatnagar, A., & Srivastava, S. K. (2005). Requirement of aldose reductase for the hyperglycemic activation of protein kinase C and formation of diacylglycerol in vascular smooth muscle cells. *Diabetes, 54*(3), 818-829.

44. Shiba, T., Inoguchi, T., Sportsman, J. R., Heath, W. F., Bursell, S., & King, G. L. (1993). Correlation of diacylglycerol level and protein kinase C activity in rat retina to retinal circulation. *Am J Physiol, 265*(5 Pt 1), E783-793.

45. Smrcka, A. V., Hepler, J. R., Brown, K. O., & Sternweis, P. C. (1991). Regulation of polyphosphoinositide-specific phospholipase C activity by purified Gq. *Science, 251*(4995), 804-807.

46. Spiegel, A. M. (1987). Signal transduction by guanine nucleotide binding proteins. *Mol Cell Endocrinol, 49*(1), 1-16.

47. Strassheim, D., Palmer, T., & Houslay, M. D. (1991). Genetically acquired diabetes: adipocyte guanine nucleotide regulatory protein expression and adenylate cyclase regulation. *Biochim Biophys Acta, 1096*(2), 121-126.

48. Strathmann, M., & Simon, M. I. (1990). G protein diversity: a distinct class of alpha subunits is present in vertebrates and invertebrates. *Proc Natl Acad Sci U S A, 87*(23), 9113-9117.

49. Strathmann, M., Wilkie, T. M., & Simon, M. I. (1989). Diversity of the G-protein family: sequences from five additional alpha subunits in the mouse. *Proc Natl Acad Sci U S A, 86*(19), 7407-7409.

50. Stryer, L., & Bourne, H. R. (1986). G proteins: a family of signal transducers. *Annu Rev Cell Biol, 2*, 391-419.

51. Tang, W. J., & Gilman, A. G. (1991). Type-specific regulation of adenylyl cyclase by G protein beta gamma subunits. *Science, 254*(5037), 1500-1503.

52. Tannous, M., Rabini, R. A., Vignini, A., Moretti, N., Fumelli, P., Zielinski, B., *et al.*. (1999). Evidence for iNOS-dependent peroxynitrite production in diabetic platelets. *Diabetologia, 42*(5), 539-544.

53. Tappia, P. S., Asemu, G., Aroutiounova, N., & Dhalla, N. S. (2004). Defective sarcolemmal phospholipase C signaling in diabetic cardiomyopathy. *Mol Cell Biochem, 261*(1-2), 193-199.

54. Taussig, R., Quarmby, L. M., & Gilman, A. G. (1993). Regulation of purified type I and type II adenylylcyclases by G protein beta gamma subunits. *J Biol Chem, 268*(1), 9-12.

55. Toro, M. J., Montoya, E., & Birnbaumer, L. (1987). Inhibitory regulation of adenylyl cyclases. Evidence inconsistent with beta gamma-complexes of Gi proteins mediating hormonal effects by interfering with activation of Gs. *Mol Endocrinol, 1*(10), 669-676.

56. Tucek, S., Michal, P., & Vlachova, V. (2001). Dual effects of muscarinic M2 receptors on the synthesis of cyclic AMP in CHO cells: background and model. *Life Sci, 68*(22-23), 2501-2510.

57. Weber, L. P., & Macleod, K. M. (1997). Influence of streptozotocin diabetes on the alpha-1 adrenoceptor and associated G proteins in rat arteries. *J Pharmacol Exp Ther, 283*(3), 1469-1478.

58. Wickman, K. D., Iniguez-Lluhl, J. A., Davenport, P. A., Taussig, R., Krapivinsky, G. B., Linder, M. E., *et al.*. (1994). Recombinant G-protein beta gamma-subunits activate the muscarinic-gated atrial potassium channel. *Nature, 368*(6468), 255-257.

59. Williams, B., Gallacher, B., Patel, H., & Orme, C. (1997). Glucose-induced protein kinase C activation regulates vascular permeability factor mRNA expression and peptide production by human vascular smooth muscle cells in vitro. *Diabetes, 46*(9), 1497-1503.

60. Williams, B., Tsai, P., & Schrier, R. W. (1992). Glucose-induced downregulation of angiotensin II and arginine vasopressin receptors in cultured rat aortic vascular smooth muscle cells. Role of protein kinase C. *J Clin Invest, 90*(5), 1992-1999.
61. Wong, Y. H., Conklin, B. R., & Bourne, H. R. (1992). Gz-mediated hormonal inhibition of cyclic AMP accumulation. *Science, 255*(5042), 339-342.
62. Yang, J. M., Cho, C. H., Kong, K. A., Jang, I. S., Kim, H. W., & Juhnn, Y. S. (1999). Increased expression of Galphaq protein in the heart of streptozotocin-induced diabetic rats. *Exp Mol Med, 31*(4), 179-184.
63. Yasuda, H., Maeda, K., Sonobe, M., Kawabata, T., Terada, M., Hisanaga, T., *et al.*. (1994). Metabolic effect of PGE1 analogue 01206.alpha CD on nerve Na(+)-K(+)-ATPase activity of rats with streptozocin-induced diabetes is mediated via cAMP: possible role of cAMP in diabetic neuropathy. *Prostaglandins, 47*(5), 367-378.
64. Zhang, Z., Apse, K., Pang, J., & Stanton, R. C. (2000). High glucose inhibits glucose-6-phosphate dehydrogenase via cAMP in aortic endothelial cells. *J Biol Chem, 275*(51), 40042-40047.

Adaptation Biology and Medicine (Volume 6: Cell Adaptations and Challenges)
Editors: P. Wang, C.-H. Kuo, N. Takeda and P.K. Singal
Copyright © 2011 Narosa Publishing House Pvt. Ltd., New Delhi, India

9. Redox Regulation of Autophagy in the Ischemic Myocardium

Dipak K. Das* **and Narasimman Gurusamy**

Cardiovascular Research Center, University of Connecticut School of Medicine, Farmington, CT 06030-1110, USA

Summary

Autophagy is an intracellular catabolic process through which long lived or damaged proteins, macromolecules or organelles are recycled using lysosomal degradation machineries. Autophagy is known to occur in several cardiac diseases including ischemic or dilated cardiomyopathy, heart failure, hypertrophy, and ischemia reperfusion injury. Oxidative stress in cellular system could induce autophagy, and oxidatively modified macromolecules and organelles can be selectively removed by autophagy. Autophagy induced under mild oxidative stress could provide a first line of protection against major damages like apoptosis and necrosis.

Keywords: Autophagy, Reactive oxygen species, Oxidative stress, Heart, Ischemia/reperfusion, Bag-1

Introduction

Autophagy (in Greek: self-digestion) is a catabolic process in which cell's own components are degraded via lysosomal machinery. In normal cells, autophagy occurs to degrade or recycle long-lived proteins, macromolecules, and organelles like mitochondria, endoplasmic reticulum, peroxisomes and ribosomes (1). Three types of autophagy are known such as macroautophagy, microautophagy, and chaperon-mediated autophagy. Macroautophagy is the most active form of autophagy and plays a major role in intracellular degradation (2). Macroautophagy (hereafter referred as autophagy) is initiated by the formation of an isolation membrane, which further gets elongated to form a double-membrane structure called autophagosome. Autophagosome surrounds portions of the cytoplasm and or organelles such as mitochondria, peroxisomes, and toxic protein aggregates (2). Autophagosome fuses with lysosome forming autophagolysosome, where the autophagic substrates along with the inner membranes of autophagosomes are degraded by the lysosomal hydrolases (2). During microautophagy, cytosolic macromolecules are locally taken up into the lysosomes by the invagination of lysosomal membrane (2). During chaperone mediated autophagy, modified proteins are delivered into lysosomes via heat shock protein 73 and the receptor lysosome-associated membrane protein type 2a (Lamp2a) (3).

*Address for Correspondence: Dipak K. Das, Ph.D., Sc.D., FAHA Professor & Director Cardiovascular Research Center University of Connecticut School of Medicine Farmington, CT 06030-1110 USA Tel: (860) 679-3687 Fax: (860) 679-4606 E-mail:DDAS@NEURON.UCHC.EDU

Reactive oxygen species (ROS) like superoxide, H2O2 are normally produced in cells either via electron leakage during mitochondrial oxidative phosphorylation or through the activation of several cellular enzymes, such as nicotinamide adenine dinucleotide phosphate (NADPH) oxidase, xanthine oxidase, etc (4). Excessive ROS can induce oxidative damage to DNA, lipids, proteins and other macromolecules. Intracellular redox state is demonstrated by the balance of oxidant production and the antioxidant capacity of the cell. At lower concentrations, ROS can serve as a second-messenger to transmit biological information, which is known as redox signaling.

Autophagy in Myocardium

In normal heart, autophagy plays an important role in homeostatsis, and serves as a catabolic energy source during starvation (5). Autophagy triggered by ischemia, could be a homeostatic mechanism preventing cell death by apoptosis and limiting the deleterious effects of chronic ischemia (6). Autophagy is shown to play distinct roles during ischemia and reperfusion, i.e. autophagy elicited during ischemia is protective, however the autophagy induced during reperfusion phase is detrimental (7). Autophagy renders a powerful protective mechanism against ischemia reperfusion injury in HL-1 cardiac cells (8). Bnip3, a mitochondrial proapoptotic Bcl-2 protein contributes to myocardial injury and triggers a protective stress response via upregulation of autophagy, which in turn helps the removal of damaged mitochondria (9). Recently, we have shown that cardiac adaptation augments autophagy in BAG-1 dependent manner, and protect the cells from ischemia reperfusion injury (10, 11). However, cardiac autophagy in various pathophysiological conditions such as in patients with dilated cardiomyopathy, in the pressure-overloaded human heart, in human patients with end stage heart failure, in failing human hearts, and in a model of human dilated cardiomyopathy are shown to induce cell death (12).

Mechanism of Autophagosome Formation

Many of the autophagic genes are conserved from yeast to humans. Autophagic genes identified in Saccharomyces cerevisiae had been denoted as Apg or Aut, but recently, Atg nomenclature has been introduced (13). Atg12 is covalently conjugated to Atg5 independent of autophagy induction (14). Atg12 is activated by Atg7 in an ATP-dependent manner (54). The carboxy-terminal glycine of Atg12 is transferred to Atg10 and Atg7 is released. Atg10 activates the conjugation of Atg12 with lysine 149 in Atg5 (Fig. 1) (15). The conjugation between Atg12 and Atg5 is irreversible, and at the same time essential for the formation of autophagosome (84). A dimer of Atg16 forms a complex with a pair of Atg12-Atg5, where Atg16 is found to bind only with Atg5, but not with Atg12. The new complex Atg12-Atg5-Atg16 is required for the elongation of the isolation membrane (16). However, the complex Atg12-Atg5-Atg16 dissociate when mature autophagosome is formed (Fig 1) (16, 17).

Another conjugation system involves Atg8, which is otherwise known as microtubule-associated protein light chain 3 (LC3) (18). Atg4, a cysteine protease cleaves the C-terminal arginine from precursor LC3 forming LC3-I, where glycine residue in the carboxy-terminal is exposed (Fig. 1) (19). Atg7 activates the glycine residue of LC3-I, and finally conjugated to phosphatidylethanolamine (LC3-II) (Fig 1) (20) . LC3-II is then transferred to the outer autophagosomal membrane via Atg5 (21). LC3-II is the best marker for the component of autophagosomal membrane. Atg4 can also convert back LC3-II into LC3-I through deconjugation of phosphatidylethanolamine from LC3-II (22). In addition to

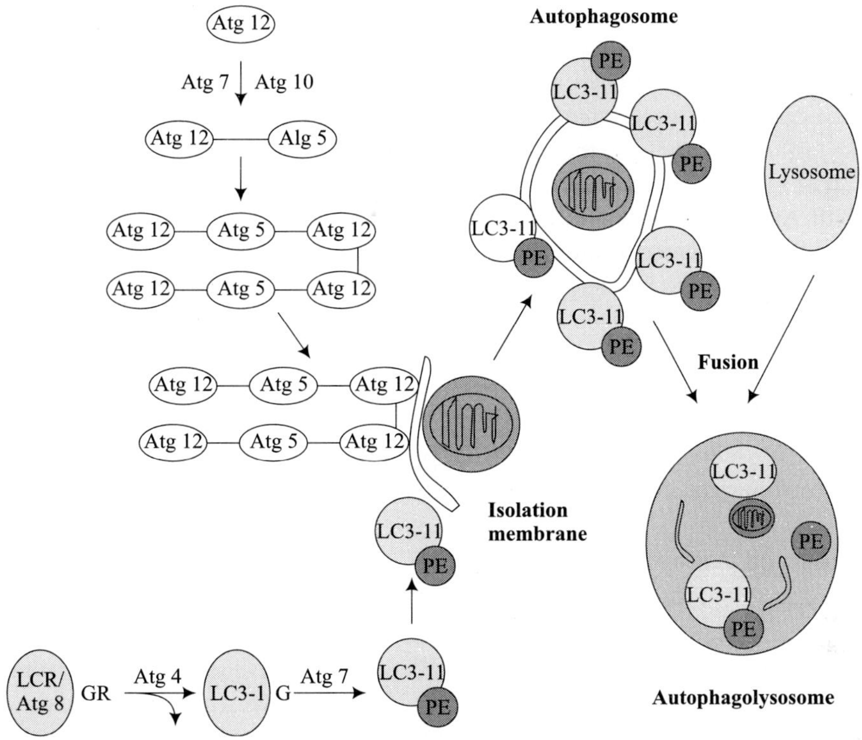

Fig. 1. Regulation of autophagosome formation by different Atgs

LC3, there are two other mammalian homologues of Atg8 such as gamma aminobutyric acid type A receptor associated protein (GABARAP) and Golgi-associated ATPase enhancer of 16 kDa (GATE-16), which have been found to be associated with autophagosomal membrane (23, 24).

When the autophagic sequestration is complete, autophagosome fuses with lysosome forming autophagolysosome; the sequestered contents along with the inner membranes of autophagosomes are degraded within lysosomes by the lysosomal hydrolases (Fig. 1) (17, 25, 26). Lysosomal receptor proteins Lamp1 and Lamp2 along with a family of Rab GTPase, play an important role in the fusion of autophgosomes with lysosomes (27).

The formation of double layered autophagosomes can be monitored by electron microscopy (28). Monodansylcadaverine, a fluorescent dye can be used to stain acidic autophagic vacuoles (29). However, LC3 is the reliable marker to study autophagosome formation. The conversion of LC3 and the turn over of LC3-II can be studied by Western blotting. Transgenic mice expressing green fluorescent protein fused to LC3 is a useful tool to study autophagic activity in vivo (30). Lysosomal receptor molecule Lamp-2a is used as a marker to study chaperone-mediated autophagy (3).

Regulators of Autophagy

Autophagy is known to be altered by several extracellular stimuli such as starvation, hormone or therapeutic treatment and intracellular stimuli such as accumulation of misfolded proteins, and invasion of microorganisms (32). Nutrient deprivation is the most generalized stimuli for macroautophagy. Insulin and amino acids like leucine have been shown to inhibit autophagy synergistically (33). Inhibiting mTOR activity with rapamycin induced the process of autophagy. TOR activation negatively regulates the association between Atg13 and Atg1. Nutrient-rich conditions activate TOR leading to hyperphosphorylation of Atg13, and thus inhibiting its association with Atg1 (34). Inactivation of TOR under starvation or rapamycin treatment causes rapid dephosphorylation of Atg13, and this promotes the binding with Atg1, leading to autophosphorylation and activation of Atg1 and induction of autophagy (35). AMP-activated protein kinase (AMPK), which serves as a general integrator of metabolic responses to changes in energy availability, is activated in response to elevations of the AMP/ATP ratio (36). Recently, it has been shown that glucose deprivation in cardiac cells and cardiac ischemia activate AMPK leading to the activation of autophagy (37). Class I phosphatidylinositol 3kinases (PI3-K) have been shown to suppress autophagy, whereas class III PI3-K activates autophagy. Beclin1/Atg6 forms a complex with Vps34, a class III PI3-K and this complex is involved in the initiation of autophagy and formation of autophagosomes (38). Inhibitors of class III PI3K like 3-methyladenine, Wortmannin, and LY294002 inhibit autophagy by interfering this pathway (39).

Oxidative Stress and Autophagy

Oxidative stress occurs when excess ROS are generated that cannot be adequately counteracted by the intrinsic antioxidant systems. A variety of cellular stresses like endoplasmic reticulum stress or mitochondrial dysfunction can induce autophagy (40, 41). Oxidative stress mediated via ROS plays an important role in the myocardial repair/remodeling (42). Oxidative stress-mediated autophagy occurs under starvation and ischemia/reperfusion conditions (43). Recently, Scherz-Shouval et al have demonstrated that ROS act as signaling molecules during starvation-induced autophagy (44), where, ROS directly oxidize a cysteine residue in the cysteine protease Atg4, which regulate the formation of autophagosome. Redox regulations of Atg4 can cause rapid activation and inactivation of this protease leading to induction of the autophagic process.

Oxidative stress enhances the formation of hydroxyl radicals within lysosomes that risks the membrane stability of iron-rich lysosomes and autophago lysosomes (2, 43). Nutrient deprivation causes an increase in the availability of ferric iron pool resulting in increased sensitivity of cells to oxidative stress leading to autophagy (45). These results indicate that the regulation of the lysosomal content of redox-active iron seems to be essential for the survival of cells both in the short and long term basis. Under mild oxidative stress conditions, chaperone-mediated autophagy selectively removes the soluble oxidized proteins from the cytosol (46). During chaperone-mediated autophagy, oxidized proteins are directly translocated into lysosomal lume.

Though autophagy is shown to be non-selective under oxidative conditions, it is shown as an effective mechanism in removing oxidatively damaged mitochondria, peroxisomes and endoplasmic reticulum (47). Oxidative damage increases the permeability and depolarization of mitochondrial membrane, leading to the initiation of programmed cell death (48). Macroautophagy protects cells from apoptosis/ necrosis by removing the damaged mitochondria at the early stage (49). This protective macroautophagy is considered to be the cross road between cell death and survival (50). Low levels of death associated protein kinase and tumor necrosis factor-related apoptosis inducing ligand may induce macroautophagy (51); however, if the activated autophagy is insufficient to remove the cellular damage, the intracellular level of the two death factors would remain high to induce cell death. It is clear from the above results that when autophagy and apoptosis are governed by the extent of oxidative stress, protective macroautophagy truly draws a borderline between cell survival and cell death.

Autophagy in the myocardium predominantly serves as a pro-survival mechanism especially nutrient deprivation and other forms of cellular stress. However, excess autophagy can destroy essential cellular components or autophagic machineries leading to cell death. Although autophagy is shown to protect the myocardium and cardiac cells during cardiac injury (8, 11, 14), the precise role of autophagy in the conditions of cardiac diseases such as heart failure, ischemic cardiomyopathy, dilated cardiomyopathy, and cardiac hypertrophy is poorly understood. Further studies are required to elucidate the role of autophagy in cardiac diseases.

Oxidative Stress Adaptation and Autophagy

Autophagy is a cellular response to stress and thus may be viewed as an adaptive phenomenon since the cells respond to stress with adaptation leading to repair. Cellular lysosomal degradation pathway of autophagy is cell's response to stress that ultimately decides whether a cell will live or die. Autophagy can be induced by a variety of stresses including metabolic [e.g., starvation), hypoxic and oxidative stresses. Oxidatively modified macromolecules often crosslink with other proteins, lipids, carbohydrates, or metals, forming a brownish pigment known as lipofuscin. Lipofuscin, a biomarker of aging, is an autofluorescent pigment that accumulates during aging in the lysosomal lumen of postmitotic cells (52). The formation of lipofuscin requires the presence of H2O2 and ferrous iron in the lysosomal lumen. H2O2 generated by mitochondria and other organelles permeates into the lumen of secondary lysosomes (52). These lysosomes contain iron derived from cellular structures undergoing autophagic degradation. Diffusion of hydrogen peroxide into the lysosomes results in Fenton-type reactions with the formation of hydroxyl radicals and ensuing peroxidation of lysosomal contents with the formation of lipofuscin (53). Increased levels of oxidative stress also were shown to accumulate another autofluorescent pigment known as carotid (54). The normal accumulation of lipofuscin in lysosomes seems to reduce the autophagic capacity of senescent postmitotic cells, which resulted in an insufficient autophagy, causing accumulation of damaged cellular components (55). Oxidative stress enhanced the formation of hydroxyl radicals within lysosomes that risks the membrane stability of iron-rich lysosomes and autophagolysosomes (55, 56). Cells experimentally loaded with lipofuscin showed an increased susceptibility to oxidative stress, leading to lysosomal membrane breakage (57). Nutrient deprivation in cells induces autophagy, which causes an increase in the availability of ferric iron pool, resulting in the increased sensitivity of cells to oxidative stress (58). The rupture of a limited number of lysosomes has been recognized as an early upstream event in many cases of apoptosis, particularly oxidative stress-induced apoptosis, whereas necrosis results from a major lysosomal break (59). Thus, the regulation of the lysosomal content of redox-active iron seems to be essential for the survival of cells in both the short and long term. Oxidative stress can affect the permeability of the lysosomal membrane by inducing the crosslinking of lysosomal membrane proteins via disulphide bonds; it results in increased lysosomal proton permeability, luminal pH, and membrane potential (60). These results show that the redox status of the lysosomal membrane thiol groups influence its permeability. Although autophagy is nonselective under oxidative conditions, it is an efficient mechanism in removing oxidatively damaged mitochondria, peroxisomes, and endoplasmic reticulum (61). Oxidative damage increases the permeability and depolarization of the mitochondrial membrane and results in leakage of intramitochondrial components such as cytochrome c, which initiates programmed cell death (62). Macroautophagy protects cells from major harm by removing the damaged mitochondria at an early stage (63). This protective macroautophagy is considered to be the crossroad between cell death and survival (64). Death-associated protein kinase and tumor necrosis factor-related apoptosis inducing ligand are the two molecules that have been shown to be involved in the activation of both autophagy and apoptosis in a caspase-dependent or -independent manner (65). At low levels, these factors may induce macroautophagy; however, if the activated autophagy is insufficient to remove the cellular damage, the intracellular level of the two death factors would remain high to induce cell death. It is clear from these results that protective macroautophagy truly draws a borderline between cell survival and cell death, when autophagy and apoptosis are governed by the extent of oxidative stress. Chaperone-mediated autophagy participates under mild oxidative-stress conditions by the removal of soluble and oxidatively modified proteins from the cytosol (66). During chaperone-mediated autophagy, oxidized substrate proteins are directly translocated into the lysosomal lumen. Oxidative modification of proteins facilitates the exposure of the lysosomal targeting motif, KFERQ, which is recognized through Hsc73 (67). The substrate-chaperone complex is translocated into the lysosomal lumen via interaction with LAMP2a, a receptor protein present in the lysosomal membrane (68). Studies have shown that the lysosomal ability to internalize substrates is enhanced during mild oxidative stress in cultured cells and in animals (69). Microautophagy refers to the formation of lysosomal membrane invaginations, which sequester the contents of cytoplasm and degrade them intralysosomally (70). Microautophagy is a constitutive mechanism for the continuous degradation of long-lived proteins inside many types of cells. However, the molecular mechanisms of microautophagy have not been well characterized. Although starvation has been shown to induce both macroautophagy and chaperone mediated autophagy, the molecular events directing the autophagic route toward macroautophagy, microautophagy, and chaperone-mediated autophagy are not well understood. It is possible that the extent of oxidative stress may play a role in this process. A genomic study by Thorpe *et al.* (71) showed that autophagic genes have been upregulated in response to protection against reactive oxygen species in Saccharomyces cerevisiae. Recently, Chen *et al.* (72) demonstrated that oxidative stress caused by treatment with H2O2 induces autophagic cell death in transformed and cancer cell lines. Inhibition of autophagosome accumulation through treatment with 3-methyl adenine and knocking down the genes involved in autophagy, such Beclin-1, Atg5, and Atg7, by treatment with respective siRNAs effectively blocked the oxidative stress-induced autophagic cell death. Inhibiting ROS generation by treatment with the ROS scavenger tiron inhibited autophagic cell death. Further, it was demonstrated that H2O2-induced autophagic cell death is independent of apoptosis (72). Yu *et al.* (73) showed that inhibition of caspase induces the degradation of catalase, leading to the accumulation of ROS and autophagic cell death. Taken together, these studies clearly indicate the role of ROS in the induction of autophagy.

Acknowledgements

This study was supported in part by NIH HL 34360, HL22559 and HL33889.

References

1. Gustafsson AB, and Gottlieb RA. Recycle or die: the role of autophagy in cardioprotection. J Mol Cell Cardiol 44: 654-661, 2008.
2. Cuervo AM. Autophagy: many paths to the same end. Mol Cell Biochem 263: 55-72, 2004.
3. Kaushik S, and Cuervo AM. Autophagy as a cell-repair mechanism: activation of chaperone-mediated autophagy during oxidative stress. Mol Aspects Med 27: 444-454, 2006.
4. Sawyer DB, Siwik DA, Xiao L, Pimentel DR, Singh K, and Colucci WS. Role of oxidative stress in myocardial hypertrophy and failure. J Mol Cell Cardiol 34: 379-388, 2002.
5. Kuma A, Hatano M, Matsui M, Yamamoto A, Nakaya H, Yoshimori T, Ohsumi Y, Tokuhisa T, and Mizushima N. The role of autophagy during the early neonatal starvation period. Nature 432: 1032-1036, 2004.
6. Yan L, Sadoshima J, Vatner DE, and Vatner SF. Autophagy: a novel protective mechanism in chronic ischemia. Cell Cycle 5: 1175-1177, 2006.
7. Matsui Y, Takagi H, Qu X, Abdellatif M, Sakoda H, Asano T, Levine B, and Sadoshima J. Distinct roles of autophagy in the heart during ischemia and reperfusion: roles of AMP-activated protein kinase and Beclin 1 in mediating autophagy. Circ Res 100: 914-922, 2007.
8. Hamacher-Brady A, Brady NR, and Gottlieb RA. Enhancing macroautophagy protects against ischemia/reperfusion injury in cardiac myocytes. J Biol Chem 281: 29776-29787, 2006.
9. Hamacher-Brady A, Brady NR, Logue SE, Sayen MR, Jinno M, Kirshenbaum LA, Gottlieb RA, and Gustafsson AB. Response to myocardial ischemia/reperfusion injury involves Bnip3 and autophagy. Cell Death Differ 14: 146-157, 2007.
10. Gurusamy N, Lekli I, Gherghiceanu M, Popescu LM, and Das DK. BAG-1 induces autophagy for cardiac cell survival. Autophagy 5: 120-121, 2009.
11. Gurusamy N, Lekli I, Gorbunov N, Gherghiceanu M, Popescu LM, and Das DK. Cardioprotection by adaptation to ischemia augments autophagy in association with BAG-1 protein. J Cell Mol Med, 2008.
12. Gurusamy N, and Das DK. Autophagy, Redox Signaling and Ventricular Remodeling. Antioxid Redox Signal, 2009.
13. Klionsky DJ, Cregg JM, Dunn WA, Jr., Emr SD, Sakai Y, Sandoval IV, Sibirny A, Subramani S, Thumm M, Veenhuis M, Ohsumi Y: A unified nomenclature for yeast autophagy-related genes. Dev Cell. 5, 539-45 (2003)
14. Mizushima N, Noda T, Yoshimori T, Tanaka Y, Ishii T, George MD, Klionsky DJ, Ohsumi M, Ohsumi Y: A protein conjugation system essential for autophagy. Nature. 395, 395-8 (1998)
15. Yorimitsu T, Klionsky DJ: Autophagy: molecular machinery for self-eating. Cell Death Differ. 12 Suppl 2, 1542-52 (2005)
16. Mizushima N, Yamamoto A, Hatano M, Kobayashi Y, Kabeya Y, Suzuki K, Tokuhisa T, Ohsumi Y, Yoshimori T: Dissection of autophagosome formation using Apg5-deficient mouse embryonic stem cells. J Cell Biol. 152, 657-68 (2001)
17. Cuervo AM: Autophagy: many paths to the same end. Mol Cell Biochem. 263, 55-72 (2004)

18. Tanida I, Ueno T, Kominami E: LC3 conjugation system in mammalian autophagy. Int J Biochem Cell Biol. 36, 2503-18 (2004)
19. Kirisako T, Baba M, Ishihara N, Miyazawa K, Ohsumi M, Yoshimori T, Noda T, Ohsumi Y: Formation process of autophagosome is traced with Apg8/Aut7p in yeast. J Cell Biol. 147, 43546 (1999)
20. Marino G, Uria JA, Puente XS, Quesada V, Bordallo J, Lopez-Otin C: Human autophagins, a family of cysteine proteinases potentially implicated in cell degradation by autophagy. J Biol Chem. 278, 3671-8 (2003)
21. Mizushima N, Yamamoto A, Hatano M, Kobayashi Y, Kabeya Y, Suzuki K, Tokuhisa T, Ohsumi Y, Yoshimori T: Dissection of autophagosome formation using Apg5-deficient mouse embryonic stem cells. J Cell Biol. 152, 657-68 (2001)
22. Kabeya Y, Mizushima N, Yamamoto A, Oshitani-Okamoto S, Ohsumi Y, Yoshimori T: LC3, GABARAP and GATE16 localize to autophagosomal membrane depending on form-II formation. J Cell Sci. 117, 2805-12 (2004)
23. Mizushima N, Ohsumi Y, Yoshimori T: Autophagosome formation in mammalian cells. Cell Struct Funct. 27, 421-9 (2002)
24. Mizushima N, Yamamoto A, Hatano M, Kobayashi Y, Kabeya Y, Suzuki K, Tokuhisa T, Ohsumi Y, Yoshimori T: Dissection of autophagosome formation using Apg5-deficient mouse embryonic stem cells. J Cell Biol. 152, 657-68 (2001)
25. Pattingre S, Espert L, Biard-Piechaczyk M, Codogno P: Regulation of macroautophagy by mTOR and Beclin 1 complexes. Biochimie. 90, 313-23 (2008)
26. Winchester BG: Lysosomal membrane proteins. Eur J Paediatr Neurol. 5 Suppl A, 11-9 (2001)
27. Jager S, Bucci C, Tanida I, Ueno T, Kominami E, Saftig P, Eskelinen EL: Role for Rab7 in maturation of late autophagic vacuoles. J Cell Sci. 117, 4837-48 (2004)
28. Mizushima N: Methods for monitoring autophagy. Int J Biochem Cell Biol. 36, 2491-502 (2004)
29. Biederbick A, Kern HF, Elsasser HP: Monodansylcadaverine (MDC) is a specific in vivo marker for autophagic vacuoles. Eur J Cell Biol. 66, 3-14 (1995)
30. Mizushima N, Yamamoto A, Matsui M, Yoshimori T, Ohsumi Y: In vivo analysis of autophagy in response to nutrient starvation using transgenic mice expressing a fluorescent autophagosome marker. Mol Biol Cell. 15, 1101-11 (2004)
31. Dice JF: Chaperone-mediated autophagy. Autophagy. 3, 295-9 (2007)
32. Codogno P, Meijer AJ: Autophagy and signaling: their role in cell survival and cell death. Cell Death Differ. 12 Suppl 2, 1509-18 (2005)
33. Blommaart EF, Luiken JJ, Blommaart PJ, van Woerkom GM, Meijer AJ: Phosphorylation of ribosomal protein S6 is inhibitory for autophagy in isolated rat hepatocytes. J Biol Chem. 270, 2320-6 (1995)
34. Scott SV, Nice DC, 3rd, Nau JJ, Weisman LS, Kamada Y, Keizer-Gunnink I, Funakoshi T, Veenhuis M, Ohsumi Y, Klionsky DJ: Apg13p and Vac8p are part of a complex of phosphoproteins that are required for cytoplasm to vacuole targeting. J Biol Chem. 275, 25840-9 (2000)

35. Kamada Y, Funakoshi T, Shintani T, Nagano K, Ohsumi M, Ohsumi Y: Tor-mediated induction of autophagy via an Apg1 protein kinase complex. J Cell Biol. 150, 1507-13 (2000)
36. Samari HR, Seglen PO: Inhibition of hepatocytic autophagy by adenosine, aminoimidazole4-carboxamide riboside, and N6-mercaptopurine riboside. Evidence for involvement of amp-activated protein kinase. J Biol Chem. 273, 23758-63 (1998)
37. Matsui Y, Takagi H, Qu X, Abdellatif M, Sakoda H, Asano T, Levine B, Sadoshima J: Distinct roles of autophagy in the heart during ischemia and reperfusion: roles of AMP-activated protein kinase and Beclin 1 in mediating autophagy. Circ Res. 100, 914-22 (2007)
38. Zeng X, Overmeyer JH, Maltese WA: Functional specificity of the mammalian BeclinVps34 PI 3-kinase complex in macroautophagy versus endocytosis and lysosomal enzyme trafficking. J Cell Sci. 119, 259-70 (2006)
39. Blommaart EF, Krause U, Schellens JP, Vreeling-Sindelarova H, Meijer AJ: The phosphatidylinositol 3-kinase inhibitors wortmannin and LY294002 inhibit autophagy in isolated rat hepatocytes. Eur J Biochem. 243, 240-6 (1997)
40. Hoyer-Hansen M, and Jaattela M. Connecting endoplasmic reticulum stress to autophagy by unfolded protein response and calcium. Cell Death Differ 14: 1576-1582, 2007.
41. Hamacher-Brady A, Brady NR, Gottlieb RA, and Gustafsson AB. Autophagy as a protective response to Bnip3-mediated apoptotic signaling in the heart. Autophagy 2: 307-309, 2006.
42. Terman A, Gustafsson B, and Brunk UT. Autophagy, organelles and ageing. J Pathol 211: 134-143, 2007.
43. Kiffin R, Bandyopadhyay U, and Cuervo AM. Oxidative stress and autophagy. Antioxid Redox Signal 8: 152-162, 2006.
44. Scherz-Shouval R, Shvets E, Fass E, Shorer H, Gil L, and Elazar Z. Reactive oxygen species are essential for autophagy and specifically regulate the activity of Atg4. Embo J 26: 1749-1760, 2007.
45. Ollinger K, and Roberg K. Nutrient deprivation of cultured rat hepatocytes increases the desferrioxamine-available iron pool and augments the sensitivity to hydrogen peroxide. J Biol Chem 272: 23707-23711, 1997.
46. Kiffin R, Christian C, Knecht E, and Cuervo AM. Activation of chaperone-mediated autophagy during oxidative stress. Mol Biol Cell 15: 4829-4840, 2004.
47. Kim I, Rodriguez-Enriquez S, and Lemasters JJ. Selective degradation of mitochondria by mitophagy. Arch Biochem Biophys 462: 245-253, 2007.
48. Assuncao Guimaraes C, and Linden R. Programmed cell deaths. Apoptosis and alternative deathstyles. Eur J Biochem 271: 1638-1650, 2004.
49. Lemasters JJ, Nieminen AL, Qian T, Trost LC, Elmore SP, Nishimura Y, Crowe RA, Cascio WE, Bradham CA, Brenner DA, and Herman B. The mitochondrial permeability transition in cell death: a common mechanism in necrosis, apoptosis and autophagy. Biochim Biophys Acta 1366: 177-196, 1998.
50. Shimizu S, Kanaseki T, Mizushima N, Mizuta T, Arakawa-Kobayashi S, Thompson CB, and Tsujimoto Y. Role of Bcl-2 family proteins in a non-apoptotic programmed cell death dependent on autophagy genes. Nat Cell Biol 6: 1221-1228, 2004.

51. Paglin S, Hollister T, Delohery T, Hackett N, McMahill M, Sphicas E, Domingo D, and Yahalom J. A novel response of cancer cells to radiation involves autophagy and formation of acidic vesicles. Cancer Res 61: 439-444, 2001.
52. Terman A and Brunk UT. Lipofuscin. Int J Biochem Cell Biol36: 1400-1404, 2004.
53. Cuervo AM. Autophagy: many paths to the same end. Mol Cell Biochem 263: 55-72, 2004.
54. Katz ML and Robison WG Jr. What is lipofuscin? definingcharacteristics and differentiation from other autofluorescentlysosomal storage bodies. Arch Gerontol Geriatr 34:169-184, 2002.
55. Cuervo AM. Autophagy: many paths to the same end. MolCell Biochem 263: 55-72, 2004.
56. Kiffin R, Bandyopadhyay U, and Cuervo AM. Oxidativestress and autophagy. Antioxid Redox Signal 8: 152-162, 2006.
57. Kihara A, Noda T, Ishihara N, and Ohsumi Y. Two distinct Vps34 phosphatidylinositol 3kinase complexes function in autophagy and carboxypeptidase Y sorting in Saccharomyces cerevisiae. J Cell Biol 152: 519-530, 2001.
58. Ollinger K and Roberg K. Nutrient deprivation of cultured rat hepatocytes increases the desferrioxamine-available iron pool and augments the sensitivity to hydrogen peroxide. J Biol Chem 272: 23707-23711, 1997.
59. Brunk UT and Terman A. Lipofuscin: mechanisms of agerelated accumulation and influence on cell function. Free Radic Biol Med 33: 611-619, 2002.
60. Wan FY, Wang YN, and Zhang GJ. The influence of oxidation of membrane thiol groups on lysosomal protopermeability. Biochem J 360: 355-362, 2001.
61. Kim I, Rodriguez-Enriquez S, and Lemasters JJ. Selective degradation of mitochondria by mitophagy. Arch Biochem Biophys 462: 245-253, 2007.
62. Assuncao Guimaraes C and Linden R. Programmed cell deaths: apoptosis and alternative deathstyles. Eur J Biochem 271: 1638-1650, 2004.
63. Lemasters JJ, Nieminen AL, Qian T, Trost LC, Elmore SP, Nishimura Y, Crowe RA, Cascio WE, Bradham CA, Brenner DA, and Herman B. The mitochondrial permeability transition in cell death: a common mechanism in necrosis, apoptosis and autophagy. Biochim Biophys Acta 1366: 177-196, 1998.
64. Shimizu S, Kanaseki T, Mizushima N, Mizuta T, Arakawa-Kobayashi S, Thompson CB, and Tsujimoto Y. Role of Bcl-2 family proteins in a non-apoptotic programmed cell death dependent on autophagy genes. Nat Cell Biol 6: 1221-1228, 2004.
65. Paglin S, Hollister T, Delohery T, Hackett N, McMahill M, Sphicas E, Domingo D, and Yahalom J. A novel response of cancer cells to radiation involves autophagy and formation of acidic vesicles. Cancer Res 61: 439-444, 2001.
66. Kiffin R, Christian C, Knecht E, and Cuervo AM. Activation of chaperone-mediated autophagy during oxidative stress. Mol Biol Cell 15: 4829-4840, 2004.
67. Dice JF. Chaperone-mediated autophagy. Autophagy 3: 295- 299, 2007.
68. Eskelinen EL. Roles of LAMP-1 and LAMP-2 in lysosome biogenesis and autophagy. Mol Aspects Med 27: 495-502, 2006.

69. Kiffin R, Christian C, Knecht E, and Cuervo AM. Activation of chaperone-mediated autophagy during oxidative stress. Mol Biol Cell 15: 4829-4840, 2004.

70. Cuervo AM. Autophagy: many paths to the same end. Mol Cell Biochem 263: 55-72, 2004.

71. Thorpe GW, Fong CS, Alic N, Higgins VJ, and Dawes IW. Cells have distinct mechanisms to maintain protection against different reactive oxygen species: oxidative-stressresponse genes. Proc Natl Acad Sci U S A 101: 6564-6569, 2004.

72. Chen Y, McMillan-Ward E, Kong J, Israels SJ, and Gibson SB. Oxidative stress induces autophagic cell death independent of apoptosis in transformed and cancer cells. Cell Death Differ 15: 171-182, 2008.

73. Yu L, Wan F, Dutta S, Welsh S, Liu Z, Freundt E, Baehrecke EH, and Lenardo M. Autophagic programmed cell death by selective catalase degradation. Proc Natl Acad Sci U S A 103: 4952-4957, 2006.

Adaptation Biology and Medicine (Volume 6: Cell Adaptations and Challenges)
Editors: P. Wang, C.-H. Kuo, N. Takeda and P.K. Singal
Copyright © 2011 Narosa Publishing House Pvt. Ltd., New Delhi, India

10. Implication of CaMKII in H_2O_2-induced Signaling Pathways and in Cardiovascular Pathophysiology

Ali Bouallegue and Ashok K. Srivastava*

Laboratory of Cell Signaling, Montreal Diabetes Research Centre, Centre de Recherche-Centre hospitalier de l'Université de Montréal (CRCHUM) and Department of Medicine, Université de Montréal, Montreal, Quebec, Canada

Abstract

Ca^{2+}/calmodulin-dependent protein kinase II (CaMKII) is a multifunctional protein kinase which has been implicated in numerous cardiovascular diseases such as cardiac hypertrophy, heart failure and hypertension, as well as in diabetes-induced vascular dysfunction. Several cellular responses of Ca^{2+} are believed to be mediated through CaMKII, and although oxidative stress has been shown to increase intracellular Ca^{2+} levels and to trigger many signaling events linked to cell growth and proliferation in vascular smooth muscle cells (VSMC), an involvement of CaMKII in modulating these signaling pathways in VSMC has not been investigated in any detail. We have shown earlier a requirement of Ca^{2+} and calmodulin in H_2O_2-induced phosphorylation of mitogen-activated protein kinase (MAPK) and protein kinase B (PKB). In addition, other studies have also reported an implication of CaMKII in H_2O_2 and other agonist-induced signaling pathways in cardiomyocytes and other cell types. Evidence has also accumulated to suggest a role of CaMKII in cardiovascular diseases. Therefore, in the present review, we will highlight the role of CaMKII in H_2O_2-induced activation MAPK and PKB pathways and its implication in cardiovascular pathophysiology.

Keywords: Calmodulin, phosphorylation, atherosclerosis, pathophysiology, superoxide dismutase, proliferation, hypertrophy

Introduction

An important role of oxidative stress in the pathogenesis of several diseases such as diabetes [1,2] and cardiovascular diseases, including hypertension and atherosclerosis, has been suggested [3-5].Oxidative stress is caused by a failure of the cellular anti-oxidant defence system to mitigate the endogeneously generated excess of reactive oxygen species (ROS).The most important ROS molecules include superoxide anion ($O_2^{-\bullet}$) and hydrogen peroxide (H_2O_2). In VSMC, NAD(P)H oxidase is one of the primary enzymes responsible for the generation of $O_2^{-\bullet}$ [6]. NAD(P)H oxidase catalyzes

*Address for Correspondence: Ashok K. Srivastava, Ph.D. Research Centre, CRCHUM- Angus Campus, 2901, Rachel East, Montreal (Quebec) H1W 4A4 Canada Tel.: (514) 890 8000 ext. 23604 Fax: (514) 412 7648
E-mail:ashok.srivastava@umontreal.ca

$O_2^{-\bullet}$ production by the one electron reduction of O_2 where NAD(P)H is the electron donor. Under physiological conditions, $O_2^{-\bullet}$ undergoes dismutation either spontaneously or by a reaction catalyzed by superoxide dismutase (SOD) to produce H_2O_2, which is more stable than $O_2^{-\bullet}$ and is a freely diffusible ROS molecule . Normally, H_2O_2 is scavenged by catalase and glutathione peroxidase to produce H_2O [8,9]. However, an aberration either in generation or scavenging of H_2O_2 or other ROS molecules leads to an upregulation of the ROS system which has been implicated in the pathophysiology of various cardiovascular diseases [3-5]. The precise mechanism by which ROS contribute to various disease states is not fully understood. However, multiple studies using H_2O_2 or other ROS molecules as experimental inducers of oxidative stress have demonstrated that these molecules activate signaling pathways that are linked with growth, proliferation and hypertrophic responses of the cells. H_2O_2 has been shown to increase the intracellular calcium concentrations in several cell types, including vascular smooth muscle cells VSMC [10] and endothelial cells [11]. H_2O_2 is also known to activate several signal transduction pathways such as mitogen-activated protein kinase (MAPK) and protein kinase B (PKB), which play critical roles in mediating the growth promoting, proliferative and survival responses of the cells [12-14]. We have demonstrated earlier that Ca^{2+} and calmodulin (CaM) are essential to tranduce ROS-induced signaling events in VSMC [15]. We have also shown recently that activation of Ca^{2+}/calmodulin dependent protein kinase II (CaMKII), a downstream effector of CaM, is essential to stimulate H_2O_2-induced activation of ERK1/2 and PKB in VSMC [16]. Since recent evidence has implicated CaMKII in the pathogenesis of heart disease [17,18], the goal of this chapter is to provide an overview on the role of CaMKII in cardiovascular disease and to highlight the involvement of CaMKII in mediating signaling events linked to proliferative and hypertrophic responses in VSMC.

CaMKII

CaMKII is a multifunctional serine /threonine protein kinase which exists as four isoforms, (α, β, γ, δ), encoded by separate genes. CaMKIIα and β are the predominant neuronal isoforms, whereas the γ and δ isoforms are expressed in diverse tissues including the heart and VSMC [19-21]. CaMKII holoenzyme is a multimeric protein which contains 12 subunits arranged in two sets of six subunits that form a stacked hexagonally shaped rings [22]. Each monomer of CaMKII contains three main regions: a N-terminal catalytic region responsible for catalyzing the phosphotransferase reaction, a regulatory region that contains Ca^{2+}/CaM binding sites and autoinhibitory domain (AID), also known as pseudosubstrate domain, and the C-terminal subunit association region responsible for assembling the multimeric holoenzyme (Fig. 1) [22]. In the absence of bound Ca^{2+}/CaM, the CaMKII is maintained in an inactive state because of an interaction of the AID with the catalytic domain of its own subunit that masks the substrate and ATP binding sites [23]. Binding of Ca^{2+}/CaM to a sequence that overlaps the AID results in a conformational change and relieves the inhibitory effect of the AID on the catalytic activity. The catalytically active CaMKII induces the autophosphorylation of Thr^{286} in case of α isoform and Thr^{287} in case of the δ isoform, which are located in the regulatory domain [22]. This region of the regulatory domain also posseses a recently identified activating oxidation site (Met^{281}/Met^{282}) [24]. Phosphorylation of Thr^{286} or oxidation of Met^{281} and Met^{282} prevents the re-association of the catalytic and regulatory domains even after dissociation of Ca^{2+}/CaM complex, and thus allows a sustained effect of CaMKII.

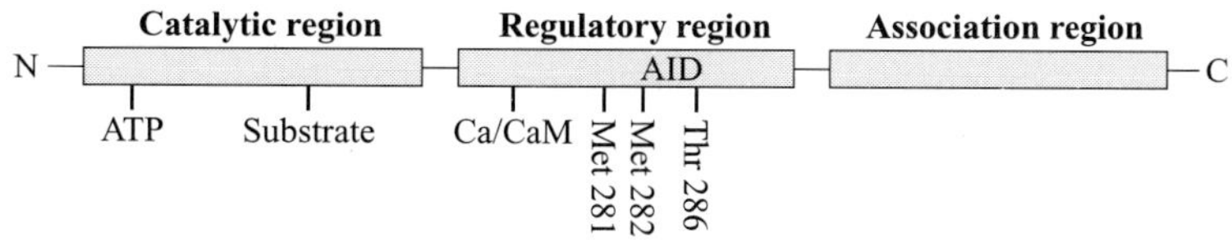

Fig. 1. Schematic representation of prototype monomeric subunit of CaMKII. Each subunit includes a N-terminal catalytic region containing ATP and substrate binding sites, followed by a regulatory region that contains Ca^{2+}/CaM binding sites, threonine (Thr) autophosphorylation sites, oxidizable methionines (Met) and autoinhibitory domain (AID) also known as pseudosubstrate site, and a C-terminal association region responsible for formation of multimeric holoenzyme.

Activation of CaMKII

Under basal conditions, CaMKII is inactive because the intramolecular binding of the catalytic domain to the regulatory domain masks substrate and in ATP binding sites on CaMKII the molecule [25-27]. However, a fraction of CaMKII is active even under this condition, because autophosphorylated $Thr^{286/287}$ is detectable even in quiescent cells [28]. The binding of Calcium bound-CaM (CaM and Ca^{2+} complex) to the regulatory domain induces a conformational change that frees the catalytic domain from its pseudosubtrate [29]. The allosteric rearrangement of CaMKII, upon binding with Ca^{2+}/CaM, allows access to the ATP binding pocket, which in turn allows CaMKII to catalyze the transfer of a phosphate donor group to downstream targets, including its sites autophosphorylation. Autophosphorylation, which occurs by an intra-holoenzyme reaction [30], has several important implications for CaMKII activity. First, $Thr^{286/287}$ autophosphorylation results in a 1,000-fold increase in the affinity for binding with CaM, a property known as "CaM trapping" [31]. Second, autophosphorylation results in the ability of the kinase to maintain catalytic activity even in the absence of CaM binding [32].

Role of CaMKII in Heart Disease

There is growing evidence for a pathophysiological role of CaMKII in cardiac hypertrophy and heart failure [17,18,33]. ET-1 has been shown to increase the activity of CaMKII and to induce hypertrophy in cardiomyocytes. Pretreatment with KN-62, a CaMK inhibitor, was able to suppress ET-1-induced cardiomyocyte hypertrophy measured by β-myosin heavy chain promoter activation and [^{3}H] phenylalanine uptake [34]. Hypertrophic responses induced by Leukemia inhibitory factor (LIF), a glycoprotein cytokine, was also attenuated by CaMKII inhibition in cardiomyocytes [35]. Furthermore, an increase in CaMKII expression [36] and activity [37] has been reported in hypertrophied myocardium from spontaneously hypertensive rats (SHR). More recently, by using hearts extracted from CaMKIIδ-null mice, a critical role of CaMKII in pathological cardiac hypertrophy and remodeling has been reported [38]. CaMKII upregulation and activation has also been shown to play a role in development and maintenance of heart failure [35]. Further evidence linking CaMKII with heart failure comes from studies in which the mice expressing an 'inhibited' form of CaMKII exhibited reversal of heart failure-associated changes [39]. The mechanism by which CaMKII induces alterations in cardiac functions appears to involve an effect on ryanodine receptor, which regulates Ca^{2+} leak from the sarcoplasmic reticulum [18]. CaMKII activation plays a role not only in heart failure and cardiac hypertrophy, but also in cardiac arrythmias. It has been reported that CaMKII is a proarrhythmic signaling molecule in cardiac hypertrophy *in vivo* [40] and its inhibition prevents development of the arrhythmogenic transient inward current [41].

Role of CaMKII in Vascular Disease

Similar to the cardiac system, an involvement of CaMKII in regulating vascular function has also been suggested. CaMKII regulates proliferation of VSMC isolated from rat thoracic aorta [42], and carotid arteries [43]. Other studies have indicated a role of CaMKIIδ_2 in cell migration [44]. siRNA-mediated suppression of CaMKIIδ_2 resulted in the inhibition of wound-induced Golgi reorganization and ERK1/2 activation, leading to inhibition of migration of VSMC [44]. In addition, by using an antisense knockdown of CaMKIIδ, an essential role of CaMKII in ERK-mediated signaling in differentiated smooth muscle cells has been suggested [45]. This notion is further supported by studies showing that pharmacological blockade of CaMKII by KN-93 improved vascular hyperplasia and hypertension in Angiotensin II (Ang II)-induced hypertensive rats [46], and normalized aberrant vascular reactivity in diabetes-induced vascular dysfunction [47,48]. Furthermore, norepinephrine and Ang II-induced arachidonic acid release, which is considered as the first step of inflammation, have been shown to be mediated by CaMKIIα activation in rabbit aortic smooth muscle cells [20,49]. More recently, CaMKII has been also found to be activated in smooth muscle cells in response to arterial injury and has been suggested as a positive regulator of proliferation and migration in the vessel wall, contributing to neointima formation and vascular remodeling [43]. Since a hightened proliferation and migration of VSMC are hallmarks of vascular disease, CaMKII has been suggested to play an important role in the pathogenesis of vascular diseases [46,47].

Role of CaMKII in Signaling Pathways

Both CaMKIIα and CaMKIIδ have been implicated in several agonist-induced signaling pathways in VSMC. A requirement of CaMKIIδ activation in ionomycin-induced phosphorylation of ERK1/2 was reported by Ginnan *et al.* [50]. They also showed that ionomycin-induced EGFR phosphorylation was dependent on CaMKII in VSMC [50]. This group also demonstrated CaMKII-dependence in ERK1/2 activation by ATP [21]. An important role of CaMKII in ERK1/2, p38 phosphorylation and actin reorganization evoked by H_2O_2 in endothelial cell was also reported by Nguyen *et al.* [13]. A similar role of CaMKIIα in mediating norepinephrine-induced arachidonic acid release via MAPK in rabbit artery VSMC has also been documented [20]. Recently, we have reported a critical role of CaMKIIα in H_2O_2-induced ERK1/2 and PKB phosphorylation in A10 VSMC [16]. In these studies, KN-93 attenuated H_2O_2-induced ERK1/2 and PKB phosphorylation in a dose-dependent fashion (Fig.2). In addition, siRNA-induced knock down of CaMKIIα also inhibited the stimulatory effect of H_2O_2 on the phosphorylation of both PKB and ERK1/2 (Fig. 3).

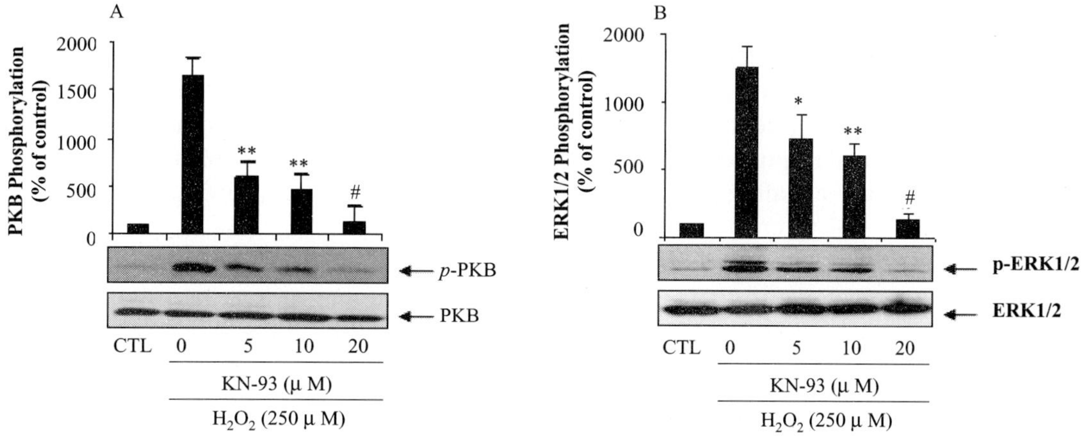

Fig. 2. Attenuation of H_2O_2-induced ERK1/2 and PKB phosphorylation by KN-93 in A-10 VSMCs. Serum-starved quiescent A-10 cells were pretreated in the absence (0) or presence of the indicated concentrations of KN-93 for 30 min followed by stimulation with 250 μM of H_2O_2 for 5 min. Cell lysates were immunoblotted with phospho-specific-Ser^{473}-PKB antibodies (A) and phospho-specific- Thr^{202}/ Tyr^{204}-ERK1/2 antibodies (B), as shown in the middle panels of each section. Blots were also analyzed for total ERK1/2 and PKB (bottom panels of each section). Top panels (bar diagrams) in each section represent average data quantified by densitometric scanning of immunoblots showing in the middle panel. Values are the means ± SE of at least 3 independent experiments and are expressed as percentage of control, taken as 100%. *P*< 0.05 considered as statistically significance versus H_2O_2 stimulation alone. * indicates that *P* < 0.05, ** indicates that *P* <0.005 and # indicates that *P* <0.0005. Adapted from Bouallegue *et al. Free Radic Biol Med* 47:858-866, 2009.

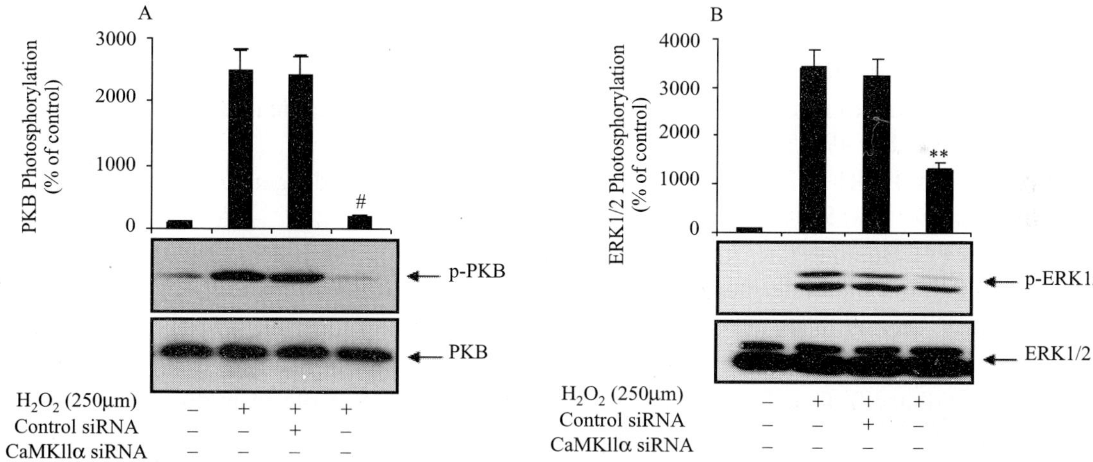

Fig. 3. Attenuation of H_2O_2-induced ERK1/2 and PKB phosphorylation by CaMKIIα siRNA in A-10 VSMCs. A-10 VSMC were transfected with CaMKII siRNA or control siRNA for 48h prior to stimulation with H_2O_2 (250 μM, 5min). Cell lysates were immunoblotted with phospho-specific-Ser^{473}-PKB antibodies (A) and phospho-specific-Thr^{202}/Tyr^{204}-ERK1/2 antibodies (B). Blots were also analyzed for total PKB and ERK1/ 2 (Bottom panels of each section). Top panels (bar diagrams) represent average data quantified by densitometric scanning of immunoblots of p-Proteins shown in the middle panel. Values are the means ± SE of at least 3 independent experiments and are expressed as percentage of control, taken as 100%. *P*< 0.05 considered as statistically significance versus H_2O_2 stimulation alone. ** indicates that *P* < 0.005 and # indicates that *P* < 0.0005. Adapted from Bouallegue *et al. Free Radic Biol Med* 47:858-866, 2009.

Conclusions

In conclusion, there is an increasing body of evidence to support a role of CaMKII in the pathogenesis of cardiovascular disease. An involvement of CaMKII in agonists as well as H_2O_2-induced MAPK and PKB pathways has also been reported. Since these pathways are critical in promoting proliferative, hypertrophic and survival responses of cardiomyocytes and VSMCs, it is possible that CaMKII, by its ability to modulate Ca^{2+} signaling proteins and MAPK/PKB pathways contributes to cardiovascular pathphysiology.

Acknowledgements

This work was supported by a grant from the Canadian Institutes of Health Research (CIHR) to AKS. AB was supported by a studentship jointly sponsored by CIHR , Canadian Hypertension Society and Pfizer Canada

References

1. Houstis N, Rosen ED, Lander ES. Reactive oxygen species have a causal role in multiple forms of insulin resistance. *Nature* **440**: 944-948, 2006
2. Jain SK. Oxidative stress and metabolic diseases: Introduction. *Pathophysiology* **13**:127-128,2006.
3. Griendling KK, Alexander RW. Oxidative stress and cardiovascular disease. *Circulation* **96**: 3264-3265, 1997.
4. Dhalla NS, Temsah RW, Netticaden T. Role of oxidative stress in cardiovascular disease. *J Hypertens* **18**: 655-673, 2000.
5. Singal PK, Khaper N, Farahmand F, Bello-Klein A. Oxidative stress in congestive heart failure. *Current Cardiaol Rep* **2**: 206-211, 2000.
6. Griendling KK, Sorescu D, Ushio-Fukai M. NAD(P)H oxidase: role in cardiovascular biology and disease. **86:** 494-501, 2000.
7. Cai H. NAD(P)H oxidase-dependent self-propagation of hydrogen peroxide and vascular disease. *Circ Res*; **96**: 818-822, 2005.
8. Schafer FQ, Buettner GR. Redox environment of the cell as viewed through the redox state of the glutathione disulfide/glutathione couple. *Free Radic Biol Med* **30**:1191-1212, 2001.
9. Droge W. Free radicals in the physiological control of cell function. *Physiol Rev* **82**: 47-95, 2002.
10. Roychoudhury S, Ghosh SK, Chakraborti T, Chakraborti S. Role of hydroxyl radical in the oxidant H2O2-mediated Ca2+ release from pulmonary smooth muscle mitochondria. *Mol Cell Biochem* **159**: 95-103, 1996.
11. Hecquet CM, Ahmmed GU, Vogel SM, Malik AB. Role of TRPM2 channel in mediating H2O2-induced Ca2+ entry and endothelial hyperpermeability. *Circ Res* **102**: 347-355, 2008.
12. Blanc A, Pandey NR, Srivastava AK. Synchronous activation of ERK 1/2, p38mapk and PKB/Akt signaling by H_2O_2 in vascular smooth muscle cells: potential involvement in vascular disease. *Int Journ Mol Med* **11**: 229-234, 2003.

13. Nguyen A, Chen P, Cai H. Role of CaMKII in hydrogen peroxide activation of ERK1/2, p38 MAPK, HSP27 and actin reorganization in endothelial cells. *FEBS Lett* **572**: 307-313, 2004.
14. Mehdi MZ, Azar ZM, Srivastava AK. Role of receptor and nonreceptor protein tyrosine kinases in H2O2-induced PKB and ERK1/2 signaling. *Cell Biochem Biophys* **47**:1-10, 2007.
15. Blanc A, Pandey NR, Srivastava AK. Distinct roles of Ca2+, calmodulin, and protein kinase C in H2O2-induced activation of ERK1/2, p38 MAPK, and protein kinase B signaling in vascular smooth muscle cells. *Antioxid Redox Signal* **6**:353-366, 2004.
16. Bouallegue A, Pandey NR, Srivastava AK. CaMKII knockdown attenuates H2O2-induced phosphorylation of ERK1/2, PKB/Akt, and IGF-1R in vascular smooth muscle cells. *Free Radic Biol Med* **47**:858-866, 2009.
17. Couchonnal LF, Anderson ME. The role of calmodulin kinase II in myocardial physiology and disease. *Physiology (Bethesda)* **23**:151-159, 2008.
18. Ling H, Zhang T, Pereira L *et al.* Requirement for Ca2+/calmodulin-dependent kinase II in the transition from pressure overload-induced cardiac hypertrophy to heart failure in mice. *J Clin Invest* **119**(5):1230-1240, 2009.
19. Tobimatsu T, Fujisawa H. Tissue-specific expression of four types of rat calmodulin-dependent protein kinase II mRNAs. *J Biol Chem* 1989; **264**:17907-17912, 1989.
20. Muthalif MM, Benter IF, Uddin MR, Malik KU. Calcium/calmodulin-dependent protein kinase IIalpha mediates activation of mitogen-activated protein kinase and cytosolic phospholipase A2 in norepinephrine-induced arachidonic acid release in rabbit aortic smooth muscle cells. *J Biol Chem* **271**:30149-30157, 1996.
21. Ginnan R, Pfleiderer PJ, Pumiglia K, Singer HA. PKC-delta and CaMKII-delta 2 mediate ATP-dependent activation of ERK1/2 in vascular smooth muscle. *Am J Physiol Cell Physiol* **286**:C1281-C1289, 2004.
22. Soderling TR. Structure and regulation of calcium/calmodulin-dependent protein kinases II and IV. *Biochim Biophys Acta* **1297**:131-138, 1996.
23. Cruzalegui FH, Kapiloff MS, Morfin JP *et al.* Regulation of intrasteric inhibition of the multifunctional calcium/calmodulin-dependent protein kinase. *Proc Natl Acad Sci U S A* **89**:12127-12131, 1992.
24. Erickson JR, Joiner ML, Guan X *et al.* A dynamic pathway for calcium-independent activation of CaMKII by methionine oxidation. *Cell* **133**:462-474, 2008.
25. Colbran RJ, Smith MK, Schworer CM, Fong YL, Soderling TR. Regulatory domain of calcium/calmodulin-dependent protein kinase II. Mechanism of inhibition and regulation by phosphorylation. *J Biol Chem* **264**: 4800-2804, 1989.
26. Kelly PT, Weinberger RP, Waxham MN. Active site-directed inhibition of Ca2+/calmodulin-dependent protein kinase type II by a bifunctional calmodulin-binding peptide. *Proc Natl Acad Sci U S A* **85**:4991-4995, 1988.
27. Smith MK, Colbran RJ, Brickey DA, Soderling TR. Functional determinants in the autoinhibitory domain of calcium/calmodulin-dependent protein kinase II. Role of His282 and multiple basic residues. *J Biol Chem* **267**: 1761-1768, 1992.

28. Huke S, Bers DM. Temporal dissociation of frequency-dependent acceleration of relaxation and protein phosphorylation by CaMKII. *J Mol Cell Cardiol*; **42**: 590-599, 2007.
29. Lou LL, Lloyd SJ, Schulman H. Activation of the multifunctional Ca2+/calmodulin-dependent protein kinase by autophosphorylation: ATP modulates production of an autonomous enzyme. *Proc Natl Acad Sci U S A* **83**: 9497-9501, 1986.
30. Kuret J, Schulman H. Mechanism of autophosphorylation of the multifunctional Ca2+/ calmodulin-dependent protein kinase. *J Biol Chem* **260**: 6427-6433, 1985.
31. Meyer T, Hanson PI, Stryer L, Schulman H. Calmodulin trapping by calcium-calmodulin-dependent protein kinase. *Science* **256**:1199-1202, 1992.
32. Schworer CM, Colbran RJ, Soderling TR. Reversible generation of a Ca2+-independent form of Ca2+(calmodulin)-dependent protein kinase II by an autophosphorylation mechanism. *J Biol Chem* **261**: 8581-8584, 1986.
33. Zhang T, Brown JH. Role of Ca2+/calmodulin-dependent protein kinase II in cardiac hypertrophy and heart failure. *Cardiovasc Res* **63**: 476-486, 2004.
34. Zhu W, Zou Y, Shiojima I *et al.* Ca2+/calmodulin-dependent kinase II and calcineurin play critical roles in endothelin-1-induced cardiomyocyte hypertrophy. *J Biol Chem* **275**:15239-15245, 2000.
35. Kato T, Sano M, Miyoshi S *et al.* Calmodulin kinases II and IV and calcineurin are involved in leukemia inhibitory factor-induced cardiac hypertrophy in rats. *Circ Res* **87**: 937-945, 2000.
36. Hagemann D, Bohlender J, Hoch B, Krause EG, Karczewski P. Expression of Ca2+/calmodulin-dependent protein kinase II delta-subunit isoforms in rats with hypertensive cardiac hypertrophy. *Mol Cell Biochem* **220**: 69-76, 2001.
37. Boknik P, Heinroth-Hoffmann I, Kirchhefer U *et al.* Enhanced protein phosphorylation in hypertensive hypertrophy. *Cardiovasc Res* **51**: 717-718, 2001.
38. Backs J, Backs T, Neef S *et al.* The delta isoform of CaM kinase II is required for pathological cardiac hypertrophy and remodeling after pressure overload. *Proc Natl Acad Sci U S A* **106**: 2342-2347, 2009.
39. Wu Y, Shintani A, Grueter C *et al.* Suppression of dynamic Ca(2+) transient responses to pacing in ventricular myocytes from mice with genetic calmodulin kinase II inhibition. *J Mol Cell Cardiol* **40**: 213-223, 2006.
40. Wu Y, Temple J, Zhang R *et al.* Calmodulin kinase II and arrhythmias in a mouse model of cardiac hypertrophy. *Circulation* **106**:1288-1293, 2002.
41. Wu Y, Roden DM, Anderson ME. Calmodulin kinase inhibition prevents development of the arrhythmogenic transient inward current. *Circ Res* **84**:906-912, 1999.
42. House SJ, Ginnan RG, Armstrong SE, Singer HA. Calcium/calmodulin-dependent protein kinase II-delta isoform regulation of vascular smooth muscle cell proliferation. *Am J Physiol Cell Physiol* **292**: C2276-C2287, 2007.
43. House SJ, Singer HA. CaMKII-delta isoform regulation of neointima formation after vascular injury. *Arterioscler Thromb Vasc Biol* **28**: 441-447, 2008.

44. Mercure MZ, Ginnan R, Singer HA. CaM kinase II delta2-dependent regulation of vascular smooth muscle cell polarization and migration. *Am J Physiol Cell Physiol* **294**: C1465-C1475, 2008.

45. Marganski WA, Gangopadhyay SS, Je HD, Gallant C, Morgan KG. Targeting of a novel Ca+2/calmodulin-dependent protein kinase II is essential for extracellular signal-regulated kinase-mediated signaling in differentiated smooth muscle cells. *Circ Res* **97**: 541-549, 2005.

46. Muthalif MM, Karzoun NA, Benter IF *et al.* Functional significance of activation of calcium/calmodulin-dependent protein kinase II in angiotensin II—induced vascular hyperplasia and hypertension. *Hypertension* **39**: 704-709, 2002.

47. Benter IF, Yousif MH, Canatan H, Akhtar S. Inhibition of Ca2+/calmodulin-dependent protein kinase II, RAS-GTPase and 20-hydroxyeicosatetraenoic acid attenuates the development of diabetes-induced vascular dysfunction in the rat carotid artery. *Pharmacol Res* **52**:252-257, 2005.

48. Yousif MH, Benter IF, Akhtar S. Inhibition of calcium/calmodulin-dependent protein kinase II normalizes diabetes-induced abnormal vascular reactivity in the rat perfused mesenteric vascular bed. *Auton Autacoid Pharmacol* **23**: 27-33, 2003.

49. Muthalif MM, Benter IF, Uddin MR, Harper JL, Malik KU. Signal transduction mechanisms involved in angiotensin-(1-7)-stimulated arachidonic acid release and prostanoid synthesis in rabbit aortic smooth muscle cells. *J Pharmacol Exp Ther* **284**: 388-398, 1998.

50. Ginnan R, Singer HA. CaM kinase II-dependent activation of tyrosine kinases and ERK1/2 in vascular smooth muscle. *Am J Physiol Cell Physiol* **282**: C754-C761, 2002.

Adaptation Biology and Medicine (Volume 6: Cell Adaptations and Challenges)
Editors: P. Wang, C.-H. Kuo, N. Takeda and P.K. Singal
Copyright © 2011 Narosa Publishing House Pvt. Ltd., New Delhi, India

11. 10 Gy Total Body Irradiation Increases Risk of Coronary Sclerosis, Degeneration of Heart Structure and Function in a Rat Model

***John E Baker, Ph.D.[1,2,3], Brian L Fish[4], Jidong Su, M.D.[1], Steven T Haworth, Ph.D.[5], Jennifer L Strande, M.D., Ph.D.[5,6], Richard A Komorowski, M.D.[7], Raymond Q Migrino, M.D.[6], Anil Doppalapudi, M.D.[5], Leanne Harmann, RDCS, RDMS, RVT[5], X Allen Li, Ph.D.[4], John W Hopewell, D.Sc.[8], John E Moulder, Ph.D.[4]**

Division of Cardiothoracic Surgery[1], Departments of Pharmacology and Toxicology[2], Medical College of Wisconsin, Milwaukee, WI, USA Children's Research Institute[3], Radiation Oncology[4], Medicine[5], Division of Cardiovascular Medicine[6], Pathology[7], Children's Hospital of Wisconsin, Milwaukee, WI, USA Department of Clinical Oncology[8], The Churchill Hospital, Oxford, United Kingdom

Abstract

Experiments were conducted to determine the impact of 10 Gy total body irradiation (TBI) or local thorax irradiation, a dose relevant to a radiological terrorist threat, on lipid and liver profile, coronary microvasculature and ventricular function. WAG/RijCmcr rats received 10 Gy TBI followed by bone marrow transplantation, or 10 Gy local thorax irradiation. Age-matched, non-irradiated rats served as controls. The lipid profile and liver enzymes, coronary vessel morphology, nitric oxide synthase (NOS) isoforms, protease activated receptor (PAR)-1 expression and fibrinogen levels were compared. Two dimensional strain echocardiography assessed global radial and circumferential strain on the heart. TBI resulted in a sustained increase in total and low density lipoprotein (LDL) cholesterol (190±8 vs. 58±6; 82±8 vs. 13±3 mg/dL, respectively). The density of small coronary arterioles was decreased by 32%. Histology revealed complete blockage of some vessels while cardiomyocytes remained normal. TBI resulted in cellular peri-arterial fibrosis whereas control hearts had symmetrical penetrating vessels with less collagen and fibroblasts. TBI resulted in a 32±4% and 28±3% decrease in endothelial NOS and inducible NOS protein respectively, and a 21±4% and 35±5% increase in fibrinogen and PAR-1 protein respectively, after 120 days. TBI reduced radial strain (19±8 vs. 46±7%) and circumferential strain (-8±3

*Address for Correspondence: John E. Baker, Ph.D. Division of Cardiothoracic Surgery Medical College of Wisconsin 8701 Watertown Plank Road Milwaukee, WI 53226 USA Tel: (414) 456-8706, Fax: (414) 453-9700 E-mail: jbaker@mcw.edu

vs. -15±3%) compared to controls. Thorax-only irradiation produced no changes over the same time frame. TBI with 10 Gy, a dose relevant to radiological terrorist threats, worsened lipid profile, injured coronary microvasculature, altered endothelial physiology and myocardial mechanics. These changes were not manifest with local thorax irradiation. Non-thoracic circulating factors may be promoting radiation-induced injury to the heart.

Keywords: vascular sclerosis, blood lipids, ventricular function, total body irradiation, thorax only irradiation, morphology, cardiovascular risk factors

Introduction

The extent to which total body irradiation (TBI), from a radiation accident or radiological terrorism, affects the cardiovascular system is poorly understood. Evidence that TBI-induced injury may be a cardiovascular risk factor comes from longitudinal studies of Japanese atomic bomb survivors. In this population, mortality from cardiovascular disease is significantly increased after more than 40 years after a single dose exposure of 1-2 Gy to the entire body[1-3].

Radiation therapy has also been linked to the development of cardiovascular disease. For example, radiation therapy for the treatment of peptic ulcers is correlated with an increased mortality from coronary heart disease[4]. Radiotherapy for breast cancer and Hodgkin's lymphoma is also associated with increased risk of cardiovascular disease[5] and myocardial infarction[6], respectively. In these clinical exposures the heart or at least part of the heart was irradiated.

Further evidence that radiation causes cardiovascular disease comes from animal studies. Irradiation of the thorax or specifically the heart in animals induced injury[7-11] following doses of ≥15 Gy. Single dose exposure to 15-60 Gy produced long term effects on cardiovascular function in the rat, resulting in morphological degeneration, mechanical dysfunction[12], damage to the endothelium [13] and increased mortality [9].

The extent to which exposure to 10 Gy TBI, a potentially survivable dose in a radiation accident or radiological terrorism event[14], results in injury to the cardiovascular system is unknown. In this setting, the entire body rather than a single organ, such as the heart, would be exposed. In such a radiological terrorism event, children would likely account for a significant portion of the population affected. Children may be more susceptible to injury from radiation. The long-term consequences of late effects in individuals exposed as children is greater than in adults and this specifically applies to the heart [15]. Studies are needed to determine whether a single dose exposure to 10 Gy TBI in the child would injure the cardiovascular system.

Evidence from the clinical use of radiation to treat children with cancer supports the induction of cardiac injury. Exposure to 10 Gy TBI, in the setting of bone marrow transplantation in children treated for leukemia, results in immediate and delayed cardiac abnormalities manifest as decreased left ventricular ejection fraction and abnormal electrocardiogram [16, 17]. This supports the notion that radiation exposure in children can lead to cardiac dysfunction during development into adulthood. However, these patients with leukemia also receive chemotherapy which may also contribute to the overall treatment-induced cardiotoxicity.

The confounding effects of chemotherapy on radiation-induced injury can be eliminated using an animal model of radiation injury. A rat model allowing survival following TBI is available[18] to study radiation-induced injury in vital organs such as the heart at a dose relevant to the radiological terror threat. It is hypothesized that a single TBI dose of 10 Gy would increase circulating lipids resulting in damage to the coronary vasculature and heart function.

Development of cardiovascular disease following TBI may result from direct or indirect causes. For instance, cardiac dysfunction may result from the direct irradiation of the heart during both TBI and local thorax irradiation. Another possibility is that cardiac dysfunction results from indirect injury in which cardiac dysfunction is only manifest following TBI and not irradiation of the thorax alone. The objectives of the present study were to determine the effect of 10 Gy, either as local thorax irradiation or TBI on lipid and liver profile, coronary vasculature and cardiac function in a rat model, and to determine if injury is a result of direct or indirect effects on the heart.

Methods and Materials

Experimental animals

Young WAG/RijCmcr male rats (n=62), 5 weeks of age, weighing 71±3g, (41% of that of young adult rats, 8-10 weeks of age, of the same strain) were maintained on sterilized rat chow in a moderate-security barrier facility at the Biomedical Resource Center of the Medical College of Wisconsin (MCW), Milwaukee, Wisconsin, USA. Animal care was in accordance with NIH guidelines. The Animal Care and Use Committee at the MCW approved all protocols.

TBI and thorax only irradiation

Rats received TBI or local thorax irradiation with a single dose of 10.0 Gy. TBI was done with a posterior-anterior field at a dose rate of 1.95 Gy/min. Thorax-only irradiation was done with parallel-opposed lateral fields at a dose rate of 1.62 Gy/min. Irradiation was with 300 kVp orthovoltage x-rays, half value layer to 1.4 mm Cu. The radiation dosimetry has been described in detail previously[19]. The TBI dose is low enough so that long-term *in vivo* studies could be carried out following bone marrow transplantation without the risk of acute radiation injury to other organ systems [18]. Non-irradiated rats, not receiving a bone marrow transplant, served as controls. Rats recovered in the barrier facility.

Bone marrow transplantation

One to 2 hr after TBI, rats were given fresh isogenic bone marrow cells using techniques previously published[18]. Rats exposed to local thorax irradiation did not receive a bone marrow transplant nor did the separate un-irradiated controls for this study.

Lipid profile and liver function

Serum was taken at the time of irradiation and then at 20, 40, 60, 80, 100 and 120 days after TBI or after 60, 120, 180 and 240 days after local thoracic irradiation and analyzed for lipid levels (total cholesterol, low density lipoprotein (LDL) cholesterol and triglycerides) and liver function (alkaline phosphatase (ALP), alanine transaminase (ALT), aspartate transaminase (AST) enzyme activity, and total bilirubin, albumin and total protein) (Marshfield Veterinary Laboratories, Marshfield, Wisconsin, USA).

Kidney function

Serum was taken at the time of irradiation and at 120 days after TBI and was analyzed for urea levels [20].

Coronary vessel imaging

After 120 days, the hearts of 6 TBI and control rats were isolated and perfused in a retrograde manner at 37°C with bicarbonate buffer [21] containing high potassium (16 mM) and low calcium (0.8 mM) at constant pressure (85 mmHg) for two minutes to wash out blood from the coronary vasculature and arrest the heart in diastole. Each heart was then perfused with contrast media and processed for micro-computed tomography as previously described [22]. The density of coronary vessels was then assessed in hearts from irradiated and non-irradiated animals.

Histology

To evaluate tissue damage at 120 days following irradiation, the entire heart (n=3) was fixed in formalin (10% v/v), embedded in paraffin, and 5mm-thick sections cut from each block and stained with hematoxylin and eosin (H&E) or Masson-trichrome according to standard methods. Ten sections from each heart were evaluated. Hearts from age-matched controls (n=3) were prepared in the same way.

Blood pressure

Systolic blood pressure was also measured non-invasively after 120 days in control and irradiated rats (n=6/group) using a tail cuff plethysmograph (Life Science Instruments, Woodland Hills, California, USA) in conscious rats.

Cardiac echocardiography

Segmental myocardial viability was assessed using two dimensional strain echocardiography in control and irradiated rats (n=6/group) after 120 days. The operator was blinded to the identity of the rat. An echocardiograph Vivid 7 (General Electric, Waukesha, Wisconsin, USA) was used with a M12L (11-MHz) linear-array transducer. Closed-chest imaging was performed in the short-axis view at the mid-LV level verified by the presence of prominent papillary muscles. The image depth was 2.5 cm, with 236 frames/second acquisition, second harmonic imaging, and with electrocardiographic gating [23].

Echocardiography image analysis

The images were processed on a workstation with EchoPAC Q analysis software (General Electric). A cardiac cycle was defined from the peak of one R wave to the peak of the following R wave. The method has been previously described [23]. The endocardial border was traced in an end-systolic frame. The software automatically selected 6 equidistant tissue-tracking regions of interest in the myocardium. The outer border was adjusted to approximate the epicardial border. The software then provided a profile of radial (myocardial deformation toward the center) and circumferential (myocardial deformation along the curvature) strain (%) with time. End systolic radial and circumferential strain was obtained for each of the 6 segments and the global strain calculated as the average. Three consecutive heart beats were measured and the average used for analysis.

Western blot analysis

The vasculature has to play an important role in maintaining vasodilation and the balance between pro- and anti-coagulation. Endothelial (e) nitric oxide synthase (NOS) and inducible (i)NOS are responsible for producing nitric oxide, a critical mediator of vasodilation. Protease activated receptor (PAR)-1, the vasculature receptor for thrombin, is responsible for mediating coagulation. The effects of radiation on these mediators of vascular function were assessed. In three irradiated and control animals, the free wall of the left ventricle was excised 120 days after TBI and snap frozen in liquid nitrogen for analysis. Western blot analysis of eNOS, iNOS, PAR-1, β-actin and glyceraldehyde 3-phosphate dehydrogenase (GAPDH) expression was carried out as previously described[21, 24]. Nitrate and nitrite levels in ventricular tissue were measured as described previously[25].

Statistical analysis

All values are expressed as the mean ± standard deviation (SD). For the lipid and liver function, densitometry and echocardiography analysis for statistical significance was determined by performing a one-way analysis of variance with Bonferroni's multiple comparison test as the post hoc test. Significance was attributed to those groups that had a P value <0.05.

Results

Total Body Irradiation

Lipid profile and liver function

From ≥ 20 day following TBI, a sustained increase in serum total cholesterol was seen for the next 40-120 days (Figure 1). LDL cholesterol progressively increased to a peak value of 82±8 mg/dL at 80 days compared with 13±3 mg/dL in unirradiated rats. There was a transient increase in triglyceride levels 40 days after TBI, which then declined to values present in unirradiated rats by 100 days (Figure 1). Serum ALP activity was consistently decreased 40-120 days following TBI. ALT activity was also decreased 100-120 days following TBI, while AST was decreased 40-60 days after TBI. Total bilirubin levels were consistently increased, with albumin and total protein levels decreased 40-120 days following TBI (Figure 2).

Kidney function

Serum urea nitrogen levels were determined 120 days following TBI as 303±31 mg/dL, compared with 22±1 mg/dL in unirradiated control rats.

Blood pressure

After 120 days, systolic blood pressure was higher in TBI treated rats (163±7 mmHg) compared with unirradiated rats (120±8 mmHg).

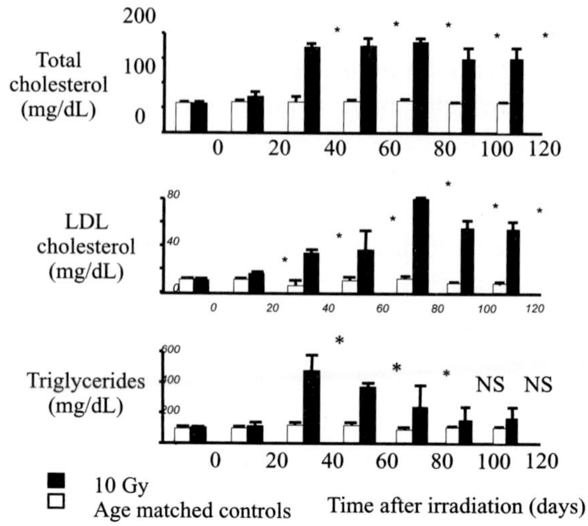

Fig. 1. Time-related changes in total cholesterol, LDL cholesterol and triglycerides following TBI. Data shown as means + SD, n=6/group. * p<0.05, 10 Gy vs controls.

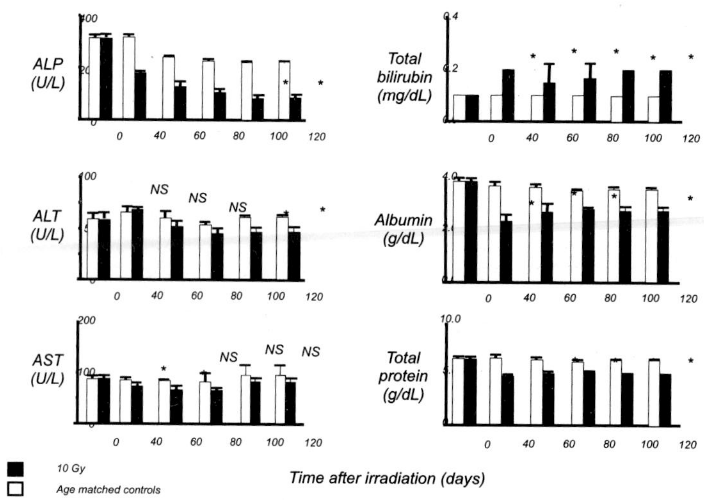

Fig. 2. Time-related changes in liver dysfunction assessed by measurements of three liver enzymes ALP, ALT and AST plus serum bilirubin, albumin and total protein. Data shown as means + SD, n=6/group. * p<0.05, 10 Gy vs controls.

Coronary vessel density

Micro computed tomography showed a reduction in the density of the smaller diameter coronary vessels (<50 μm diameter) at 120 days after TBI (Figure 3A). The diameter of the epicardial arteries were unaffected by TBI. Three dimensional reconstruction of the coronary network revealed a decrease in the smaller diameter penetrating vessels in 32% of fields examined at random compared with the controls (Figure 3B).

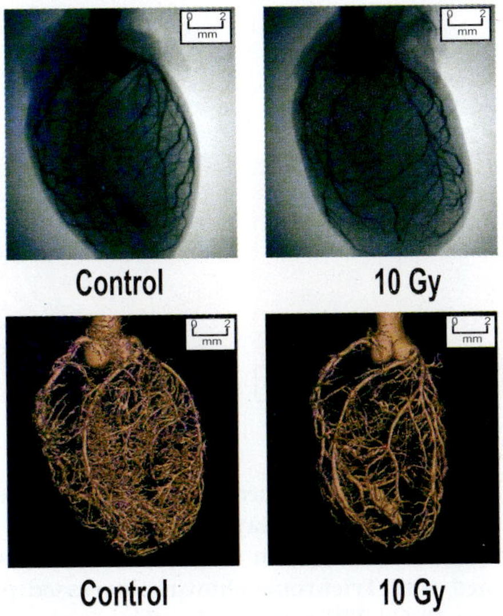

Fig. 3. Coronary vessel changes. A. Micro computerized tomography images of isolated rat heart with contrast agent perfused through the coronary arteries. Representative image of a heart obtained 120 days after exposure to 10 Gy compared with an age matched control. The limit for visualization of coronary vessel diameter is 4 μm. B. Three dimensional micro computed tomography reconstruction of coronary network showing density of coronary vessels at 120 days following 10 Gy TBI compared with an age matched control. TBI resulted in a decrease in the density of the smaller diameter coronary vessels (<50 μm). The density of large diameter coronary vessels was unchanged following 10 Gy TBI.

Cardiac histology

The most striking histological lesion seen in hearts 120 days following TBI was the sclerosis of intramural coronary vessels. Affected vessels had partial to complete luminal sclerosis due to concentric laminar thickening of the vessel walls with accumulation of amphophilic matrix material between layers of hyperplastic and vacuolated smooth myocytes. The cardiomyocytes from TBI treated hearts remained normal in appearance (Figure 4A). Trichrome staining revealed peri-arterial fibrosis and irregular collagen deposition around the penetrating coronary vessels of irradiated hearts. Control hearts had symmetrical penetrating vessels with less collagen and fibrosis (Figure 4B).

Cardiac function

Following TBI, rats had significantly reduced global radial (Figure 5) and circumferential (Figure 5) strain compared to controls as assessed by echocardiography after 120 days. Thus a single exposure to 10 Gy TBI was associated with the development of risk factors for cardiovascular disease, cardiac tissue degeneration and cardiac dysfunction.

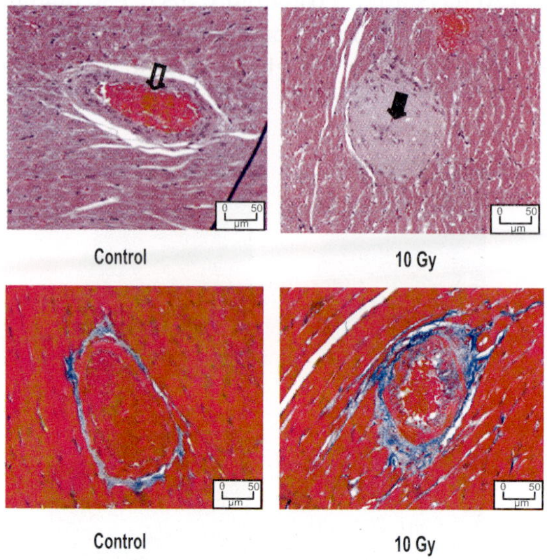

Fig. 4. Morphological changes A. Heart sections stained with H&E show vessel lumen completely blocked (➔) as a result of myointimal proliferation 120 days after 10 Gy TBI. Three hearts were studied in each group. The lumen of a comparable vessel in an age matched heart is patent and contains red blood cells (⇨). B. Heart section stained with Trichrome showing increased peri-arterial fibrosis in small caliber coronary vessel 120 days after 10 Gy TBI compared with a comparable vessel in an age matched control. Three hearts were studied in each group. Fibrosis appears as blue using trichrome staining.

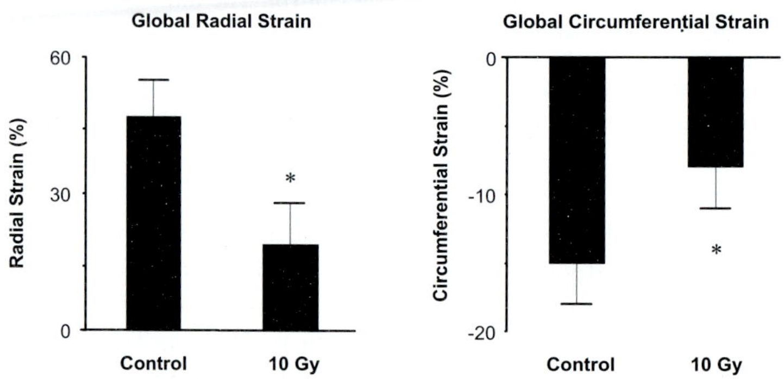

Fig. 5. Changes in ventricular function at 120 days after 10 Gy TBI compared with age matched controls. Data shown as means + SD, n=6/group. * $p<0.05$, 10 Gy vs age matched control.

Nitric oxide synthase and nitric oxide production

There was a 27% decrease in eNOS and a 29% decrease in iNOS protein after 120 days (Figure 6A). Nitric oxide levels were measured in the heart to determine whether TBI impaired NOS function. Nitrite and nitrate are stable oxidation products of nitric oxide and are used as an index of nitric oxide activity. Nitrite plus nitrate content in hearts from rats exposed to 10 Gy TBI was 20% less than in hearts from control rats after 120 days (Figure 6B).

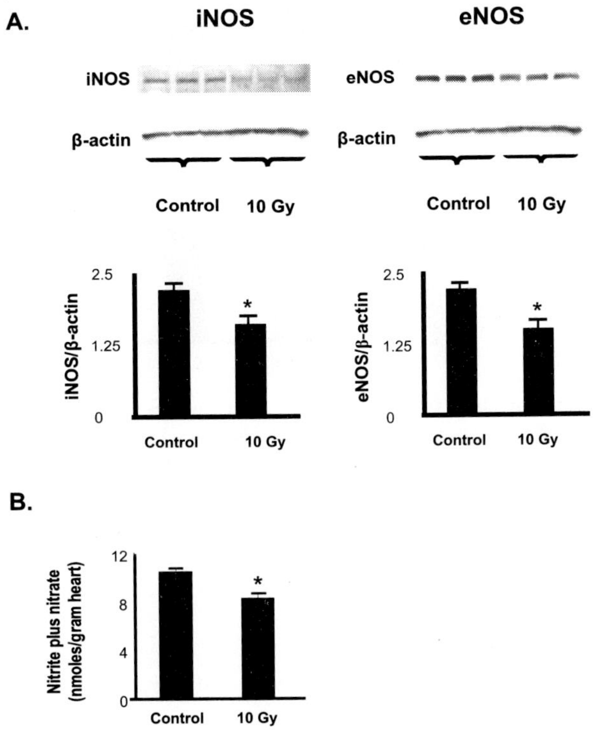

Fig. 6. Variation in iNOS and eNOS protein content in heart 120 days following exposure to 10 Gy TBI compared with age matched controls. A. Western analysis of iNOS and eNOS expression in heart homogenates. B. Changes in nitrite plus nitrate content in heart 120 days after TBI. Data shown as means + SD, n=3/group. * = $p<0.05$, 10 Gy vs aged matched control.

Circulating fibrinogen levels and expression of PAR-1 in heart

TBI resulted in a 30% increase in expression of PAR-1 protein relative to controls (Figure 7A). Plasma levels of fibrinogen were elevated 21% 120 days after 10 Gy TBI (Figure 7B).

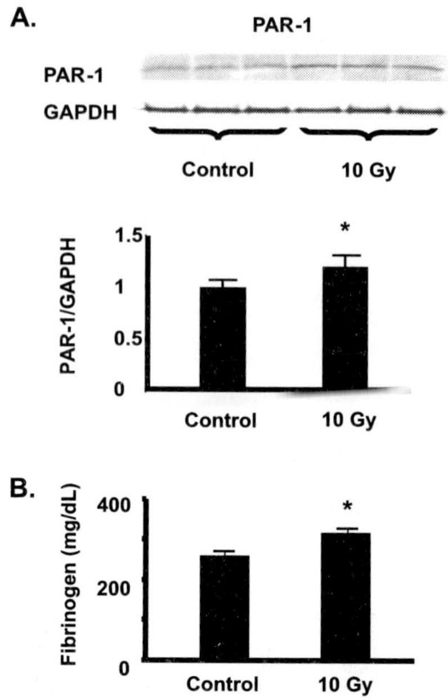

Fig. 7. Changes in coagulant function following TBI. A. Changes in the expression of PAR-1 in heart homogenates 120 days after exposure to 10 Gy TBI compared with controls. Data shown as means + SD, n=3/group. B. Changes in fibrinogen levels in the plasma 120 days after exposure to 10 Gy TBI compared to aged matched controls. Data shown as means + SD, n=6/group. * p<0.05, 10 Gy vs age matched control.

Local Thorax Irradiation

Lipid profile and liver function

There were no significant changes observed over the 60-240 days period investigated (Figure 8).

Nitric oxide synthase, fibrinogen and PAR-1 expression

There was no change in expression of eNOS or iNOS protein, nitrite plus nitrate content, fibrinogen or PAR-1 after 240 days (results not shown).

Coronary vessel density and cardiac histology

Histologic examination after 240 days revealed a normal coronary vasculature and cardiomyocyte structure in local thorax-irradiated rats compared with non-irradiated controls (results not shown). There was no structural injury arising from thorax only irradiation.

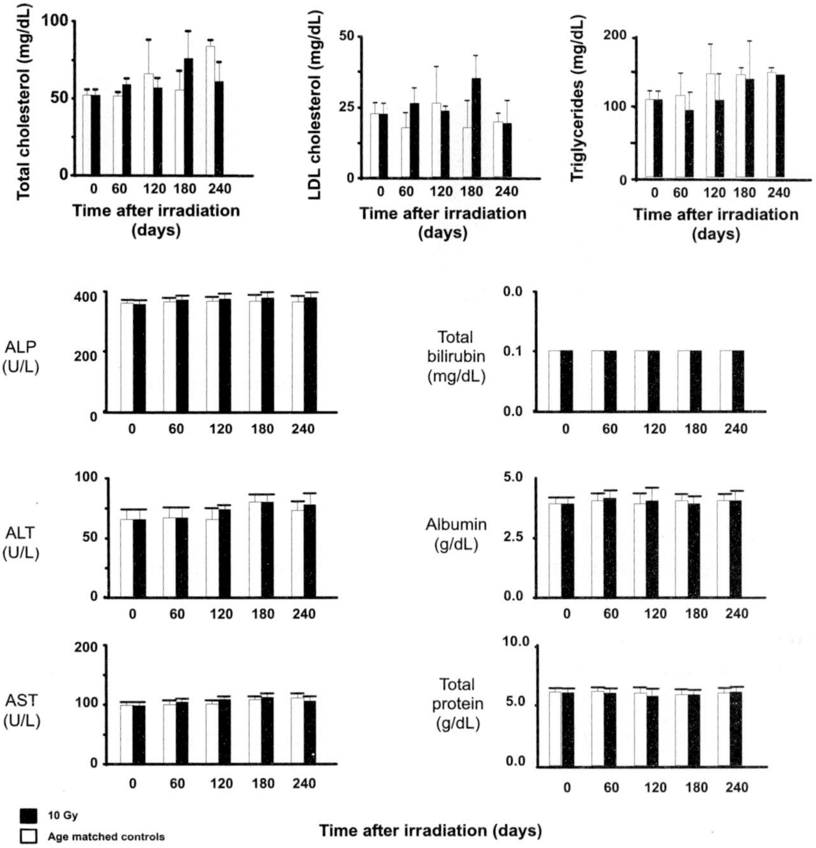

Fig. 8. Time related changes in blood lipid and liver function after local thorax irradiation. Data shown as means + SD, n=6/group.

Cardiac function

Global radial and circumferential strains were not affected at 240 days after local irradiation of the thorax (Figure 9).

Discussion

The results of the present study show that a single TBI exposure with a dose of 10 Gy results in a time-dependent increase in serum total cholesterol, LDL cholesterol, and triglycerides, all of which are biomarkers for the increased risk for cardiovascular disease. Hypercholesterolemia is associated with morphological injury to the vascular endothelium resulting in stenosis, decreased density of the smaller diameter coronary vessels, and a decrease in ventricular function at 120 days following TBI, manifest functionally as a decline in global radial and circumferential strain. TBI decreased expression of eNOS and iNOS and increased expression of PAR-1 and fibrinogen in the first 120 days after TBI. The dose used in the present study, in young 5 week old rats is considerably less than doses of 15-60 Gy, previously shown by others [7-11, 26], to cause injury to the heart following local thorax irradiation in adult

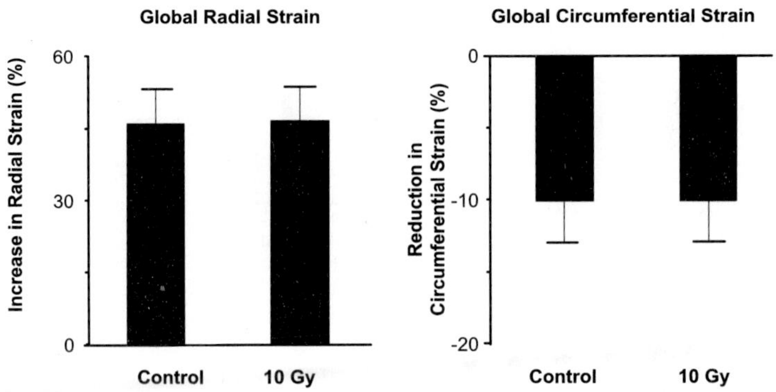

Fig. 9. Ventricular function 240 days after local thorax irradiation with 10 Gy compared to age matched controls. Data shown as means + SD, n=6/group.

rats. The study also demonstrated that 10 Gy, local thorax irradiation, does not result in hypercholesterolemia or hypertriglyceridemia, liver enzyme dysfunction or cardiac morphological and ventricular dysfunction even after 240 days.

There is an urgent need to understand the risk of injury to vital organs, such as the heart, following a radiological terrorist attack or nuclear accident, define the mechanisms underlying the injury, and devise treatment strategies. A study published in the *British Medical Journal* [27] estimated that a small 12.5-kiloton nuclear bomb detonated in New York City would cause 50,000 deaths immediately and several hundred thousand cases of radiation-induced sickness. Children would account for a significant portion of this population. An attack on or an accident at a nuclear power plant or waste storage facility poses another radiological threat. An equal concern is the vulnerability of spent fuel pools that can contain 20 to 30 times more radioactive material than a reactor core and are in buildings not nearly as strong as those that house the reactors.

To study the effects of radiation in an animal model closer to potential pediatric patients than in previous studies on the heart, rats were used at the earliest age when they could live independently from the mother. This feature of the experimental design avoids any potential confounding effects arising from nutrition and care that could originate from the mother. WAG/RijCmcr rats are weaned at around 4 weeks of age so that by 5 weeks of age the offspring can exist independently. At 5 weeks of age rats weight 71±3g, 41% of the young adult body weight, with well developed hearts, livers and lung, and thus they are likely to be comparable with pre pubescent humans. There is relatively little information about tissue responses to radiation in immature vs. mature rats and what information does exist is conflicting as it is in humans; however, the consequences of late effects developing in exposed children have obviously raised special concerns. To the knowledge of the present authors, no other organ systems in the rat have been evaluated with respect to changes in tolerance between immature and mature animals. The present study is possibly the first to evaluate tolerance to TBI in the immature rat.

The present study shows that risk factors for cardiovascular disease are not increased 60 – 240 days after local thorax irradiation with 10 Gy. In contrast, local irradiation of the heart/thorax using higher doses of 15-60 Gy [7-11, 26] does result in delayed injury to the heart in the same time frame. The present findings suggest that injury to the heart caused by 10 Gy TBI is indirect and is likely caused by dysfunction in other organs resulting in the export of systemic factors that contribute to coronary sclerosis and cardiac ventricular dysfunction. For instance, radiation injury to the kidneys results in proteinuria and hypertension [28]. It is well known that proteinuria is associated with hypercholesterolemia as well as low serum albumin and total protein [29]. In addition, untreated hypertension is known to contribute to coronary artery sclerosis. Furthermore, serum urea nitrogen is elevated following TBI indicating renal injury is present. Therefore, the detrimental effects on liver lipid metabolism and cardiac function in this study may be a result of radiation-induced renal failure.

The present study describes a persistent increase in total cholesterol and LDL cholesterol with a transient elevation in triglycerides after TBI. Previous studies in rats have described a transient increase in blood lipids in response to 4-8 Gy TBI [30, 31]. This transient increase in plasma cholesterol levels in the rat was associated with increased liver cholesterol synthesis following irradiation [32, 33]. Elevated serum cholesterol levels would be expected in radiation nephropathy because of the associated proteinuria that adversely affects hepatic lipoprotein synthesis resulting in hypercholesterolemia. Based on AST and ALT levels, as enzyme markers for hepatocyte health, liver function remains fairly stable following irradiation. Most likely, the low serum albumin and total protein may be a result of proteinuria, since if it were due to decreased liver synthesis, marked elevations in AST and ALT would be expected, which was not seen.

Hepatic damage, as well as renal damage may be occurring after TBI, as serum albumin levels are decreased after 9.5 Gy TBI [34]. Hypoalbuminemia is a symptom of hepatic damage but can also be caused by renal damage. The present study shows that liver function is compromised over the 120 days following TBI, as manifest by increased bilirubin levels and decreased activity of the liver enzymes ALP, ALT and AST (Figure 2). The liver is also damaged by local irradiation [35, 36]. However, the present study does contrast with the previously observed decline in activity of ALT but not AST in liver homogenates at 85-90 days following local irradiation of rat liver with 10 Gy [37]. Difference in experimental design, the strain of rat studied and the endpoints used may account for the apparent divergence in outcomes in the present study vs. that of Geraci and Mariano. Albumin and total protein levels are also decreased at 120 days following TBI. However, it is not known if local kidney or liver irradiation increases risk factors for cardiovascular disease. Further studies are needed to determine whether local irradiation of these organs results in increased risk factors for cardiovascular disease, decreased capillary density, and impaired ventricular function to an equivalent level that is present following TBI.

The study also shows that 10 Gy TBI decreases the density of the smaller caliber coronary vessels and confirms previous observations of radiation-induced loss in small vessel density in the heart [38]. Nitric oxide plays a central role in vascular biology, and hence changes in enzyme function will likely have a major impact on endothelial cell and vascular physiology following TBI. It has been shown in the present study that TBI decreases protein levels for constitutive NOS (eNOS), inducible NOS isoforms and nitric oxide generation in the heart. However, the underlying mechanisms are unknown.

TBI results in an increase in plasma fibrinogen levels at 120 days after TBI. Fibrinogen has a plasma half-life of about 4 days. This finding therefore suggests the stimulus for increased production of fibrinogen is still present at 120 days after 10 Gy TBI. TBI also increases expression of PAR-1, the thrombin receptor, in the heart. Activation of PAR-1 may result in proliferation of collagen by increased activation of p42/44 mitogen activated protein kinase (MAPK), which can then cause fibrosis. The role the PAR plays in mediating radiation-induced endothelial dysfunction and the progression to fibrosis in the intestine has also been shown [39]. Taken together with the present findings of increased PAR-1 expression in the heart, further investigations as to the role the PAR plays in radiation-induced injury to the heart are warranted. Radiation also exerts a pro-thrombotic effect to increase deposition of von Willebrand factor [13, 40]. The present findings support the notion that the rat is in a hypercoagulable state 120 days after TBI, most likely as a result of stress response from the liver. Furthermore, clinical and epidemiological studies demonstrate a consistent association between elevated fibrinogen levels and increased risk of atherosclerotic vascular disease. It is proposed that TBI results in an imbalance between procoagulant and anticoagulant function of the vasculature resulting in coronary sclerosis, fibrosis and ventricular dysfunction.

It was also shown in this study that a single 10 Gy dose of TBI results in hypertension by 120 days after irradiation that is associated with hyperlipidemia. It has previously been shown that radiation-induced hypertension over this time course is mediated by the kidney [41]. Further studies are needed to determine the role of the kidney in radiation-induced abnormal lipid metabolism.

Limitations of the study

The present study was not able to define the sequence of events following TBI that result in cardiac injury. The time course over which cardiac dysfunction is manifest in relation to the appearance of elevated lipid levels, hypertension, and occlusion of the smaller caliber coronary vessels needs to be determined and whether this is associated with renal dysfunction. It has also been shown that TBI, but not local thorax irradiation, decreases cardiac function as indicated by a decreased strain pattern, but that overall ejection fraction is preserved. Cardiac strain is a very early predictor of heart failure but is still under clinical investigation. Cardiac dysfunction may continue to deteriorate into a loss of ejection fraction in both groups if followed for a longer period of time.

Conclusion

10 Gy TBI worsened the blood lipid profile in a rat model, resulted in coronary microvasculature injury, altered endothelial physiology and myocardial mechanics. Injury to the heart following TBI may be indirect with non-thoracic organs contributing to coronary sclerosis and ventricular dysfunction. The mechanisms underlying TBI-induced injury to the heart need to be understood and treatments developed to decrease the risk of developing cardiovascular disease after accidental or intentional exposure.

Acknowledgements

This work was supported in part by cooperative agreement AI067734 and grants HL54075 and AI080363 from the National Institutes of Health. The secretarial support of Mary Lynne Koenig, technical support of Eric Jensen, DVM, Mary Lou Mader and Amy Irving, Vladimir Semenko, PhD for dosimetry studies and stimulating discussions with Eric P. Cohen, MD are gratefully acknowledged.

References

1. McGale P, Darby SC. Low doses of ionizing radiation and circulatory diseases: a systematic review of the published epidemiological evidence. *Radiat Res.* 2005;163(3):247-257.

2. Preston DL, Shimizu Y, Pierce DA, Suyama A, Mabuchi K. Studies of mortality of atomic bomb survivors. Report 13: Solid cancer and noncancer disease mortality: 1950-1997. *Radiat Res.* 2003;160(4):381-407.
3. Yamada M, Wong FL, Fujiwara S, Akahoshi M, Suzuki G. Noncancer disease incidence in atomic bomb survivors, 1958-1998. *Radiat Res.* 2004;161(6):622-632.
4. Carr ZA, Land CE, Kleinerman RA, Weinstock RW, Stovall M, Griem ML, Mabuchi K. Coronary heart disease after radiotherapy for peptic ulcer disease. *Int J Radiat Oncol Biol Phys.* 2005;61(3):842-850.
5. Hooning MJ, Botma A, Aleman BM, Baaijens MH, Bartelink H, Klijn JG, Taylor CW, van Leeuwen FE. Long-term risk of cardiovascular disease in 10-year survivors of breast cancer. *J Natl Cancer Inst.* 2007;99(5):365-375.
6. Swerdlow AJ, Higgins CD, Smith P, Cunningham D, Hancock BW, Horwich A, Hoskin PJ, Lister A, Radford JA, Rohatiner AZ, Linch DC. Myocardial infarction mortality risk after treatment for Hodgkin disease: a collaborative British cohort study. *J Natl Cancer Inst.* 2007;99(3):206-214.
7. Cilliers GD, Harper IS, Lochner A. Radiation-induced changes in the ultrastructure and mechanical function of the rat heart. *Radiother Oncol.* 1989;16(4):311-326.
8. Fajardo LF, Stewart JR. Experimental radiation-induced heart disease. I. Light microscopic studies. *Am J Pathol.* 1970;59(2):299-316.
9. Lauk S, Kiszel Z, Buschmann J, Trott KR. Radiation-induced heart disease in rats. *Int J Radiat Oncol Biol Phys.* 1985;11(4):801-808.
10. Yeung TK, Hopewell JW. Effects of single doses of radiation on cardiac function in the rat. *Radiother Oncol.* 1985;3(4):339-345.
11. Yeung TK, Lauk S, Simmonds RH, Hopewell JW, Trott KR. Morphological and functional changes in the rat heart after X irradiation: strain differences. *Radiat Res.* 1989;119(3):489-499.
12. Wondergem J, van der Laarse A, van Ravels FJ, van Wermeskerken AM, Verhoeve HR, de Graaf BW, Leer JW. In vitro assessment of cardiac performance after irradiation using an isolated working rat heart preparation. *Int J Radiat Biol.* 1991;59(4):1053-1068.
13. Boerma M, Kruse JJ, van Loenen M, Klein HR, Bart CI, Zurcher C, Wondergem J. Increased deposition of von Willebrand factor in the rat heart after local ionizing irradiation. *Strahlenther Onkol.* 2004;180(2):109-116.
14. Coleman CN, Blakely WF, Fike JR, MacVittie TJ, Metting NF, Mitchell JB, Moulder JE, Preston RJ, Seed TM, Stone HB, Tofilon PJ, Wong RS. Molecular and cellular biology of moderate-dose (1-10 Gy) radiation and potential mechanisms of radiation protection: report of a workshop at Bethesda, Maryland, December 17-18, 2001. *Radiat Res.* 2003;159(6):812-834.
15. Shankar SM, Marina N, Hudson MM, Hodgson DC, Adams MJ, Landier W, Bhatia S, Meeske K, Chen MH, Kinahan KE, Steinberger J, Rosenthal D. Monitoring for cardiovascular disease in survivors of childhood cancer: report from the Cardiovascular Disease Task Force of the Children's Oncology Group. *Pediatrics.* 2008;121(2):e387-396.

16. Eames GM, Crosson J, Steinberger J, Steinbuch M, Krabill K, Bass J, Ramsay NK, Neglia JP. Cardiovascular function in children following bone marrow transplant: a cross-sectional study. *Bone Marrow Transplant.* 1997;19(1):61-66.
17. Pihkala J, Saarinen UM, Lundstrom U, Salmo M, Virkola K, Virtanen K, Siimes MA, Pesonen E. Effects of bone marrow transplantation on myocardial function in children. *Bone Marrow Transplant.* 1994;13(2):149-155.
18. Moulder JE, Fish BL. Late toxicity of total body irradiation with bone marrow transplantation in a rat model. *Int J Radiat Oncol Biol Phys.* 1989;16(6):1501-1509.
19. Cohen EP, Fish BL, Sharma M, Li XA, Moulder JE. Role of the angiotensin II type-2 receptor in radiation nephropathy. *Transl Res.* 2007;150(2):106-115.
20. Datta PK, Moulder JE, Fish BL, Cohen EP, Lianos EA. TGF-beta 1 production in radiation nephropathy: role of angiotensin II. *Int J Radiat Biol.* 1999;75(4):473-479.
21. Baker J, Su J, Hsu A, Shi Y, Zhao M, Xu H, Eis A, Komorowski R, Wang J, Fu X, Konorev E, Jensen E, Tweddell J, Gross G. Human thrombopoietin reduces myocardial infarct size, apoptosis and stunning following ischemia/reperfusion in rats *Cardiovasc Res.* 2008;77:44-53.
22. Karau KL, Molthen RC, Dhyani A, Haworth ST, Hanger CC, Roerig DL, Johnson RH, Dawson CA. Pulmonary arterial morphometry from microfocal X-ray computed tomography. *Am J Physiol Heart Circ Physiol.* 2001;281(6):H2747-2756.
23. Migrino RQ, Zhu X, Pajewski N, Brahmbhatt T, Hoffmann R, Zhao M. Assessment of segmental myocardial viability using regional 2-dimensional strain echocardiography. *J Am Soc Echocardiogr.* 2007;20(4):342-351.
24. Strande J, Hsu A, Su J, Fu X, Gross G, Baker JE. SCH 79797, a selective PAR-1 antagonist, limits myocardial ischemia-reperfusion injury in rat hearts. *Basic Res Cardiol.* 2007;102:350-358.
25. Shi Y, Hutchins W, Ogawa H, Chang C-C, Pritchard Jr K, Zhang C, Khampang P, Lazar J, Jacob HJ, Rafiee P, Baker JE. Increased Resistance to Myocardial Ischemia in the Brown Norway vs. Dahl S Rat: Role of Nitric Oxide Synthase and Hsp90. *J Mol Cell Cardiol.* 2005;38(7):625-635.
26. Lauk S. Strain differences in the radiation response of the rat heart. *Radiother Oncol.* 1986;5(4):333-335.
27. Helfand I, Forrow L, Tiwari J. Nuclear terrorism. *BMJ.* 2002;324(7333):356-359.
28. Moulder JE, Fish BL, Cohen EP. Impact of angiotensin II type 2 receptor blockade on experimental radiation nephropathy. *Radiat Res.* 2004;161(3):312-317.
29. Wheeler DC, Bernard DB. Lipid abnormalities in the nephrotic syndrome: causes, consequences, and treatment. *Am J Kidney Dis.* 1994;23(3):331-346.
30. Ahlers I, Misurova E, Praslicka M, Tigranyan RA. Biochemical changes in rats flown on board the Cosmos 690 biosatellite. *Life Sci Space Res.* 1976;14:185-188.
31. Feurgard C, Bayle D, Guezingar F, Serougne C, Mazur A, Lutton C, Aigueperse J, Gourmelon P, Mathe D. Effects of ionizing radiation (neutrons/gamma rays) on plasma lipids and lipoproteins in rats. *Radiat Res.* 1998;150(1):43-51.

32. Gould RG, Bell VL, Lilly EH. Stimulation of cholesterol biosynthesis from acetate in rat liver and adrenals by whole body x-irradiation. *Am J Physiol.* 1959;196(6):1231-1237.
33. Shelley RN. The effect of ionizing radiations on cholesterol levels of adrenals, liver, and plasma. *Radiat Res.* 1966;29(4):608-613.
34. Moulder JE, Fish BL, Holcenberg JS, Sun GX. Hepatic function and drug pharmacokinetics after total body irradiation plus bone marrow transplant. *Int J Radiat Oncol Biol Phys.* 1990;19(6):1389-1396.
35. Dawson LA, Normolle D, Balter JM, McGinn CJ, Lawrence TS, Ten Haken RK. Analysis of radiation-induced liver disease using the Lyman NTCP model. *Int J Radiat Oncol Biol Phys.* 2002;53(4):810-821.
36. Geraci JP, Mariano MS, Jackson KL, Taylor DA, Still ER. Effects of dexamethasone on late radiation injury following partial-body and local organ exposures. *Radiat Res.* 1992;129(1):61-70.
37. Geraci JP, Mariano MS. Radiation hepatology of the rat: parenchymal and nonparenchymal cell injury. *Radiat Res.* 1993;136(2):205-213.
38. Lauk S. Endothelial alkaline phosphatase activity loss as an early stage in the development of radiation-induced heart disease in rats. *Radiat Res.* 1987;110:118.
39. Wang J, Boerma M, Fu Q, Hauer-Jensen M. Significance of endothelial dysfunction in the pathogenesis of early and delayed radiation enteropathy. *World J Gastroenterol.* 2007;13(22):3047-3055.
40. Verheij M, Dewit LG, van Mourik JA. The effect of ionizing radiation on endothelial tissue factor activity and its cellular localization. *Thromb Haemost.* 1995;73(5):894-895.
41. Cohen EP, Robbins ME. Radiation nephropathy. *Semin Nephrol.* 2003;23(5):486-499.

Adaptation Biology and Medicine (Volume 6: Cell Adaptations and Challenges)
Editors: P. Wang, C.-H. Kuo, N. Takeda and P.K. Singal
Copyright © 2011 Narosa Publishing House Pvt. Ltd., New Delhi, India

12. Hypoxia Inducible Factor (HIF-1α), HSPs, Antioxidant Enzymes and Membrane Resistance to ROS in Endurance Exercise Performance after Adaptive Hypoxic Preconditioning

Tatyana G. Sazontova, Antonina V. Bolotova, Irina V. Bedareva, Nadezhda V. Kostina, Andrey R. Yurasov, and Yuriy V. Arkhipenko

Faculty of Fundamental Medicine, M.V. Lomonosov Moscow State University, Moscow, Russia

Abstract

Reactive oxygen species (ROS) can not only be damaging but can also induce the synthesis of defense system proteins of the cell as adaptive reaction to stress, hypoxia, etc. This is an overview of some modern concepts as well as the authors' original data on the redox signaling pathways, particularly, at single session of hypoxic preconditioning. Initiation of redox signal system results in induction of transcription factors, which activate genes responsible for cellular defense systems - peroxyredoxines, heat shock proteins, hemoxygenase (HOx), antioxidant enzymes and other urgent response proteins and increase of membrane resistance to ROS. The most significant among transcription factors reacting to hypoxia is HIF-1α. We analyzed components of the redox-signaling pathways in the heart and liver after single exposure during 1h to a) continuous hypoxia, 10% O_2; b) 5-min inhalation of a gas mixture with 10% O_2, - hypoxia and 3-min intervals of normoxia; (H/N) or c) hypoxia, 10% O_2 - hyperoxia, 30%, (H/H) with or without acute physical load. Thereafter we compared the dynamics of HIF-1 α levels after H/H with the dynamics of urgent protein levels (inducible proteins: HOx-1, HSP72 - by Western-blot using monoclonal antibodies and second antibodies with peroxidase label; antioxidant defense enzymes - by spectrophotometry). The findings indicate a tissue specificity of induction of HIF-1α and urgent response defense proteins: the response to hypoxia, physical load and exposure to H/H was fastest and maximal in the heart, compared with liver. It is important that an increase in HIF-1α level was paralleled by expression of HSP family and antioxidant enzymes activity: there was direct correlation of HIF-1α content with superoxide dismutase activity, HOx-1 and HSP72 levels in the heart and HIF-1α content with HSP72 level in the liver. The stability of membrane structures of different organs are maintained by activation of redox signaling after H/H and compensatory synthesis of defense proteins. The preliminary single session of H/H is the most adequate, as both physical endurance raises, and free radical processes by acute physical load are compensated. Thus, from tested methods of preconditioning the new preconditioning regime - single session of hypoxia-hyperoxia is the optimal mode for increase of physical endurance in stressogenic conditions.

*Address for Correspondence: T.G. Sazontova Ph.D., D.Sc. or Yu.V.Arkhipenko, Ph.D., D.Sc. Faculty of Fundamental Medicine M.V. Lomonosov Moscow State University Lomonosovsky pr., 31, korp. 5 Moscow 119192, Russia Tel./fax: +7(495)932-8814 E-mail: arkhipenko@fbm.msu.ru

Keywords: Redox signaling pathways, preconditioning, adaptation, hypoxia, hyperoxia, reactive oxygen species, HIF-1α, antioxidant enzymes, heat shock proteins, heart, liver

Introduction

Participation of hypoxic and stress conditions in pathogenesis of various diseases is mediated by increase in reactive oxygen species (ROS) level and puts a problem of increase in resistance of an organism to such damaging factors. Exogenic antioxidants may be used as mode of protection, however their application is not always effective due to suppression of endogenic own antioxidant systems of an organism [Sazontova *et al.* 1987; Belykh *et al.* 1992; Mosconi *et al.* 1988; Arkhipenko & Sazontova, 1995; Jaeschke, 1995; Fehrenbach, Northoff, 2001]. Thus, the crucial role in effective protection from ROS should be sought in the formation of intrinsic endogenous protective systems including, in particular, a multi component antioxidant protective system rather than in the use of exogenous antioxidants.

Therefore the methods of limitation of excessive activation of free radical processes are perspective, because they can induce the synthesis of its own protective system, for example after activation of redox signaling [Semenza G.L., 1999; Chandel N.S., *et al.* 2000] during single session as preconditioning or repeated sessions as adaptation to changing oxygen level under normobaric conditions by hypoxia or hyperoxia.

Essentially, similar mechanisms are realized both at adaptation and at single session of hypoxic preconditioning. However there are some differences between them:

(a) at single session despite the synthesis of protective systems, the stage of full compensation of free radical generation cannot be reached;

(b) synthesis of protective systems can be excessive more than necessary for compensation. In other words, the price of single preconditioning session is often more than the price of adaptation.

It is known that acute exhausting physical load leads to activation of free radical oxidation with the subsequent development of ROS-dependent damages. Therefore preliminary single or repeated sessions of oxygen level variations, through activation of redox signaling and synthesis of protective systems, apparently can be effective in limitation of free radical processes, activated by physical load.

To solve the problem "how to increase the working capacity for sportsmen and physical endurance in stress situations", we have performed the cycle of investigations directed to development of new technology based on sessions of preconditioning or adaptation to different oxygen levels at acute physical load. The first question was - which mechanisms participate in possible protective effect of different kinds of hypoxic preconditioning.

Role Ros in Protective Effects of Single Session of Hypoxic Preconditioning

ROS formation is a constant physiological process, and ways of formation ROS are numerous, for example, electron transporting chain in mitochondria, auto oxidation of catecholamine, output iron ions (Fe^{2+}) during destruction of hemoproteins, activation of NADH- ? NADPH-oxidase, etc. In a cell, there is a set of antioxidant systems, and the main factors are antioxidant enzymes. They maintain the physiological balance between the level of prooxidants and antioxidant processes. The shift of balance, for example, by ROS signal, leads to the usual response of a cell - synthesis of protective systems.

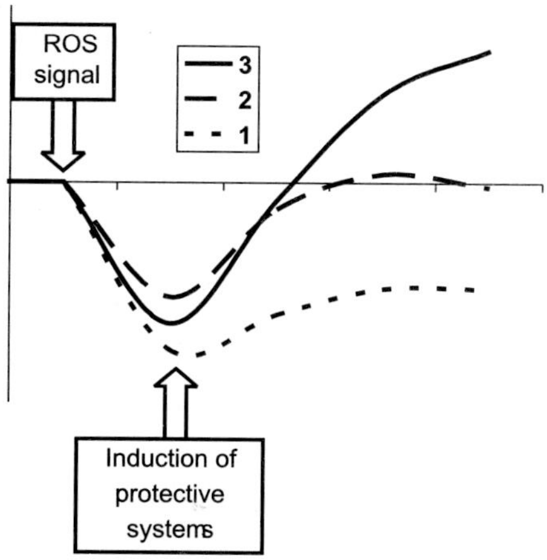

Fig. 1. ROS signal-induced shift of physiological prooxidant/antioxidant balance and synthesis of protective cell systems [Sazontova T.G., 2002].

There are different kinds of cell responses to ROS-signal (Fig.1):

1. The first is - de-compensation, high level of ROS with following damage of cell structures. - line 1
2. Another answer - compensation of ROS signal - line 2
3. And the third is - increase in level of protective system. This answer usually realized under preconditioning or adaptive session - line 3

How does it occur? In recent years, the role of the redox-signaling system and ROS-dependent intracellular system to transfer the external signal to the cell nucleus with the subsequent initiation of proteins synthesis is actively being studied [Semenza G.L., 1999; Chandel N.S., *et al.* 2000]. The major consequence of initiation of redox-signaling systems is activation of transcription factors HIF-1α and HIF-3α [Wiener C.M. *et al.* 1996; Semenza G.L., 1999; Sazontova T.G. *et al.* 2007]. This process of inducing protective proteins often promotes resistance of an organism. Now it is known that more than 60 genes activate HIF-1α. The basic proteins of the answer to the ROS-signal at stress, hypoxia, and ischemia are antioxidant enzymes, HSP family proteins, Fe-regulators, enzymes of reparation and peroxiredoxins [Graven K.K. *et al.* 1993; Ryter S.W. *et al.* 2000; Zhukova A.G. *et al.* 2004]. As a result, the redox signaling system leads to saturation of a cell by the molecules raising its protection against damaging effects. These processes exist both under preconditioning and adaptive sessions.

The First and Second Window of Preconditioning

There are two different protective effects of preconditioning - the first and second window of preconditioning. In preconditioning, there are two periods of protective effect - early, during the first 3 hours, and late, during 24 hours after preconditioning session.

It is important to note that depending on the duration after the preconditioning session, different protective mechanisms are switched on. The first stage of the protection is characterized by the mobilization of already existing reserves of the cell, in particular antioxidants enzymes. Within 12-24 hours after preconditioning session, there is activation of expression of different protective genes, for example, synthesis of heat shock proteins - Hsp72, and increase of their level are in relation to thermal stress [Qian Y.-Z., Bernardo N.L., 1997].

Table 1 shows the three peculiarities of different kinds preconditioning investigations:

Table 1 Protective effects of I and II preconditioning windows (I / II w), protective cross-effects are underlined.

Author	Nature of influence	Received effect
Zucchi R. *et al.* 1995; Bosetti F. *et al.* 2000	Ischemia, isolated rat heart (3 × 3 min)	Stability to ischemia – I w
Yamashita N. *et al.* 1994	Hypoxia in cardiomyocyte cultures (1h)	Stability to hypoxia – I w
Leeuwenburgh C. *et al.* 1994, 2001	Acute physical load, rats	Resistance to cold – I w
Levshina I.P. *et al.* 1994	Hypobaric oxygenation (1 àtm, 1h)	Stability to stress – I w
Wiese A.G. *et al.* 1995; 2009	Oxidizing stress by H_2O_2 to liver cells	Decrease in intensity of ROS processes – I / II w
Hart B.A. *et al.* 1995	Cd 8,9 мкM, n = 3 in culture of lung cells - IMR-90	Resistance to oxidizing stress by H_2O_2 – I / II w
Das D. *et al.* 1997; 2008	Ischemia 5 min, reperfusion 10 min, n = 3	Role of tyrosine kinase in II w
Zager R.A. *et al.* 1995; 2009	Occlusion of kidney arteries	Stability to ischemia – II w
Ashraf M., 1997; 2007	Ischemia 5 min, reperfusion 10 min, n = 4	Effect I / II w on Ca-level and on PK C – II w
Samoilov M.O. *et al.* 2007	Hypoxia 10 min – n = 3	Allay depression – II w

First, one can use very different factors as applied preconditioning, mainly of hypoxic and ischemic nature with subsequent restoration of oxygen level.

Secondly, it is visible, that at the initial stages of development the effects of the first window of preconditioning has been studied, in 2-3 hours after influence.

A follow up study on the effects of both the first and second window of preconditioning was conducted. But by present time the main part of work is dedicated to the study of protective effects of the second window of preconditioning, in 24 hours after influence.

Thirdly, it is important, that among many works on preconditioning in which protective action was realized by the factor of the lowered intensity, there are works on the realization of cross effects - they are underlined in the Table. So, preconditioning by means of physical activities raises resistance to a cold, and precondition by hypoxia removes depressions and application by cadmium increases the stability to oxidizing stress.

The aim of this work consists in the studying of cross protective effects of the second window of hypoxic and hyperoxic adaptive preconditioning sessions on physical endurance, intensity of free radical processes, activity of antioxidant enzymes and the level of HIF-1α and HSP family of proteins.

Methods

Experimental Groups

Experiments were performed on male Wistar rats weighing 250 g. Some kind of preconditioning sessions have been applied:

1. By continuous hypoxia
2. By intermittent hypoxia in 2 regimes: a) hypoxia-normoxia and b) a new way - hypoxia-hyperoxia.

Efficiency of these preconditioning sessions in the increase of physical endurance was controlled. The intermittent regimes with 3-5 min intervals of hypoxia, normoxia and hyperoxia were used.

The experimental animals were divided into 7 groups (9 animals in each group):

1. **C** - control.
2. **PhL** - acute physical load (exhausting swimming with 5% from weight, at 26°C).
3. **Hypo + PhL** - acute physical load 24 hours later of single session of continuous hypoxia (64-min inhalation of a gas mixture containing 8% O_2).
4. **H/N + PhL** - acute physical load 24 hours later of single session of interval of normobaric training (5-min inhalation of a gas mixture containing 10% O_2 with 3-min intervals of normoxia, 64 min).
5. **H/H + PhL** - acute physical load 24 hours after single session of interval of normobaric training (5-min inhalation of a gas mixture containing 10% O_2 with 3-min intervals of hyperoxia, 30% O_2; 64 min).
6. **H/H, 3h** - 3 hours after single session of interval of normobaric training (5-min inhalation of a gas mixture containing 10% O_2 with 3-min intervals of hyperoxia, 30% O_2; 64 min).
7. **H/H, 24h** - 24 hours after single session of interval of normobaric training (5-min inhalation of a gas mixture containing 10% O_2 with 3-min intervals of hyperoxia, 30% O_2; 64 min).

The air mixtures with low and high content of O_2 were obtained with the help of a device (developed in co-operation of our Faculty and "Metaks" Company under management of Dr. Alexander I. Kostin) operating on the membrane gas partition principle with on-line computer control of O_2 levels in isolated animal cage.

Homogenate Preparation

The animals were decapitated after urethane i/p injection on the next day after cessation of the training course or 2 hours after acute physical load. The hearts, brains and livers were extracted, frozen in liquid nitrogen and homogenized in a Teflon-glass homogenizer in 8 (heart and brain tissue) and 12 (liver tissue) volumes of a medium containing 30 mM Trizma@ and 100 mM NaCI pH 7.4.

Antioxidant Enzymes Assays

Catalase activity (mcmol H_2O_2/min/mg protein) was determined by the method of Luck [1963] at 240 nm. SOD activity (a.u./mg protein) was measured spectrophotometrically by the rate of inhibition of superoxide radical formation in the xanthine-xanthine oxidase system [Beauchamp, Fridovich, 1971] at 560 nm by formazan formation from tetranitrotetrazolium blue. Glutathione reductase activity (mcmol NADPH/min/mg protein) was determined by NADPH consumption in the presence of GSSG [Beutler, 1984].

Lipid Peroxidation

The resistance of heart, brain and liver tissues to free radical oxidation was studied in vitro by following the oxidation in ROS-inducing systems. Freshly prepared diluted (1:5) homogenates (exactly 3 mg protein) were incubated at 37°C in 20 mM Trizma@ (pH 7.0), 100 mM NaCI, $FeSO_4^{2+}$ (10-50 nmoles/mg) and 0.3 mM sodium ascorbate. The concentrations of ROS-induced oxidation products were determined by following the characteristic absorption spectra using the method of Ohkawa *et al.* [1979], in the modification of Kikugawa *et al.* [1992].

Protein Concentration

Protein concentration was calculated from the amplitude of the fourth derivative of the absorption spectrum around 240-320 nm using the method of Padros *et al.* [1984].

Western-blot Analysis

HIF-1α, HSP70 and HSP32 or hemoxygenase-1 (HO-l) were determined in the cytoplasmic fractions of the heart, brain and liver. The tissues were homogenized in a Teflon-glass homogenizer (4°C) in a lyzing buffer which contained 50 mM Tris-HCl, 5 mM EDTA, 1 mM DTT and 1% Triton X-100 pH 7.5 (buffer:tissue ratio = 6: 1, by weight). The homogenate was filtered through six layers of gauze and centrifuged for 30 min at 12000 g (4°C). The proteins were separated in 10% polyacrylamide gel and transferred to a PVDF membrane by electroelution (overnight). Preincubation of Western blots was carried out in TBST supplemented with 1 % Tween-20 containing 5% (w/v) defatted milk (1 h). The samples prepared from the thymuses of rats and from H35 hepatoma cells subject to heat shock (41.5°C, 30 min) were used as a positive control. The Western blots were incubated in the presence of first polyclonal antibodies against HSP32 (Stressgen) (1: 1000) and first monoclonal antibodies (Stressgen, Canada; Santa Cruz, USA) against HSP70 in the same dilution for 1 h, washed and incubated in the presence of second antibodies (Jackson Immuno Research) conjugated with horseradish peroxidase (1:2000) for 1h. Detection of HSP32 and HSP70 was performed in a reaction with ECL reagents (Kodak film).

Statistical Analysis

Experimental results were analyzed statistically using STATISTICA 6.0 software according to biomedical statistical recommendations [Platonov, 2000] and presented as medians. Treatment of experimental results was carried out on the basis of parametric and nonparametric statistic criteria. Independent groups were compared using the Mann-Whitney nonparametric U-test.

Results and Discussion

Figure 2 shows results of duration of swimming depending on a kind of preconditioning session. It is shown, that even single session of changing in oxygen level increases the physical endurance. The duration of swimming in group with acute physical load followed by single preconditioning session of hypoxia-hyperoxia was 42% higher than in group of swimming without preliminary preconditioning.

In contrast, in group with precondition by hypoxia-normoxia the increase in physical endurance was not observed.

It is interesting, that preconditioning by means of continuous hypoxia, unlike to hypoxia-normoxia, considerably raises duration of exhausting physical load (Fig. 2).

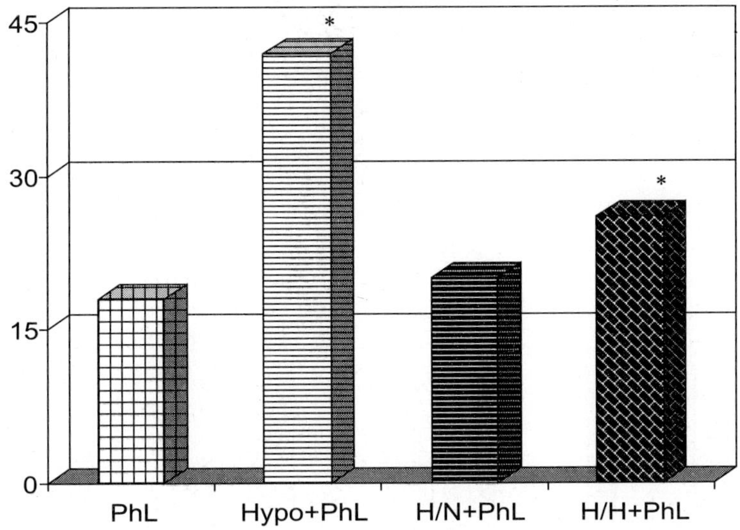

Fig. 2. Effect of different methods of preconditioning (continuous hypoxia, hypoxia/normoxia, hypoxia/hyperoxia) to endurance of exhaustive swimming.
Ordinate - time in min. * - P<0.01.

Therefore the question is, whether "long-swimming" animals receive stress of greater intensity or not. Experiments on measuring the intensity of ROS-dependent processes answered to this question. The intensity of free radical oxidation (on 12 and 25 minute of induction in vitro, Fig.3) in group with acute physical load was higher when compared to control by 68 % and 52 %.

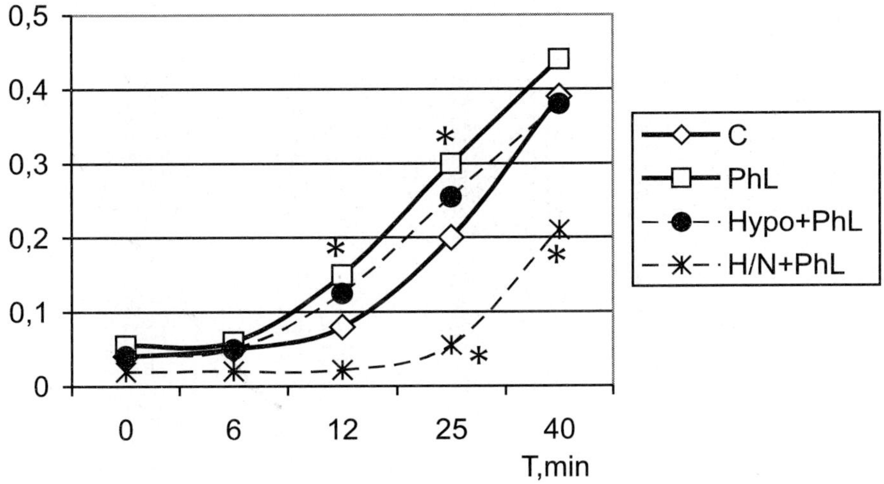

Fig. 3. Increase of thiobarbituric acid reactive substances concentration during incubation of rat liver homogenate. Preconditioning with hypoxia/normoxia. * - P<0.01.

Additionally, the activity of protective antioxidant enzymes (Fig. 4) considerably increases. As a result of exhausting physical load, the activity of catalase and GR is increased by 16 % and 34 % (P<0.01).

(a) Gutathione reductase:

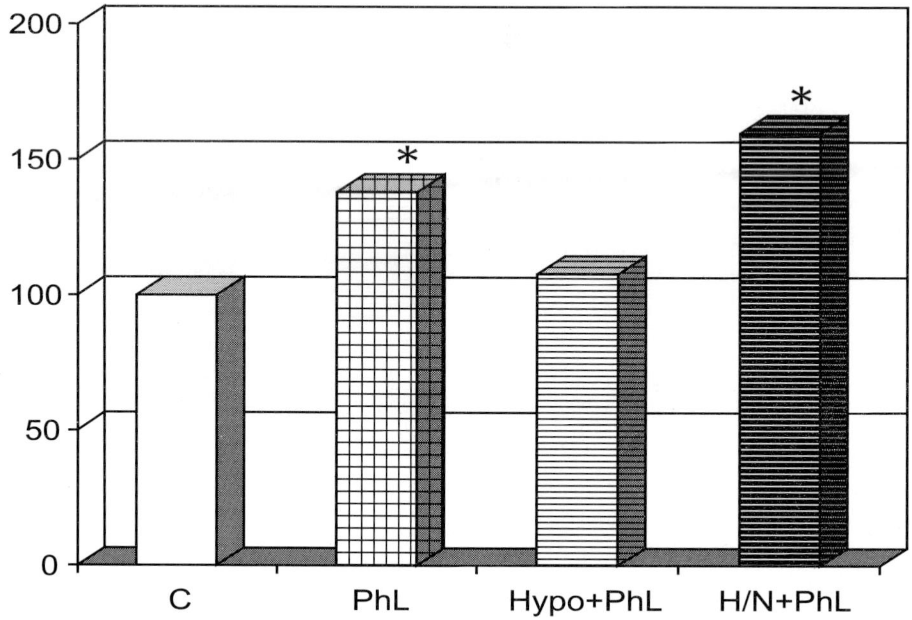

(b) Catalase:

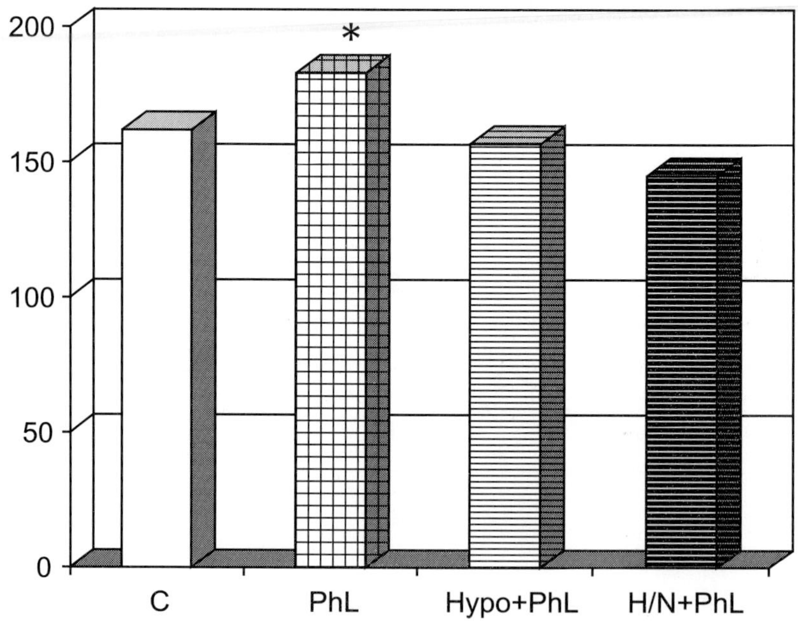

Fig. contd..

Fig. contd..

(c) Superoxide dismutase:

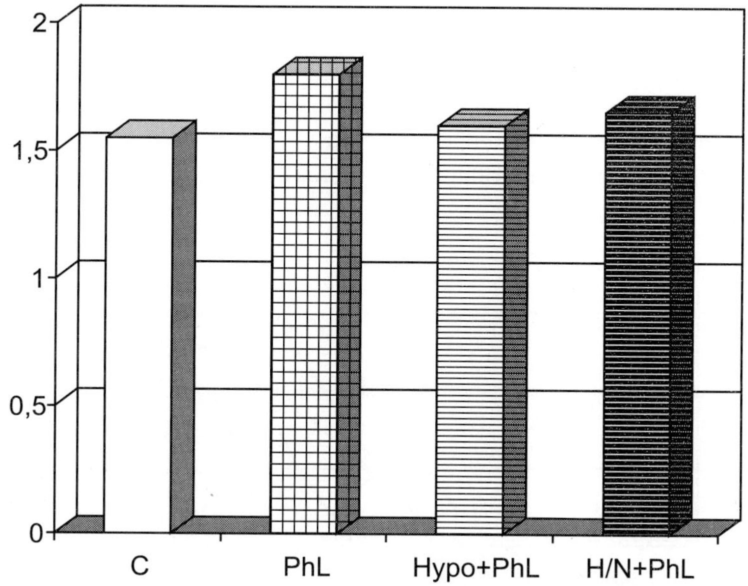

Fig. 4. Activities of gutathione reductase, mcmol NADPH/min/mg protein (A), catalase, mcmol H_2O_2/min/mg protein (B) and superoxide dismutase, a.u./mg protein (C) in rat liver. Designations are in section "Materials and Methods". * - P<0.01.

Exhausting physical load leads to development of oxidative stress. The increased ROS-signal, despite intensification of free radical processes in the cellular membranes, simultaneously induces a number of protective components leading to its partial compensation. Nevertheless, the increased level of antioxidant protection is not capable to compensate for the high intensity of the ROS-signal in the case of physical load.

Preconditioning in the form of a preliminary single session of hypoxia-normoxia limits the intensity of the oxidizing stress induced by physical load. It is visible (Fig. 3) that the intensity of free radical oxidation at physical load after hypoxia-normoxia is considerably lower, than at physical load without precondition than in control. Such decrease in intensity of free radical processes at physical load after single session of hypoxia-normoxia is accompanied by great increase in activity of GR in 59% (Fig.4) which probably contributes to suppression of the ROS-signal.

It is important that despite similar effect of restriction of physical load-induced oxidizing stress by single preconditioning session of hypoxia-normoxia exists, it is not reflected in duration of exhausting swimming, and all effect of the preliminary single preconditioning session of hypoxia-normoxia is to decrease the intensity of the ROS-signal caused physical load, only.

On the other hand, the session of continuous hypoxia within 1 hour leads to an increase of physical endurance, but nevertheless does not reduce intensity of free radical signal and does not compensate for the ROS-processes in physical load (Fig. 2 and 3).

Therefore the new method of preconditioning by single session of hypoxia-hyperoxia has been applied for the more efficiency of precondition.

First, it is visible, that preconditioning session by hypoxia-hyperoxia passes in compensation mode, because the intensity of free radical processes and the level of antioxidant enzymes do not exceed above control level at the second window of precondition in 24 h (Fig. 5 and 6). The main result has been received by the analysis of the effect of single preconditioning session of hypoxia-hyperoxia on ROS-processes and protective systems at physical load. The single preconditioning session of hypoxia-hyperoxia at the stage of the second window suppresses physical load-induced oxidizing stress, leading to restriction of the raised level of free radical processes. Besides, the activity of antioxidant systems is normalized and comes back to control level.

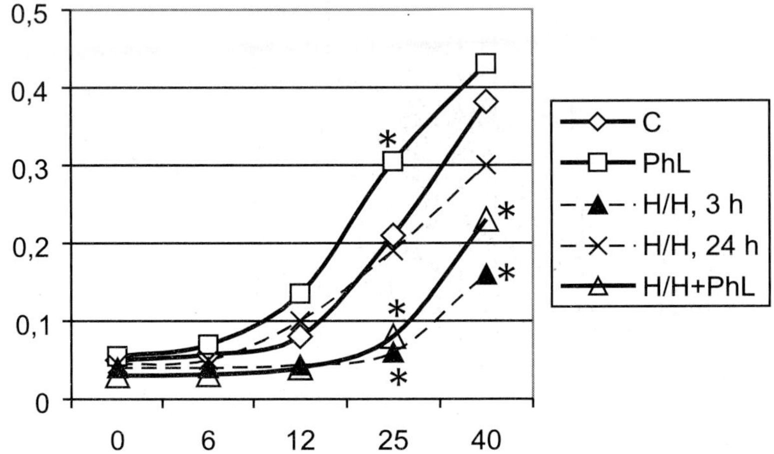

Fig. 5. Increase of TBARS concentration during incubation of rat liver homogenate. Preconditioning with hypoxia/hyperoxia.* -P<0.01.

(a) Gutathione reductase:

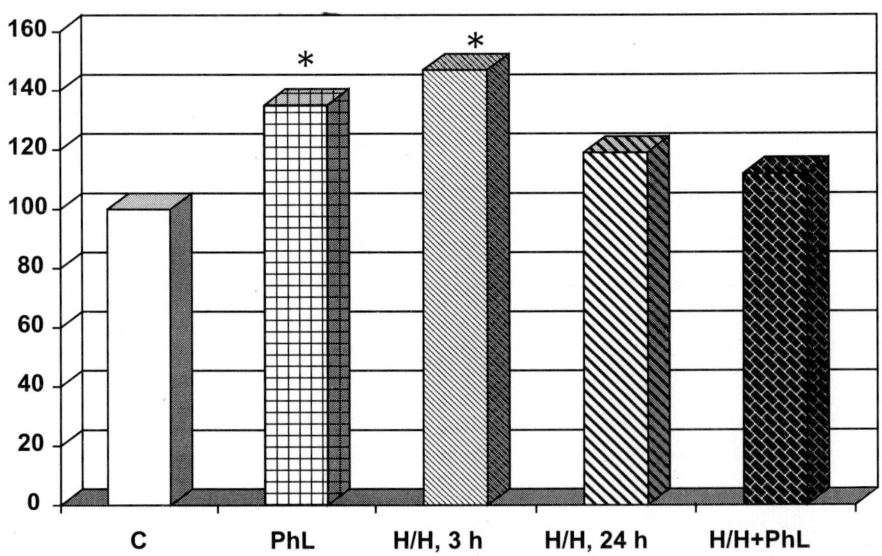

Fig. contd..

Fig. contd..

(b) Catalase:

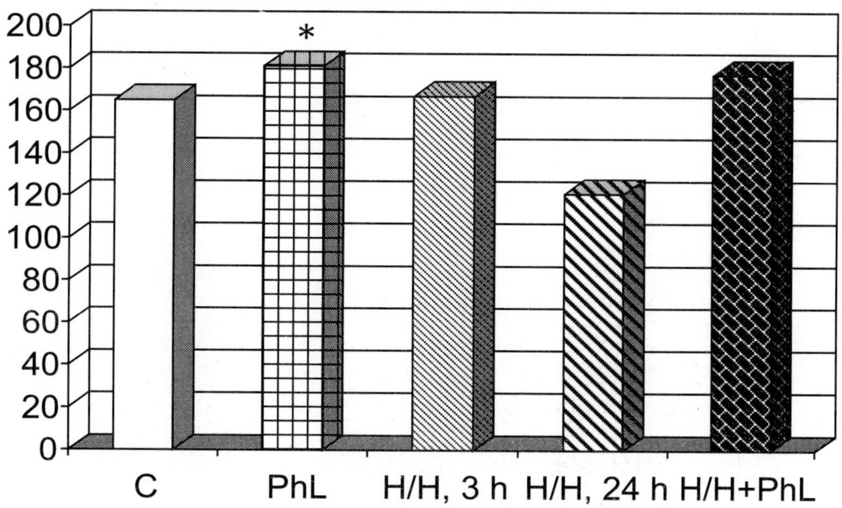

(c) Superoxide dismutase:

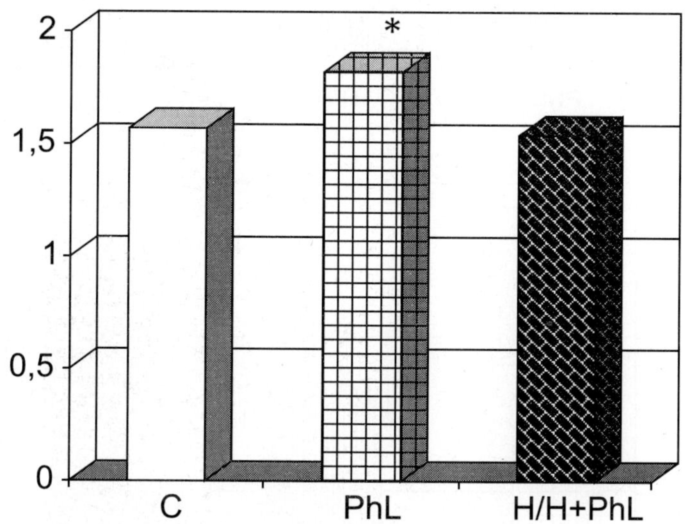

Fig. 6. Activities of gutathione reductase, mcmol NADPH/min/mg protein (A), catalase, mcmol H_2O_2/min/mg protein (B) and superoxide dismutase, a.u./mg protein (C) in rat liver. Preconditioning with hypoxia/ hyperoxia. Designations are in section "Materials and Methods". * - P<0.01.

Similar normalization is shown for HSP family - inducible forms of proteins - HOx-1 and stress-protein HSP72, responsible for reparation of ROS-damaging molecules. If physical load took place after preliminary single preconditioning session of hypoxia-hyperoxia, an induction of these stress proteins is considerably lower, than at physical load without precondition (Fig.7).

Therefore we have correlated change of a level of components of redox signaling systems - the hypoxia induced transcription factor - HIF-1α and HSP family of proteins in heart and liver at physical load, taken separately and after single preconditioning session of hypoxia-hyperoxia.

(a) HIF-lα

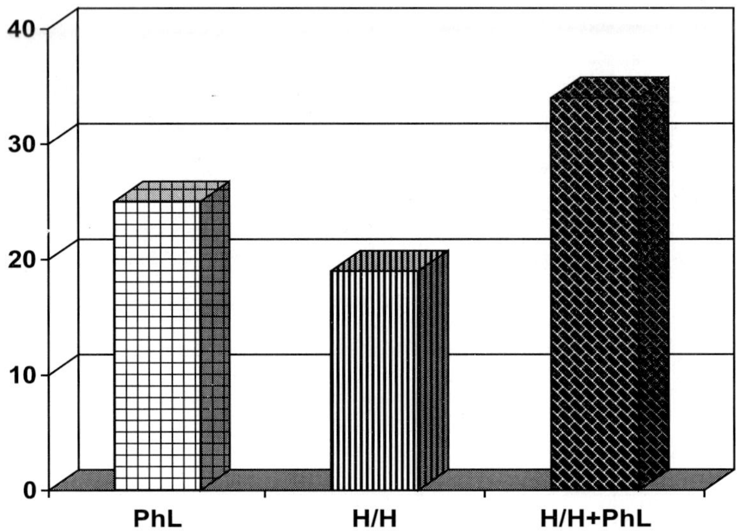

(b) HSP72

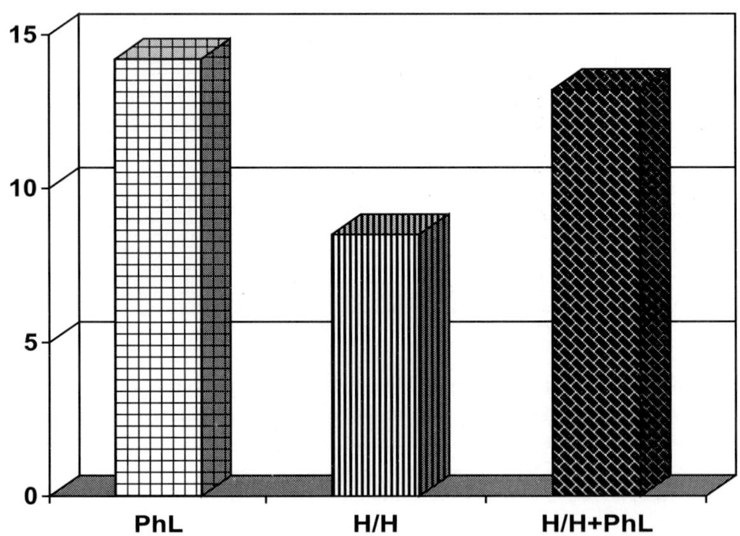

Fig. contd..

Fig. contd..

(c) HOx-1

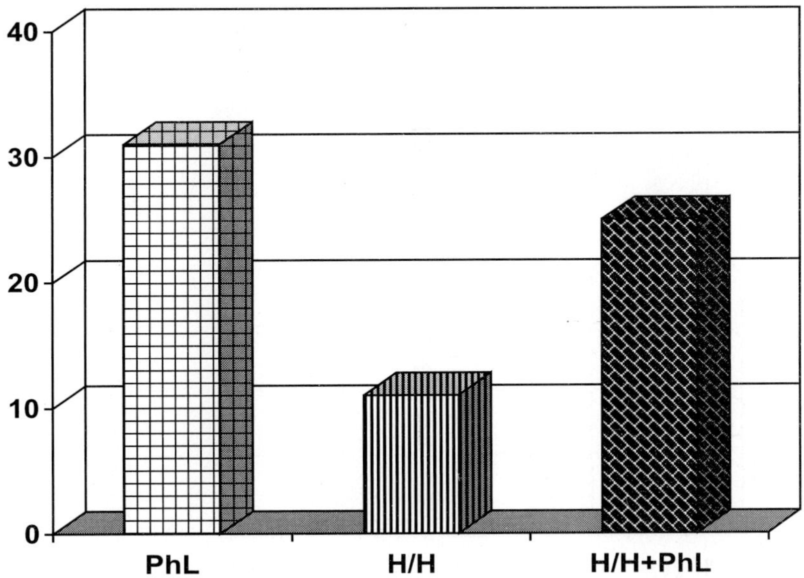

Fig. 7. Levels of HIF-1α (A), HSP72 (B) and HOx-1 (C) in rat heart. Preconditioning by hypoxia/hyperoxia. Designations are in section "Materials and Methods". * - P<0.01.

We have studied the level of HIF-1α in heart and liver. Fig.8 shows an image of Western blot film. In liver when compared with the control, the level of HIF-1α is increased both after physical load and single preconditioning session of hypoxia-hyperoxia. Similar activation is observed in heart, too.

There is important difference in the induction of HIF-1α between heart and liver after single preconditioning session of hypoxia-hyperoxia (Fig.9).

In heart, the induction of HIF-1α occurs much earlier, than in liver and already through 3h exceeds the control level by 2.5 times. In liver HIF-1 α level starts to increase only 24h after physical load. Similar relationships have been found by us earlier (Fig.10), but on different model - continuous hypoxia with 8 % O2: early induction of HIF-α a in heart, but not in liver.

Conclusions

Thus, by summarizing the results, it is possible to conclude, that the comparison of efficiency of application of various kinds of preconditioning has shown, that:

1. Continuous hypoxia, unlike periodic hypoxia-normoxia, raises physical endurance, but does not compensate for a high level of free radical processes, and thus demonstrates the high cost of physical endurance.
2. Preconditioning session by periodic hypoxia-normoxia compensates the free radical processes, but does not raise physical endurance.

A Liver:

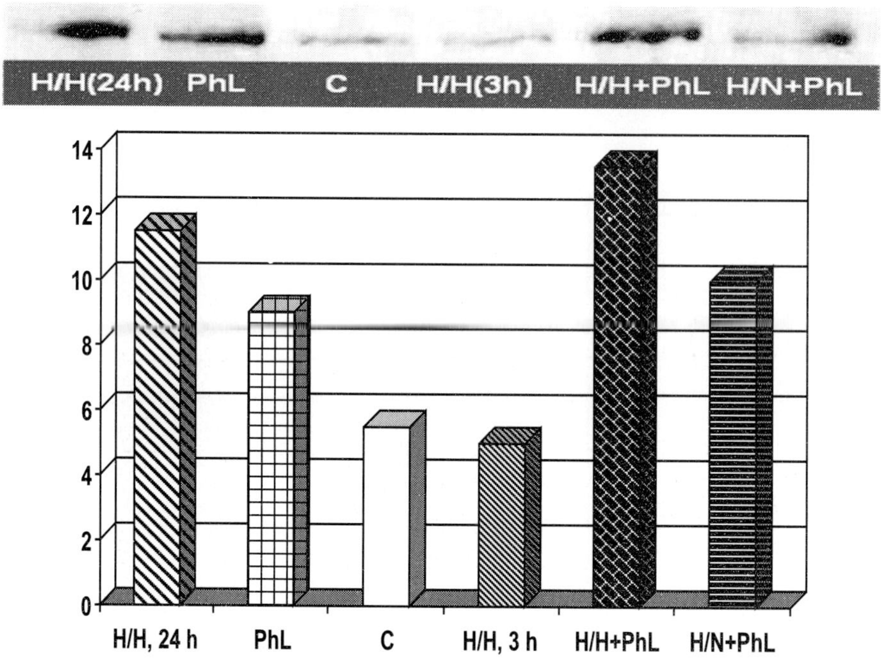

B Heart:

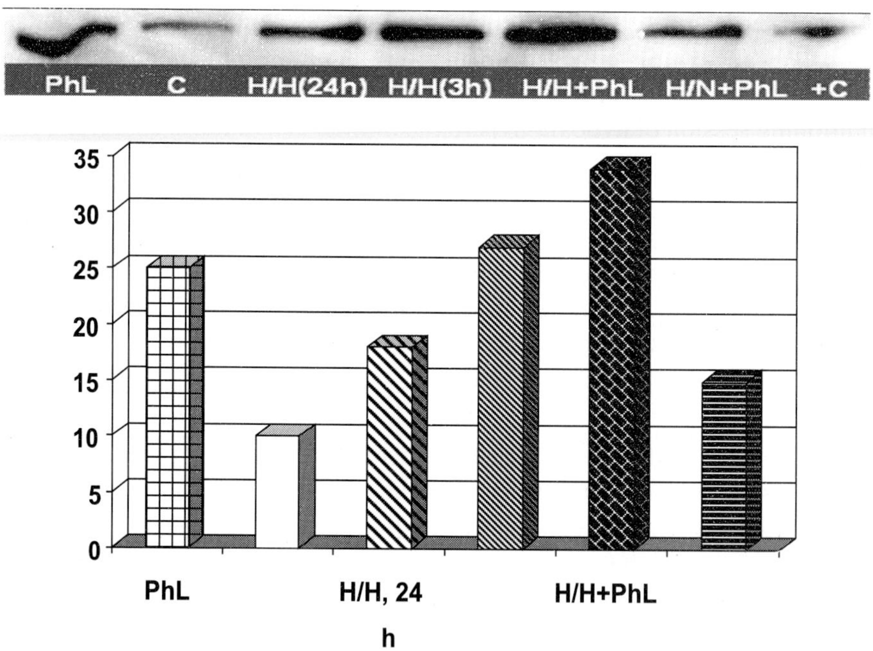

Fig. 8. Immunoblotting of liver (A) and heart (B) HIF-1a after different ways of preconditioning and physical load by swimming. Ordinate - relative optical density units.

A Heart:

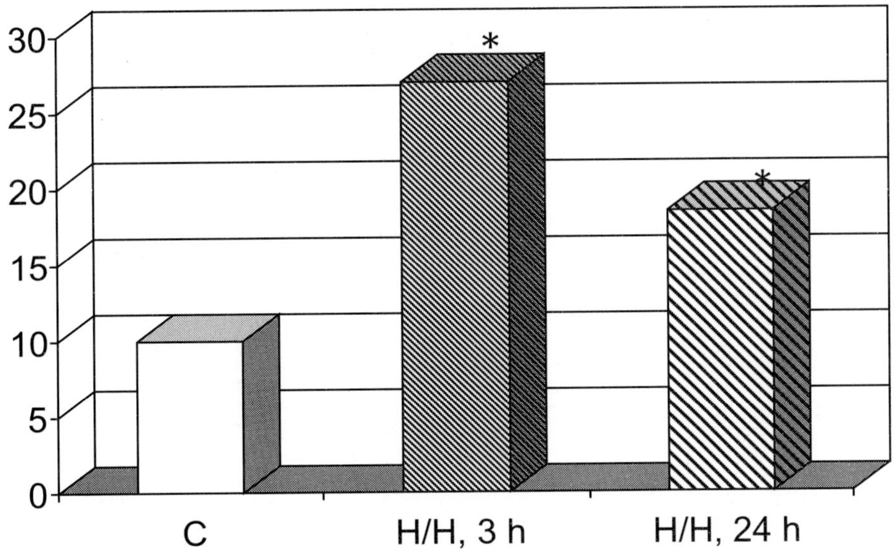

B Liver:

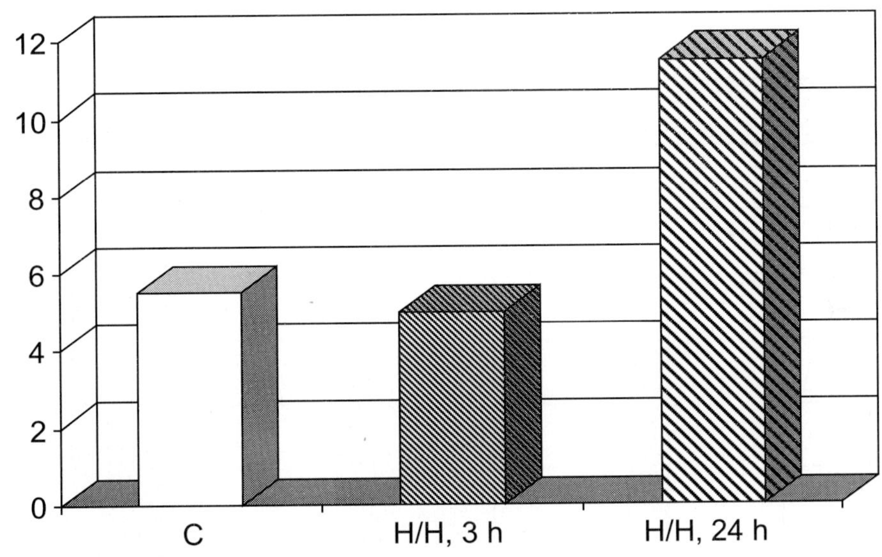

Fig. 9. HIF-1α induction in heart (A) and liver (B) after 3 and 24 h of 1 h-preconditioning with hypoxia/hyperoxia.

A. Heart:

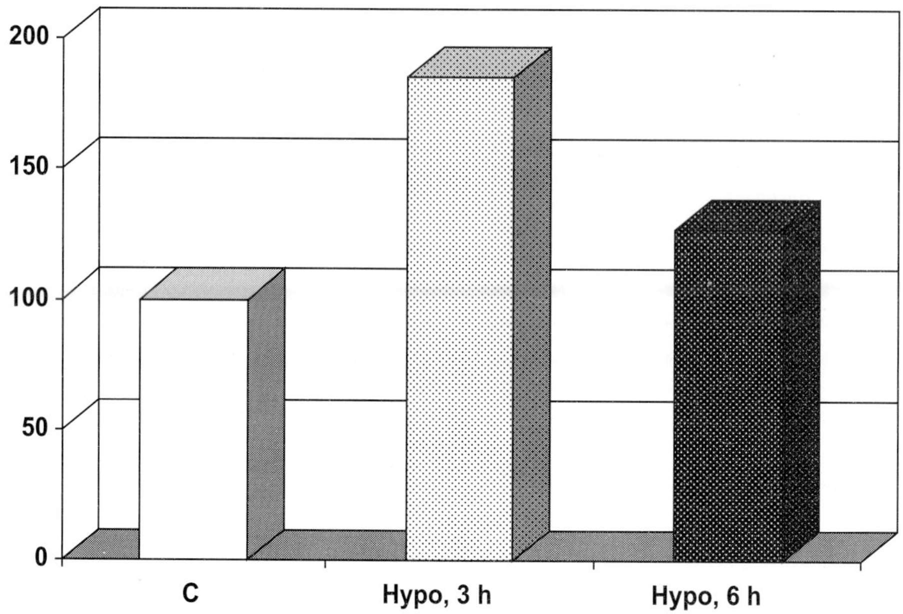

B. Liver:

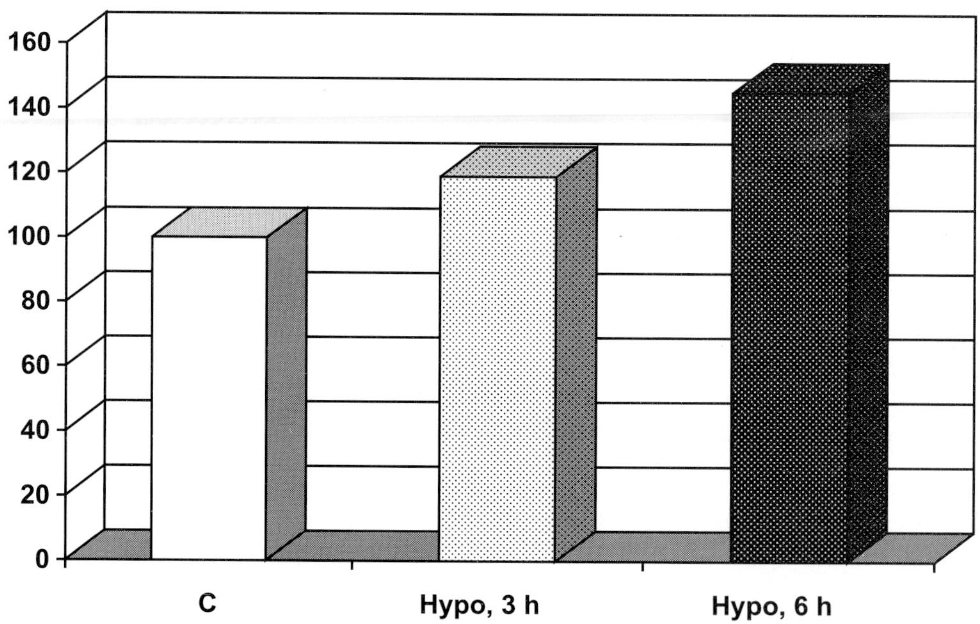

Fig. 10. Expression of HIF-1α in heart (A) and liver (B) 3 and 6 h after acute hypoxia (8% O2, 2 h).

3. The preliminary preconditioning session of hypoxia-hyperoxia is the most adequate, as only at hypoxia-hyperoxia both physical endurance raises, and free radical processes are compensated.

Thus, the preliminary single preconditioning session of hypoxia-hyperoxia is capable to raise physical endurance and compensate free radical signal and normalize a level of protective systems. This kind of preconditioning - hypoxia-hyperoxia - has low "price". From three tested methods of preconditioning, the single session of hypoxia-hyperoxia is the optimal mode for the increase of physical endurance in stressogenic conditions.

References

1. Arkhipenko YuV and Sazontova TG. Mechanisms of the cardioprotective effect of a diet enriched with n-3 polyunsaturated fatty acids. Pathophysiology. 1995;2:131-140.
2. Ashraf M. Calcium, a trigger for cardioprotective adaptive responses to myocardial ischemia. In: Abstracts of the Fifth World Congress of the International Society for Adaptive Medicine (ISAM), Framingham, USA, 1997; 28.
3. Beauchamp C and Fridovich I. Superoxide dismutase: Improved assay and an assay applicable to acrylamide gels. Analyt Biochem. 1971;44: 276-287.
4. Belykh AG, Gukasov VM and Chukaev SA. Free radical oxidation during normobaric hypoxia. Physiol J. 1992;38:73-76 [in Russian].
5. Beutler E. Red cell metabolism. A manual of biochemical methods. Orlando, Fl.: Gune and Stratton, Inc. 1984; 188P.
6. Bosetti F, Yu G, Zucchi R, Ronca-Testoni S and Solaini G. Myocardial ischemic preconditioning and mitochondrial F1F0-ATPase activity. Mol Cell Biochem. 2000;215: 31-37.
7. Chandel NS and Schumacker PT. Cellular oxygen sensing by mitochondria: old questions, new insight. J Appl Physiol. 2000;88: 1880-1889.
8. Das D. Tyrosine kinase - P38MAP kinases - MAPKAP kinase 2 signaling in ischemic preconditioning leads to the transcription of HSP27 and HSP70: a mechanism for second window of protection. In: Abstracts of the Fifth World Congress of the International Society for Adaptive Medicine (ISAM), Framingham, USA, 1997; 13.
9. Das M and Das DK. Molecular mechanism of preconditioning. IUBMB Life. 2008;60: 199-203.
10. Fehrenbach E and Northoff H. Free radicals, exercise, apoptosis, and heat shock proteins. Exerc Immunol. 2001;7:66-89.
11. Graven K.K., Zimmerman L.H., Dickson E.W., Weinhouse G.L., Farber H.W. Endothelial cell hypoxia associated proteins are cell and stress specific. J. Cell. Physiol. 1993;157: ?3: 544-554.
12. Hart BA, Prabhu RM, Eneman JD, Durieuxlu CC, Janssen AML and Borm PJA. Oxidant resistance of cadmium-adapted human lung fibroblasts. Toxicology 1995;98: 1-13.
13. Jaeschke H. Mechanisms of oxidant stress-induced acute tissue injury. Proc Soc Exp Biol Med. 1995;209:104-111.

14. Kikugawa K, Kojima T, Yamaki S and Kosugi H. Interpretation of the thiobarbituric acid reactivity of rat liver and brain homogenates in the presence of ferric ion and ethylenediaminetetraacetic acid. Analyt Biochem. 1992;202: 249-255.
15. Leeuwenburgh C, Fiebig R, Chandwaney R and Ji LL. Aging and exercise training in skeletal muscle: Responses of glutathione and antioxidant enzyme systems. Am J Physiol. 1994;267: R439-R445.
16. Lyovshina IP, Lazareva NA, Mikhalyov SL, Koloskov YuB, Alekseeva NN, Demurov EA and Gulyaeva NV. Influence of different protocols of hypobaric oxygenation and stress on free radical reactions in rat brain. Bull Exp Biol Med. 1994;118: 212-216.
17. Luck H. Catalase. In: Bergmeyer HU. ed., Methods of enzymatic analysis, New York, Verlag-Chemie, Academic Press, 1963:Vol.4, 885-888.
18. Mosconi C, Colli S, Medini L, Stragliotto E, Maderta P, Tremoli E and Galli C. Vitamin E influences the effects of fish oil on fatty acids and eicosanoid production in plasma and circulating cells in the rat. Biochem Pharmacol. 1988;37: 3415-3421.
19. Ohkawa H, Ohishi N and Yagi K. Assay for lipid peroxides in animal tissues by thiobarbituric acid reaction. Analyt Biochem. 1979;95: 351-358.
20. Padros E, Dunach M, Morros A, Sabes M and Manosa J. 4-th-derivative spectrophotometry of proteins. Trends Biochem Sci. 1984;9: 508-510.
21. Platonov AE. Statistical analysis in medicine and biology: Tasks, terms, logic, computer methods. Moscow, Russian Academy of Med. Sci. Publishers, 2000: 51P. [in Russian]
22. Qian Y-Z and Bernardo NL. Ischemic preconditioning does not produce the second window of protection in rat despite increased expression of 72 kilodalton heat shock protein. In: Abstracts of the Fifth World Congress of the International Society for Adaptive Medicine (ISAM), Framingham, USA, 1997: 35.
23. Ryter SW and Tyrrell RM. The heme synthesis and degradation pathway: role in oxidant sensitivity. Free Radic Biol Med. 2000;8: 289-309.
24. Sazontova TG and Arkhipenko YuV. The significance of prooxidants and antioxidants, equivalent participants of metabolism. Pathol Physiol Experim Ther. 2007; No.3: 2-18. [in Russian]
25. Sazontova TG, Arkhipenko YuV and Meerson FZ. Increase in activities of heart antioxidant protection enzymes during adaptation of rats to short-term stress exposures. Bull Exp Biol Med. 1987;103: 411-413.
26. Sazontova TG, Tkatchouk EN and Kolmykova SN. Comparative analysis of peroxidation and antioxidant enzyme activities in rats adapted to different regimes of normobaric hypoxia. Hypoxia Med J. 1994;4: 4-7.
27. Semenza GL. Perspectives on oxygen sensing. Cell. 1999;98: 281-284.
28. Wiener C.M., Booth G., Semenza G.L. In vivo expression of mRNAs encoding hypoxia-inducible factor 1. Biochem. Biophys. Res. Commun. 1996; 225: 485-488.
29. Wiese AG, Pacifici RE and Davies KJA. Transient adaptation to oxidative stress in mammalian cells. Arch Biochem Biophys. 1995;318: 231-240.

30. Yamashita N, Nishida M, Hoshida S, Kuzuya T, Hori M, Taniguchi N, Kamada T and Tada M. Induction of manganese superoxide dismutase in rat cardiac myocytes increases tolerance to hypoxia 24 hours after preconditioning. J Clin Invest. 1994;94: 2193-2199.
31. Zager RA, Burkhart KM and Gmur DJ. Postischemic proximal tubular resistance to oxidant stress and Ca2+ ionophore-induced attack - Implications for reperfusion injury. Lab Invest. 1995;72: 592-600.
32. Zager RA and Johnson AC. Renal ischemia-reperfusion injury upregulates histone-modifying enzyme systems and alters histone expression at proinflammatory/profibrotic genes. J Physiol Renal Physiol. 2009;296: F1032-F1041.
33. Zhukova A.G., Sazontova T.G. Heme oxygenase: function, regulation, biological role. Hypox.Med.J;2004: ? 3-4: 30-43.
34. Zucchi R, Ronca-Testoni S, Di Napoli P, Gallina S, Bosco G, Ronca G, Calafiore AM, Mariani M and Barsotti A. Sarcoplasmic reticulum calcium uptake in human myocardium subjected to ischemia and reperfusion during cardiac surgery. J Mol Cell Cardiol. 1996;28: 1693-1701.

Adaptation Biology and Medicine (Volume 6: Cell Adaptations and Challenges)
Editors: P. Wang, C.-H. Kuo, N. Takeda and P.K. Singal
Copyright © 2011 Narosa Publishing House Pvt. Ltd., New Delhi, India

13. Human Adaptation to Intermittent Hypoxia: Effects on Hematopoietic Stem Cells and Immune Function

Tatiana Serebrovska[1*], Igor Nikolsky[2], Vadim Ishchuk[3]

[1]Bogomoletz Institute of Physiology, Kiev, Ukraine

[2]Institute of Genetic and Regenerative Medicine, Academy of Medical Sciences, Kiev, Ukraine

[3]State Institute of Gerontology, Academy of Medical Sciences, Kiev, Ukraine

Abstract

It has been proven by Ukrainian and Russian scientists that intermittent hypoxia training (IHT) improves the functioning of respiratory, cardiovascular and immune systems, increases hematopoiesis, stimulates clearing of the respiratory tract of microscopic particles, reduces negative effects of ionizing irradiation and increases body tolerance to various toxins. The purpose of this study was to define the IHT effects on human hematopoietic stem cells (HSC), various factors of natural resistance and main humoral and cellular components of adaptive immunity in peripheral blood. 10 healthy subjects (25-35 years old) were observed twice before and twice after IHT (normobaric rebreathing technique up to SaO_2 = 80-82 %, 5 min hypoxia, 5 min break, 4/day, 15 days). The content of HSC (enumerated as $CD45^+CD34^+$ bright cells) was studied according to standard direct immunofluorescence techniques. Phagocyte function and bactericide activity of neutrophils, the content of immunoglobulins IgM, IgG, IgA, circulating immune complexes (CIC) were determined. It was firstly shown that IHT causes the decrease in circulating HSC count in healthy subjects in a week after IHT. As a result of IHT, the enhancement of natural resistance system occurs, including complement activation, platelet number augmentation, the increase in phagocytic capacity of neutrophils with a decrease in their spontaneous function and an increase in reserve bactericidity. In such mode IHT doesn't affect the red blood cell status as well as erythropoietin concentration. Thus, humoral and cellular factors of non-specific resistance are sensitive to such regime of IHT. Their activity increases significantly. It leads to the enhancement of natural protection of the organism. A decrease in HSC number in peripheral blood is probably associated with the change of their migration capacity. However, it remains unclear whether there is an inhibition of HSC migration into circulation or an activation of HSC escape from circulation. In both cases tissues could accumulate more HSC which in turn could enhance hematopoiesis and general regenerative potential. It assures the HSC transdifferentiation capacity, plasticity and intercellular cooperation.

Keywords: Hematopoietic stem cells; intermittent hypoxia; immunofluorescence; neutrophils; immunoglobulins

[*]Address for Correspondence: Dr. Tatiana V. Serebrovska Principal Researcher of the Bogomoletz Institute of Physiology Bogomoletz str., 4 Kiev, 01024, Ukraine Email: sereb@biph.kiev.ua

Introduction

One of the main mechanisms of adaptation to hypoxia is the response of stem cells. But until recently, the effects of oxygen on the functioning of human stem cells have been largely ignored.

Stem cells and progenitor (precursor) cells represent the minor cell populations which are initial elements forming various cell and tissue types during embryogenesis and postnatal ontogenesis (Weissman *et al.* 2001). They are characterized by the ability to self-renewal and retention of multipotency in different degree, i.e. the ability to differentiate into various cell types. The term "embryonic stem cells" is used towards the elements derived from the inner mass of blastocyst and possessing properties of totipotency, i.e. the ability to develop into a whole organism. The term "postembryonic stem cells" (fetal cells and cells at subsequent postembryonic ontogenetic stages) is served for the designation of regional stem cells because of their exclusive pluripotency, i.e. the ability to provide a variety of cellular elements mainly within one or two germ layers. Under certain conditions these cells could undergo the transdifferentiation and generate nonrelevant cell types.

Stem cells are also found in adults. They maintain and repair the body and are located in the various tissues of an organism. Such stem cells have the potential to take on different characteristics in order to form the cells of a limited number of tissues. They are called multipotent.

Hematopoetic stem cells (HSCs) are one of the earliest elements in well-organized chain of cell proliferation, differentiation and migration which leads to formation of all blood cells. Besides, HSCs have the ability for transdifferentiation and can give rise to different cell types such as hepatic oval cells (Lagasse *et al.* 2000), neurons (Mezey & Chandross, 2000), skeletal muscle cells (Bossolasco *et al.* 2004), cardiac muscle and vascular endothelium cells (Jackson *et al.* 2001). One of the most important features of HSCs is their ability to migrate and recirculate in bloodstream thereby supporting the pool of stem cells in different places of an organism in a quantity sufficient for the reparation processes. Stress and enhanced level of corticosteroids lead to a decrease in recirculating HSCs pool and the reduction of steroids quantity alone is accompanied by the increase in HSCs. (Bezin *et al.* 1975).

It was observed long ago that the stem cells were able to respond immediately to hypoxia (Murphy & Lord, 1973). Later investigations have shown that hypoxia inhibits the proliferation of $CD34^+$-cells and preserves the capacity for colony-forming-cell (CFC) generation in secondary cultures (pre-CFC) (Desplat V. *et al.* 2002), even independently in the presence of growth factors (Sun *et al.* 2000). Different oxygen pressures and time of hypoxic exposure could play important and varying roles in the development of stem cells.

HSCs and their proliferating progenitors are naturally distributed along pO_2 gradient, with the HSCs occupying the most hypoxic niches (Olive *et al.* 2002). Different oxygen pressures and time of hypoxic exposure could play important and divergent roles in cell development. Transcription factors and cytokines clearly influence these processes. Recent work has revealed an important link between the factors that are involved in regulating stem and progenitor cell behaviour and the hypoxia-inducible factors, which provide a molecular framework for the hypoxic control of differentiation and cell fate (Wang *et al.* 2007). Some authors demonstrated that hematopoiesis is improved ex vivo, when the cultured cells are exposed to 1-3% oxygen saturation (Cipolleschi, 1993; Danet, 2003). But there are few evidences about hypoxia effects on human HSCs (Ciulla *et al.* 2005; Mancuso *et al.* 2008). Very scant information about the intermittent hypoxia effects on stem cells that was obtained generally in cell culture models reveals that intermittent hypoxia at certain duration and intensity is a more potent trigger of transcription activation than constant hypoxia (Zhu *et al.* 2005; Zhao *et al.* 2008; Tang *et al.* 2009).

This study was designed to define the effects of intermittent hypoxia training (IHT) on human hematopoietic stem cells, various factors of natural resistance and main humoral and cellular components of adaptive immunity in peripheral blood.

Methods

This work was officially approved and authorized by the Ethics Committee for Human Experiments of the Bogomoletz Institute of Physiology.

Subjects

Ten healthy male volunteers (age 30.9±0.6 yr, weight 76.2±2.8 kg, height 176.9±1.8 cm) participated in the study and gave their consent. All subjects were sea level residents and non-smokers. Most subjects consumed 1-3 alcoholic drinks per week. The subjects were in good health with no evidence of cardiovascular, pulmonary or immunological disorders.

Experimental protocol

To ascertain that all changes registered in this investigation result exactly from IHT and not from casual fluctuations, we examined all subjects originally twice before IHT start (in two weeks and one day). Totally, four examinations have been performed: **I** - two weeks before IHT; **II** - the day before IHT; **III** - the day after two-week course of IHT; **IY** - one week after two-week course of IHT. Venous blood was drawn from the median antecubital vein during the early morning with the subjects fasting to provide hematology assessment.

IHT protocol

IHT was performed in the morning, from 10 a.m. till 12 p.m., 2 hours after light breakfast. With the subjects in a sitting position, normobaric, isocapnic hypoxia was induced for 5 min, 4 times per day for 14 days with a device Hypoxytron (Serebrovska, 1995; Serebrovska *et al.* 2009). The four periods of hypoxia were separated by three 5-min periods of room air inspiration. Initial inspired gas composition (F_i) was 20.9% O_2 and 79.1% N_2. During the first 1-1.5 min of rebreathing FiO_2 fell progressively by body utilization until FiO_2 of 10% was reached. Then O_2 was added gradually to the Hypoxotron to maintain FiO_2 at 10% during the remaining 3.5-4 min. The final SaO_2 was about 80-82 %. $P_{ET}CO_2$ was maintained at the initial pre-test pressure for each subject, typically 38-40 mm Hg, throughout the training period. Subjects easily endured the hypoxia periods without any side-effects. ECG and ventilation were continuously monitored during IHT, and arterial pressure was measured at 2 min intervals.

Blood analysis

The content of $CD45^+34^+$-cells was studied according to standard direct immunofluorescence techniques with flow-cytometer FACSAria® (Becton Dickinson & Co.) by ISHAGE protocol (Gratama *et al.* 2001). Rather than using CD45 only as a gating reagent for leukocytes, this protocol also takes advantage of the dim CD45 expression by the $CD34^+$, SSC^{low} HPC.

The content of IgM, IgG, IgA was determined by radial immunodiffusion using monospecific antisera against human IgM, IgG and IgA (Manchini *et al.* 1964). Circulating immune complexes (CIC) were determined by the measurement of CIC after precipitation of serum in 3.75% polyethylene glycol 6000 by precipitate density photometry (Haskova *et al.* 1977). The erythropoetin (EPO) content in serum was determined by two-site ELISA (Biomerica). Interferon-γ (IFN-γ) was analyzed using standard ELISA kits (BioSource International, Camarillo, CA) according to the manufacturer's instructions. The content of hemoglobin in the peripheral blood was studied by routine hemoglobin cyanide method. The content of erythrocytes, reticulocytes, platelets and leucocytes in peripheral blood was studied by routine microscopic analysis of blood smears. Phagocyte function of neutrophils was determined by its ability to phagocyte the Staphylococcus aureus-cells. Bactericide activity of neutrophils was determinate in the test of the nitroblue tetrazolium restoration (NBT-test).

Statistical analyses

All values are expressed as means ± SE. Treatment means were compared statistically with the ANOVA test. Correlations were identified by the least square method and expressed as linear regression slopes. Differences were considered significant for P< 0.05.

Results

All examined parameters did not differ significantly during 2 weeks preceding to IHT course. Therefore, all changes after IHT are to be considered as direct effects of IHT.

$CD45^+34^+$-cells in peripheral blood are found in very small quantity (vary from 0,8 to 5,8 cells/mkl, mean value = 1,6±0,2/mkl). As a result of IHT, the level of $CD45^+34^+$ showed the trend to increase during training period and decrease significantly (by 27%) in a 1 week after IHT (Fig. 1).

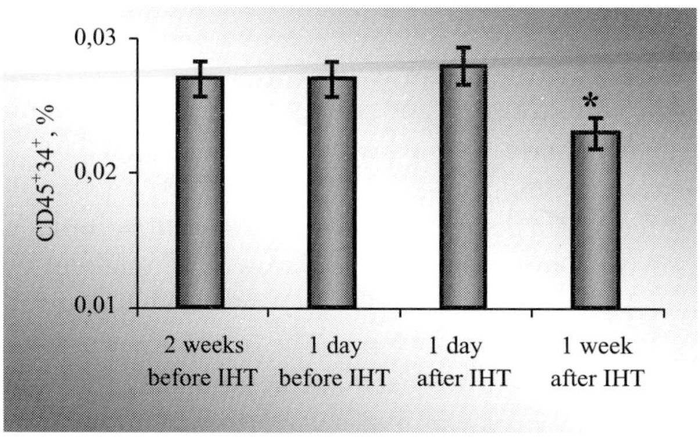

Fig. 1. Content of $CD45^+34^+$-cells in human peripheral blood under IHT
– p < 0.05 vs. 1 day before IHT

Red blood cell indices did not change under IHT according to average values. A trend to an increase in erythropoietin content (by 13 %) was observed. Substantial rise of platelets (by 31%) was registered which continued to increase in I week after IHT (Fig. 2).

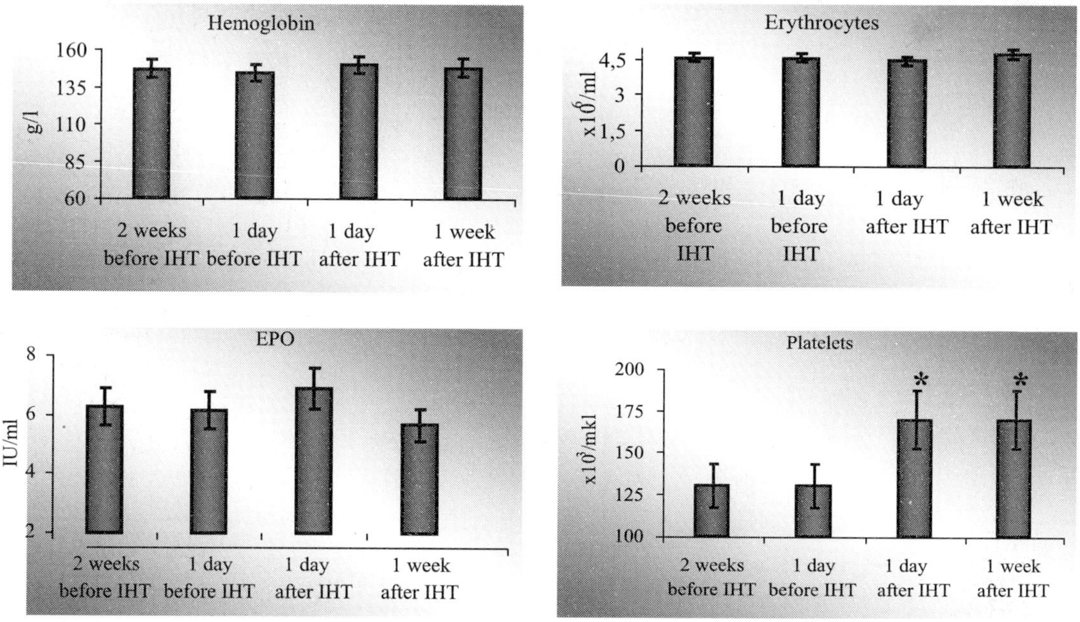

Fig. 2. Content of hemoglobin, erythrocytes, erythropoietin and platelets in human peripheral blood under IHT.

– $p < 0.05$ vs. 1 day before IHT

Simultaneously the activating processes in non-specific immune resistance occur: a) complement activity substantially increases (by 46% at the end of the training and by 51% in 1 week after IHT); b) phagocytic activity of neutriphils augments significantly, spontaneous bactericidal action decreases in a week after IHT and reserve bactericidity increases (by 47% just after IHT and by 62% in a week after IHT); c) level of IgG and IgA increases considerably (by 18% and 24%, respectively) in I week after IHT; d) content of IFN-γ increased by the end of training course and kept higher level at least during 1 week (Fig. 3, 4). Throughout the study we didn't observe changes in the concentration of IgM and CIC.

According to our data we found that IHT, as performed in this investigation, causes a decrease in circulating HSC count (enumerated as $CD45^{+}CD34^{+}$ bright cells) in healthy subjects in a week after training. Besides, IHT affected both cellular and humoral immunity. Changes in cellular immunity consist in an increase in phagocytic activity of neutriphils, decrease in their spontaneous bactericidity and an increase in reserve bactericidity. Changes in humoral components are manifested in an enhancement of complement activity and the level of immunoglobulins. Red blood cells indices, hemoglobin content as well as EPO concentration did not show significant shifts, but just the level of platelets increased. Most parameters sensitive to intermittent hypoxia manifest their major modifications in one week after the course of training.

Discussion

This investigation is a first attempt to elucidate how intermittent hypoxia influences circulating hemopoietic stem cells and some factors of non-specific immune resistance in healthy subjects.

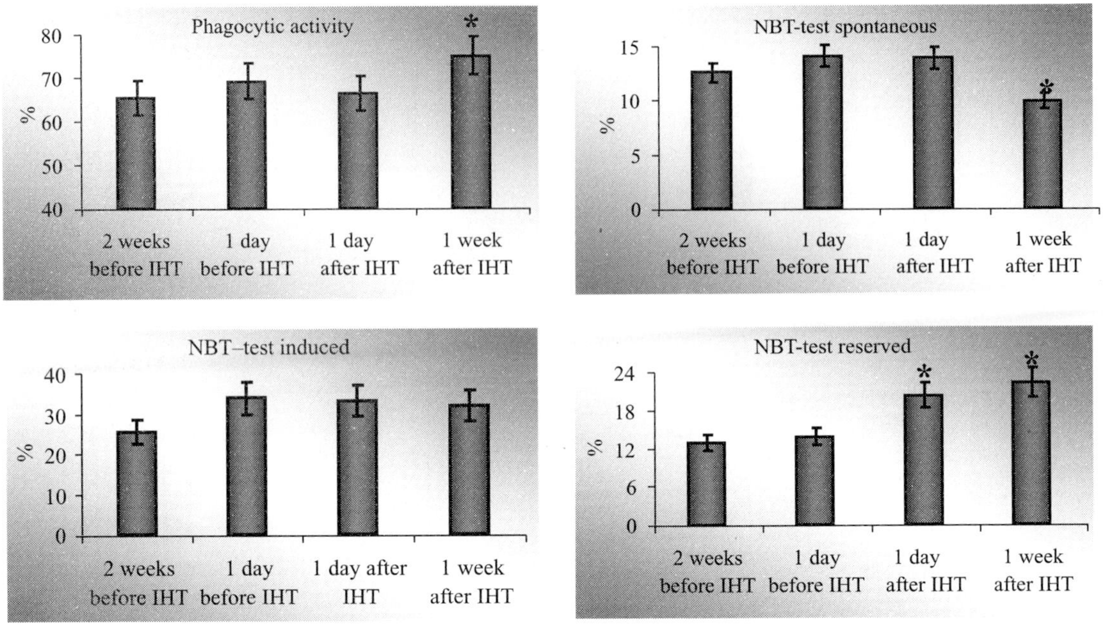

Fig. 3. Phagocytic and bactericidal activity of neutrophils in human peripheral blood under IHT
**** – p < 0.05 vs. before IHT***

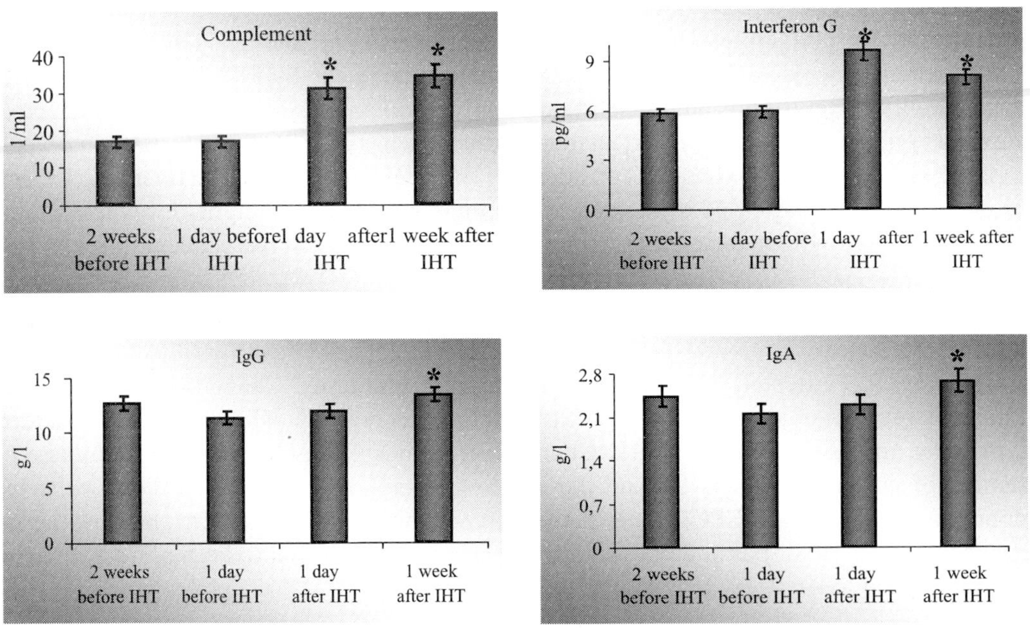

Fig. 4. The content of complement IFN-γ and immunoglobulins in human peripheral blood under IHT
**** – p < 0.05 vs. before IHT***

Hypoxia is an optimal environment for maintenance and functioning of various stem cell types (Weissman *et al.* 2001; Fan *et al.* 2007). It effectively maintains biological characteristics of $CD34^+$ cells by keeping lower intracellular ROS levels by regulating NADPH oxidase. It promotes the maintenance of stem cells in a quiescent state with a necessary rate of self-renewal. There are certain regulatory mechanisms of the maintenance of hypoxic conditions in stem cell microenvironment named niches. This microenvironment influences essentially the stem cell proliferation and differentiation and protects the cells against toxic and mutagenic effects of free oxygen-derived radicals. The interested reader is referred to more extensive recent scientific reviews (Eliasson & Jönsson, 2009; Ivanovic, 2009; Nikolsky & Serebrovskaya, 2009; Panchision, 2009).

An additional exogenous hypoxic impact could lead to the activation of stem cell system. Theiss *et al.* (2008) presented the study showing that 1-week exposure of healthy volunteers to moderate altitude (1700 m) with physical activity leads to increased level of circulating progenitor cells. Ciulla *et al.* (2005) reported their observations of one healthy subject after a trek in the Himalayas that also showed an increase in the number of mature endothelial progenitor cells and in the endothelial colony-forming unit capacity. Both of these effects were completely reverted at sea level, 45 days after the subject's trek. But other investigation (Mancuso *et al.* 2008) provided at 3000m altitude, has shown that after 12 days of sojourn at this altitude, circulating HSC, endothelial cells and progenitors count significantly decreased in mountain trekkers.

On the contrary, we could not find any published evidences on the effects of intermittent hypoxia on human circulating HSC. Most information was obtained in-situ or from cell culture models. For example, in a study of adult rats exposed to hypobaric hypoxia (corresponding to altitudes of 3,000 m and 5,000 m) for 4 hours per day during 2 weeks, the cells labelled with 5-bromo-2-deoxyuridine-5-monophosphate (BrdU) grew up in subventricular zone (SVZ) and dentate gyrus. The number of BrdU-labeled cells in the SVZ returned to normal level 4 weeks following the end of intermittent hypoxia. However, the number of BrdU-labelled cells increased two-fold in dentate gyrus in 4 weeks after intermittent hypoxia exposure (Zhu *et al.* 2005). Based on these findings, the authors concluded that intermittent hypoxia facilitates neural stem cell (NSC) proliferation in situ, and that newly divided cells in SVZ and dentate gyrus respond to hypoxia differently. In other words, proliferation of NSC, possibly, facilitates adaptation changes following the exposure to intermittent hypoxia.

Other studies on cell culture models showed that, with certain duration and intensity, intermittent hypoxia is a more powerful factor for triggering transcriptional activation than continuous hypoxia. Nanduri & Prabhakar (2007) summarized that intermittent hypoxia activates HIF-1, immediate early gene c-fos, activator protein-1, nuclear factor kappa-B, and cAMP-response-element-binding protein. It also induces the expression of proteins associated with neuron survival and apoptosis as well as post-translation modification of proteins, resulting in enhanced biological activity. The comparison between continuous and intermittent hypoxia showed pronounced differences in both the kinetics of protein kinase activation and the types of activated protein kinase.

More recent paper (Panchision, 2009) demonstrates that at lower oxygen tensions, HIF-1a facilitates signal transduction pathways that promote self-renewal and inhibits pathways that promote NSC differentiation or apoptosis.

Processes of HSC migration, both homing into hematopoietic organs and sequestration in the circulation, are regulated by chemokines and cytokines (Potocnik *et al.* 2000; Muench *et al.* 2000; Krstiæ *et al.* 2009). In this investigation we registered an enhancement of some immunoglobulins level without significant changes in erythropoietin content. Other papers also reported that IHT did not accelerate erythropoiesis despite the increase in serum EPO (Gore, 2006; Abellan *et al.* 2007). In another investigation (Villa, 2005) similar to our IHT regimen, revealed rather similar results: EPO concentration had a tendency to an increase with no effect on reticulocytes or erythrocyte count.

Number of circulating HSC could be increased or reduced by several methods, for example by using corticosteroids (Petrov *et al.* 1981; Berardi *et al.* 1990). It is important to note that there is a significant inter-individual variability in humans and inter-strain variability in mice in the ability to mobilize hematopoietic stem and progenitor cells, suggesting that there is genetic regulation of mobilization (Geiger *et al.* 2004). In our study we observed the tendency to an increase in circulating HSC just after 2-weeks of IHT and substantial decrease in a week after IHT.

Our investigation has shown that humoral and cellular factors of non-specific resistance are sensitive to short periods of hypoxic exposures. It leads to an enhancement of natural protection for an organism. A decrease in HSC number in peripheral blood is probably associated with the change of their migration capacity. However, it remains unclear whether there is an inhibition of HSC migration into circulation or an activation of HSC escape from circulation. In both cases tissues could accumulate more HSC which in turn could enhance hematopoiesis and general regenerative potential. It assures the HSC transdifferentiation capacity, plasticity and intercellular cooperation.

Conclusion

IHT, as shown in this investigation, causes a decrease in the content of hematopoetic stem cells in human peripheral blood in a week after IHT. IHT affects both cellular and humoral immunity. Changes in cellular immunity consist of an increase in phagocytic activity of neutriphils, decrease in their spontaneous bactericidity and an increase in reserve bactericidity. It indicates the growing cell phagocytic activity and higher state of readiness to the fight against an infection. Changes in humoral components confirm this conclusion, especially an increase in complement activity and the level of immunoglobulins. It leads to an enhancement of natural protection of an organism. Most parameters sensitive to intermittent hypoxia manifest their major modifications in one week after the course of training.

Acknowledgements

This work was supported by STCU grant #4299. The authors are grateful to Prof. O.V.Korkushko and Dr. V.B.Shatilo (State Institute of Gerontology, Kiev) for helpful suggestions and assistance. Many thanks to the staff of the Institute of Genetic and Regenerative Medicine, Kiev, (particularly Dr. Valentyna Nikolska, Dr. Lubov Taranukha, Mrs. Svitlana Halytska) for their high-skilled work.

References

1. Abellan R, Ventura R, Remacha AF, Rodríguez FA, Pascual JA, Segura J. Intermittent hypoxia exposure in a hypobaric chamber and erythropoietin abuse interpretation. *J Sports Sci.* 2007;25(11):1241-1250.

2. Berardi A, Geraci L, Quaglietta AM, Di Bartolomeo G, Dragani A. Pharmacological mobilization of haemopoietic progenitor cells in human peripheral blood. *Haematologica.* 1990;75 Suppl 1:15-17.
3. Bezin GI, Khaitov RM, Moroz BB, Petrov RV, Romashko OO. The factors controlling stem cell recirculation. II. ACTH-induced inhibition of migration of hemopoietic stem cells. *Blood.* 1975;46(1):79-84.
4. Bossolasco P, Corti S, Strazzer S, Borsotti C, Del Bo R, Fortunato F, Salani S, Quirici N, Bertolini F, Gobbi A, Deliliers GL, Pietro Comi G, Soligo D. Skeletal muscle differentiation potential of human adult bone marrow cells. *Exp Cell Res.* 2004 ,15;295(1):66-78.
5. Cipolleschi MG, Dello Sbarba P, Olivotto M. The role of hypoxia in the maintenance of hematopoietic stem cells. *Blood.* 1993;82(7):2031-2037.
6. Ciulla MM, Giorgetti A, Lazzari L, Cortiana M, Silvestris I, Annoni G, De Asmundis C, Fiore AV, Montelatici E, Paliotti R, Magrini F, Rebulla P, Cortelezzi A. High-altitude trekking in the Himalayas increases the activity of circulating endothelial cells. *Am J Hematol.* 2005;79(1):76-78.
7. Danet GH, Pan Y, Luongo JL, Bonnet DA, Simon MC. Expansion of human SCID-repopulating cells under hypoxic conditions. *J Clin Invest.* 2003 ;112(1):126-135.
8. Desplat V, Faucher JL, Mahon FX, Dello Sbarba P, Praloran V, Ivanovic Z. Hypoxia modifies proliferation and differentiation of CD34($^{+}$) CML cells. *Stem Cells.* 2002;20(4):347-354.
9. Eliasson P, Jönsson JI. The hematopoietic stem cell niche: Low in oxygen but a nice place to be. *J Cell Physiol.* 2009 Sep 1.
10. Fan J, Cai H, Tan WS. Role of the plasma membrane ROS-generating NADPH oxidase in CD34+ progenitor cells preservation by hypoxia. *J Biotechnol.* 2007;130(4):455-462.
11. Geiger H, Szilvassy SJ, Ragland P, Van Zant G. Genetic analysis of progenitor cell mobilization by granulocyte colony-stimulating factor: verification and mechanisms for loci on murine chromosomes 2 and 11. *Exp Hematol.* 2004; 32(1):60-67.
12. Gore CJ, Rodríguez FA, Truijens MJ, Townsend NE, Stray-Gundersen J, Levine BD. Increased serum erythropoietin but not red cell production after 4 wk of intermittent hypobaric hypoxia (4,000-5,500 m). *J Appl Physiol.* 2006;101(5):1386-1393.
13. Gratama JW, Sutherland DR, Keeney M, Papa S. Flow cytometric enumeration and immunophenotyping of hemetopoetic stem and progenitor cells. *J of Biol Regul and Homeostatic Agents.* 2001; 15(1): 14-22.
14. Hasková V, Kaslík J, Matl I, Matejcková M. New technique of circulating immunocomplex estimation in human sera (author's transl). *Cas Lek Cesk.* 1977;116(14):436-437 [In Czech].
15. Ivanovic Z. Hypoxia or in situ normoxia: The stem cell paradigm. *J Cell Physiol.* 2009;219(2):271-275.
16. Jackson KA, Majka SM, Wang H, Pocius J, Hartley CJ, Majesky MW, Entman ML, Michael LH, Hirschi KK, Goodell MA. Regeneration of ischemic cardiac muscle and vascular endothelium by adult stem cells. *J Clin Invest.* 2001;107(11):1395-1402.

17. Krstiæ A, Vlaski M, Hammoud M, Chevaleyre J, Duchez P, Jovèiæ G, Bugarski D, Milenkoviæ P, Bourin P, Boiron JM, Praloran V, Ivanoviæ Z. Low O 2 concentrations enhance the positive effect of IL-17 on the maintenance of erythroid progenitors during co-culture of $CD34^+$ and mesenchymal stem cells. *Eur Cytokine Netw.* 2009;20(1):10-16.
18. Lagasse E, Connors H, Al-Dhalimy M, Reitsma M, Dohse M, Osborne L, Wang X, Finegold M, Weissman IL, Grompe M. Purified hematopoietic stem cells can differentiate into hepatocytes in vivo. *Nat Med.* 2000; 6(11):1229-1234.
19. Manchini G, Vaerman J-P, Carbonera AO, Heremans JF. A simple radial-diffusion method for the immunological quantitation of protein. In Peeters N. (Ed.) Proceedings of the biological fluids, Amsterdam, L., N.Y.: Elsevier, 1964:370-379.
20. Mancuso P, Peccatori F, Rocca A, Calleri A, Antoniotti P, Rabascio C, Saronni L, Zorzino L, Sandri MT, Zubani A, Bertolini F. Circulating endothelial cell number and viability are reduced by exposure to high altitude. *Endothelium.* 2008; 15(1):53-58.
21. Mezey E, Chandross KJ. Bone marrow: a possible alternative source of cells in the adult nervous system. *Eur J Pharmacol.* 2000; 405(1-3):297-302.
22. Muench MO, Humeau L, Paek B, Ohkubo T, Lanier LL, Albanese CT, Bárcena A. Differential effects of interleukin-3, interleukin-7, interleukin 15, and granulocyte-macrophage colony-stimulating factor in the generation of natural killer and B cells from primitive human fetal liver progenitors. *Exp Hematol.* 2000; 28(8):961-973.
23. Murphy MJ Jr, Lord BI. Hematopoietic Stem Cell Regulation. I. Acute Effects of Hypoxic-Hypoxia on CFU Kinetics. *Blood.* 1973; 42 (1):81-87.
24. Nanduri J, Prabhakar NR. Cellular mechanisms associated with intermittent hypoxia. *Essays Biochem.* 2007; 43:91-104.
25. Nikolsky IS, Serebrovskaya TV. Hypoxia and Stem Cells. In: Intermittent Hypoxia: From *Molecular Mechanisms to Clinical Applications*, editors: Lei Xi & Tatiana V. Serebrovskaya, Nova Science Publishers, 2009, Chapter 22:469-487.
26. Olive PL, Luo CM, Banath JP. Local hypoxia is produced at sites of intratumour injection. *Br J Cancer.* 2002; 86:429-435.
27. Panchision DM . The role of oxygen in regulating neural stem cells in development and disease. *J Cell Physiol.* 2009; 220(3):562-568.
28. Petrov RV, Kchaitov RM, Manko VM, Michailova AA. Control and regulation of the immune response. – Leningrad: Medicina, 1981. – 311p.
29. Potocnik AJ, Brakebusch C, Fassler R. Fetal and adult hematopoietic stem cells require (beta) I integrin function for colonizing fetal liver, spleen and bone marrow. *Immunology.* 2000;12(6):653-663.
30. Serebrovska TV. Method for nonspecific body resistance increasing by means of intermittent hypoxic influences "Hypotron". Author's Certificate PA #32, 06 Dec 1995.
31. Serebrovska TV, Lopata VA, Roy VV, Roitman EM. Device for breathing with hypoxic mixtures "Hypoxytron". Patent ¹ 44179, ÌÏÊ À61Ì 16/00; Ukraine, 25.09.2009.
32. Tang YL, Zhu W, Cheng M, Chen L, Zhang J, Sun T, Kishore R, Phillips MI, Losordo DW, Qin G. Hypoxic preconditioning enhances the benefit of cardiac progenitor cell therapy for

treatment of myocardial infarction by inducing CXCR4 expression. *Circ Res.* 2009; 104(10):1209-1216.

33. Theiss HD, Adam M, Greie S, Schobersberger W, Humpeler E, Franz WM. Increased levels of circulating progenitor cells after 1-week sojourn at moderate altitude (Austrian Moderate Altitude Study II, AMAS II). *Respir Physiol Neurobiol.* 2008; 160(2):232-238.
34. Villa JG, Lucía A, Marroyo JA, Avila C, Jiménez F, Garcia-López J, Earnest CP, Córdova A. Does intermittent hypoxia increase erythropoiesis in professional cyclists during a 3-week race? *Can J Appl Physiol.* 2005; 30(1):61-73.
35. Wang JS, Lin HY, Cheng ML, Wong MK. Chronic intermittent hypoxia modulates eosinophil- and neutrophil-platelet aggregation and inflammatory cytokine secretion caused by strenuous exercise in men. *J Appl Physiol.* 2007; 103(1):305-314.
36. Weissman IL, Anderson DJ, Gage F. Stem and progenitor cells: origins, phenotypes, lineage commitments, and transdifferentiations. *Annu Rev Cell Dev Biol.* 2001; 17:387-403.
37. Zhao T, Zhang CP, Liu ZH, Wu LY, Huang X, Wu HT, Xiong L, Wang X, Wang XM, Zhu LL, Fan M. Hypoxia-driven proliferation of embryonic neural stem/progenitor cells—role of hypoxia-inducible transcription factor-1alpha. *FEBS J.* 2008;275(8):1824-1834.
38. Zhu LL, Zhao T, Li HS, Zhao H, Wu LY, Ding AS, Fan WH, Fan M. Neurogenesis in the adult rat brain after intermittent hypoxia. *Brain Res.* 2005; 1055(1-2):1-6.

Adaptation Biology and Medicine (Volume 6: Cell Adaptations and Challenges)
Editors: P. Wang, C.-H. Kuo, N. Takeda and P.K. Singal
Copyright © 2011 Narosa Publishing House Pvt. Ltd., New Delhi, India

14. Non-Invasive Measurement of Skeletal Muscle Microvascular Blood Flow using Photoplethysmography: Experimental Studies on the Human Leg

Qiuxia Zhang*

Department of Orthopaedics, University of Gothenburg, Sahlgrenska University Hospital, Gothenburg, Sweden

Abstract

Muscle function depends on blood flow at the microvascular level since all exchange of oxygen, nutrients, fluids and waste products occurs at this level. Blood flow changes in skeletal muscle tissue at the microvascular level are of great interest in both health and disease. This article reviews a new non-invasive photoplethysmography (PPG) method for continuous assessment of relative changes in local skeletal muscle microvascular blood flow (MBF) in the human leg. Experimental conditions expected to impede or increase local muscle blood flow include arterial occlusion, muscle contractions, and increased intramuscular pressure. PPG technique enables non-invasive, continuous assessment of local muscle altered MBF in response to arterial occlusion, muscle contractions, and increased intramuscular pressure. The PPG may be a promising non-invasive diagnostic tool for direct measurement of local skeletal muscle MBF under various physiological and pathological conditions.

Keywords: photoplethysmography, skeletal muscle blood flow, perfusion pressure, intramuscular pressure, venous obstruction, muscle exercise, non-invasive, limb elevation, hyperaemia.

Introduction

Measurements of local skeletal muscle microvascular blood flow (MBF) are of particular importance, because they provide valuable information about the function and regulation of the circulatory system. Different methods measure different levels in the circulatory system, such as limb blood flow, whole muscle blood flow, and local microcirculation. The magnitude of muscle blood flow has been estimated from measurements of total limb flow by using venous occlusion plethysmography. The blood flow measured using this technique represents total flow to various tissues (McCully &

---

*Address for Correspondence: Qiuxia Zhang, PhD Department of Orthopaedics Sahlgrenska University Hospital Per Dubbsgatan 14, 6th floor 413 46, Göteborg, Sweden
Email: qiuxia.zhang@orthop.gu.se +46 31 3426616 office +46 31 416924 fax

Posner 1995), so the technique cannot measure local skeletal muscle blood flow directly. Blood flow over the limb can also be determined using the ultrasound Doppler, which measures arterial inflow to a human limb during intermittent static contraction. However, to avoid insonation failures, the exercise mode must be such that the limb and the artery are in a fixed position. The technique is expensive, technically demanding and has a significant learning curve (Rådegran 1999). The local ^{133}Xenon washout technique introduced by Lassen and co-workers has been the most often used for determining local blood flow in skeletal muscle (Lassen *et al.* 1964). Recent developments in laser Doppler flowmetry (LDF) now allow for the assessment of blood flow in deeper tissues by means of a single-fibre probe. With this technique, the laser light is continuously guided via an optical fibre inserted through the subcutaneous tissue into the muscle (Salerud & Öberg 1987). This technique has also been applied to human muscle for measuring continuous changes in local muscle blood flow in relation to current muscle work (Kvernebo *et al.* 1990; Hoffmann *et al.* 1995; Jensen *et al.* 1995; Larsson *et al.* 1995). However, the local trauma caused by using an invasive method can complicate the flow measurements. Until recently, direct measurements of local skeletal muscle blood flow in humans have been limited, owing to the absence of suitable methods (Jensen *et al.* 1995; Hughson & Tschakovsky 1999).

This article reviews a new non-invasive photoplethysmography (PPG) method for continuous assessment of relative changes in local skeletal muscle microvascular blood flow in the human leg. Experimental conditions expected to impede or increase local muscle blood flow include arterial occlusion, muscle contractions, and increased intramuscular pressure.

Photoplethysmography

PPG is a non-invasive optical technique that has been used mainly for monitoring blood perfusion in skin (Challoner 1979; Kamal *et al.* 1989; Lindberg & Öberg 1991). Hertzman and Spealman illustrated changes in blood volume produced by exercise, cold and the Valsalva manoeuvre in 1937. They contributed greatly to the understanding of the PPG technique in their earlier work (Hertzman & Spealman 1937). Hertzman called his apparatus a "photoelectric plethysmography" in 1938 (Hertzman 1938). Since then numerous photoplethysmographic measuring devices have been developed for various applications.

The principle on which PPG is based is simple, although the underlying detailed optical mechanisms remain unknown (Kamal *et al.* 1989). PPG light from a light emitting diode (LED) is directed toward the skin and absorbed and scattered in the tissue. A small amount of this light is received by a photodetector placed adjacent to the LED. Altered blood flow and blood volume in the underlying tissue are related to variations in the photodetector signal (Challoner 1979; Kamal *et al.* 1989). Light penetration into the tissue increases primarily with increasing wavelength (Anderson & Parrish 1981). The light penetration depth also depends on the optical geometry of the PPG probe. An increasing distance between the light source and the photodetector enables to image vascular bed at different depths in human tissue (Fridolin *et al.* 2000; Fridolin & Lindberg 2000). The PPG signal combines AC and DC components. The PPG_{AC} component correlates with heart rate and depends on pulsatile blood flow changes due to varying lumen diameter of the vessel (Graaf 1993; Lindberg and Oberg 1993). An increase in amplitude of the PPG_{AC} signal indicates increased microvascular blood flow and the PPG_{DC} component of the signal reflects altered total blood volume within the underlying tissue (Challoner 1979).

A modified PPG instrument with a custom-designed PPG probe (Department of Biomedical Engineering, Linköping University, Sweden) has been developed for the purpose of measuring relative changes in skeletal muscle blood perfusion (Sandberg *et al.* 2005a; Zhang *et al.* 2001a, b; 2004). The PPG probe consists of three photodetectors and six light sources (light emitting diodes LEDs). Four LEDs emitted light of a wavelength of 560 nm (green light) and two LEDs emitted light in the near-infrared region of 880 nm. All components are placed in a special pattern and embedded in dark-coloured silicone. The centre to centre distance between the LEDs and the photodetectors is 3.5 mm and 20 mm for wavelengths 560 and 880 nm, respectively. Placing the green LED 3.5 mm from the photodetector allows for superficial penetration of light to monitor microvascular flow of the skin. Whereas, a long distance (20 mm) between the near-infrared LED and a photodetector enables deep penetration of light to measure skeletal muscle microvascular flow. In a previous study (Sandberg *et al.* 2005a), the depth of the near-infrared light penetration by the PPG instrument was evaluated by inserting an optic fiber into the anterior tibial muscle of the human leg and connecting to an Optical Power Meter for recording the radiant power of radiation in the muscle, and the radiant power measured in the muscle tissue indicates that the near-infrared light penetrates down to at least a vascular depth of 13.0 mm from the skin surface. The distance between the skin surface and muscle fascia averaged 5.4 mm (range 2.2-13.8) at the middle of the anterior tibial muscle in 43 subjects as measured by ultrasound (Sandberg *et al.* 2005a).

Limitations of the PPG method, such as possible interference of high adipose tissue thickness and influence of the motion artefact must be taken into consideration. Light penetration depends strictly upon the source-detector distance and the subcutaneous adipose tissue thickness. Since light is highly scattered in tissue, a detector positioned on the surface of the skin can measure reflections from a range of depths. Those reflections are variously absorbed, depending on whether the light encounters weakly or highly absorbing tissue. A thick skin or adipose tissue prevents some light passing through the muscle tissue. In the absence of any blood volume changes, the signal level will be determined by the muscle type, skin type, probe positioning, local temperature, surrounding light and of course the geometry and sensitivity of the sensor itself. In addition, the amplitude of the PPG signal cannot be calibrated. Therefore, only relative changes in blood flow in an individual can be measured. Furthermore, one of the difficulties with the clinical use of a PPG device remains that of motion artefacts that have a deteriorating effect of the measurements and restrict study design. The method cannot measure MBF during dynamic exercise.

Measurements of Skeletal Muscle Microvascular Blood Flow by PPG

Post-occlusion and Post-exercise Hyperaemia

Post-occlusive reactive hyperaemia is the well-known physiological reaction following reperfusion after arterial occlusion (Bayliss 1902). The longer the period of occlusion, the greater the increase in blood flow above pre-occlusion levels. In addition, blood flow increases many-fold in the active muscles after both intense static and dynamic exercise (Laughlin 1987). The increased blood flow is maintained throughout the period of increased metabolic activity. In a previous study, PPG monitored the altered MBF in the human anterior tibial muscle following arterial occlusion and after muscle contractions (Zhang *et al.* 2001a). Single-fibre LDF was used as a reference. After arterial occlusion lasting three minutes, reactive hyperaemia was observed in muscle tissue using both PPG and single-fibre LDF. The post-occlusive reactive hyperaemia was 150% by PPG and 182% by LDF (compared with baseline=100%). The post-exercise hyperaemia was 150% by PPG and 169% by LDF after 1 min of maximal static contraction (compared with baseline=100%). After 1 min of maximal concentric contractions, the post-exercise hyperaemia was 158% by PPG and 170% by LDF (compared with baseline=100%). These observations indicate that post-occlusion reactive hyperaemia and post-exercise hyperaemia can be measured in human anterior muscle by using a non-invasive PPG technique.

MBF Responses to Limb Elevation and Increased Intramuscular Pressure IMP

The ability to measure how abnormally elevated IMP affects local muscle blood flow is of clinical interest. To investigate the relationship between blood flow and increased IMP, both parameters should be measured in the same muscle. PPG may offer the possibility of measuring changes in local muscle blood flow in relation to increased tissue pressure in the same muscle.

In 1975, Ashton studied the effects of external compression by inflatable plastic splints on limb blood flow in human subjects (Ashton 1975). She demonstrated a significant reduction in blood flow by the application of pressure at 40 mmHg and elevation of the limb 9 to 35 cm above heart level. By applying venous stasis to a casted leg, abnormally increased pressure in the anterior compartment of the human leg was induced (Styf *et al.* 1998; Wiger *et al.* 1998; Wiger *et al.* 2000). Furthermore, by elevating the limb, perfusion pressure decreased to about 25 mmHg and neuromuscular dysfunction evolved. The same experimental human model was used to improve the understanding of the effects of increased IMP on muscle blood flow and neuromuscular function (Zhang *et al.* 2001b). MBF was measured in the anterior muscle of the human leg by PPG and found that MBF decreased significantly by 50% in the elevated leg and by 42% in the leg at heart level when the IMP was maintained at 40 mmHg. Perfusion pressure of the anterior compartment decreased from 42 to 17 mmHg (60%) in the elevated leg and from 65 to 43 mmHg (34%) in the leg kept at heart level. These findings are in agreement with results of others, which show that IMP values between 30 and 50 mmHg and perfusion pressure below 30 mmHg may impair normal function (Ashton 1975; Hargens *et al.* 1979; Heppenstall *et al.* 1988; Matsen *et al.* 1979; Styf & Wiger 1998; Wiger & Styf 1998; Wiger *et al.* 2000).

MBF at Rest after Exercise Under Normal and Elevated IMP Conditions

A non-invasive tool for monitoring muscle blood flow before and after exercise is helpful in studying the pathophysiology of exercise-induced pain in the leg and impaired skeletal muscle perfusion, such as chronic compartment syndrome. It has been shown that the chronic compartment syndrome is caused by abnormally elevated IMP during exercise and at rest after exercise (Styf 1989).

In a previous study (Zhang *et al.* 2004), PPG was used to measure the relative changes in the anterior tibial muscle MBF in response to exercise when IMP was elevated (the test leg) or when IMP was normal (the control leg). The elevated IMP was induced by applying venous obstruction of 40 mmHg or 65 mmHg (thigh tourniquet) of a casted leg in healthy subjects. They find that MBF is significantly lower in the leg exposed to elevated IMP than in the control leg exposed to normal IMP at rest after exercise. This means that the increased IMP causes a decrease in MBF at rest after exercise. The lower increase in muscle blood flow at intramuscular pressures of 39.4 mmHg and 58.3 mmHg at rest after exercise compared to conditions of normal intramuscular pressure may be explained by the arteriovenous pressure difference. In addition, increased IMP may compress the vessels at the capillary level, leading to enhanced hyperaemia after exercise. Styf and Körner reported that muscle blood flow decreased significantly during exercise when muscle relaxation pressure was increased to 42.6 mmHg, in patients who had chronic compartment syndrome (Styf *et al.* 1987). Delayed post-exercise hyperaemia was found, when IMP was 74.1 mmHg, in chronic compartment syndrome patients one minute after exercise (Abraham *et al.* 1998). Mohler and co-workers measured tissue oxygenation of exercised skeletal muscle with near-infrared spectroscopy in patients who had chronic compartment syndrome of the leg (Mohler *et al.* 1997). They reported greater relative deoxgenation during exercise as well as delayed reoxygenation after exercise. These findings are probably a consequence of impaired muscle blood flow due to elevated IMP, and are consistent with the results obtained from the PPG measurements when abnormally elevated intramuscular pressure impedes muscle blood flow at rest after exercise. PPG seems to be a suitable method to measure changes in local muscle MBF caused by abnormally elevated IMP at rest after exercise.

Other applications of PPG to study skeletal muscle MBF also have been reported. For example, PPG studies of different patterns of blood flow response in the trapezius muscle following needle stimulation (acupuncture) between healthy subjects and patients with fibromyalgia and work-related trapezius myalgia were reported (Sandberg *et al.* 2003, 2004, 2005b). In another study, Sandberg *et al.* (2007) used PPG to examine the effects of transcutaneous electrical nerve stimulation on local blood flow in the trapezius muscle and overlying skin. These observations indicate the potentiality of the PPG method for non-invasive measurement of local muscle perfusion.

Summary

In summary, PPG, with the aid of a custom-designed probe, enables non-invasive, continuous assessment of relative changes in local muscle MBF in response to, for example, arterial occlusion, muscle contractions, and increased intramuscular pressure. The PPG technique is a promising non-invasive diagnostic tool for direct measurement of local muscle MBF under clinical conditions with and without abnormally elevated IMP.

References

1. Abraham P, Leftheriotis G and Saumet JL. Laser Doppler flowmetry in the diagnosis of chronic compartment syndrome. *J Bone Joint Surg Br*. 1998;80: 365-369.
2. Anderson RR and Parrish JA. The optics of human skin. *J Invest Dermatol*. 1981;77: 13-19.
3. Ashton H. The effect of increased tissue pressure on blood flow. *Clin Orthop*. 1975;113: 15-26.
4. Bayliss WM. On the local reactions of the arterial wall to changes of internal pressure. *J Physiol*. 1902;28: 220-231.
5. Challoner AVJ. Photoelectric plethysmography for estimating cutanneous blood flow. In: Rolfe P. eds., *Non-invasive Physiological Measurements*. Academic, London, 1979:Vol. 1, 127-151.
6. Fridolin I, Hansson K and Lindberg LG. Optical non-invasive technique for vessel imaging: II. A simplified photon diffusion analysis. *Phys Med Biol*. 2000;45: 3779-3792.

7. Fridolin I and Lindberg LG. Optical non-invasive technique for vessel imaging: I. Experimental results. *Phys Med Biol*. 2000;45: 3765-3778.
8. Graaf R, Dassel A, Koeolink M, de Mul F, Aarnoudse J and Zilstra W. Optical properties of human dermis in votro and in vivo. *Appl Opt*. 1993;32: 435-447.
9. Hargens AR, Romine JS, Sipe JC, Evans KL, Mubarak SJ and Akeson WH. Peripheral nerve-conduction block by high muscle-compartment pressure. *J Bone Joint Surg Am*. 1979;61: 192-200.
10. Heppenstall RB, Sapega AA, Scott R, Shenton D, Park YS, Maris J and Chance B. The compartment syndrome. An experimental and clinical study of muscular energy metabolism using phosphorus nuclear magnetic resonance spectroscopy. *Clin Orthop*. 1988;226: 138-155.
11. Hertzman AB. The blood supply of various skin areas as estimated by the photoelectric plethysmograph. *Am J Physiol*. 1938;124: 328.
12. Hertzman AB and Spealman CR. Observations on the finger volume pulse recorded photoelectrically. *Am J Physiol*. 1937;119: 334.
13. Hoffmann U, Uckay I, Fischer M, Wen S, Franzeck UK and Bollinger A. Simultaneous assessment of muscle and skin blood fluxes with the laser-Doppler technique. *Int J Microcirc Clin Exp*. 1995;15: 53-59.
14. Hughson RL and Tschakovsky ME. Cardiovascular dynamics at the onset of exercise. *Med Sci Sports Exerc*. 1999;31: 1005-1010.
15. Jensen BR, Sjogaard G, Bornmyr S, Arborelius M and Jorgensen K. Intramuscular laser-Doppler flowmetry in the supraspinatus muscle during isometric contractions. *Eur J Appl Physiol Occup Physiol*. 1995;71: 373-378.
16. Kamal AA, Harness JB, Irving G and Mearns AJ. Skin photoplethysmography—a review. *Comput Methods Programs Biomed*. 1989;28: 257-269.
17. Kvernebo K, Staxrud LE and Salerud EG. Assessment of human muscle blood perfusion with single-fiber laser Doppler flowmetry. *Microvasc Res*. 1990;39: 376-385.
18. Larsson SE, Cai H, Zhang Q, Larsson R and Öberg PA. Measurement by laser-Doppler flowmetry of microcirculation in lower leg muscle at different blood fluxes in relation to electromyographically determined contraction and accumulated fatigue. *Eur J Appl Physiol Occup Physiol*. 1995;70: 288-293.
19. Lassen NA, Lindbjerb J and Munch O. Measurement of blood flow skeletal muscle by intramuscular injection of xenon-133. *Lancet*. 1964;28: 686-689.
20. Laughlin MH. Skeletal muscle blood flow capacity: role of muscle pump in exercise hyperemia. *Am J Physiol*. 1987;253: H993-1004.
21. Lindberg LG and Öberg PA. Photoplethysmography. Part 2. Influence of light source wavelength. *Med Biol Eng Comput*. 1991;29: 48-54.
22. Lindberg LG and Oberg PA. Optical properties of blood in motion. *Opt Eng*. 1993;32: 253-258.
23. Matsen FA 3rd, Krugmire RB and King RV. Nicolas Andry Award. Increased tissue pressure and its effects on muscle oxygenation in level and elevated human limbs. *Clin Orthop*. 1979;144: 311-320.

24. McCully KK and Posner J. D. The application of blood flow measurements to the study of aging muscle. *J Gerontol A Biol Sci Med Sci.* 1995;50: 130-136.
25. Mohler LR, Styf JR, Pedowitz RA, Hargens AR and Gershuni DH. Intramuscular deoxygenation during exercise in patients who have chronic anterior compartment syndrome of the leg. *J Bone Joint Surg Am.* 1997;79: 844-849.
26. Rådegran G. Limb and skeletal muscle blood flow measurements at rest and during exercise in human subjects. *Proc Nutr Soc.* 1999;58: 887-898.
27. Salerud, EG and Öberg, PA. Single-fiber laser Doppler flowmetry. A method for deep tissue perfusion measurements. *Med Biol Eng Comput.* 1987;25: 329-334.
28. Sandberg ML, Sandberg MK andDahl J. Blood flow changes in the trapezius muscle and overlying skin following transcutaneous electrical nerve stimulation. *Phys Ther.* 2007; 87(8): 1047-1055.
29. Sandberg M, Zhang Q, Styf J, Gerdle B and Lindberg LG. Non-invasive monitoring of muscle blood perfusion by photoplethysmography: evaluation of a new application. *Acta Physiol Scand.* 2005a;183(4): 335-343.
30. Sandberg M, Larsson B, Lindberg LG and Gerdle B. Different patterns of blood flow response in the trapezius muscle following needle stimulation (acupuncture) between healthy subjects and patients with fibromyalgia and work-related trapezius myalgia. *Eur J Pain.* 2005b; 9(5): 497-510.
31. Sandberg M, Lindberg LG and Gerdle B. Peripheral effects of needle stimulation (acupuncture) on skin and muscle blood flow in fibromyalgia. *Eur J Pain.* 2004;8(2): 163-1671.
32. Sandberg M, Lundeberg T, Lindberg LG and Gerdle B. *Eur J Appl Physiol.* 2003;90(1-2): 114-9.
33. Styf J. Chronic exercise-induced pain in the anterior aspect of the lower leg. An overview of diagnosis. *Sports Med.* 1989;7: 331-339.
34. Styf J, Körner L and Suurkula M. Intramuscular pressure and muscle blood flow during exercise in chronic compartment syndrome. *J Bone Joint Surg Br.* 1987;69: 301-305.
35. Styf J and Wiger P. Abnormally increased intramuscular pressure in human legs: comparison of two experimental models. *J Trauma.* 1998;45: 133-139.
36. Wiger P and Styf JR. Effects of limb elevation on abnormally increased intramuscular pressure, blood perfusion pressure, and foot sensation: an experimental study in humans. *J Orthop Trauma.* 1998;12: 343-347.
37. Wiger P, Zhang Q and Styf J. The effects of limb elevation and increased intramuscular pressure on nerve and muscle function in the human leg. *Eur J Appl Physiol.* 2000;83: 84-88.
38. Zhang Q and Styf J. Abnormally elevated intramuscular pressure impairs muscle blood flow at rest after exercise. *Scand J Med Sci Sports.* 2004;14(4): 215-220.
39. Zhang Q, Lindberg LG, Kadefors R and Styf J. (2001a) A non-invasive measure of changes in blood flow in the human anterior tibial muscle. *Eur J Appl Physiol.* 2001a;84: 448-452.
40. Zhang Q, Styf J and Lindberg LG. Effects of limb elevation and increased intramuscular pressure on human tibialis anterior muscle blood flow. *Eur J Appl Physiol.* 2001b;85: 567-571.

Adaptation Biology and Medicine (Volume 6: Cell Adaptations and Challenges)
Editors: P. Wang, C.-H. Kuo, N. Takeda and P.K. Singal
Copyright © 2011 Narosa Publishing House Pvt. Ltd., New Delhi, India

15. Adaptative Response of Lower Extremity Veins to the Iliac Valve Regurgitation Disease

Otoni Moreira Gomes[1*], Amarjit Arneja[2] and Rohit K. Singal[3]

[1]São Francisco de Assis Truth is Jesus Cardiovascular Foundation, Belo Horizonte, Brazil; [2]Department of Internal Medicine, [3]Cardiac Sciences, Faculty of Medicine, University of Manitoba, Winnipeg, Canada

Summary

Background: Varicose veins of the lower extremities is one of the most common circulatory diseases but there is little information on the comparative incidence of the regurgitation of femoral and iliac valves. The objective of this study was to analyze the occurrence of selective valve regurgitation in the external iliac vein with or without an association with insufficiency of the valves in the femoral or great saphenous veins.

Methods: The venous system of the lower limbs of thirty patients aged between 21 and 75 years old (mean 39 years old) with varicose veins in the lower extremities were studied with Color Flow Duplex Scanning aiming to investigate the regurgitation in the iliac, femoral, popliteal and great saphenous veins.

Results: Iliac regurgitation greater than or equal to 26.0 cm/s present in 19 limbs was isolated (without reflux in the great saphenous ostium valve) in only one (5.3%) case. In cases where iliac regurgitation was less than 26.0 cm/s, regurgitation in the great saphenous vein ostium was absent in four (14.3%) limbs. Considering the total of 60 lower extremities studied, the incidence of iliac regurgitation was seen in 47 (78.3%) cases, and dysfunction of the saphenous ostium valve in 45 (75%) cases. Otherwise, saphenous ostium valve reflux without iliac regurgitation occurred only in three (6.3%) cases, with only one case above 25.9 cm/s.

Conclusion: Anomalies of valvular function in iliac veins present significant correlation with the intensity of insufficiency in the great saphenous vein ostium valve and with the gravity of the varicose veins in the lower extremities.

Keywords: Varicose Veins. Iliac Vein, surgery. Iliac Vein, pathology. Venous Insufficiency

Introduction

Varicose veins of the lower extremities are one of the most common circulatory diseases. Two percent of the population experiences severe form of venous insufficiency [Callam 1992]. The disease carries significant morbidity that necessitates frequent hospitalizations, absence from work, recurrent

*Address for Correspondence: Otoni Moreira Gomes São Francisco de Assis Truth is Jesus Cardiovascular Foundation, Rua José do Patrocínio, 522 - 31525160 Belo Horizonte - MG - Brazil
E-mail: gomes@servcor.com

debilitating symptoms despite treatment and changes in lifestyle [Mayberry *et al.* 1991]. Venous valvular incompetence is the most common cause of lower extremity chronic venous insufficiency which leads to skin changes, leg ulcerations, and deep vein thrombosis [Kistner *et al.* 1995]. Venous ulcerations in the gaiter area of legs occurs as a consequence of inabated persistent venous insufficiency. This is due to valvular deficiency of superficial perforator or deep veins alone or in combination [Rosales *et al.* 2008]. Valvular reconstruction [Tripathi *et al.* 2004] as well as stenting [Antoniou *et al.* 2008; de Borst *et al.* 2003] has been used to address the underlying anatomic pathology. However, the associated comparative incidence of femoral and iliac valve regurgitation has not been examined in great detail. In order to better understand the adaptative response of veins in the lower extremities to regurgitant disease of the iliac valve, we must consider the great physiologic importance of this valve. The following differentiation between the venous valves is based in their morphology and pressure stress resistance: (1) Atrioventricular tricuspid valve, (2) Pulmonary artery valve, (3) Iliac vein valve (Fig. 1), (4) Deep femoral vein valve, (5) Great saphenous vein valve and (6) Other small venous valves in the circulatory system.

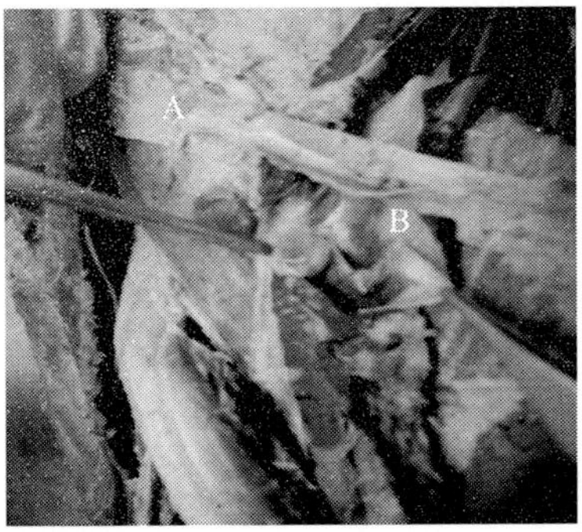

Fig. 1. A-Inguinal ligament; B- Tricuspid iliac valve

Lower limb varicose veins were classified with the following criteria: Class 1 - micro and minimal varicose veins; sclerotherapy responsive; Class 2 - small varicose veins needing surgery for esthetic purposes, Class 3 - large varicose veins with formal surgery indication and Class 4 - complicated varicose veins (Fig. 2)

Methods

The occurrence of selective valve regurgitation in the external iliac vein associated with or without insufficiency of the valves in the femoral or in the great saphenous veins was studied. The venous system of the lower limbs of thirty patients aged between 21 and 75 years of age (mean 39 years old), with varicose veins in the lower extremities were studied with Color Flow Duplex Scanning (Esaote array; 7.5 MHZ transducer; bidimensional, pulsed doppler and collor flow techniques) aiming to investigate the regurgitation in the iliac, femoral, popliteal and great saphenous veins. Six of the patients were male (20%). The results were statistically considered based on the Analysis of Variance and t Test for significance level of =0.05.

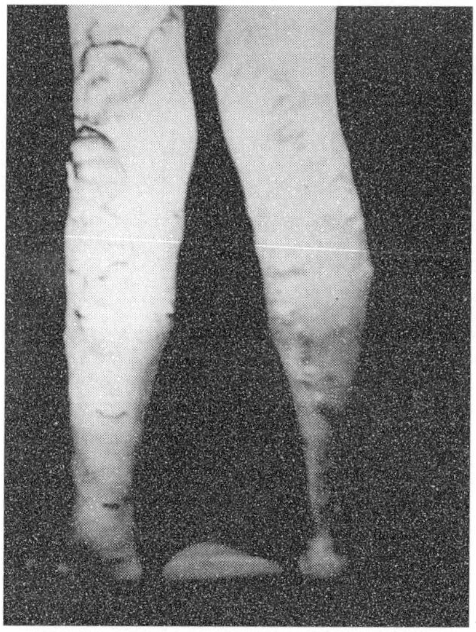

Fig. 2. Class 4 complicated lower limb varicose veins

Results

The reflux intensity (peak velocity) was registered during effort from sudden and voluntary coughing, with the patient standing. Considering maximum value of three consecutive measures the following reflux classification was employed: Class 0 – absent; Class 1 – less than 9.0 cm/s; Class 2 – 9.0 to 17.0 cm/s; Class 3 - 18.0 to 25.0 cm/s; Class 4 – 26.0 to 34.0 cm/s; Class 5 – 35.0 to 42.0 cm/s; Class 6 - greater than 42.0 cm/s (Figs. 3, 4)

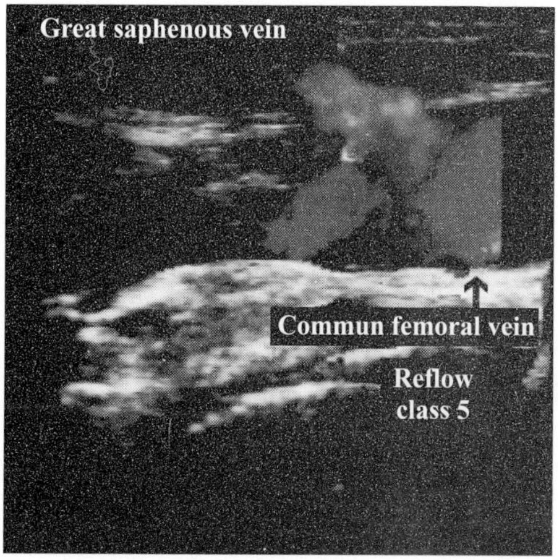

Fig. 3. Reflux classification 5

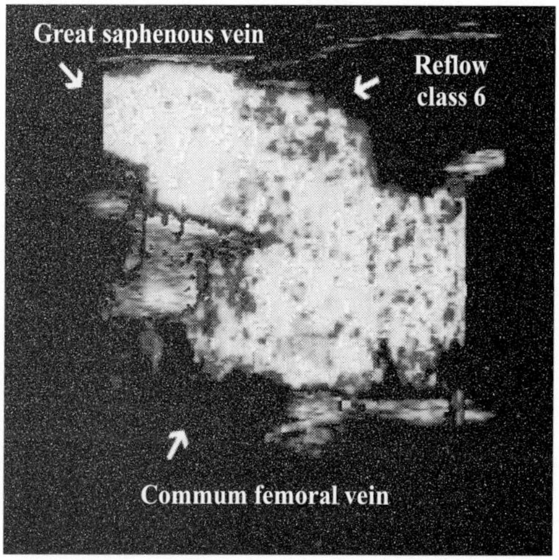

Fig. 4. Reflux classification 6

Iliac regurgitation = to 26.0 cm/s (present in 19 limbs) occurred in isolation without reflux in the great saphenous ostium valve in only one (5.3%) case. In cases of iliac regurgitation < 26.0 cm/s, absence of regurgitation in the great saphenous vein ostium occurred in four (14.3%) limbs. Considering the total of 60 lower extremities studied, it was observed that the incidence of iliac regurgitation was 78.3% (47 cases), and incidence of dysfunction of the saphenous ostium valve in 75% (45 cases). Otherwise, saphenous ostium valve reflux without iliac regurgitation occurred in only 6.3% (3) of cases, with only one case above 25.9 cm/s (Table 1).

Table 1. Number of patients with regurgitation incidence in the iliac and great saphenous vein

Regurgitation		Absent	= 26.0cm/s	> 26.0cm/s
ILIAC	RLL	6	16	8
	LLL	7	15	8
G Saph V	RLL	8	12	10
	LLL	7	11	12

RLL and LLL: Right and Left Lower Limbs

The most important result regarding adaptative response of lower limbs veins to the iliac venous valve regurgitation can be observed in Figure 5, confirming 82.35 % incidence of class 3 varicose veins and 11.70% in class 4 with regurgitation values >26.0cm/sec. Under 26.0cm/sec regurgitation 65.11% are in class 1 and no case of class 4 venous disease.

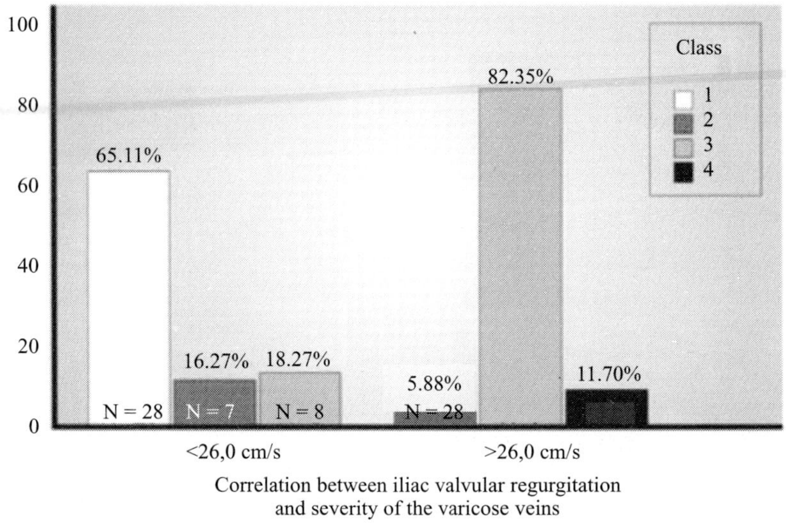

Fig. 5.

Discussion

The existence of valves in the venous system was described in the sixteenth century by Fabrício Acquapendente from the Anatomy School of Vesálio; figuring consistently in the majority of anatomy, angiology and cardiovascular books since that time. Notably, however, the difference in structure of the iliac valve, which is found at the start of the external iliac vein and is the strongest and most developed valve in all of the venous systems has not been highlighted. Here we choose to denominate the iliac valve as the valve that is at the start of the external iliac vein; stressing its special morphologic structure, greater consistency [Eger and Casper 1943; Powell and Lynn 1951; Basmajian 1952] and for the possible higher frequency of its presence or absence in the pathology of venous diseases of lower limbs. Other valves in the iliac venous segment are considered here as iliac valve accessories.

The position of the iliac valve is also important. To better guarantee its function, it is surrounded by resistant osteofibrosis structures, mainly represented by the inguinal ligament, anteriorly and by the ilio-psoas and pectineal muscles over the hips posteriorly. In these conditions the separation of its two cuspids is more difficult and the result is valvular failure with significant reflux. This is important since the iliac valve supports the complete weight of the column of blood of a standing individual, and because accessory iliac valves are rare. There are no valves in the inferior vena cava, with only the rudimentary Eustachian valve at the junction of the right atrium. Below the iliac valve there is the ostial valve of the great saphenous vein and the common femoral vein, at the transition with the superficial femoral vein. Since the deep femoral veins do not have valves, the valves in the iliac vein represent an important protection against endothelial stress in supporting all the hydrostatic venous blood pressure. Among the cardiac valvular diseases; stenosis, failure, double dysfunction, atresia and agenesis have been described. Iliac valvular disease has two main forms, which are valve failure and agenesis, and these diseases are more common in accessory iliac valves that are more fragile.

Iliac valve regurgitation was observed in 78.3% of the patients, all of whom suffered from lower limb varicose veins of varying degrees. In contrast, saphenous ostial regurgitation without iliac reflux occurred only in three cases (6.3%), with only one case (2.1%) greater than 25.9 cm/s. This favors a theory of the 'domino effect' for the non-infection and non-traumatic venous degeneration in lower limbs. Congenital malformation of the iliac valve and its assessories creates a venous pressure hammer on the ostial valve of the great saphenous vein and on the valve at the junction of the common and superficial femoral veins, which are structurally weaker and less protected by supporting structures [Gomes and Gomes 2002; Gomes 2005]. Mainly, the great saphenous valve is inclined to dilatation and regurgitation. The valve at the start of the common femoral vein possesses a different structural constitution, is protected by musculo-aponeurotic structures of the thigh and it is more prevalent than the iliac vein valve, giving greater protection for the popliteal vein and its direct branches. This fact justifies the presence of reflux in the popliteal vein in only two patients (3.3%) of the study group, with the most accentuated refluxes in the iliac valve. Thus, regurgitation of the iliac valve may promote distention of the common femoral vein and the great saphenous ostium, thereby progressively deteriorating their valves. When the great saphenous valve has a better structure and resistance, the tension acts with a greater intensity on the ostial valve of the superficial femoral vein, since it is more resistant and reduces the progression of the disease. On the other hand, the common femoral vein transmits the stress of the pressure to its branches having valves with little consistency.

Consequently, collateral venous dilations are developed, damaging the communicating perforating valves and this degeneration propagates to the great and lesser saphenous vein segments in the legs. This pathophysiologic condition favors the explanation of clinical symptoms of paresthesia, the sensation of weight and orthostatic edema in patients without varicose veins and without reflux in the great saphenous vein, as all the tension is exerted on the region of drainage of the deep femoral vein and make the centripetal drainage of the communicators difficult. Although the varicose veins depend on other primary or secondary factors, which may depend on the constitutional resistance of the venous walls and the surrounding tissues which restrict dilation, two important studies support the previously mentioned domino theory. The most common histopathologic alteration in varicose veins is proliferation-type endothelial degeneration and fibrosis of the intima, as has been explained by the continuous endothelial pressure stress [Garrido 1987]; and demonstrated iliofemoral regurgitation in patients who are with symptoms of varicose veins [Ristow et al 1979]. In a general analysis, it can be concluded that the anomalies of valvular function in iliac veins represent the determining factor with the greatest correlation with failure of the ostial valve of the great saphenous vein and with the gravity of varicose veins of the lower limbs.

Furthermore, it can be inferred that future epidemiological studies in different age ranges may indicate the patients with a greater risk to develop venous diseases of the lower limbs, as in pregnancy and during the development of cardiomyopathies, permitting an earlier preventative treatment. New frontiers are opened for the treatment of patients evolving with more severe venous stasis of lower limbs too. The evidence in this study of bilateral iliac venous regurgitation raises the therapeutic hypothesis for the implantation of stents with valves in the iliac vein or the infrarenal inferior vena cava, by endovascular surgery or by a direct surgical procedure. This kind of research is already being performed with the implantation of porcine aortic cuspid valvular segments preserved in glutaraldehyde, in the infra-renal vena cava of dogs with promising initial results. [Gomes et al 2002]

Thus far, beyond the current observed incidence of valvular dysfunction in the iliac veins of 78.3% of lower limbs with varicose veins and the direct relationship between the intensity of the venous reflux and the gravity of the varicose veins, other possibilities have not been demonstrated.

References

1. Antoniou GA, Georgiadis GS, Souftas VD, Deftereos SP, Lazarides MK. Reversal of deep vein reflux after successful stenting in a patient with venous hypertension after thigh access graft creation. Vasc Endovasc Surg Dec/Jan 2008; 6: 547-550.
2. Basmajian JV. The distribution of valves in the femoral external iliac and common iliac veins and their relationship to varicose veins. Surg Gynecol Obstet 1952; 95:537- 45.
3. Callam, MJ. Prevalence of chronic leg ulceration and severe chronic venous disease in Western countries. Phlebology 1992; 7(Suppl.1): 6-12.
4. de Borst GJ, Teijink JAW, Patterson M, Quijano TC, Moll FL. A percutaneous approach to deep venous valve insufficiency with a new self-expanding venous frame valve. J Endovasc Ther 2003; 10: 341-349.
5. Eger SA, Casper SI. The etiology of varicose veins from an anatomic aspect, based on dissection of 38 adult cadavers. JAMA 1943; 123: 148-58.

6. Garrido MBM. Patologia e diagnóstico de varizes dos membros inferiores. In: Maffei FHA, Lastória S, Yoshida WB, Rollo HA. Doenças vasculares periféricas. Rio de Janeiro: MEDSI, 1987.
7. Gomes OM, Gomes ES. Iliac valve regurgitation disease: implications for the surgical approach of the venous insufficiency in the lower extremities. Rev Bras Cir Cardiovasc 2002; 17(4): 339-344
8. Gomes OM, Zandim R, Rodrigues JBT, Miranda ED, Barros D, Oliveira DM - Implante de Válvula Heteróloga na Veia Cava Inferior Infra-Renal para Tratamento da Doença de Regurgitação Valvar Venosa Ilíaca: Estudo Experimental. Rev Bras Cardiovasc 2002; 17(4):367-369
9. Gomes OM. Fisiologia cardiovascular aplicada, Belo Horizonte: Edicor, 2005 www.servcor.com PDF / HTTML Down Load
10. Kistner RL, Eklof B, Masuda EM. Deep venous valve construction. Cardiovasc Surg 1995; 3: 129-140.
11. Mayberry JC, Moneta GL, Taylor LM, *et al.* Fifteen-year results of ambulatory compression therapy for chronic venous ulcers. Surgery 1991; 109: 575-581.
12. McMurrich JP. The valves of the iliac vein. Br Med J 1906; 2: 1699-700.
13. Powell T, Lynn RB. The valves of the external iliac femoral and upper third of the popliteal veins. Surg Gynecol Obstet 1951; 92: 453-63.
14. Ristow AVB, Arruda AM, Albuquerque JT, Medina AL. Varizes primárias: 10 anos de experiência com o tratamento cirúrgico. Rev Assoc Med Bras 1979; 25: 216-8.
15. Rosales A, Jorgensen JJ, Slagsvold CE, Stranden E, Risum O, Kroese AJ. Venous valve reconstruction in patients with secondary chronic venous insufficiency. Eur J Vasc Endovasc Surg 2008; 36: 466-472.
16. Tripathi R, Sieunarine K, Abbas M, Durrani N. Deep venous valve reconstruction for non-healing leg ulcers: Techniques and results. ANZ J Surg 2004; 74: 34-39.

Adaptation Biology and Medicine (Volume 6: Cell Adaptations and Challenges)
Editors: P. Wang, C.-H. Kuo, N. Takeda and P.K. Singal
Copyright © 2011 Narosa Publishing House Pvt. Ltd., New Delhi, India

16. Alveolar Macrophages Phenotype: Adaptive Programming and Disorders in Pulmonary Diseases

S.V. Lyamina, E.A. Monastyrskaya, O.A. Borodovitsina, D.A. Belousova, E.B. Manukhina and I.Y. Malyshev*

Department of Adaptive Biology and Medicine, Institute of General Pathology and Pathophysiology, Russian Academy of Medical Sciences, Moscow, Russia

Summary

Macrophages are the main cells of the innate immunity providing immideate reaction to the penetration of xenogenous pathogenic agents in the organism and participating in the startup and realization of adaptive immune reactions. There are no other certain systems for monitoring biological parameters as this one. Any disorders in macrophages function lead to chronic inflammation, autoimmune diseases, and can promote tumordevelopment.

Keywords: Macrophages, microbes, neutrophils, chemokines, autocrine

Introduction

Every day during life the organism is exposed to different physical, chemical and biological factors, such as temperature, pressure, light, sounds, odors, food taste and innumerable quantity of pathogens inside and outside the organism. It is vitally important for the organism to keep up with changes of the internal and external environment. That's why during evolution, special sense organs with specialized sensor cells such as eyes, ears, nose and tongue were developed. Evolution followed the way of maximum specialization of organs and sensor cells to accept different physical and chemical factors, e.g. light, sound, odor and taste. However evolution did not create any special organs or regions for acceptance of biological factors and found another way. Special sensor cells detecting pathogens in the organism present in each and every organ and tissue of the body. Macrophages are the key cells among the above mentioned ones. Macrophages constitute from 10 to 15% of the total number of cells in every organ of the body.

*Address for Correspondence: Dr. Igor Y. Malyshev Professor and Head, Laboratory of Stress and Adaptation, Department of Adaptive Biology and Medicine, Institute of General Pathology and Pathophysiology, Russian Academy of Medical Sciences Baltijskaya Street 8, Moscow 125315, Russia, Email: igor.malyshev@mtu-net.ru

Macrophages originate from monocytes initially developing from CD34$^+$ (CD-cluster antigen) myeloid progenitor cells of bone marrow. Monocytes are circulating in blood flow and spread in different tissues where they mature in different types of resident macrophages (18). During monocyte transformation to macrophages the following structural and functional changes are observed (19):

- apparent increase of cell size (up to 25-50 μm), and amplification of chondriosomes in cytoplasm, pinocytotic vesicles, lysosomes and enlargement of Goldgi complex;
- transformation of plasma membrane with considerable number of folds formation, increasing quantity of microvilli and receptors to IgG and component C3b of complement system;
- activation of respiratory and lysosome enzymes and simultaneous decrease of peroxidase count;
- strengthening of mobility, metabolic activity, adhesion properties, ability to pinocytosis and phagocytosis, and general increase of microbicidal properties;
- change of sensitivity to hormones.

Macrophages originated from monocytes in tissues reside in an inactive condition which is characterised by low consumption of oxygen, low speed of protein synthesis and moderate synthesis of cytokines. Depending on the degree of maturity, location and activation by antigenes or lymphocytes, macrophages are characterized by certain structural and functional heterogeneity. Resident tissue macrophage can quickly react under the influence of environment stimuli by changing the character of genes (21). Such quick reaction usually observed during inflammation or tissue damage is called macrophages activation, and is also characterized by increasing the synthesis of cytokines, chemokines and other inflammation mediators which, in turn, promote attraction of new macrophages (27).

Alveolar macrophages of the lungs first contact the air, and accordingly are the first cells to perceive action of environmental biological factors. Nearly 10000 litres of air containing various pathogenic microorganisms and particles pass through lungs of the person during the day. Macrophages are the central branch of the innate immune response and their reaction on pathogenes determines development of adaptive immunity. So the following important consequence is observed: macrophages are the core cellular transductors of biological signal, whether it is a pathogenic microorganism or pollen of plants and other allergens on immune system of an organism. Thanks to immune system, potential pathogenic agents are recognized and removed from an organism (25). It is clear that quality and adequacy of biological signal transduction through alveolar macrophages will define probability of development of pulmonary diseases and their outcomes. Therefore it is obvious that macrophages deserve very close attention while studying pulmonary diseases and searches of new therapeutic approaches.

In respiratory ways and lungs of a healthy person 95 % of leukocytes compiles alveolar macrophages, about 1-4 % - lymphocytes and nearby 1 % - neutrophils.

Alveolar macrophages are the basic cells providing startup of the innate immune system in lungs. Before analyzing the role of these macrophages in the development of pulmonary diseases, it is necessary to remind shortly at least a role of macrophages in the immunity.

Role of Macrophages in Immune Reactions of an Organism: M1/M2 Macrophages and the Th1/Th2 Paradigm

Immune defense of the organism is provided by mechanisms of innate and adaptive immunity. Cell division of innate immunity consists of hemopoietic cells, mast cells, basophils, monocytes, dendrite cells and macrophages. Adaptive immunity includes CD4+ and CD8+ T cells, regulatory T cells and B cells. Macrophages play a critical signaling role in immunity as invasion of an alien agent induces activation of macrophages and release of cytokines. Macrophages distinguish endocellular pathogenes, such as bacteria, viruses, extracellular pathogenes, such as parasites, protozoa, fungi or helminthes, thanks to presence at microorganisms pathogen-associated molecular patterns (PAMPs), such as LPS, and at macrophages - pattern recognition receptors (PRR), such as Toll-Like Receptors (TLRs) and Nucleotide Oligomerization Domain (NOD) receptors. During the interaction of PAMPs of endocellular microbes with macrophages PRR macrophages produce proinflammatory cytokines IL-12, TNF-α, IL-18 and chemokines (Figure 1). Chemokines attract to inflammatory focus natural killers (NK), neutrophils and naive T-lymfocytes (Th0). Macrophage IL-12, TNF-α and IL-18 interact with natural NK to increase secretion of IFN-γ by these cells. IL-12 and IL-18 provide autocrine stimulation of IFN-γ production in macrophages. IFN-γ even more stimulates production of IL-12 and TNF-α and enhances phagocyte and bactericide properties of macrophages.

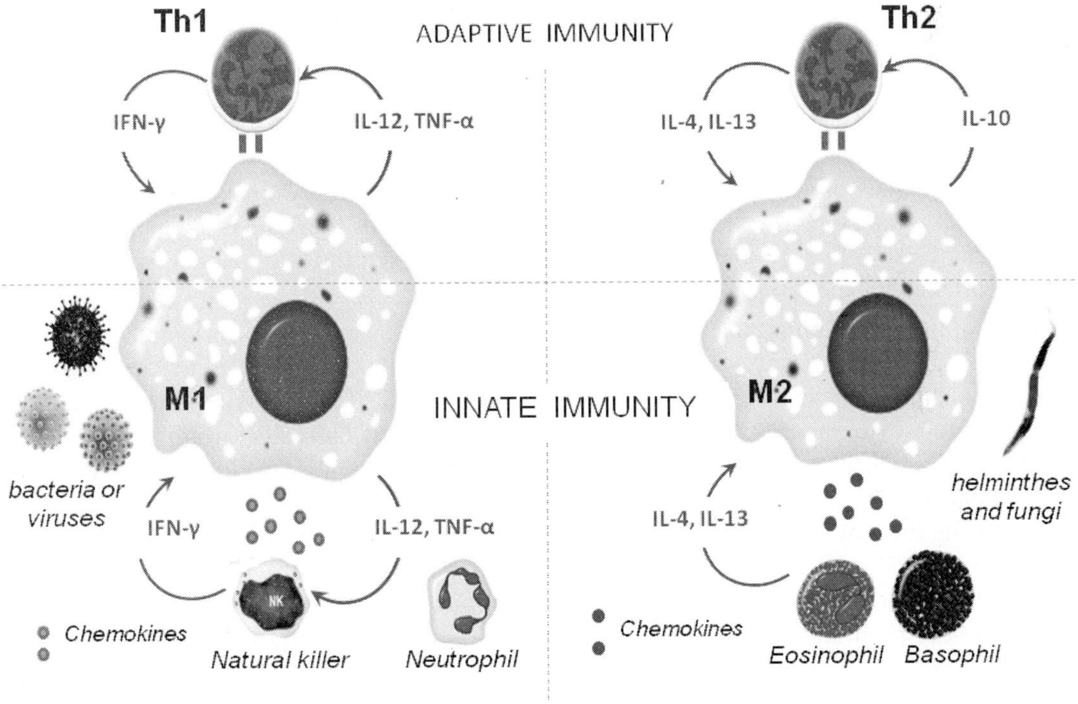

Fig. 1. Role of macrophages in immune responses of the body: M1/M2 macrophages and the Th1/Th2 paradigm.

If PAMPs of extracellular pathogens, such as fungi or helminths co-operate with macrophages, they release antiinflammatory cytokines, such as IL-10 and IL-13 and other chemokines. Chemokines attract the cells producing IL-4 and IL-13, such as eosinophils and basophiles, as well as naive T-lymphocytes. IL-4 and IL-13 stimulate even more macrophages to release IL-10. IL-10 reduces antimicrobial properties of macrophages by inhibiting the release of pro-inflammatory cytokines, reactive oxygen and nitrogen species.

Interaction of macrophages with PAMPs of intracellular microbes, attraction of IFN-γ producing cells and release of IL-12 by macrophages, or alternative interaction of macrophages with PAMPs of extracellular pathogens, attraction of IL-4 and IL-13 producing cells and secretion of IL-10 by macrophages, represents the first wave of alternative activation of macrophages and marks the development of the innate immune response. The innate immune response represents initial and rather nonspecific response of the immune system and distinguishes the most often seen microorganisms.

The phenotype of macrophages formed during interaction with intracellular microbes and/or IFN - γ is named as classical or M1 phenotype. Phenotype macrophages acquire during interaction with extracellular pathogens and/or IL-4 and IL-13 is specified as alternative or M2 phenotype.

Thus depending on the nature of pathogens (intracellular microbes or extracellular parasites) even at the stage of the innate immunity, adaptive alternative changing/programming of macrophages phenotype takes place (12).

Along with these macrophages, the development of the specific adaptive immune response is predetermined (28). For successful eradication of the pathogen, an adaptive response may develop by either cell (Th1) or humoral (Th2) type. There are two types of signals carrying information on expediency of Th1 or Th2 type of response activation. The first signal type forms due to dendrite cells and macrophages presenting antigens for Th0 cells (CD4+ T-cells). The second type of signals is provided by cytokines produced during the innate response. On these signals the differentiation of "naive" Th0-cells can be pointed to Th1 or Th2. Genetic predisposition, age of the patient and/or the involved organs can influence the trend of differentiation. Often the way passed by Th0-cell is of crucial importance for the disease outcome (35).

Antigens of intracellular microbes, M1 phenotype of macrophages and pro-inflammatory cytokines IL-1, TNF-α, IL-12 and IFN-γ, potentiate development of progenitor cells Th0 to Th1 and suppress the Th2 response (Figure 1). Antigens of extracellular invaders, M2 phenotype of macrophages and anti-inflammatory cytokines IL-10 and IL-4 shift differentiation of Th0 cells towards Th2 and inhibits Th1 proliferation. Differentiation of T-cells in Th1 or Th2 marks the development of the specific adaptive response. Th1 response counteracts pathogenic intracellular microorganisms at the expense of strengthening the macrophages activity directed on destruction of pathogens, production of IFN-γ, IL-2, IL-3, GM-CSF and stimulation of cytotoxic T-lymphocytes proliferation. In Th2 response, neutralization of extracellular parasites is achieved by production of IL-4, IL-5, IL-6, IL-9, IL-10, IL-13, stimulation of proliferation and differentiation of B-lymphocytes and the release of antibodies of different classes.

Th1 or Th2 cells produce their own specific set of cytokines, Th1 or Th2, respectively. Th1 cytokines and first of all IFN-γ, influence macrophages and even more polarise them towards M1 phenotype (19). Th2 cytokines and first of all IL-4 and IL-13 influence macrophages and even more polarise them towards M2 phenotype (19). The influence (effect) of Th1 or Th2 cytokines on macrophages results in the second wave of alternative adaptive activation of macrophages. Th-dependent activation of macrophages is provided by expression of ligands CD40 on the surface of activated Th-lymphocytes (19). These proteins get in touch with corresponding surface receptors of macrophages and provide "physical" affinity of macrophages and Th cells and quick contact of Th cytokines with macrophages, respectively.

Thus during the immune response, alternative two-level activation of macrophages and T cells differentiation are interdependent with positive feedback: M1 by means of presenting antigen and secretion of IL-12 shifts differentiation of Th0 cells towards Th1, and Th1 cells at the expense of secretion of IFN-γ even more strengthen polarisation of macrophages towards M1 phenotype.

Similarly by means of the presence of antigen and secretion of IL-10 M2, shifts differentiation of Th0 cells towards Th2, and Th2 at the expense of IL-4 and IL-13 releases yet more enforces of polarisation of macrophages towards M2 phenotype.

It is interesting that M1 and M2 phenotypes differ morphologically in cell culture (Figure 2).The effector functions of M1 and M2 macrophage phenotypes also differ strongly.

Classically activated M1 macrophage undergoes some changes which significantly increase the efficiency of its abilities to control pathogens and enhance immune response. M1 macrophages induced by bacteria and viruses, inflammatory molecules such as lipopolysaccharide (LPS) and IFN-γ or granulocyte macrophage colony-stimulating factor (GM-CSF) secrete significant amount of proinflammatory cytokines, such as IL-1beta, IL-6, IL-15, IL-18, TNF-α, IL-12, IL-23, macrophage inflammatory protein 1 (MIP-1), and low levels of IL-10. M1 starts to produce large amounts of reactive oxygen species (ROS) and generates reactive nitrogen species, such as nitric oxide (NO), via expression of inducible nitric oxide synthase (iNOS). ROS and NO are responsible for cytotoxic activity of activated macrophages. M1 also secretes the specific set of chemokines: CCL15/HCC-2, CCL20/MIP-3-α, and CXCL13/BCA-1, and the angiostatic INFgamma-responsive chemokines CXCL9/Mig, CXCL10/IP-10, and CXCL11/I-TAC (12, 27). Moreover, surface expression of CD40 and tumor necrosis factor a (TNF-α) receptors increase. Macrophage begins to produce TNF-α extensively, which has autocrine action on receptors on macrophage surface and enhances its activation. Also there is enhanced expression of main histocompatibility complex class II (MHC II) molecules and B7 molecules (costimulator participating in antigen production process) on macrophage surface (19). Functionally, M1 cells are potent effector cells integrated into Th1 response, which kill intracellular microorganisms and tumor cells. This increase of microbicidal and antitumor activity is mediated by different mechanisms including restriction of iron and other nutrients for microorganisms, acidification of the phagosome, synthesis of ROI, release of nitric oxide (NO), pro-inflammatory cytokines production and activation of Th1 adaptive response. M1 macrophage displays anti-angiogenic properties.

M2 macrophages are induced by extracellular parasites, exposure to certain stimuli, such as IL-4, IL-13, as well as immune complexes in combination with IL-lbeta or LPS, IL-10, transforming growth factor beta (TGF-ß), or glucocorticoids. The M2 macrophages express high levels of anti-inflammatory molecules, such as IL-10 and IL-1 decoy receptor (IL-1ra), and low levels of pro-inflammatory cytokines, such as IL-12 and IL-23. M2 metabolizes arginine through arginase rather than iNOS. The iNOS activity is blocked by arginase by various mechanisms, which include competing for the arginine substrate needed for NO production. Arginine metabolism of M2 phenotype is shifted towards ornithine and polyamines production by arginase activity (26). M2 phenotype expresses large amount of mannose receptors, scavenger receptors and galactose-type receptors. In contrast to M1, M2 macrophages tune the inflammatory responses, limiting pro-inflammatory cytokine production and render harmless extracellular parasites due to promotion of Th2 responses and, in the first place, allergic reactions of reaginic type and immune reactions under parasitic infections. M2 macrophages are characterized by enhanced phagocytic activity and attract eosinophils to the inflammatory site. By synthesizing a lot of growth factors, M2 macrophages are also promoting angiogenesis, tissue remodeling and repair (27), and promote tumor growth.

Comparative characteristics of polarization stimuli, cytokine and other bioactive substances secretion and main cellular markers of M1 and M2 macrophages are shown in Tables 1-3.

Table 1. Stimuli for M1 and M2 polarization

M1	M2
- Th1 cytokines (IFN-γ, TNF-α)	- Th2 cytokines (IL-4, IL-13)
- Pathogen-associated molecular patterns (LPS, lipoproteins, dsRNA, lipoteic acid)	- immunocomplexes with IL-1ß
	- IL-10, TGF-ß
- Different gram-positive and gram-negative microorganisms, cytomegalovirus	- PPAR-g agonists (peroxisome proliferator-activated receptor gamma)
- Heat shock proteins, extracellular matrix components, high mobility group box 1 (HMGB1)	- Sometimes intracellular pathogens (Coxiella burnetii, Leismania)
	- vitamin D3
	- hormones (glucocorticoids)
	- apoptotic cells

Table 2. Cytokine and bioactive substances secretion by M1 and M2 phenotypes

M1	M2
- iNOS induction (NO and peroxynitrite production)	- IL-10
	- transforming growth factor b (TGF-β)
- ROS synthesis	- mannose receptors exprassion
- pro-inflammatory cytokines (IL-1ß, IL-6, TNF-α, IL-12 and IL-23, IL-15)	- arginase-1 (urea, ornithine and proline production), do not produce NO
- chemokines: CXCL10 (IP-10), CCL8 (MCP-2), CCL15 (MIP-5), CCL19 (MIP3ß), CCL20 (MIP-3α) and CXCL13 (MCP-4)	- fibronectin
	- chemokines: CCL17 (TARC), CCL13 (MCP-4), CCL14 (HCC-1), CCL23 (MPIF-1) and CCL26 (eotaxin-3)

Lipid metabolism also substantially differs between M1 and M2 macrophages. Macrophage maturation is accompanied by suppression of PG-endoperoxide synthases (PTGS1 and PTGS2), arachidonate-5-lipoxygenase (ALOX5) and leukotriene A4 hydrolase. After classical M1 activation predictable significant induction of cyclooxygenase 2 type (COX-2) occurs, which leads to enhanced synthesis of arachidonic acid metabolites as it happens in inflammatory conditions; however it is

Table 3. Cellular surface markers for M1 and M2 phenotypes

M1	M2
B (CD80)	M130 (CD163)
B7.2 (CD86)	CD206 (MRC1, mannose receptor)
CCR7 (MCP-3)	FceRII (CD23)
CXCL10 (IP-10)	Nucleotide receptors (GPR86, GPR105,
TLR-2	P2Y8, P2Y11 and P2Y12)
TLR-4	C-type lectin-like repector Dectin-1
Fcγ RIII (CD16)	DC-SIGN (CD209)
Fcγ RII (CD32)	DCIR (CLECSF6)
LAM-1 (CD62)	CLACSF13
IL-1R1	FIZZ1, ST2 (mouse)
IL-7R (CD127)	Scavenger receptors (SR-A, M60)
IL-2R (alpha chain)	CXCR4, fusin (CD184)
IL-15R	(alpha chain) TRAIL
IL-17R (CTLA8) (CDw217)	IL-1Rα

accompanied by further suppression of constitutively active cyclooxygenase 1 type (COX-1), leukotriene A4 hydrolase, tromboxane A synthase 1 and ALOX5. After alternative M2 activation, promotion of arachidonate-15-lipoxygenase and COX-1 occurs (26).

Also M1 macrophages express sphingosine kinase whereas M2 macrophages express ceramide kinase, which are sphingomyelin cycle enzymes. Sphingomyelin cycle is the universal secondary messenger for apoptotic signaling, which helps to regulate essential cellular functions such as proliferation, phagocytosis, differentiation, angiogenesis, and chemotaxis. Ceramide plays a critical role in the regulation of cellular processes by sphingomyelin signal pathway. It acts as a secondary messenger, inhibits proteinkinase C and suppresses proliferation, modulates protein phosphorylation, phospholipase A2 activity and potentially inducts apoptosis 0(2). However ceramide produced by enzymatic hydrolysis of sphingomyelin displays proapoptotic effect (by inducing mitochondria destabilization, activating Fas etc.). Sphingosine generated from ceramide by ceramidase activity inhibits proteinkinase C, activates some other kinases and regulates large repertoire of enzymes participating in extracellular signaling, and therefore shows antiapoptotic action. Derivate of sphingosine, which is produced by sphingosinekinase activity, sphingosine-1-phosphate (S1P) also acts as antiapoptotic agent, either via receptors or intracellulary. Pro-inflammatory cytokines stimulate activity of sphingomyelinase, but not ceramidase, which leads to ceramide accumulation, whereas growth factors also stimulate ceramidase and sphingosinkinase which lead to enhanced S1P production. By this mechanism there is accumulation of proapoptotic ceramide in M1 macrophages and antiapoptotic sphingosine and S1P in M2 macrophages (2).

In addition, M1 polarization leads to enhanced expression of ABCA1 which reduces tissue cholesterol and group of apolipoproteins participating in cholesterol transport (26).

By now there are only scarce 7data about intracellular mechanisms of M1 and M2 macrophage polarization presented in literature. One of the intracellular enzymes which appears to participate in macrophage polarization is SH2-containing inositol-5'-phosphatase (SHIP). It represses the proliferation, survival, and activation of hematopoietic cells, in large part by translocating to membranes following extracellular stimulation and hydrolysing the phosphatidylinositol-3-kinase (PI3K)-generated second messenger PI-3,4,5-P3 (PIP3) to PI-3,4-P2. SHIP-/- mice have, as a result, an increased number of monocyte/macrophages because their progenitors display enhanced survival and proliferation, as well as more rapid differentiation. Interestingly, SHIP-/- mice do not display lipopolysaccharide (LPS)- or CpG oligonucleotide-induced tolerance because this blunting of inflammatory mediator production is contingent upon LPS- and CpG-induced upregulation of SHIP in their macrophages. This upregulation is mediated by the production of autocrine-acting TGFbeta which is induced by the MyD88-dependent pathway. The increased levels of SHIP then inhibit both MyD88-dependent and independent signaling. Peritoneal and alveolar macrophages from SHIP knockout mice produce less NO because they have constitutively high arginase 1 levels and this enzyme competes with inducible nitric oxide synthase (iNOS) for the substrate L-arginine cleaving it to ornithine and urea (39).

In addition, it has been shown that macrophage stimulation with anti-inflammatory cytokines (IL-10 and TGF-β) leads to rapid phosphorylation/activation of AMP-dependent proteinkinase (AMPK) whereas stimulation of macrophages by LPS leads to its deactivation. Supression of AMPK expression causes substantial rise of LPS-induced cytokines (TNF-α, IL-6 and COX-2) mRNA concentration. After transfection of macrophages with constitutively active AMPK, there was a decrease of IL-6 and TNF-α and an increase in IL-10 production. Moreover, AMPK reduces IkB-α degradation and enhances Akt, accompanied by suppression of glycogen synthetase ß kinase and activation of CREB. Therefore AMPK leads intracellular signaling in macrophage cells towards anti-inflammatory polarization (37).

It has been shown that NFkB plays an important role in classical macrophage M1 polarization. Thus mice M2 phenotype macrophages which are found in tumor tissue display reduced NFkB activity due to enhanced nuclear expression of p50 NFkB-inhibitory homodimer. They show poor activation by classical activatiors (LPS, CD40L, TNF-α and IL-1ß) (36).

Some of the possible intracellular signaling pathways which activate during macrophage polarization by M1 and M2 phenotype are shown on the Figure 2.

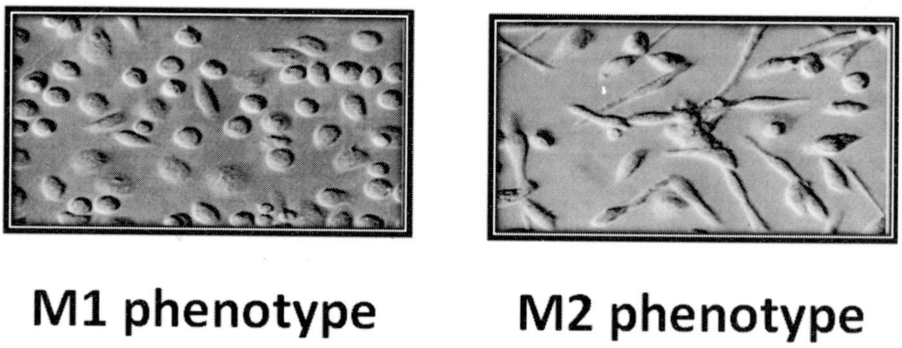

Fig. 2. Macrophage phenotypes M1 and M2 differ morphologically.

While the classical M1 phenotype of activated macrophages is well-described in literature, the data concerning M2 phenotype are somehow inconsistent. Alternatively activated M2 macrophages are often subdivided into 3 groups: M2a, M2b and M2c (27).

Appearance of M2a macrophages was described based on IL-4 or IL-13 action. Such macrophages participate in activation of Th2 reactions. It has been shown particularly that they play the main role in helminthic infections. Cells with such phenotype regulate Th reactions towards Th2 type (mainly by suppression of Th1 reactions), participate in wound healing caused by helminthes, connective tissue growth and help to attract eosinophils to the lesion site (probably by producing leukotriene B4 and Yml which are described in murine macrophages) (21).

M2b phenotype was described based on combined immunocomplex and IL-1ß or LPS action, and macrophage polarization was connected with Toll-like receptors (TLRs) and IL-1R activation. It was shown that macrophages of M2b phenotype participate in suppression and regulation of inflammatory and immune reactions and promote Th2 activation (27).

Appearance of M2c macrophages was described based on IL-10, TGF-ß or glucocorticoids action. Such macrophages activate extracellular matrix synthesis and participate in remodeling processes. M2a and M2b macrophages mostly display anti-inflammatory activity. M2c macrophages have more similarities with M1 macrophages except for expression of IL-10 instead of pro-inflammatory cytokines (20). Comparative characteristic of these three groups of M2 macrophages according to described classification is shown in Table 4.

Table 4. Subdivisions of M2 phenotype macrophages

	M2a	M2b	M2c
Polarization stimuli	IL-4, IL-13	Immunocomplexes + LPS or IL-1ß	IL-10, TGF-ß, glucocorticoids
Functions	Activation of Th2 reactions, attraction of eopinophils, connective tissue growth	Suppression and regulation of inflammatory and immune reactions, activation of Th2 reactions	Remodelling activity, extracellular matrix synthesis

Another classification of macrophage phenotypes which is described in literature marks out three different groups: classically activated macrophages (Ca-M), alternatively activated macrophages (AA-M) and so-called M-II macrophages which are polarized in presence of immunocomplexes (9). Ca-M are prototypical immune effector cells which kill intracellular pathogens via the production of oxygen and nitrogen radicals and pro-inflammatory cytokines. They correspond to M1 phenotype. AA-M arise in response to the Th2 cytokines IL-4 and/or IL-13, are associated with parasitic diseases and may contribute to the production of the extracellular matrix. A third population of activated macrophages was generated by activating macrophages in the presence of immune complexes. They display different patterns of cytokine production: the cross-linking of FcγRs during activation resulting in an abrogation of IL-12 production and strong induction of IL-10. IL-10 is a potent anti-inflammatory mediator. Such macrophages promote maturation of Th2 cells acting as antigen-presentating cells

(APC). Macrophages of the same phenotype were isolated from tumor masses and induced by glucocorticoids.

There has been evidence of detailed study of metabolic and functional properties of these three groups of macrophages in mice. Study of cytokine profile showed that Ca-M produced relatively high levels of IL-12 but low levels of IL-l0; M- II produced high levels of IL-10 and low levels of IL-12. AA-M did not secrete detectable levels of IL-12 or IL-10. Cytokine production by AA-M was not significantly different from unstimulated macrophages. The stimulation of IL-4-primed cells with LPS resulted in an increase in the production of IL-10, which remained substantially below that of M-II.

There were also substantial differences in arginine metabolism between these groups of macrophages. Ca-M and M-II produced relatively high levels of NO, whereas the AA-M produced virtually no NO. On the contrary arginase activity was high in AA-M but was virtually absent in Ca-M and M-II. Antigen-performance was highest in M-II phenotype macrophages: despite IL-10 synthesis, they had the highest expression of CD25 and CD69, and less expression of CD62L. Ca-M expresses the same markers of cellular activation but in lower amounts. AA-M display minimal capacity for T-cells activation similar to unstimulated macrophages. Also AA-M were poor in driving T-cell proliferation, even after stimulation with LPS, whereas Ca-M and M-II did induce active proliferation. T cells activated by Ca-M produced relatively high levels of IFN-γ and less IL-4, whereas T cells activated by M-II produced relatively high levels of IL-4 and significantly reduced levels of IFN-γ. T cells activated by M-II, but not Ca-M, produced high levels of IL-10. AA-M produced high levels of proline by arginase activity which promotes extracellular matrix synthesis (9). The characteristic features of phenotype macrophage described is presented in Table 5.

Table 5. Comparison of three types murine macrophages

	Ca-M	**AA-M**	**M-II**
Stimuli	LPS	IL-4/IL-13	Immune complexes, glucocorticoids
Cytokines	High IL-12, low IL-10	Low IL-12 and IL-10	High IL-10, low IL-12
Arginine metabolism	NO	Ornithine and urea	NO
Biomarker	iNOS, IL-12	Arginase, Fizz1 (mice)	iNOS, IL-10
Antigen presentation ability	Enhanced MHC II and CD86 expression	Not enhanced MHC II and CD86 expression	Maximal MHC II and CD86 expression
Acting on T cells	Potent induction of T cells proliferation, Th1 differentiation	Minimal even after LPS	Potent induction of T cells proliferation, Th2 differentiation

In M1 and M2 macrophage classification, M2 group combines AA-M and M-II because they do not produce IL-12 in contrast to M1 macrophages. However because of the difference between AA-M and M-II cells, probably M1 and M2 classification is a simplification of actual situation in organism.

Sometimes functional macrophage phenotype in different pathologies is different from the classifications described. Thus in streptococcal infection, there is unusual macrophage activation with some signs either of classical activation (pro-inflammatory cytokines and co-stimulating factor synthesis) or alternative activation (such as IL-1Rα and IL-10 synthesis). It is accompanied by activation of arginase expression, but not iNOS expression, and bactericide activity of such macrophages is mediated mostly by ROS. These macrophages have enhanced expression of genes encoding metallothioneines (Mt1 and Mt2), thioredoxin reductase (Txnrd1) and superoxide dismutase 2 (Sod2), which act as traps for reactive oxygen species and participate in free radicals detoxication. Also there is activation of endosomal prostaglandin synthase 2 (PTGS2) expression which can participate in pro-inflammatory prostaglandin production. In addition, these cells possess enhanced expression of cytoprotective molecules (of Gadd45 family, which promote DNA and Tnfaip3 recovery and act as antiapoptotic inhibitor of caspase 8); probably they help to maintain homeostasis in stress conditions induced by streptococcal infection. Therefore in streptococcal infection, macrophages possess mixed M1/M2 phenotype or form mixed population of different phenotypes (12).

Activated macrophages with different phenotype from standard M1 or M2 phenotype were found in some other pathologies. In cytomegalovirus infection there were M1 macrophages with spectrum of produced chemokines resembling mixed M1/M2 phenotype (6). In malignant tumor growth, mixed M1/M2 macrophage phenotype was observed (42). In adipose tissue in obesity, special phenotype of macrophages was described, which characterized M2-like surface receptors expression but they produced high levels of pro-inflammatory cytokines (TNF-α, IL-6, IL-1, MCP-1 and MIP-1α), which may cause development of insulin resistance (45).

Therefore under contemporary literature data, classification of activated macrophages in organism on two phenotypes, M1 and M2, appear to be rather relative. It seems likely that macrophages in organism form a continuum of functional states, one pole of which makes up macrophages actively stimulating inflammation and another pole forms macrophages stimulating tissue regeneration after suppression of inflammatory reaction. Non-activating macrophages found in healthy undamaged tissue appear to be between these two poles (Figure 4) (13, 14).

Maturation of macrophage phenotype is determined not only by nature of factors affecting cell and by their combinations, but also by the consistency of their action. Thus it was shown that prior treatment of macrophages with IFN-γ several hours before LPS stimulation forced cells to produce high levels of TNF-α (15) and changed cell reaction on IL-10 in comparison to macrophages which were not treated with IFN-γ (16). Therefore the reaction of macrophages on stimulus persisted even after polarization changed. It means that despite prior polarization and degree of differentiation, macrophages retain an ability to respond adequately on different new stimuli. Such ability has been named the plasticity of macrophage phenotype (13, 41). In compliance with the conception of plasticity, macrophages are considered as cells which are capable to change their phenotype in frames of continuous phenotypes according to changes in microenvironment. For example macrophages are capable to respond adequately on non-specific pathogen-associated molecular patterns (PAMP) such as LPS and muramildipeptide, regardless of their phenotype. Also pro-inflammatory active macrophages retain an ability to respond to anti-inflammatory stimuli by reducing their pro-inflammatory activity (13).

Essentially continuum or phenotypical plasticity of macrophages reflect their high adaptation potential to different changes of environment with simultaneous reprogramming to required effector and secretory reactions.

Taking into consideration the above data, it appears to be convenient and reasonable to classify macrophages to M1 and M2 phenotypes which should be considered as two polar forms of existence of active macrophages in organism. However we need to take into account the existence of intermediate and mixed phenotypes and, in studying behavior of macrophage in different environment and in concrete pathologies, information about functional activity, expression of certain markers and metabolic features appear to be useful for the determination of the direction of immune response in pathology, the role of macrophages in pathogenesis and the influence on macrophage phenotype on the whole course of inflammatory and immune reaction.

Macrophages in Pulmonary Diseases

Alveolar macrophages represent a heterogeneous population of different phenotypes at different stages of activation and differentiation. These macrophages are also capable to change to a certain phenotype (26). The process of alternative transformation of macrophage phenotype to secretory activity has been defined as "re-programming" (44). The phenotypic polarisation of macrophages, i.e. the ratio of macrophage phenotypes M1/M2, can essentially change in pathological conditions and pulmonary diseases. This, in turn, can lead to prevalence of one or another way of T-cells differentiation (Th1/Th2). In one case the change of macrophages phenotype polarisation represents adequate "therapeutic" response and provides recovery, otherwise the response becomes inadequate "pathogenic" and can promote progression of disease and even lead to fatality. Really, it has been shown that in various pulmonary inflammatory diseases, acute and chronic, essential changes of the immune response are observed, and in particular, change of cytosis in airways and change of effector and secretory functions of alveolar macrophages.

Immune Response Changes in Chronic Obstructive Pulmonary Disease (COPD)

Predominant quantity of alveolar macrophages among the cells of broncho-alveolar lavage fluid is observed in COPD patients (4). However, phagocytic activity is decreased. These data were confirmed by Sandra Hodge *et al.* in addition have shown that in patients with COPD who had given up smoking or do not smoke, the phagocytic activity of alveolar macrophages was slightly more than seen in smokers with COPD (17). It was established that the decrease of phagocytic ability of macrophages is connected with the influence of hypoxia and pyrexia on the inflammation (10).

The same factors (hypoxia and pyrexia) determine the increase of secretary function of alveolar macrophages, and, mainly, the hypersecretion of proinflammatory mediators (23). It is established that initiating of cytokine release in COPD is also connected with direct influence of cigarette smoke, namely: smoke enhanced expression of IL-8 gene and its release by epithelial bronchial cells, intensified the release of TNF-alpha and IL-6 by alveolar macrophages, and oxidants of cigarette smoke cause release of proinflammatory cytokines - IL-1 and IL-8 by macrophages and epithelial cells (7). Besides, in the sputum of COPD patients, an increase in the levels of IL-6, IL-1ß, TNF- and IL-8 was observed, in comparison with healthy individuals (Kolling *et al.* 2001) and exacerbation of the disease was marked.

A number of studies for definition of changes in superficial M1 and M2 macrophage markers in COPD patients were carried out in view of the fact that increase of proinflammatory cytokine production can reflect changes of macrophage phenotype towards M1 phenotype. It has really been proved that in comparison with healthy individuals alveolar macrophages of COPD patients express M1 phenotype and track the adaptive immune response across the path of cell-mediated reactions (Th1) (24, 33).

Immune Response Changes in Bronchial Asthma

In bronchial asthma as well as in COPD population of airway cells we can also see the predominance of alveolar macrophages with the decreased phagocytic function. According to the studies conducted by Petrova *et al.* (2005) and other authors the decrease of phagocytic index of alveolar macrophages correlated with the severity of bronchial asthma (30). Similar data were received by Anne M. Fitzpatrick et al. (2008) in patients with bronchial asthma (11).

In the definition of superficial markers of expression of M1 and M2 phenotype in bronchial asthma, unlike COPD, alveolar macrophages of M2 phenotype are included (40). Nowadays, Th2-type of the immune response is considered to be the central, molecular mechanism, determining the development of bronchial asthma (43).

The results of comparative analysis of cytokine profiles in bronchial asthma and COPD are presented in Table 6. The pattern of cytokine production confirms that the immune response in COPD mainly develops as Th1 way, whereas in bronchial asthma - as Th2.

Table 6. Cytokine profile in bronchial asthma and COPD (7)

	Bronchial asthma	COPD
IL-4 (Th2 cytokine)	++	±
IL-5 (Th2 cytokine)	++	±
IL-10 (Th2 cytokine)	+ or -	?
IL-13 (Th2 cytokine)	+	?
IL-8 (Th1 cytokine)	±	++
IL-1ß (Th1 cytokine)	?	+
TNF- α (Th1 cytokine)	+	+
TGF-ß (Th2 cytokine)	+	+

Immune Response Changes in Sarcoidosis

Sarcoidosis nowadays is considered to be one of the most abundant interstitial lung diseases with unknown origin (38). Clinical implications of sarcoidosis are nonhomogeneous and essentially depend on ethnic and geographical location of patients, their sex, age and professional activities (1, 3, 8, 31). According to the recent studies specific alveolar macrophages in sarcoidosis produce increased amount of proinflammatory cytolines (especially, TGF - α, IL-2, IFN-γ). Apart from the above-listed factors, phase of the disease and overexpression of the surface markers peculiar to phenotype M1 was noted (34, 45, 46). According to these data the evolution of immune response in sarcoidosis originates as Th1 response.

Immune Response Changes in Pneumonia

Pneumonia is a group of the acute inflammatory diseases caused by various infectious agents. In noncomplicated disease, bacteria or viruses induce overproduction of Th1 cytokines in cell-mediated response (TNF, IL-6, IL-12, IL-1 ß) and chemokines (CCL2, CCL5 and CXCL8). In the meantime predominant polarisation of alveolar macrophages in M1 phenotype is observed (5). It results in quick development of innate and the following cell-mediated Th1 adaptive immune response assuring successful eradication of pathogenic bacteria and viruses.

However, in some cases, polarisation of macrophages in M1 phenotype can become uncontrollable, e.g. Escherichia coli or some types of Streptococcus. This can result in serious disorders of immune response, tissue damage or multiple organ failure (5, 29, 32).

Usually M1 phenotype formation is a "therapeutic" immune reaction in response to development of acute bacterial or virus infections. However some bacterial agents such as Salmonella, Mycobacterium, Coxiella burnetii are able to disturb the process of polarisation of alveolar macrophages in M1 phenotype or stimulate M2-polarization (5). M2 macrophages and Th2 immune response possess substantially smaller bacterial growth-inhibitory activity as compared to M1 macrophages and Th1 immune response. In the infectious process, the re-programming of macrophages in M2 phenotype on the one hand probably is a part of strategy of survival for pathogenic microorganisms, and on the other hand forms pathogenic branches of chronic infectious diseases with a great number of complications leading to fatal cases.

Generally, conception of M1/M2 macrophages polarization and Th1/Th2 paradigm allow us to formulate at least two significant trends in the clinical picture of pulmonary diseases:

In the first place, phenotyping of pulmonary macrophages, i.e. definition of their phenotype (M1/M2), can become an informative diagnostic and predictive tool and the efficacy endpoint of drug therapy in the management of pulmonary diseases;

In the second place, targeted re-programming of alveolar macrophages phenotype for the purpose of correction of broken immune response in lungs can become an extremely effective therapeutic way of management in various pulmonology.

The work is executed with support from the grant of the Russian Foundation for Basic Research № 09-04-00976 and grant of Federal Target Program "Scientific, academic and teaching staff of innovative Russia" government contract № 02.740.11.5033

References

1. Ainslie G.M., Bateman E.D. The clinical spectrum and natural history of pulmonary sarcoidosis in Southern Africa. Abstract book of 7th WASOG Congress in Stockholm June 16-19 2002., Abstr. N 20
2. Alesenko AV, Bugrova AE, Dudnik LN, Korobko VG, Shingarova LN. The role of sphingomyelin cycle and tumor necrosis factor alpha in toxic effect of beta-amyloid peptide. Zh Nevrol Psikhiatr Im S S Korsakova. 2004; 104(3):55-61 (in Russian).
3. Almonacid C, Girón RM, Pérez-Arellano JL, et al. Sarcoidosis. En: Villasante C (editor). Enfermedades respiratorias. Madrid: Aula. Médica, 2002; 599-610

4. Barnes PJ. Alveolar Macrophages as Orchestrators of COPD. J Chronic obstructive pulmonary disease, 2004, Vol.1: 59-70
5. Benoit M, Desnues B and Mege JL. Macrophage Polarization in Bacterial Infections. J Immunol, 2008, 181:3733 -3739
6. Chan G, Bivins-Smith ER, Smith MS, Smith PM, Yurochko AD Transcriptome analysis reveals human cytomegalovirus reprograms monocyte differentiation toward an M1 macrophage. J Immunol., 2008, 181(1): 698-711.
7. Chung KF. Cytokines in chronic obstructive pulmonary disease. Eur Respir J 2001; 18:50S-59S
8. Edmondstone WM; Wilson AG. Sarcoidosis in Caucasians, Blacks and Asians in London. Br J Dis Chest 1985, 79(1):27-36
9. Edwards JP, Zhang X, Frauwirth KA, and Mosser DM. Biochemical and functional characterization of three activated macrophage populations. J Leukoc Biol., 2006, 80: 1298-1307.
10. Ferrara F, D'Adda D, Falchi M, Dall'Asta L The macrophagic activity of patients affected by pneumonia or chronic obstructive pulmonary disease. Int J Tissue React. 1996;18(4-6):109-14,
11. Fitzpatrick A, Holguin F, Teague W, Brown L. Alveolar macrophage phagocytosis is impaired in children with poorly controlled asthma. J Allergy Clin Immunol, 2008, 121(6): 1372-1378
12. Goldmann O, von Kockritz-Blickwede M, Holtje C, Chhatwal GS, Geffers R, and Medina E. Transcriptome Analysis of Murine Macrophages in Response to Infection with Streptococcus pyogenes Reveals an Unusual Activation Program. Infect and Immun., 2007, 4148-4157.
13. Gratchev A, Kzhyshkowska J, Kothe K, Muller-Molinet I, Kannookadan S, Utikal J, Goerdt S. Mphi1 and Mphi2 can be re-polarized by Th2 or Th1 cytokines, respectively, and respond to exogenous danger signals. Immunobiology., 2006, 211: 473-486.
14. Gratchev A, Kzhyshkowska J, Utikal J, Goerdt S. Interleukin-4 and dexamethasone counterregulate extracellular matrix remodelling and phagocytosis in type-2 macrophages. Scand J Immunol., 2005, 61: 10-17.
15. Hayes MP, Freeman SL, Donnelly RP. IFN-gamma priming of monocytes enhances LPS-induced TNF production by augmenting both transcription and MRNA stability. Cytokine., 1995, 7(5): 427-435.
16. Herrero C, Hu X, Li WP, Samuels S, Sharif MN, Kotenko S, Ivashkiv LB. Reprogramming of IL-10 activity and signaling by IFN-gamma. J. Immunol., 2003, 171(10): 5034-5041.
17. Hodge S, Hodge G, Ahern J, Jersmann H, Holmes M, Reynolds PN. Smoking alters alveolar macrophage recognition and phagocytic ability: implications in chronic obstructive pulmonary disease. Am J Respir Cell Mol Biol., 2007, Vol. 37: 748-755
18. Hume DA, Ross IL, Himes SR, Sasmono RT, Wells CA, and Ravasi T. The mononuclear phagocyte system revisited. J Leukoc Biol., 2002, 72: 621-627.
19. Janeway CA, Travers P, Walport M, Shlomchik M. Immunobiology. The immune system in health and disease. - Garland Science Publishing, 2005.

20. Kreider T, Anthony RM, Urban JrJF, and Gause WC. Alternatively activated macrophages in helminth infections. Curr. Opin. Immunol., 2007, 19(4):448-453.
21. Lang R, Patel D, Morris JJ, Rutschman RL, Murray PJ. Shaping gene expression in activated and resting primary macrophages by IL-10. J. Immunol., 2002, 169: 2253-2263.
22. Liang MD, Bagchi A, Warren HS, Tehan MM, Trigilio JA, Beasley-Topliffe LK, Tesini BL, Lazzaroni JC, Fenton MJ, Hellman J. 2005. Bacterial peptidoglycan-associated lipoprotein: a naturally occurring Toll-like receptor 2 agonist that is shed into serum and has synergy with lipopolysaccharide. J. Infect. Dis. 191: 939-948;
23. Lobanova E.M., Taganovich A.D. Stimulated alveolar macrophages phagocytic activity in the hypothermic and hypoxic conditions. Medical Journal of Belarus, 2005, No 1 (11). P. 43-46
24. Löfdahl M, Wahlström J, and Sköld CM. Different inflammatory cell pattern and macrophage phenotype in chronic obstructive pulmonary disease patients, smokers and non-smokers. Clin Exp Immunol. 2006 145(3): 428-437
25. Martin TR and Frevert CW Innate Immunity in the Lungs. Proceed Am Thorac Soc 2005, 2: 403-411
26. Martinez FO, Gordon S, Locati M, and Mantovani A. Transcriptional Profiling of the Human Monocyte-to-Macrophage Differentiation and Polarization: New Molecules and Patterns of Gene Expression. J. Immunol., 2006, 177: 7303-7311.
27. Martinez FO, Sica A, Mantovani A, Locati M. Macrophage activation and polarization. Front. Biosci., 2008, 1(13): 453-461.
28. Medzhitov R, Janeway CA Jr. Innate immunity: impact on the adaptive immune response. Curr Opin Immunol 1997; 9:4-9
29. O'Reilly M, Newcomb DE, Remick D. Endotoxin, sepsis, and the primrose path. Shock, 1999, 12: 411-420.
30. Petrova O.V. Immunoligical disorders in moderate and severe bronchial asthma and the possibility of their corrections by roncoleukin. Immunology of the Ural, 2005, No 1 (4): P. 149-150
31. Pietinalho A., Hiraga Y., Hosoda Y., Lofroos A.B., Yamaguchi M., Selroos O. The frequency of sarcoidosis in Finland and Hokkaido, Japan. A comparative epidemiological study. Sarcoidosis, 1995, 12(1):61-67
32. Pinheiro da Silva F, Aloulou M, Skurnik D, Benhamou M, Andremont A, Velasco IT, Chiamolera M, Verbeek JS, Launay P, Monteiro RC. CD16 promotes Escherichia coli sepsis through an FcR gamma inhibitory pathway that prevents phagocytosis and facilitates inflammation. Nat Med. 2007, 13(11):1368-74.
33. Pons AR, Noguera A, Blanquer D, Sauleda J, Pons J, Agustí AG. Phenotypic characterisation of alveolar macrophages and peripheral blood monocytes in COPD. Eur Respir J. 2005, 25(4):647-52
34. Prasse A, Georges CG, Biller H, Hamm H, Matthys H, Luttmann W, Virchow JJC. Th1 cytokine pattern in sarcoidosis is expressed by bronchoalveolar CD4+ and CD8+ T cells. Clin Exper Immunol, 2000, 122: 241-248

35. Rottiers P, Dooms H, Fiers W and Grooten J Macrophages Induce Cellular Immunity by Activating Th1 Cell Responses and Suppressing Th2 Cell Responses, Marjory Desmedt, J Immunol, 1998, 160: 5300-5308

36. Saccani A, Schioppa T, Porta C, Biswas SK, Nebuloni M, Vago L, Bottazzi B, Colombo MP, Mantovani A, and Sica A. p50 Nuclear Factor-KB Overexpression in Tumor-Associated Macrophages Inhibits M1 Inflammatory Responses and Antitumor Resistance. Cancer Res., 2006, 66(23): 11432-11440.

37. Sag D, Carling D, Stout RD, Suttles J. Adenosine 5'-monophosphate-activated protein kinase promotes macrophage polarization to an anti-inflammatory functional phenotype. J Immunol., 2008, 181(12): 8633-8641.

38. Shmelev E.I. Differential diagnostics of interstitial pulmonary diseases. Consilium medicum, 2003, V. 5, No 4: P. 176-181

39. Sly L.M., Ho V., Antignano F., Ruschmann J., Hamilton M., Lam V., Rauh M.J., Krystal G. The role of SHIP in macrophages. Front Biosci, 2007,.1(12): 2836-2848.

40. St-Laurent J, Turmel V, Boulet L-P, and Bissonnette E. Alveolar Macrophage Subpopulations in Bronchoalveolar Lavage and Induced Sputum of Asthmatic and Control Subjects. J Asthma 2009, 46: 1-8

41. Stout RD and Suttles J. Functional plasticity of macrophages: reversible adaptation to changing microenvironments. J Leukoc Biol, 2004, 76(3): 50.

42. Umemura N, Saio M, Suwa T, Kitoh Y, Bai J, Nonaka K, Ouyang GF, Okada M., Balazs M., Adany R., Shibata T., Takami T. Tumor-infiltrating myeloid-derived suppressor cells are pleiotropic-inflamed monocytes/macrophages that bear M1- and M2-type characteristics. J Leukoc Biol. 2008, 83(5): 1136-1144.

43. Woodruff PG, Modrek B, Choy DF, Jia G, Abbas AR, Ellwanger A, Arron JR, Koth LL, Fahy JV. T-helper type 2-driven inflammation defines major subphenotypes of asthma. Am J Respir Crit Care Med. 2009, 180(5):388-95

44. ZhangX., Morrison D.C. J. Exp. Med. 1993. Vol. 177, N 2. P. 511-516

45. Zeyda M., Farmer D., Todoric J., Aszmann O., Speiser M., Györi G., Zlabinger G.J., Stulnig T.M. Human adipose tissue macrophages are of an anti-inflammatory phenotype but capable of excessive pro-inflammatory mediator production. Int J Obes (Lond). 2007, 31(9): 1420-1428.

46. Zheng L, Teschler H, Guzman J, Hübner K, Striz I, Costabel U. Alveolar macrophage TNF-alpha release and BAL cell phenotypes in sarcoidosis. Am J Respir Crit Care Med. 1995, 152(3):1061-6.

47. Zissel G, Prasse A, Müller-Quernheim J. Sarcoidosis--immunopathogenetic concepts. Semin Respir Crit Care Med. 2007, 28(1):3-14.

Adaptation Biology and Medicine (Volume 6: Cell Adaptations and Challenges)
Editors: P. Wang, C.-H. Kuo, N. Takeda and P.K. Singal
Copyright © 2011 Narosa Publishing House Pvt. Ltd., New Delhi, India

17. Lung Surfactant System: Effects of Gravity and Restraint Stress

Irina G. Bryndina[1], Natalia N.Vasilieva[1], Pavel N.Gerasimov[1], and Alan R.Hargens[2]

[1]Department of Pathophysiology, Izhevsk State Medical Academy, Izhevsk, 426034 Russia

[2]Department of Orthopaedic Surgery, University of California, San Diego, CA 92103-8894 USA

Abstract

In chronic experiments (10 days) on male non-line albino rats, the surface active properties of pulmonary surfactant, total phospholipids, phosphatidylcholine and lysophosphatidylcholine content in bronchoalveolar lavage fluid were explored in four stress procedures: restraint in narrow boxes in orthostatic and antiorthostatic positions (8 hours daily), immobilization on the desk in prone and supine position (2 hours daily). The results of experiments show that various forms of prolonged stress reduce surfactant activity. The most significant increase of minimal and maximal surface tension is in restrain with 30° head down tilt, without significant changes of total phospholipids. The disturbance of surfactant properties occurs in both cases with the high and low amounts of total phospholipids and their most active fraction phosphatidylcholine. The phosphatidylcholine / lysophosphatidylcholine ratio decreased in all forms of prolonged stress studied. Our results suggest that changes of pulmonary surfactant properties induced by stress depend on high phospholipase activity. It is a component of lipid triad which includes activation of lipases, phospholipases and lipid peroxidation. A similar adaptation of the pulmonary surfactant system may occur in actual microgravity.

Keywords: simulated microgravity, stress, lung, surfactant

Introduction

The dependence of many body functions on normal Earth gravity has been formed during phylogenesis. The gravitation diminution or the changing of its vector can cause the disorientation of functional systems and elicit moderate stress. The influence of microgravity on the lungs causes the haemodynamic overload because of the shift of fluid media in organism (12, 11, 9). Also, short-term and prolonged immobilization disturbs the functional status of the pulmonary surfactant (13, 1). The complex of morphological changes in lung after acute immobilization stress is described by (10) and

*Address for Correspondence: Prof. Irina G. Bryndina Department of Pathophysiology Izhevsk State Medical Academy Kommunarov St. 281 Izhevsk, UR 426034 Russia E-mail: bryndina@udm.net

named “stressed lung”. This complex includes interstitial edema, multiple hemorrhages, red blood cell appearance in alveolar space and decreased lipids on the alveolar surface.

The purpose of this research is to study the characteristics of the surfactant system of lungs on microgravity stress model in comparison with other stress influences. We hypothesized that all four stress procedures in rats (restraint in narrow boxes in orthostatic and antiorthostatic positions, immobilization on the desk in prone and supine position) reduce surfactant activity in the lung as compared to normal control animals.

Methods

Chronic experiments were performed on 54 adult male non-line albino rats (3 month age, weighing 180-240 g, 6-14 per group). The IRB Committee at the Izhevsk State Medical Academy approved all procedures. The immobilization was made by fixing an animal to the desk in supine (n=14) or prone (n=10) position for 2 hours daily during 10 days. Two groups of rats were immobilized by placement into narrow boxes (restraint system) on even surface (n=10) or by placing in antiorthostatic position at the angle of 30° to the surface for 8 hours during 10 days (n=6). Normal, ambulating rats (n=14) were used as a control group. The status of the pulmonary surfactant was estimated according to the parameters of surface activity of the bronchoalveolar lavage fluid, according to the total phospholipids and cholesterol amount and fractional composition of the phospholipids. Minimal and maximal surface tension were measured using modified methods of Wilhelmy and du Noüy in condition of compression (to 25% surface) or extension (to 100%) of surfactant films in Langmuir trough (3). Also index of the film stability was calculated (2). Principal phospholipid content and its sum (total phospholipids) were estimated according to (8) using the thin layer chromatography on silica gel plates (Merck, Germany) with evaluation of spots by UV light densitometer (Sorbfil, Russia). Gravimetric index (pulmonary coefficient) was calculated as the percentage of wet lung mass to body mass. Stress was confirmed by the growth of relative suprarenal gland mass. Data were analyzed using standard Microsoft Excel program. The Student t test, with a $p < 0.05$ significance level, was used to identify significant differences between groups.

Results

A decrease of the surface activity of the pulmonary surfactant occurred in all cases compared to normal control rats. The most significant changes were registered in the microgravity stress model: the minimal surface tension was 73% increased and maximal 49% increased in comparison with the control group while the stability index was 25% decreased (Fig. 1). All animals of these groups had clearly marked cyanosis and pulmonary hemorrhage. Lung coefficient also was significantly increased only in the case of restraint with head down tilt (Fig. 2).

Animals which were placed in narrow boxes had threefold amount of phospholipids. In rats with the model of gravity stress, there was a twofold amount of phospholipids, but these changes were not significant because 2/3 of animals had an elevated amount of phospholipids (1/3 more than twice) and 1/3 of animals had more than 5 times decreased amount of phospholipids. It is important to note that animals with low content of phospholipids had low indices of the relative mass of adrenal glands. In both experiments (with immobilization in narrow boxes and gravity stress model) there were 31% and 47% reductions of cholesterol respectively in comparison with control rats (Fig. 3).

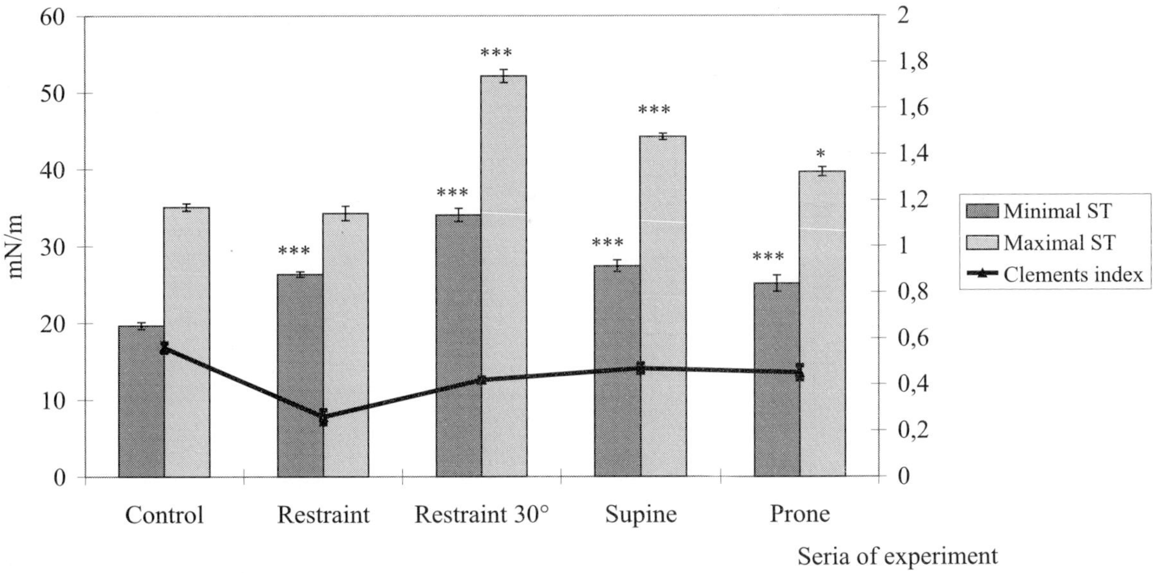

Fig. 1. Surface tension of pulmonary surfactant films in multiple restraint in narrow boxes and immobilization in supine and prone position. Values are means ± SE.
Denotes * - p<0.05, ** - p<0.01, * - p<0.001 between control and stress**

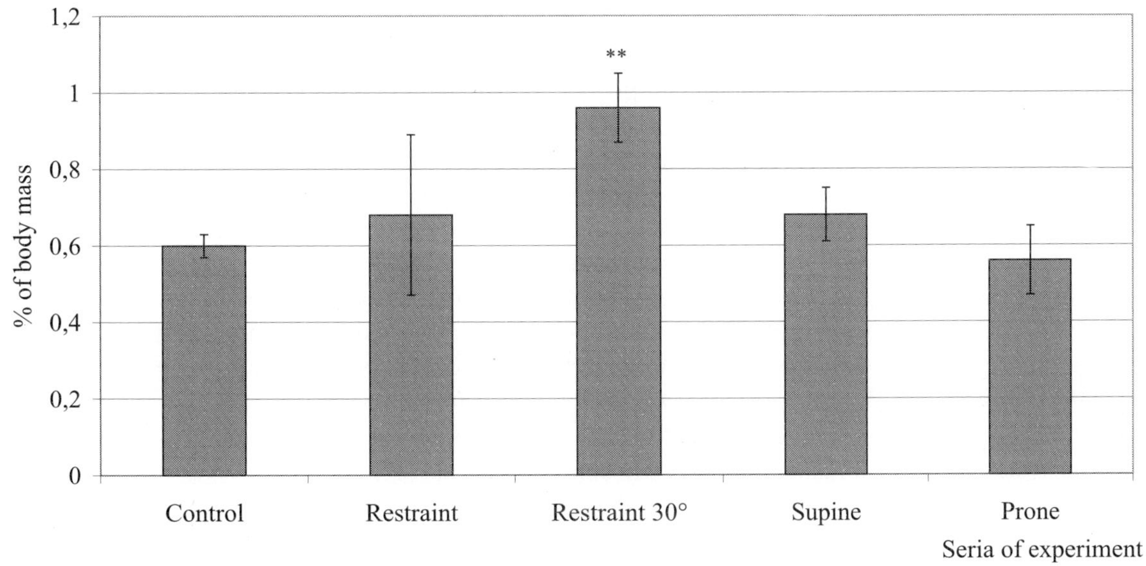

Fig. 2. Lung coefficient in multiple restraint in narrow boxes and immobilization in supine and prone position. Values are means ± SE.
Denotes * - p<0.05, ** - p<0.01, * - p<0.001 between control and stress**

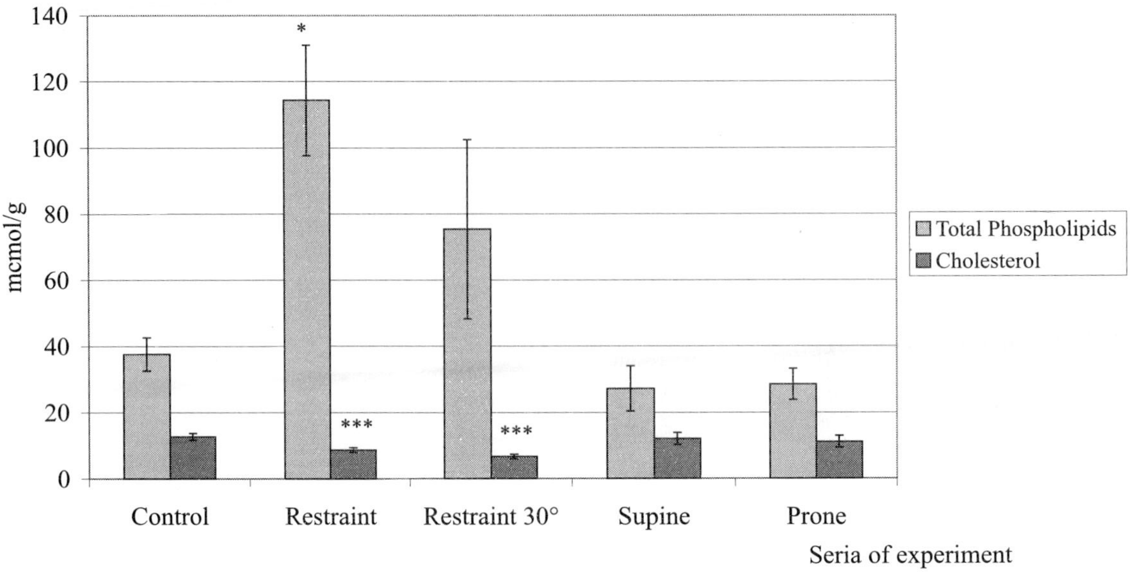

Fig. 3. Total phospholipids and cholesterol content in bronchoalveolar lavage fluid in multiple restraint in narrow boxes and immobilization in supine and prone position. Values are means ± SE. Denotes * - p<0.05, ** - p<0.01, * - p<0.001 between control and stress**

Studying the fractional composition of alveolar phospholipids, the following findings were evident: repeated immobilization of animals in narrow boxes led to growth of phosphatidylcholine amount (58% higher in comparison with control data), the growth of lysophosphatidylcholine was more significant – 23 times as high as in comparison with the non-stressed rats, the phosphatidylcholine/ lysophosphatidylcholine ratio was 5 times lower. It led to instability of surfactant and probably played the leading part in deterioration of surface activity of lungs in stress. The same changes were revealed in gravity stress model: higher level of lysophosphatidylcholine, the growth of phosphatidylcholine amount and the lowering of lysophosphatidylcholine/phosphatidylcholine index. Unlike the experiments with the motion limitation, the rigid immobilization wasn't accompanied by the growth of total phospholipids in both cases (supine or prone position) but the tendency to the diminution of them occurred. The lysophosphatidylcholine content also increased (18 times more) but the phosphatidylcholine content decreased more then 3 times; phosphatidylcholine/lysophosphatidylcholine index became 39 times less in comparison with control rats (Fig. 4).

In all cases the enlarge of adrenal gland was detected and the most significant growth of adrenal gland mass was 48% in animals immobilized in supine position (Fig. 5).

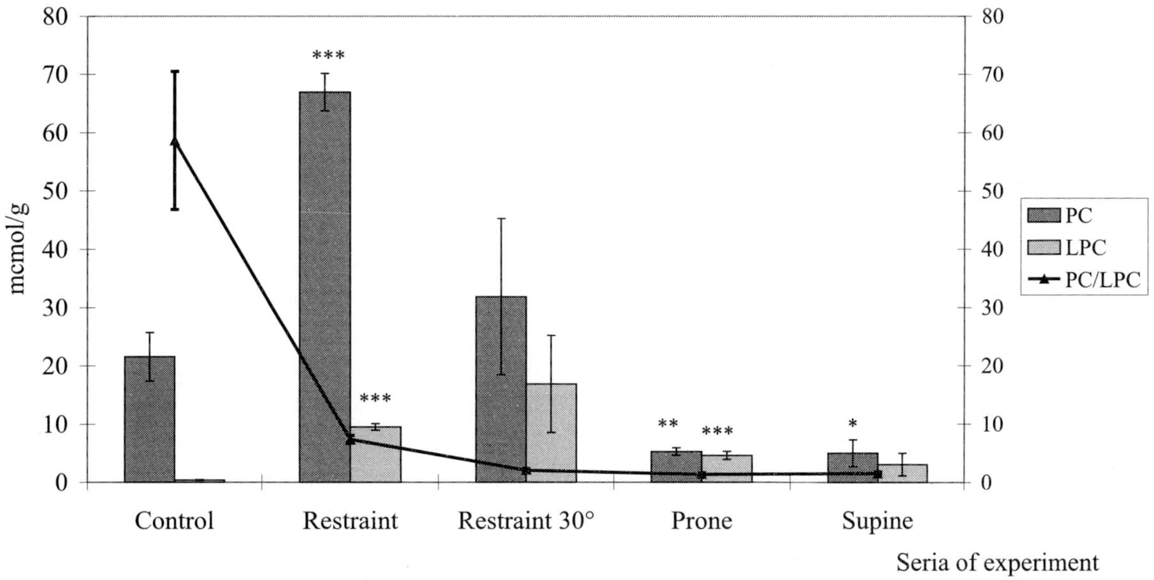

Fig. 4. Phosphatidylcholine and lysophosphatidylcholine content in bronchoalveolar lavage fluid in multiple restraint and immobilization in prone and supine position. Values are means ± SE. Denotes * - p<0.05, ** - p<0.01, * - p<0.001 between control and stress**

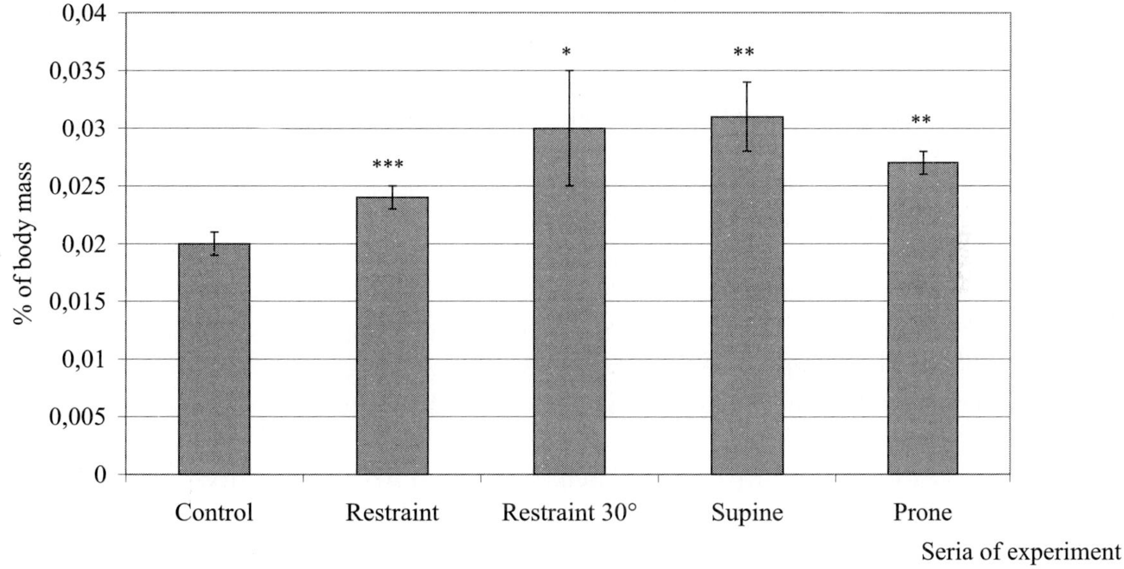

Fig. 5. Relative mass of adrenal glands in multiple restraint in narrow boxes and immobilization in supine and prone position. Values are means ± SE. Denotes * - p<0.05, ** - p<0.01, * - p<0.001 between control and stress**

Discussion

The results of this research document that all kinds of prolonged stress, used in the current experiments, lowers surfactant activity. The most significant increase of minimal and maximal surface tension is in restraint with head down tilt. We suggest that this result is caused by the shift of body fluid in head direction and alteration of lung haemodynamics on the background of another stress effects presenting in different stress models. It is interesting that disturbance of surfactant properties occurs in both cases with the high and low amounts of phospholipids and their most active fraction - phosphatidylcholine. The phosphatidylcholine/lysophosphatidylcholine ratio is decreased in all series. This finding indicates that the conversion of phosphatidylcholine to lysophosphatidylcholine is intensified through phospholipase hydrolysis. Stress is known to be characterized by the triad which includes activation of lipases, phospholipases and lipid peroxidation (7). The products of these reactions show the detergent action on cell membrane and surfactant complex, covering the alveoli. According to (4, 5), the products of free radical reactions play an important part in damaging surfactant in stress (both acute and chronic). The importance of phospholipase mechanism is documented by our experiments. These two mechanisms may play dominant roles in dysfunction of surfactant and deterioration of the structures of alveolar barrier associated with stress (6).

In summary, prolonged immobilization in narrow boxes, fixation on the desk and changing the vector of gravitation, all cause deterioration of the surface-active properties of pulmonary surfactant and restraint with head down tilt produces the greatest changes.

References

1. Bryndina IG and Danilov GE. Substance P as a factor enhancing resistance of the surfactant lung system to chronic immobilization stress. *Ross Fiziol Zh Im IM Sechenova.* 2002;88: 84-89.
2. Clements JA, Hustead RF, Johnson RP and Gribetz I. Pulmonary surface tension and alveolar stability. *J Appl Physiol.* 1961;16: 444-450.
3. Hills BA and Barrow RE. An 'engine' phenomenon displayed by monolayers of a pulmonary surfactant cycled to steady state. *Phys Med Biol.* 1984,29: 1399-1408.
4. Kovacheva S and Ribarov SR. Pulmonary xanthine oxidase activity of rats exposed to prolonged immobilization stress. *Gen Physiol Biophys.* 1998;17: 377-384.
5. Kovacheva S and Ribarov SR. Lipid peroxidation in lung of rat stressed by immobilization: effects of vitamin E supplementation. *Lung* 1995;173: 255-263.
6. Kovacheva-Ivanova S and Savov J. Immobilization and ultrastructure of alveolar type II cells. *Acta Physiol Pharmacol Bulg.* 1994;20: 1-8.
7. Meerson FZ, Kagan VE, Kozlov YuP, Belkina LM and Arkhipenko YuV. The role of lipid peroxidation in pathogenesis of ischemic damage and the antioxidant protection of the heart. *Basic Res Cardiol.* 1982;77: 465-485.
8. Pokrovskii EA and Kargapolov AV. Modification of the thin-layer chromatography method of phospholipid determination. *Lab Delo.* 1972;6: 337-341.
9. Prisk GK. Microgravity and the lung. *J Appl Physiol.* 2000;89: 385-396.

10. Rozova EV, Seredenko MM and Meerson FZ. The electron microscopic characteristics of stress lung. *Biull Eksp Biol Med.* 1989;108: 735-738.
11. Verbanck S, Larsson H., Linnarsson D., Prisk GK, West JB and Paiva M. Pulmonary tissue volume, cardiac output, and diffusing capacity in sustained microgravity. *J Appl Physiol.* 1997;83: 810-816.
12. West JB and Dollery CT. Distribution of blood flow and ventilation-perfusion ratio in the lung, measured with radioactive CO_2. *J Appl Physiol.* 1960;15: 405-410.
13. Yanev E, Momchilova-Pankova A, Markovska T, Koumanov K, Kenarov P., McGuigan FJ and Nicolov N. Influence of immobilization stress on the phospholipid composition of alveolar surfactant and lungs in rats. *Pavlov J Biol Sci.* 1990;25: 25-28.

Adaptation Biology and Medicine (Volume 6: Cell Adaptations and Challenges)
Editors: P. Wang, C.-H. Kuo, N. Takeda and P.K. Singal
Copyright © 2011 Narosa Publishing House Pvt. Ltd., New Delhi, India

18. The Thioredoxin and Thioredoxin Reductase System: Overexpression in Thyroid Cancer

David T. Lincoln[1*], Pawan K. Singal[2], Matthew Burge[3], Fatma Al-Yatama[4], Fawziah M.A. Mohammed[4], Anwar G. Al-Banaw[4], Maie Al-Bader[5] and Fred Sinowatz[6]

In Memory of Moyra Lincoln, who lost her battle with Cancer

[1]Independent Research Foundation, Chapel Hill, QLD 4069, Australia; [2]Institute of Cardiovascular Sciences, University of Manitoba, Winnipeg, Canada; [3]Cancer Care Services, Royal Brisbane Hospital, Herston, QLD. 4029, Australia, [4]Department of Medical Laboratory Sciences, Faculty of Allied Health Sciences, Kuwait University, Kuwait; [5]Department of Physiology, Faculty of Medicine, Kuwait University, Kuwait; [6]Institute of Veterinary Anatomy, University of Munich, Germany.

Abstract

Thyroid cancer is the second most common malignancy following breast cancer in Arab females. Thioredoxin (TRX) is a small multi-functional redox protein with both intracellular and extracellular functions. The protein exists in either a reduced form (thioredoxin-SH_2) or an oxidized form (thioredoxin-S_2). TRX acts like an enhancement for growth factors and stimulates the growth of cancer cells. In this study of thyroid neoplasms, involving 116 female and 60 male patients, expression of TRX and TRX-R was studied using purified mouse anti-human TRX monoclonal antibody and anti-human TRX-R antiserum from rabbits, respectively. In order to delineate tumour cell growth, proliferating cell nuclear antigen (PCNA) polyclonal antibody was used. Compared to normal thyroid tissue, expression of TRX and TRX-R was increased in the cytoplasm and nuclei of thyroid cancer cells. Furthermore, TRX expression correlated with that of TRX-R. Of the 176 thyroid neoplasms investigated, overexpression of TRX-R was found in different types of neoplasms. The majority of carcinomas showed a correlation between strongly positive TRX and TRX-R expression and a number of positive lymph nodes. In addition, there was a positive correlation with neoplastic cellular proliferation, as measured by PCNA. This indicates that increased TRX and TRX-R expression may be associated with tumourigenesis by acting as an autocrine growth stimulus. This study suggests that TRX immunoreactivity in thyroid tumours is a function of malignancy and cancer progression. In addition, secreted TRX can also act as an extracellular growth factor for both normal and tumour cells and enhance the sensitivity of the cells. Furthermore, this

*Address for Correspondence: Dr. David T. Lincoln, Independent Research Foundation, Chapel Hill, QLD 4069, Australia; Tel: (+61) 7 3878 2417; E-mail: docdtl@yahoo.com.au

study emphasizes the potential benefits of anti-TRX/TRX-R agents in cancer therapeutics in the treatment of thyroid cancers.

Keywords: Thyroid gland neoplasm, thioredoxin, thioredoxin reductase, redox activity, cellular proliferation.

Introduction

Thyroid cancer is a common endocrine malignancy with an incidence of approximately 0.5-10 cases per 100,000 (Sedliarou *et al.* 2007) and represents 0.5-1% of human cancers. According to WHO guidelines, it is classified as follicular, papillary, anaplastic or medullary carcinoma and 80-90% are highly differentiated tumours (follicular and papillary) with good prognosis and five year survival of 95%. The anaplastic tumours are undifferentiated with a poor outcome. Thyroid cancer differs from many other neoplasms and that incidence and type vary widely in various geographical areas and it has a wide spectrum of clinical behaviour and therapeutic responsiveness. Papillary carcinoma (PTC) is the most common malignant tumour, accounting for 72% of all thyroid cancers in the United States, while 17% had follicular carcinoma. Anaplastic and medullary carcinoma represent 4% and 3% of the cases respectively, whereas lymphoma and Hurtle cell carcinoma represent 1% each (Hundahl *et al.* 1998). Follicular thyroid carcinoma (FTC) and anaplastic thyroid carcinoma (ATC) are more frequent in endemic goiter regions than in goiter free areas (Bakiri *et al.* 1998). It is postulated that goitrous thyroid is a precursor lesion to the development of malignant thyroid diseases (Sugg *et al.* 1998) and Madhavan and Othman (1996) reported that 36.8% of surgically resected goitrous thyroid enlargements harboured PTC. Although FTC contributes a lesser percentage, it is necessary to have a clear-cut differentiation between PTC and FTC and clinical management and prognosis depend much on the diagnostic reliability of histopathological examination of the suspicious thyroid lesion.

Thyroid cancer and the majority of benign thyroid diseases (BTD) (e.g. nodule/adenoma, non-endemic goitre, hypo- and hyperthyroidism, Graves disease, thyroiditis) occur 3-5 times more frequently among women. This female predominance, greatest during reproductive ages, is observed in all geographical areas and ethnic groups. Evidence from epidemiological and laboratory studies suggest that reproductive factors and patterns may influence, or contribute to, the risk of BTD and thyroid cancer in women. Most cancers found during pregnancy are differentiated and have an excellent prognosis. Nevertheless, they may persue a different clinical course when develop in pregnant women. Thyroid nodules which develop in these patients appear to have a greater risk of malignancy and should be evaluated aggressively.

Thyroid cancer is the second most common cancer (after breast) among women in Kuwait, accounting for 8.1% of all cancers among Arab women. The average annual incidence rate (per 100,000) is 5.3 and the age-standardized rate is 7.7 in Kuwaiti women (Kuwait Cancer Registry, unpublished data). Similarly high relative frequency and rates of thyroid cancer among women have also been reported from other countries in the Gulf region. Data from Oman, Qatar, Saudi Arabia and the United Arab Emirates show that thyroid cancer is also the second most common cancer among women in these countries, accounting for 7.7%, 10.9%, 8.8% and 12.4% of all female cancers, respectively (Memon *et al.* 2004).

Exposure to high-dose of ionising radiation, particularly in childhood and adolescence, and previous hyperplastic thyroid disease (i.e. nodule/adenoma, goitre) are two well established risk factors for thyroid cancer. Radiation induces fibrosis within the capsule and decreases thyroid function (Ozawa, 2007). From the Chernobyl nuclear accident, it is known that radiation exposure is associated with an increased incidence of circulating antithyroid antibodies, inducing hypofunction of the thyroid. A relation between hypothyroidism and circulating antithyroid antibodies (e.g. thyroid peroxidase and thyroglobulin antibodies) has been found and may precede the development of hypothyroidism (Farahati, 2000). Thyroid hormone influences the cell cycle of both normal and neoplastic cells and has a permissive role in the growth of certain solid tumours (Hercbergs, 2006).Thyroid hormone receptors are transcription factors and act in a ligand-inducible manner by binding to DNA response elements in the regulatory regions of target genes. Potentially, THR mutations can recast gene expression profile in a cell, resulting in a tendency to tumourigenesis (Kim *et al.* 2005).

The mammalian thioredoxins (TRX) are a family of redox proteins that undergo NADPH-dependent reduction by thioredoxin reductase (TRX-R) and in turn reduce oxidized cysteine groups of proteins. The two main thioredoxins are thioredoxin-1 (TRX-1), a cytosolic and nuclear form, and thioredoxin-2 (TRX-2), found only as mitochondrial form. Human TRX-1 is a small multifunctional redox protein with both intracellular and extracellular functions. The protein exists in a reduced form (thioredoxin-SH_2). Oxidation of thioredoxin occurs via disulphide substrate giving rise to thioredoxin cysteine disulphide ($TRX(SH)_2 + X\text{-}S_2 = TRXS_2$). TRX-1 is a 104 amino acid protein, predominantly found in the cytosol. TRX-2 is found in the mitochondria and is a 166 amino acid protein containing the two cysteine residues (Nakamura *et al.* 2000).

The mechanism of TRX action involves a transient, mixed disulphide intermediate and a fast thiol-disulhide exchange in a hydrophobic environment. Thioredoxin is able to act as a powerful protein disulphide oxireductase through the reversible oxidation of its active dithiol centre and through this action it is able to affect the disulphide-dependent structural changes in other molecules, causing the activation or inhibition of them. Thioredoxin performs many biological actions including the supply of reducing equivalents to thioredoxins peroxidases and ribonucleotide reductase, the regulation of transcription factor activity, and the regulation of enzyme activity by heterodimer formation. Thioredoxin stimulates cell growth and is an inhibitor of apoptosis and may play a role in a variety of human diseases, including cancer. An increased level of thioredoxin has been reported in several human tumours, where it is associated with aggressive tumour growth (Al-Yatama *et al.* 2008; Al-Banaw *et al.* 2008; Lincoln *et al.* 2003; Lincoln *et al.* 2005; Lincoln *et al.* 2008; Mohammed *et al.* 2008).

An investigation of the role of TRX is not complete without consideration of the role of thioredoxin reductase, the enzyme able to reduce the active site of TRX. Mammalian thioredoxin reductases are homodimeric, flavin adenine dinucleotide containing proteins with a C-terminal selenocysteine residue. The cys residues of the conserved redox catalytic site of mammalian thioredoxin reductases –Cys-Val-Asn-Val-Gly-Cys-, undergo reversible oxidation reduction in much the same way as TRX. The selenocysteine is essential for the activity of mammalian thioredoxin reductases. Two full-length human thioredoxin reductases have been cloned, a 54.4 kDa thioredoxin reductase-1, which is predominantly cytosolic (Gasdaska *et al.* 1999), and a 56.2 kD thioredoxin reductase-2, which has a 33 amino acid N-terminal extension identified as a mitochondrial import sequence (Mustacich and Powis, 2000). Thioredoxin reductase is able to inhibit the release of cytochrome C from the mitochondria to the cytosol, preventing hydrogen peroxide accumulation in the cells, thus inhibiting the cell signal to apoptosis, since hydrogen peroxide accentuates the apoptotic process. From these findings one can conclude that thioredoxin reductase inhibits cytochrome C release and accelerates cancer development. Here we investigate the expression of thioredoxin and thioredoxin reductase system in neoplasms of the human thyroid gland.

Materials and Methods

Tumour Procurement and Preparation

Thyroid tumour tissues were obtained from the Mubarek Al-Kaber Hospital, Jabriya, Al-Amiri Hospital, Safat and Cancer Control Centre, Shuwaikh, Kuwait, Arabian Gulf. The age of the patients ranged from 16 to70 years with medium age of 36 years. The following thyroid neoplasms, obtained from 117 female and 59 male patients, were investigated: benign colloid nodule (n=15), colloid goiter (n=14), multinodular goiter (n=21), papillary oncocytic neoplasm (n=10), follicular adenoma (n=16), follicular carcinoma (n=24), invasive follicular carcinoma (n=19), papillary carcinoma (n=48) and medullary carcinoma (n=9). Tissues were fixed in phosphate-buffered 4% paraformaldehyde (ph 7.4) and processed for paraffin embedding. Sections (5 ìm thickness) were stained with haematoxylin-eosin or immunostaining procedures. Tissue location adjacent to tumour were used as controls.

Thioredoxin and Thioredoxin-reductase Immunohistochemistry

Mouse monoclonal anti-human thioredoxin (TRX) antibody 2B1 and rabbit monoclonal anti-human thioredoxin reductase (TRX-R) were prepared. The presence of thioredoxin and thioredoxin reductase was demonstrated by the streptavidin-biotin horseradish peroxidase complex (ABC) technique. Sections were deparaffinised and endogenous peroxidase was blocked by application of 3% hydrogen peroxide in Tris buffer for 5 min at room temperature (RT), followed by rinsing in Tris buffer. Non-specific protein binding was blocked by incubation with 20% normal goat serum in PBS containing 1% bovine serum albumin (BSA). Positive controls were prepared using paraffin wax embedded tissue sections with proven good reactivity. Negative controls were performed by a) omission of the primary antibody; b) substituting the respective primary antibody by non-immune normal mouse serum at different concentrations. Sections obtained from locations adjacent to the tumours, were used as normal control tissues. All incubations were carried out in humidified chambers to prevent evaporation.

Proliferative Indices and Quantitative Analysis

Immunohistochemical analysis of proliferating cell nuclear antigen (PCNA) was used to demonstrate levels of cellular proliferation of thyroid tumours by the peroxidase, anti-peroxidase (PAP) method. Tumour sections were scanned in the entire area with a counting grid and for each field the number of points in coincidence with the area to be measured was recorded (Point count method). The quantitative analysis of TRX, TRX-R and PCNA was done by calculating the percentage of stained cells in 10 low power magnification fields or 16 high power fields using a Zeiss microscope connected to a computer using image analysis software. To ensure reproducible and objective assessment of staining, 10 representative areas, each containing 1000 tumour cells, were observed under high-power field (objective lens, X 40) in a vertical section taken from the centre of the lesion. The proportion of positive cells was expressed as percentage of total cells counted.

Classification of Thyroid Malignancy

Histological differentiation of various thyroid swellings and the identification of the characteristic features of the malignancy were performed by applying standard WHO histological classification for thyroid cancers. According to the WHO classification, malignant tumours of the thyroid are subdivided into thyroid-specific, which are unique to the thyroid (papillary, follicular and medullary carcinoma), and tumours commonly found in other organs, but which still have some particular features when they occur in the thyroid gland (e.g. lymphoma, some types of sarcoma). Thyroid-specific tumours are derived from follicular cells (papillary and follicular carcinoma), and from parafollicular, calcitonin-producing C-cells (medullary carcinoma). Among the thyroid malignancies, papillary carcinoma is the most frequent type of follicular-cell derived carcinoma. Subtypes of papillary carcinoma are papillary microsarcoma, encapsulated variant, follicular variant, diffuse sclerosing variant and oxyphilic cell type. The variants of follicular carcinoma are: minimally invasive (encapsulated), widely invasive, oxyphilic (oncocytic cell type) cell type and clear cell variants.

Clinico-and Histopathological Diagnostic Evaluation of Thyroid Tumours

Histological examination and diagnosis of papillary thyroid carcinoma (PTC) is simple but has a limited value in tumours with follicular pattern, thus resulting in a preoperative diagnostic challenge. In such situations, immunohistochemistry may be of great value. The established prognostic factors of age, tumour size, prevalence of lymph node metastasis and histological classification of the carcinoma are of prognostic significance for survival in long term follow-up. The main criteria for the diagnosis of PTC is the orcurrence of ground-glass nuclei. These nuclei are enlarged, round to oval structures with pale karyoplasm condensing continously to the nuclear membrane due to cytoplasmic pseudo-inclusions. The nuclei are densely arranged and often overlap each other. Papillary carcinoma usually has papillae in its structure which are delicate stalks of epithelial cells situated on basal membranes covering stromal fibers and thin capillaries. Often the tumours contain round laminated calcifications (Svetislaw, 2003). The follicular variant of papillary carcinoma (PTCFC) has the characteristic nuclei of papillary carcinoma but has an almost totally follicular architecture.

The natural history of follicular thyroid cancer is relatively slow, and metastases usually involve bone and lung (Safar, 1999). Their differential diagnosis, based on specific morphological criteria, is problematic in thyroid pathology and various diagnostic/prognostic markers are often used. The diagnosis of follicular carcinoma is based on the histopathological demonstration of infiltrative growth. There are two criteria: 1) true infiltration of the venous vessels outside the tumour capsule, and 2) fungus-like infiltration through the tumour capsule into the surrounding parenchyma by capsular infiltration, or better extra-capsular extension.

Medullary (C-cell) thyroid carcinoma is characteristically composed of solid nests and infiltrating formations of polygonal or spindle-shaped cells. Amyloid deposits within the stroma are found in about half of the tumours. Medullary thyroid carcinoma (MTC) is a calcitonin-producing tumour of the parafollicular C-cells, accounting for 5-10% of all thyroid tumours. MTC may occur sporadically, in a familial form without associated endocrinopathies or as part of multiple endocrine neoplasia type 2A or 2B with an autosomal dominant inheritance. The clinical varieties are associated with distinct types of mutations in the RET proto-oncogene. The only effective treatment is the early and total surgical removal of all neoplastic tissue. The prognosis of patients with advanced MTC, with unresectable or distant metastases is poor, and chemotherapy or irradiation is of no significant value.

Undifferentiated (anaplastic) carcinoma displays cytological polymorphism and histological dedifferentiation. Histopathologically, this tumour reveals solid sheets of highly anaplastic cells or spindle cells with morphologically sarcoma-like areas and frequent appearance of giant cells. Most tumours show large areas of necrosis and haemorrhage. Some cases of anaplastic carcinoma have remmants of pre-existing differentiated (papillary or follicular) carcinoma, indicating a dedifferentiation (Passler *et al.* 1999).

Results

Association with Reproductive Factors

There was an association between age at first and last pregnancy and risk of thyroid cancer. Young age at first pregnancy seemed to have a protective effect, as the odds ratios were consistently below unity for ages 16-24 years. Women who had their last pregnancy at ages > 30 years were at a significantly increased risk compared to those who had their last pregnancy at earlier ages. There was also a significant trend in risk with increasing age at last pregnancy. With regard to parity, the odds ratios were consistently above unity for women who have more than two live births. Women with > five births were at a modestly increased risk compared to those who had fewer children. There was also some indication of an increasing trend in risk with increasing number of live births (p-trend=0.08). In contrast, miscarriage (spontaneous abortions) seemed to have a protective effect on the incidence of thyroid cancer as the odds ratios were consistently below unity, compared to the reference category. Papillary carcinoma was the commonest (58%) of the tumours, followed by follicular (20%) and anaplastic (11%). Papillary and follicular carcinomas were more frequent during the third and fourth decades of life. Anaplastic tumours were commonest during the seventh and eight decades. Goitre was the commonest clinical presentation (81%). All cases with confirmed histology were staged according to the TNM system recommended by the International Union Against Cancer.

Expression of Thioredoxin (TRX)

Compared to normal thyroid tissue cells, intensity of thioredoxin immunoreactivity in the neoplasms of the thyroid was increased in both the nuclei and cytoplasm in the majority of cases investigated. Table 1 shows the clinicopathological characteristics and thioredoxin expression, obtained from counting a total of 1000 cells in each of 10 representative fields from each case of thyroid neoplasm investigated. Intensity of staining as a measure of TRX expression was present to varying degree in all tumour specimens and consistently showed prominent immunoreactivity in malignant follicular-, papillary- and invasive follicular carcinoma tumour cells and to a lesser extend in the cells of the benign neoplasms. Expression of the TRX was predominantly cytoplasmic, but strong nuclear immunoreactivity was also apparent in many instances (Figures 1 – 4). Increased expression of TRX immunoreactivity was found in 15.1% of the 15 benign colloid nodule cases, in 38.0% of colloid goiters, in 43.4% of the 21 cases multinodular goiters, in 41.7% of the papillary oncocytic neoplasms, in 36.0% of the follicular adenomas cases, in 65.2% of the 24 follicular carcinomas cases, in 72.8% of the 48 cases of papillary carcinomas, in 69.5% of all 19 cases of invasive follicular carcinomas and in 79.6% of all medullary carcinomas investigated. The strong immunoreactivity exhibited with the anti-human thioredoxin antibody 2B1 is concordant with the specificity of this antibody. Tissue sections incubated with 20%

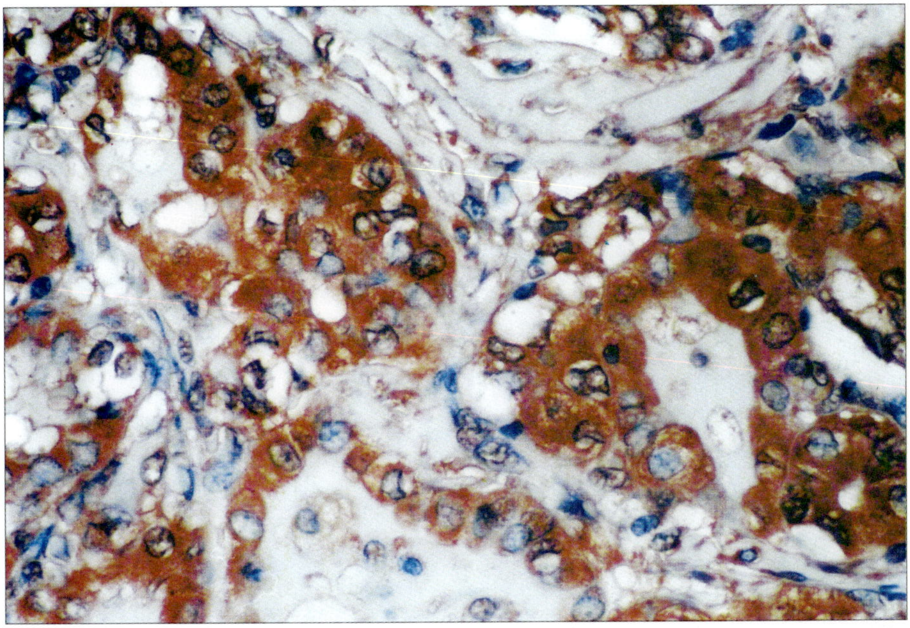

Fig. 1. Thioredoxin (TRX) expression in a papillary carcinoma of the thyroid from a 56 year old female patient (2647/98), using MAb 6F3 and visualisation with 3-amino-9-ethylcarbazole.

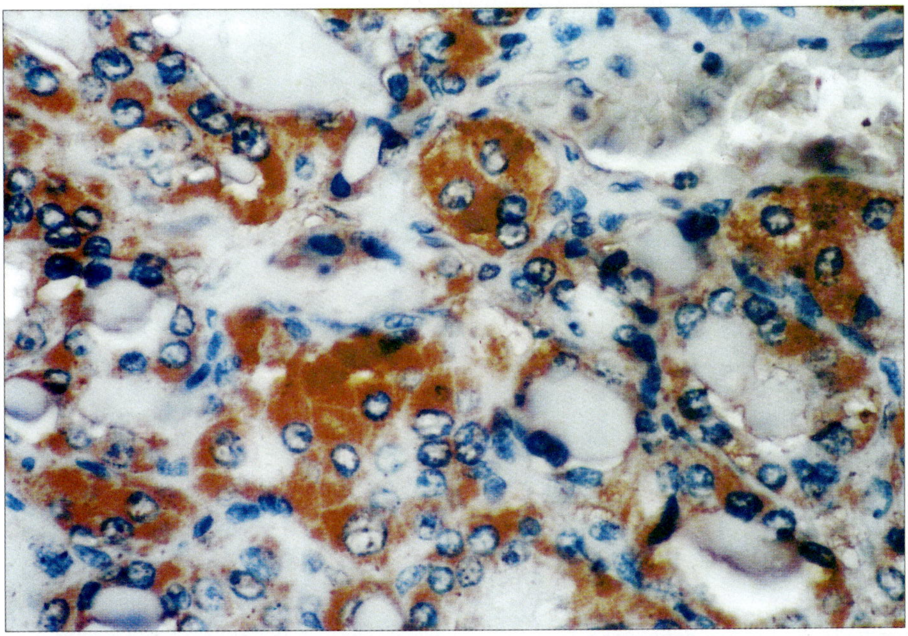

Fig. 2. Thioredoxin (TRX) expression in a follicular carcinoma of the thyroid from a 60 years old male patient (2395/95), using MAb 6F3 and visualization with 3-amino-9-ethylcarbazole.

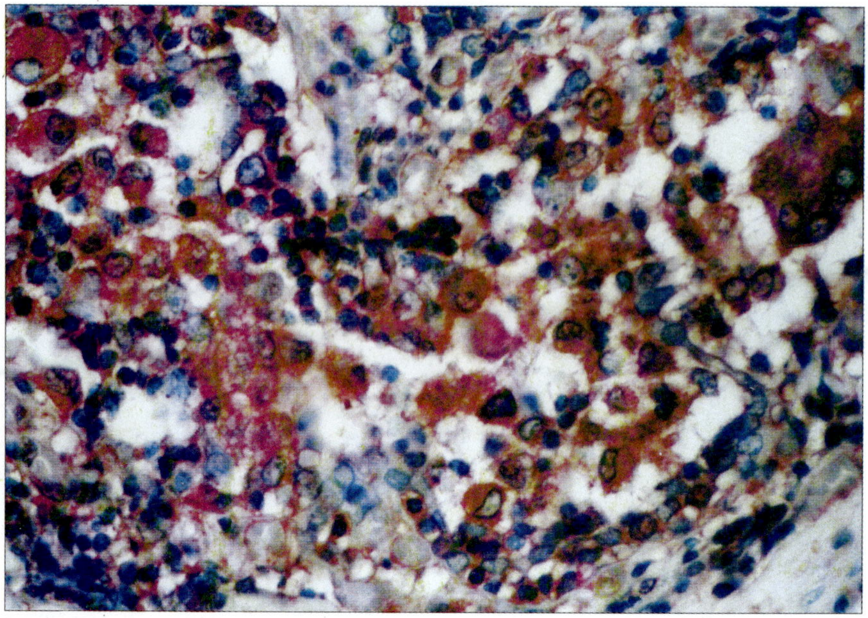

Fig. 3. Thioredoxin (TRX) expression in a medullary carcinoma of the thyroid from a 39 years old female patient (1925/98 B1), using MAb 6F3 and visualisation with 3-amino-9ethyl-carbazole.

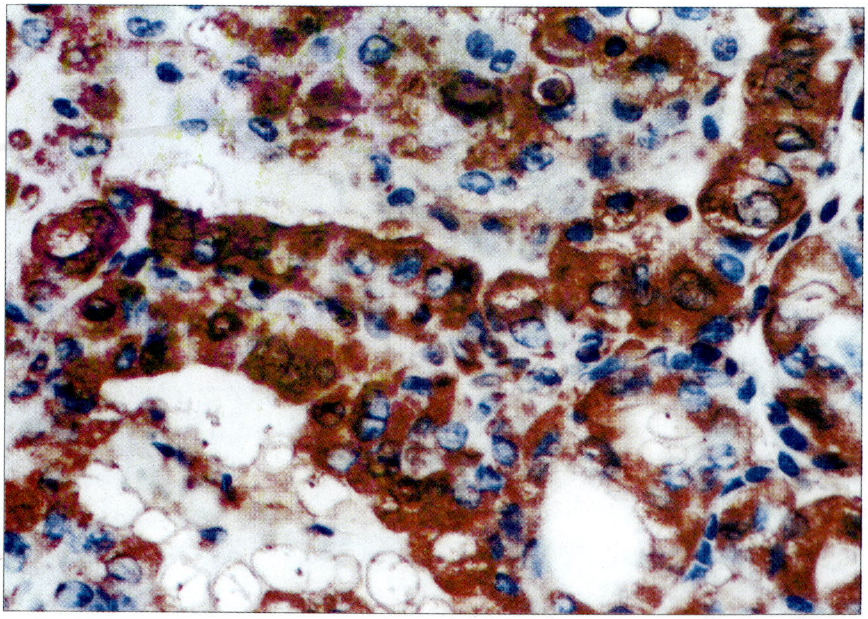

Fig. 4. Thioredoxin Reductase expression in papillary carcinoma of the thyroid from a 56 year old female patient (166/02 A2-3). Rabbit anti-human thioredoxin reductase (TR) gamma G fraction. Visualisation with 3-amino-9 ethylcarbazole.

Table 1. Expression of Thioredoxin (TRX), Thioredoxin Reductase (TR) and Proliferating Cell Nuclear Antigen (PCNA) in Neoplasm's of the Human Thyroid Gland

TumourType	Number of patient	PCNA* % +ve	TRX* % +ve	TRX-R* % +ve
Benign colloid nodule	15	5.9	15.1	12.4
Colloid goiter	14	24.7	38..0	40.9
Multinodular goiter	21	44.3	43.4	45.1
Papillary oncocytic neoplasm	10	15.9	41.7	36.6
Follicular adenoma	16	39.0	36.0	32.8
Follicular carcinoma	24	47.8	65.2	58.0
Invasive follicular carcinoma	19	58.5	69.5	71.6
Papillary carcinoma	48	51.2	72.8	65.7
Medullary carcinoma	9	61.1	79.6	74.3

*Compared to corresponding normal tissue.

normal serum, without the primary monoclonal antibody or with unrelated primary antibodies of identical isotype at the same concentration as the TRX antibody, did not display detectable immunoreactivity.

Expression of Thioredoxin-reductase (TRX-R)

Increased levels of TRX immunoreactivity positively correlated with thyoredoxin reductase (TRX-R) expression and localization. This enzyme, involved in the reduction of thioredoxin, was also highly expressed in cells of thyroid neoplasms, reflecting that a large amount of its thioredoxin substrate is also present. Of the 166 cases of thyroid neoplasms investigated, overexpression of TRX-R was found in 12.4% of benign colloid, in 40.9% of colloid goiters, in 45.1% of multinodular goiters, in 36.6% of the papillary oncocytic neoplasms, in 32.8% of follicular adenomas tumour cells, in 58.0% of the follicular carcinoma cells, in 65.7% of the papillary carcinomas, in 71.6% of invasive follicular thyroid carcinomas and in 74.3% of medullary carcinoma tumour cells investigated. In the case of the invasive follicular carcinomas, the majority of cases showed a correlation between strongly positive thioredoxin and thioredoxin reductase expression, and number of positive lymph nodes.

Proliferating Cell nuclear Antigen (PCNA)

Table 1 shows the percentage of proliferating cell nuclear antigen (PCNA) obtained from counting a total of 1000 cells in each of ten representative fields from each tumour specimen. Overnight application of primary anti-proliferating cell nuclear antigen (PCNA) at a 1:50 dilution at 4°C overnight in humidified chambers revealed distinct nuclear labelling and low background with diaminobenzidine (DAB) as the visualizing agent. Somewhat superior staining results were achieved with 3-amino-9-ethyl carbazole (AEC) as the visualising chromogen. Proliferating cell nuclear antigen immunoreactivity was present in all cases investigated and the expression of PCNA immunopositive tumour cell nuclei varied from being more than 90% negative to a positive labelling of the great majority, with a variable proportion of stained cells in different thyroid tumours. A distinct to strongly positive staining intensity was present in the nuclei of these cells. Increased expression of PCNA immunoreactivity was found in 5.9% of the 15 of the benign colloid nodule cases, in 24.7% of all colloid goiters, in 44.3% of all 21 cases multinodular goiters investigated, in 15.9% of all papillary oncocytic neoplasms investigated, in 39.9% of all the follicular adenomas cases, in 47.8% of all 24 cases of follicular carcinomas cases investigated, in 51.2% of all 48 cases of papillary carcinomas, in 58.5% of all 19 cases of invasive follicular thyroid carcinomas and in 61.1% of all medullary carcinomas investigated. The nuclei of normal tissue cells, including fibroblasts, vascular endothelial cells, adipocytes, smooth muscle cells, lymphocytes, neutrophills, Mast cells and plasma cells were mostly negative.

Discussion

Problems in the Diagnosis of Thyroid Malignancy

Thyroid carcinoma is a malignant representative of those originating from endocrine organs, arising from normal thyroid follicular cells. Follicular carcinoma is believed to originate from pre-existing follicular adenoma. On the other hand, no precursor lesions of papillary carcinoma have been identified. These carcinomas are generally slow growing with an excellent prognosis. However, once these carcinomas show anaplastic transformation (dedifferentiation), they display the most rapidly progressive character of all human malignancies. Cellular proliferation is a prominent parameter for evaluating the biological characteristics of carcinomas, and studies of positive and negative regulators of the cell cycle have been performed in various human tumours (Itoh *et al.* 2003; Lincoln *et al.* 2007; Sinowatz *et al.* 2008). The morphological characteristics betweem folliculr adenoma and follicular carcinoma are similar and hence follicular carcinoma is believed to occur from benign adenoma. Problems with observer variations can lead to low diagnostic reproduceability, especially in the diagnosis of FTC (Franc *et al.* 2003). Papillary microcarcinoma may occur in a familial form and these tumours show more aggressive clinical behaviour than sporadic cases (Lupoli *et al.* 1999). The oxyphilic-cell type of papillary carcinoma causes diagnostic problems because it obscures the pattern of ground-glass nuclei. The follicular variant of papillary thyroid carcinoma still behaves biologically as papillary carcinoma, provided it meets the nuclear criteria for diagnosis of papillary cancers. The true follicular carcinoma, in contrast, lacks these nuclear features, frequently demonstrates capsular and vascular invasion, and has a less favourable prognosis. Thus, the two need to be properly differentiated with regard to their prognosis and management.

Clinicopathological Markers for Thyroid Tumours

Two main well established discriminatory immunohistochemical markers for tumours with follicular and parafollicular origin are thyroglobulin and calcitonin, respectively. Thyroglobulin is present in more than 95% of papillary and follicular carcinomas. Poorly differentiated carcinomas contain less thyroglobulin than do more differentiated tumours. Anaplastic carcinomas are most immunogenic for thyroglobulin. The great majority of medullary carcinomas are positive for calcitonin. The vast majority of papillary carcinomas are cytokeratin type 19 immunopositive. However, cytokeratin 19 is also present in other thyroid tumour types (Svetislaw, 2003). Thyroid transcription factor-1 (TTF-1) and gal-3 (Ordonez, 2000) are also some of the potential candidates as tumour markers in differentiating thyroid malignancies that are difficult to confirm in routine histopathology.

There is direct clinical evidence for the role of hormonal factors in thyroid cancer to suggest that hormonal and biochemical changes related to reproductive events and patterns could be relevant to the etiology of benign thyroid dysfuntiion or disease and neoplasia. Elevated levels of thyroid stimulating hormone (TSH) are associated with thyroid hyperplasia and elevate the risk of neoplastic transformation. Increased secretion of TSH occurs during puberty, pregnancy, delivery and puerperium, as well as while using oral contraceptives (Burrow, 1993). In healthy pregnant women, there is a progressive increase in TSH levels, basal metabolic rate and thyroid volume (pseudo-goitre) throughout pregnancy. Serum human chorionic gonadotrophin, with its marked elevation in early pregnancy, may directly stimulate the thyroid gland through its TSH-like activity (Mann and Hoermann, 1993).

Aetiology of Thyroid Neoplasm in the Middle East

The high frequency of thyroid neoplams in the Middle East population provides a unique opportunity to study the aetiology of this disease. Familial aggregation of thyroid cancer (as well as BTD) could be ascribed to several factors, such as incidence of the disease in the general population, prevalence of BTD, genetic predisposition, exposure, or increased susceptibility to shared environmental factors, gene-environment interaction, and socioeconomic/cultural factors. Kuwait and several other Arab countries in the Gulf region, with relatively high frequency and rates of thyroid cancer, also have high birth and total fertility rates (approximately 5-6 children per woman). For example, in 1998, the total fertility rates in Kuwait, Oman, Qatar, Saudi Arabia and the United Arab Emirates were 5.2, 6.5, 5.1, 5.0 and 4.9 respectively (Economic and Social Commission for Western Asia, United Nations, 1999, 10). Other reproductive health patterns in these countries include high prevalence of consanguinity, young age at first marriage and child bearing, short intervals between births, and older ages at last birth (Kuwait Ministry of Health). This reproductive pattern and culture may, in part, account for the high frequency and familial aggregation of BTD and thyroid cancer among Arab women in the Middle East. Additional environmental causes of thyroid disease in these populations may include high levels of iodine and dietary habits (Memon *et al.* 2004).

Expression of Thioredoxin-thioredoxin Reductase

A number of cancer cells are known to secrete thioredoxin in varied amounts (Lincoln *et al.* 2003; 2005; Al-Yatama *et al.* 2008). The results from this immunohistochemical analysis show that both thioredoxin (TRX) and thioredoxin-reductase (TRX-R), compared to their normal tissue counterparts, are both overexpressed in the nuclei and cytoplasm of the majority of thyroid neoplasms (Table 1). Overexpression of thioredoxin indicates the possible involvement of thioredoxin in the process of oncogenesis. In cancer cells, the secreted thioredoxin tends to sensitize the cells to growth factors, produced by the cancer cells themselves (Gasdaska *et al.* 1995), by increasing the potentiating effect of growth factors, making the cells more susceptible to them and therefore leading to increased cell growth. Thioredoxin therefore enhances growth factors and stimulates the growth of neoplastic cells. In MCF-7 breast cancer cells the effects of IL-2 were increased up to 1000 fold and that of basic fibroblast growth factor up to 100 fold, providing evidence that thioredoxin potentiates the growth stimulation of growth factors (Powis *et al.* 1997). Plasma and serum levels of thioredoxin in normal healthy individuals are between 0.8 to 6.6 nM. This level is greatly increased in tumours and can be used as a marker (Miyazaki *et al.* 1998).

The thioredoxin amino acid sequence tends to be identical to the adult T-cell leukemia derived growth factor secreted by HTLV-1. This growth factor is able to induce the expression of the IL-2 receptor and to stimulate the growth of lymphoid cells by making them more sensitive to factors secreted by the cells themselves (Yodoi and Tursa, 1991). Thioredoxin has been shown to act as a chemotactic protein causing the migration of neutrophils, monocytes and T-cells with a potency similar to known chemokines, including IL-8. Moreover, it has some cytokine like activities such as regulating eosinophil migration and promoting lymphoid cell growth. It differs only from other chemokines in one respect, being G-protein independent (Bertin *et al.* 1999). Extracellular thioredoxin participates in the regulation of cell to cell contact and contributes to redox regulation in the cell extracellular space.

Thioredoxin reductase (TRX-R), the enzyme involved in the reduction of thioredoxin, was also present in large amount in the thyroid cancer tissues, reflecting that a large amount of its substrate (thioredoxin) is also present. Reduced thioredoxin acts as a cofactor for ribonucleotide reductase which catalyzes the transfer of ribonucleotides to deoxribonucleotides for DNA synthesis. It is involved in gene transcription and modulates gene regulation by acting directly on several transcription factors mediating redox control of critical thiol groups, including regulation of NF-kB transcription complex important for tumourigenesis and several other transcription factors, such as TFIIIC, glucocorticoid and estrogen receptors (Mathews *et al.* 1992).

Reduced thioredoxin acts as a structural component by binding with high affinity to DNA polymerase to form a complex, which in turn gives the enzyme its high processing capacity and makes an inhibitory complex with apoptosis signaling kinase 1 (ASK1), providing the means of redox regulation of apoptosis (Sailoh *et al.* 1998). When the oxidative stress rises in the cell, the antioxidant effect of thioredoxin comes into action, reducing the stress (Nakamura *et al.* 1997). A rise in cellular oxidative stress causes induction of apoptosis, whereas reduced oxidative stress inhibits apoptosis. The rise in oxidative stress, inducing apoptosis, is lowered by thioredoxin, triggering a fall in oxidative stress (inhibition of apoptosis). Moreover, overexpression of thioredoxin peroxidase, acting as antioxidant, is also related to apoptosis, the overexpression tending to reduce H_2O_2 levels, causing inhibition of the induction of apoptosis (Zhang *et al.* 1997). Thioredoxin reductase is a selenoenzyme, containing a selenocycline residue. Selenium is an essential factor of thioredoxin reductase and increased selenium incorporation translates to increased enzyme activity. Increase in the thioredoxin reductase activity shows no effect on cell growth, while a reduction in thioredoxin reductase activity, resulting in the reduction of the reduced form of thioredoxin and therefore a reduction in reducing equivalents due to inhibition, causes decreased cell growth (Powis and Mustacich, 2000). Thioredoxin reductase plays an essential role in reducing the oxidized form of thioredoxin ($TrxS_2$) to the reduced form ($Trx(SH)_2$), thereby providing the reducing equivalents, which in turn play a vital role in several key processes known to be the functions of reduced thioredoxin. Reduced thioredoxin increases cell growth by increasing the supply of reducing equivalents for DNA synthesis, resulting in cell proliferation and activation of transcription factors that regulate cell growth, thereby increasing the sensitivity of cells to growth factors and cytokines (Powis and Mustacich, 2000).

Thioredoxin and thioredoxin reductase are widely distributed, but vary considerably between different cell types. For instance, hepatocytes, epithelial cells, secretory cells, plasma cells and neurons display high contents of immunoreactive thioredoxin and thioredoxin reductase, while mesenchymal cells of connective tissues, bone, muscle and fibrous tissue, in contrast, show low or no immunoreactivity. Thioredoxin and thioredoxin reductase have been identified in the cytoplasm, nuclei, cell membrane and extra-cellularly. In the cytoplasm they exhibit antioxidant and cofactor activity; in the nucleus they are involved in the regulation of transcription factor activity; in the mitochondria they associate with isoenzymes and extracelluarly they stimulate cell growth and chemotactic activity.

This immunohistochemical investigation of a variety of thyroid neoplasms shows that thioredoxin is increased in more than half of the tumours, compared to levels in the corresponding normal tissue. The results indicate that tumour cells use thioredoxin as an autocrine growth stimulate. Overexpression of thioredoxin indicates the possible involvement of thioredoxin in the process of oncogenesis. Occurrence of oxidative stress within the cell induces thioredoxin reductase release, which, due to its anti-apoptotic activity, causes inhibition of apoptosis, resulting in abnormal cell proliferation which can intiate cancer development. The thioredoxin levels have been reported to be positively correlated ($p<0.001$) with cell proliferation measured by nuclear proliferation antigen and a highly significant negative correlation with apoptosis measured by the terminal deoxynucleotidyl transferase assay (Powis and Montfort, 2001). In a immunohistochemical study of thioredoxin-1 levels in human colon cancer, thioredoxin-1 protein was not increased compared to normal colonic mucosa in precancerous colon polyps but was increased sixfold in primary colon cancers and almost ninefold in metastatic colon tumours in adjacent lymph glands (Gallegos *et al.* 2000). Furthermore, high levels of thioredoxin-1 in the tumour appear to be associated with decreased patient survival.

Conclusion

We propose that one function of extracellular TRX secreted by cancer cells is to facilitate the mechanisms which give expression to the invasive phentoype. While a normal physiological example of this function is the implantation of embryos in the uterus, it is likely that cancer cells exploit this function of TRX to enhance their invasive capacity. This argument is supported by the following facts: (1) Thioredoxin (TRX) and thioredoxin-reductase (TRX-R), compared to normal tissues, are both overexpressed in the nuclei and cytoplasm of the majority of thyroid neoplasms; (2) TXR and TXR-R are localized in the cytoplasm, nuclei, cell membrane and extra-cellularly. In the cytoplasm they exhibit antioxidant and cofactor activity; in the nucleus they are involved in the regulation of transcription factor activity; in the mitochondria they associate with isoenzymes and extracelluarly they stimulate cell growth and chemotactic activity; (3) Compared to levels in the corresponding normal tissue, thioredoxin is increased in more than half of all thyroid neoplams investigated. The results indicate that thyroid tumour cells use thioredoxin as an autocrine growth stimulate. Overexpression of thioredoxin indicates the possible involvement in the process of oncogenesis; (4) In tumour cells, the secreted thioredoxin tends to sensitize the cells to growth factors, produced by tumour cells themselves, by increasing the potentiating of the growth factors, making the cells more susceptible to them and therefore leading to increased cell growth. Thioredoxin therefore enhances growth factors and stimulates the growth of neoplastic cells; (5) Thioredoxin reductase (TRX-R), the enzyme involved in the reduction of thioredoxin, was also present in large amount in thyroid neoplasms, reflecting that a large amount of its substrate (thioredoxin) is also present; and (6) Occurrence of oxidative stress within the cell induces thioredoxin reductase release, which, due to its anti-apoptotic activity, causes inhibition of apoptosis, resulting in abnormal cell proliferation which can intiate cancer development.

Acknowledgements

Dr. Pawan Singal is the holder of the Naranjan Dhalla Chair in Cardiovascular Research supported by the St. Boniface Hospital and Research Foundation.

References

1. Al-Banaw AG and Lincoln DT. Expression of growth hormone receptors in neoplasms of the prostate. Anticancer Res 2008; 28: 3188-3389.
2. Al-Banaw AG, Sinowatz F and Lincoln DT. Endothelial cell proliferation and angiogenesis in vascular tumours. Anticancer Res 2008; 28: 3189-3190.
3. Al-Yatama F, Al-Bader M, Mohammed FMA, Al-Banaw A G and Lincoln DT. The thioredoxin-thioredoxin reductase system: Overexpression in thyroid cancer. Anticancer Res 2008; 28: 3192-3193.
4. Bakiri F, Djemli FK and Mokrane VA. The relative roles of endemic goiter and socioeconomic development status in the prognosis of thyroid carcinoma. Cancer. 1998; 82: 1146-1153.
5. Bertin R, Howard O, Dang H, Oppenheim J, Bizzarric C, Sergi R, Caselli G, Pagliei S, Rominen B and Ghazzi P. Thioredoxin, a redox enzyme released in infection and inflammation, is a unique chemoattractant for neutrophills, monocytes and T cells. J Exp Med 1999; 189: 1783-1789.
6. Burrow G. Thyroid function and hyperfunction during gestation. Endocr Rev 1993; 14: 194-202.
7. Farahati J. Inverse association between age at time of radiation exposure and extent of disease in cases of radiation-induced childhood thyroid carcinoma in Belarus. Cancer 2000; 88:1470-1476.
8. Franc B, De La Salmoniere P, Lange P, Hoang C, Louvel A, de Roquancourt A, Vilde F, Hejblum G, Chevret S and Chastang C. Interobserver and intraobserver reproducibility in the histopathology of follicular thyroid carcinoma. Human Pathol. 2003; 34: 1092-1100.
9. Gallegos A, Raffel J, Bhattacharyya A and Powis G. Increased expression of thioredoxin in human primary and metastatic colon cancer. Am Assoc Cancer Res 2000; 41: 189.
10. Gasdaska JR, Berggren M and Powis G. Cell growth stimulation by the redox protein, thioredoxin occurs by a novel helper mechanism. Cell Growth Differ 1995; 6: 1643-1650.
12. Gasdaska JR, Harney JW, Gasdaska PY, Powis G and Berry MJ. Regulation of human thioredoxin reductase expression and activity by 3'-untranslated region selenocysteine insertion sequence and mRNA instability elements. J Biol Chem 1999; 274: 25379-25385.
13. Hercbergs A and Davis P. Hypothyroidism in the patient with cancer: how much thyroid supplementation is "safe"? Otolaryngol Head Neck Surg 2006; 133:625-626.
14. Hundahl SA, Fleming ID, Fremgen AM and Menck HR. A National Cancer Data Base report on 53,856 cases of thyroid carcinoma treated in the U.S., 1985-1995. Cancer 1998; 83: 2638-2648.
15. Itoh Y, Yoshida H, Urono T, Nakano K, Takamura Y, Miya A, Kobayashi K, Yokozawa T and Matsuzuka F. Decreased expression of cyclin G2 is significantly linked to the malignant transformation of papillary carcinoma of the thyroid. Anticancer Res 2003; 23:2335-2338.

16. Kim C, Vasko V, Kato Y, Kruhlak M, Saji M and Cheng S. AKT activation promotes metastasis in a mouse model of follicular thyroid carcinoma. Endocrinology 2005; 146: 4456-4463.
17. Lincoln DT, Emadi E M, Tonissen KF and Clarke FM. The thioredoxin-thioredoxin reductase system: Over-expression in human cancer. Anticancer Res 2003; 23: 2425-2434.
18. Lincoln DT, Singal PK, Emadi E M, Al-Khaledi MM, Mohammed FMA, Al-Yatama F, Tonissen KF and Clarke FM. The thioredoxin-thioredoxin reductase system adapts to human cancer growth by overexpression. Adaptation Biology and Medicine, Volume 4: Current Concepts, 2005; Chapter 8: 95-111.
19. Lincoln DT, Singal PK, Al-Banaw AG. Growth hormone in vascular pathology: Neovascularization and expression of receptors is associated with cellular proliferation. Anticancer Res 2007; 27: 4201-4218.
20. Lincoln DT, Al-Yatama, F, Mohammed FMA and Al-Banaw AG. Breast cancer and growth hormone receptor expression in young Arab women. Anticancer Res 2008; 28: 3383-3384.
21. Lincoln DT and Al-Banaw AG. Expression of growth hormone receptors in human lymphomas. Anticancer Res 2008; 28: 3384-3385.
22. Lincoln DT, Al-Yatama, F, Mohammed FMA and Al-Banaw AG. Histopathological aspects and risk factors of breast cancer in young Arab women. Anticancer Res 2008; 28: 3385-3386.
23. Lupoli G, Vitale G, Caraglia M, Fittipaldi MR, Abbruzzese A, Tagliaferri P and Bianco AR. Familial papillary thyroid microcarcinoma: A new clinical entity. Lancet 1999; 353: 637-639.
24. Madhaven M and Othman NH, Spectrum of thyroid diseases in Hospital University Sains Malaysia: A study of 300 consecutive cases. Mal J Med Sci 1996; 3: 58.
25. Mann K and Hoermann R. Thyroid stimulation by placental factors. J Endocrinol Invest 1993; 16: 378-384.
26. Mathews J and Wakasugi N. Thioredoxin regulates the DNA binding activity of NF-kappa B by the reduction of disulphide bond involving cystein 62. Nucl Acids res 1992; 20: 3821-3830.
27. Memon A, Berrington de Gonzalez, A, Luqmani Y and Suresh A. Family history of benign thyroid disease and cancer and risk of thyroid cancer. European J Cancer 2004; 40:754-760.
28. Miyazaki K, Noda S, Okada S, Hagiwara Y, Miyata M, Sakurabayashi I, Yamaguchi N, Sugimura T, Terada M and Wakasugi H. Elevated serum level of thioredoxin in patients with hepatocellular carcinoma. Biotherapy 1998; 11: 277-288.
29. Mohammed FMA, Al-Yatama F, Al-Banaw AG and Lincoln DT. Overexpression of HER-2/NEU proto-oncogene in breast carcinomas of young Arab women. Anticancer Res 2008; 28: 3410-3411.
30. Mustacich D and Powis G. Thioredoxin reductase. Biochem J 2000; 346: 1-8.
31. Nakamura H, Nakamura K and Yodi J. Redox regulation of cellular activation. Ann Rev Immunol 1997; 15: 353-369.
32. Nakamura H, Bai J, Nishinaka Y, Ueda S, Sasada T, Ohshio G, Imamura M, Takabayashi A, Yamaoka Y and Yodoi J. Expression of thioredoxin and glutaredoxin, redox-regulating proteins, in pancreatic cancer. Cancer Detect Prev 2000; 24: 53-60.

33. Ordonez NG. Thyroid transcription factor-1 is marker for of lung and thyroid carcinomas. Adv Anat Pathol 2000; 7: 123-127.

34. Ozawa H. Hypothyroidism after radiotherapy for patients with head and neck cancer. Am J Otolaryngol 2007; 28:46-49.

35. Passler C, Scheuba C, Prager G, Kaserer K, Flores JA, Vierhapper H and Niederle B. Anaplastic thyroid carcinoma (ATC): A retrospective analysis. Langenbecks Arch Surg; 1999; 384: 284-293.

36. Powis G, Gasdaska J, Gasdaska P, Berggren M, Kirkpatrick D, Engman L, Cotgreave I, Angulo M and Baker A. Selenium and the thioredoxin redox system: effects on cell growth and death. Oncol Res 1997; 9: 303-312.

37. Powis G and Mustacich D. Thioredoxin Reductase. Biochem J 2000; 346: 1-8.

38. Powis G and Montfort W. Properties and biological activities of thioredoxins. Ann Rev Biophys Biomol Struct 2001; 30: 421-455.

39. Safar AM, Atamira PM and Stuart KE. Synchronous brain metastasis in follicular thyroid cancer after environmental radiation accident. Anticancer Research 1999; 19: 2028.

40. Sailoh M, Nishito H, Fuj M, Takeda K, Tobiume K and Sawada Y. Mammalian thioredoxin is a direct inhibitor of apoptosis, signal-regulating kinase. EMBO J 1998; 17: 2596-2606.

41. Sedliarou I, Matsuse M, Saenko V, Rogounovitch T, Nakazawa Y, Mitsutake N, Namba H and Nagayama Y. Overexpression of wild-type THR-beta 1 suppresses the growth and invasiveness of human papillary thyroid cancer cells. Anticancer Research 2007; 27: 3999-4010.

42. Sinowatz F, Kenngot R, Lincoln DT, Al-Banaw AG. Role of prostate stem cells in development and cancer. Anticancer Res 2008; 28: 3487-3488.

43. Sugg S, Ezzat S, Rosen I, Freeman J and Asa S. Distinct multiple RET/PTC gene rearrangements in multifocal papillary thyroid neoplasia. J Clin Endocrinol Metab 1998; 83: 4116-4122.

44. Svetislaw T. Histopathological and immunohistochemical features of thyroid carcinoma. Arch Oncol 2003; 11: 173-174.

45. Yodai J and Tursa T. ADF, a growth promoting factor derived from adult T-cell leukemia and homologus to thioredoxin, involvement in lymphocyte immotion by HTLV-1 and EBV. Adv Cancer Res 1991; 57: 381-411.

46. Zhang P, Liv B, Kang S, Seo M and Rhee S. Thioredoxin peroxidase is a novel inhibitor of apoptosis with a reaction distinction from that of Bcl-2. J Biol Chem 1997; 272: 30615-30618.

Adaptation Biology and Medicine (Volume 6: Cell Adaptations and Challenges)
Editors: P. Wang, C.-H. Kuo, N. Takeda and P.K. Singal
Copyright © 2011 Narosa Publishing House Pvt. Ltd., New Delhi, India

19. The Signal Function of Succinate and Free Radicals in Mechanisms of Preconditioning and Long-term Adaptation to Hypoxia

L.D. Lukyanova*, E.L. Germanova and Yu. I. Kirova

Institute of General Pathology and Pathophysiology,
Russian Academia Medical Sciences, Moscow, Russia

Abstract

A bioenergetic mechanism for development of urgent and long-term adaptation to hypoxia is considered. Hypoxia induces reprogramming of respiratory chain function and switching from oxidation of NAD-related substrates (complex I) to succinate oxidation (complex II). Transient, reversible, and compensatory activation of respiratory chain complex II are a major mechanisms of urgent adaptation to hypoxia necessary for (1) succinate- related energy synthesis in conditions of oxygen deficiency and formation of urgent resistance in the body; (2) succinate- related stabilization of HIF-1 and initiation of its transcriptional activity related with formation of long-term adaptation; and (3) succinate-related activation of a succinate-specific receptor GPR91. Therefore succinate is a signaling molecule, and its effects are realized at three levels in hypoxia, intramitochondrial, intracellular and intercellular. In this setting, succinate shows antihypoxic and antiradical activities. Suppression of free radical processes in early phase of hypoxia potentiates the effects of succinate and facilitates the development of urgent adaptation to hypoxia. Tactics and strategy for the antihypoxic defense and development of antihypoxants with energotropic action are considered.

Keywords: Hypoxia, adaptation, energetic metabolism, enzyme complexes I and II, hypoxia inducible factor

Abbreviations

AHBH, acute hypobaric hypoxia; ATP, adenosine triphosphate; CD, conjugated dienes; GPR91, succinate-related guanine nucleotide - binding protein-coupled receptor; GSH, reduced glutathione; GSSG, oxidized glutathione; HR, high-resistance rats; HIF, hypoxia-inducible factor; HSP, heat shock proteins; INH, interval normobaric hypoxia; FAD, flavine adenine nucleotides; LP, lipid peroxidation; LOOH, lipid hydroperoxides; LR, low-resistance rats; NBH, normobaric hypoxia; NAD, nicotine adenine nucleotides; SDH, succinate dehydrogenase; ROS, reactive oxygen species; TBARS, thiobarbituric acid reactive substances.

*Address for Correspondence: Dr. Ludmila D. Lukyanova Professor and Head Department of Bioenergetics and Hypoxia Problems Institute of General Pathology and Pathophysiology Russian Academy of Medical Sciences Baltijskaya Street 8 Moscow 125315, Russia Email: ldlukyanova@gmail.com.

Introduction

Participation of mitochondria in essential regulatory processes determining intracellular, intercellular and systemic interactions is presently discussed [7, 9-11, 15-17, 20, 21, 29, 50-56, 60].

At the same time it appears that functioning of the respiratory chain substrate region differs in the conditions of oxygen shortage and normoxia and its performance is reprogrammed in early hypoxia. As a result, the electron-transport function of mitochondrial enzyme complex I depends on oxidation of NAD-related substrates [49-53]. An alternative pathway of electron transport in respiratory chain through the complex II takes a particular role in energy mechanism. Succinate, a tricarbonic cycle intermediator, is a substrate for succinate dehydrogenase, the major enzyme of respiratory chain complex II. It is presently shown that this oxidation pathway plays a leading role in three most important functional aspects of vital activity of the body under hypoxia: (1) energy production and maintenance of energy-dependent processes; (2) initiation of urgent adaptive responses necessary for subsequent formation of long-term adaptation; (3) receptor function and cell-to-cell signaling, that is, systemic regulation.

Taken together these facts allow considering succinate a signal molecule.

1. Succinate Signaling Pathway in Mitochondrial Energy Synthesis in Hypoxia

In normoxic conditions, performance of the respiratory chain is usually related with oxidation of NAD-related substrates, the major suppliers of reduction equivalents to the respiratory chain through mitochondrial complex I. Its contribution, as estimated by oxygen consumption, may reach 55-65% in intact cells [51-54]. Nevertheless, 25-30% of mitochondrial respiration under these conditions is attributed to oxidation of succinate (complex II) [51-54] where the level of mitochondrial matrix is very low (0.2-0.4 mM) [34, 41, 46]. The ratio between two oxidation pathways depends primarily on properties of the principal enzyme complexes I and II. Kinetic characteristics of two major enzymes of these complexes, NAD-ubiquinon oxidoreductase and succinate dehydrogenase (SDH), are tissue-specific and differ in intact animals with different sensitivity to hypoxia [18-22, 49-57].

For example, Vmax and Km values for NADH- ubiquinon oxidoreductase are significantly (twice) higher in the neocortex of high-resistance (HR) than low-resistance (LR) rats under normoxic conditions, which indicates a lower oxidative capability of the enzyme in LR rat brain. This enzyme becomes saturated faster with a substrate (NADH) and slowly oxidizes it, which may result in lower activity and faster inactivation of complex I in this instance. This fact determines dissimilar oxidative abilities of the NAD-dependent region in the brain respiratory chain as well as the ratio of complex I/ complex II activities in these two types of animals [18, 19, 50, 51, 53, 55, 56].

Observations on the lower efficiency of complex I performance are confirmed by our previous data on peculiarities of oxidative phosphorylation in neocortical mitochondria from intact HR and LR rats. Despite the original similarity in the efficiency of oxidative phosphorylation, the neocortex of LR rats is characterized by higher rate of phosphorylating respiration and greater tension of energy-producing processes than in HR rats. Moreover neocortical mitochondria of LR rats oxidize NAD-related substrates at a maximum velocity of electron transport in the respiratory chain; increased functional load may result in exhaustion of the respiratory chain reserve capacity, which does not happen in HR rats [19, 50, 51, 54].

Differences in baseline activities of complex I and complex II are also evident in the myocardium of HR and LR rats [50, 51]. In this instance, however, the contribution of NAD-dependent pathway of substrate oxidation to energy metabolism of HR myocardium was much smaller than that of LR in normoxic conditions. The baseline activity of succinate oxidation, which reflects the baseline complex II status, is, on the contrary, higher in HR than LR myocardium. This is probably the cause for the higher resistance of HR myocardium in early hypoxia: myocardial contractility suffers later and less severely in HR than in LR animals exposed to hypoxia. Furthermore the impairment of myocardial contractility correlates with a smaller decrease in myocardial ATP [50, 51].

Existence of differences between HR and LR rats in kinetic properties of tissue respiratory chain enzymes suggests that this phenomenon is genetically predetermined, and energy metabolism is one of the leading factors defining formation of individual resistance in the body.

Hypoxic environments induce **regulatory reprograming** of the respiratory chain substrate region, **including reversible depression of complex I function and compensatory activation of complex II** and succinate oxidation, with increased contribution of the latter to respiration and energy production. When substrate is abundant, the contribution of succinate-dependent oxidation to total respiration may reach 70-80%.

Hypoxia-induced reprogramming of the performance of the respiratory chain substrate region is the first of three phases in the development of mitochondrial disturbances in hypoxia, which we have described 25 years ago as correlating with phase changes observed at the systemic level in hypoxia [49-53]. At present time, a great wealth of experimental confirmations is available concerning the reversible inactivation of complex I electron transport function in hypoxic conditions, which persists and even increases in the post-hypoxic period (first 30 min-2 hours of reoxygenation) [1, 2, 14, 25, 27, 49-51, 53, 56, 59, 66-70, 74, 87, 88].

Many studies documenting a special role of succinate in tissue oxidative metabolism at early phase of hypoxia are available from literature. For example, in 1966, Goldberg *et al.* found that after 30 sec of global brain ischemia, succinate level increased 1.5 times in the setting of reduced concentrations of certain NAD-dependent substrates [28]. Similar data were obtained also by other investigators. In the first 30 minutes of anoxic incubation, formation of succinate increases by an order of magnitude reaching 4-7 mM and continues to grow for the first 30 minutes of reoxygenation [34, 36, 41, 83].

Simultaneously, hypoxia is associated with activation of succinate dehydrogenase [3, 13, 31, 82] and succinate oxidation [31, 34, 37-39, 42-44, 49-57, 83, 87], and with increased contribution of the latter to respiration and energy production [13, 14, 16, 42, 44]. The sharp enhancement of succinate-dependent respiration during reoxygenation allowed us to term this condition a "succinate-dependent post-hypoxic activation" [50].

Therefore hypoxic changes are associated with succinate accumulation in tissues and blood, where succinate level increases by 2 orders of magnitude and with suppressed oxidation of NAD-related substrates. Kinetic properties of both enzymes change simultaneously (fig.1). A hypoxic exposure carried out in any regimen and at any oxygenation (acute hypobaric hypoxia, at about 4% concentration in inhaled air; interval normobaric hypoxia, at 10% oxygen) results in opposite, more pronounced changes in Km values for both enzymes: increased Km values for complex I and decreased Km values for complex II. This process reflects the impaired efficiency of enzyme performance in the first instance and its enhanced efficiency in the second instance.

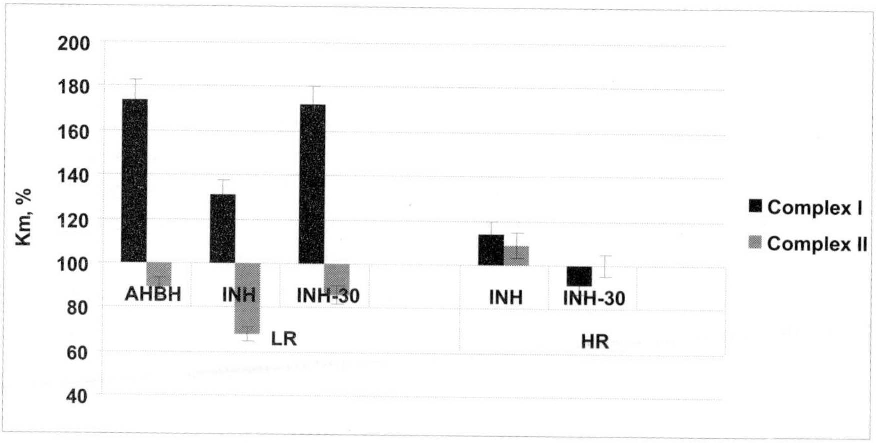

Fig. 1. Effects of different hypoxic regimens on Km values (% of control) for enzymes of complex I (NADH-ubiquinon oxidoreductase) and complex II (succinate-cytochrome c reductase) in rat cortex 2 hours after a hypoxic exposure (1h).

AHBH, acute hypobaric hypoxia (time of survival to the second agonal breath at the critical altitude (11,500, 4% O_2);

INH, interval normobaric hypoxia training consisted of alternating periods of passive breathing of a hypoxic gas mixture at constant pressure containing 10% O_2 (5 min) and breathing (3 min) atmospheric air (21% O_2). These sessions were repeated 6-7 times (a training cycle) and lasted for about one hour in total.

INH-30, interval normobaric hypoxia training consisted of alternating periods of passive breathing of a hypoxic gas mixture at constant pressure containing 10% O_2 (5 min) and breathing (3 min) 30% O_2. LR, low-resistance rats; HR, high-resistance rats.

However it should be noted that pheno- and genotypic peculiarities of the body response to hypoxia are essential in this instance as well; these changes are relatively more pronounced in the brain of LR rats and are not statistically significant for the neocortex of HR rats.

Therefore switching pathways of mitochondrial substrate oxidation occurs in different forms of hypoxia and activation of succinate oxidation constitutes a regulatory and compensatory mechanism, which is realized in conditions of oxygen shortage in most tissues (brain, myocardium, liver, kidneys, lymphocytes) [42, 44, 50, 51]. This mechanism prevents or attenuates disorders in ATP synthesis and exerts a normalizing effect on adenylate pool [19, 21, 22, 36-39, 42, 50, 54, 83, 87, 88] and vital functions of the body, stabilizes and normalizes pH to eliminate hypoxia-related acidosis [58], and increases the body resistance to oxygen shortage [50, 52, 53, 55, 56].

Hypoxia-induced switching of metabolic pathways from oxidation of NAD-related substrates to succinate oxidation is due to several causes; 1) Succinate dehydrogenase sensitivity (mitochondrial complex II) to the redox status of pyridine nucleotides, iron-sulphur complex I and coenzyme Q. Ability of succinate dehydrogenase to become activated when the degree of its reduction grows in hypoxia and is inhibited when it is oxidized [30, 31]. 2) Restriction of oxaloacetic acid inhibition, which suppresses succinate-dependent respiration, since oxaloacetate, a SDH competitive inhibitor, oxidizes to malate and exerts less inhibitory action on the enzyme in hypoxia [42, 44, 58]. 3) Kinetic advantages of succinate (FAD-related substrate) oxidation under the conditions of high reduction of NADH (hypoxia) over NAD-related substrates due to the fact that flavins remain in a more oxidized state in these conditions. 4) Ability of succinate-dependent oxidation to maintain a high velocity of reducing equivalent delivery to the respiratory chain and thereby provide maximum output of energy-rich compound per unit time. For a biological system, it is more important than a high coefficient of efficiency estimated by the number of phosphorylation points, which decrease in this case [58]. Therefore the succinate oxidase pathway (complex II) has kinetic advantages and is energetically more efficient in hypoxic conditions.

If such switching does not occur in hypoxia, the unbalanced complex I dysfunction becomes associated with sharp de-energizing (decreased membrane potential; ATP loss and changes in the adenine nucleotide pool; and disturbed respiration related to oxidation of NAD- related substrates, electron donors for complex I) [18, 19, 21, 22, 50-56]. All these phenomena precede changes in other functional and metabolic parameters that control cell vital activity including condensation of mitochondrial matrix; disturbance of calcium and potassium homeostasis; loss of oxygen sensitivity; disordered expression of mitochondrial genome; appearance of mitochondria-derived reactive oxygen species; CoQ loss; initiation of cytochrome c exit to the intermembrane space; apoptosis; impaired capability of cells for adaptation to low pO_2. In this process, the lactate/pyruvate ratio increases; cell redox potential changes; and metabolic acidosis develops. Also different energy-dependent processes become affected (for example, electrogenic function of excitable cells, anabolic processes such as urea synthesis and stage 2 biotransformation reactions, etc.) [11, 14, 21, 38, 50, 55, 58].

Depending on tissue-specific peculiarities of energy metabolism, pathways of endogenous succinate formation are different in hypoxia. However all these pathways are based on aspartate- and glutamate-dependent aminotransferase reactions, substrate-level α-ketoglutarate phosphorylation during the conversion of α-ketoglutarate to succinate by α-ketoglutarate dehydrogenese, reversal of tricarbonic cycle and oxidative phosphorylation reactions (substrate-level phosphorylation in citric acid cycle via 2-oxoglutarate dehydrogenase and fumarate reductase reactions) [10, 28, 34, 36, 37, 43, 58, 60, 65, 75, 83, 86].

However the rate of endogenous succinate formation in hypoxia may be apparently not sufficient for complete compensation of the energy deficiency. This may explain the fact that exogenous succinate or succinate-containing medications exert a pronounced antihypoxic effect in the setting of oxygen shortage. They expand the intracellular ATP pool and prevent disorders of energy metabolism (energotropic effect of succinate). For instance, a single injection of the succinate-containing compound proxipin, a 3-hydroxypyridine derivative, 15 minutes prior to exposure of rats to an extreme, life-incompatible simulated altitude (11,500 m, 2-4% O2) prevented the loss of high-energy compounds (ATP and creatine phosphate) in rat neocortex (Fig. 2). Furthermore animal tolerance of acute hypoxia also increases in this process [50-55].

The possibility of using exogenous succinate as an energotropic drug brings up a question of its ability to penetrate cell membranes. Indeed the transfer of drug through biological membranes is obviously difficult in normoxic conditions. Contrastingly, in hypoxia, the increased succinate formation in mitochondria is accompanied by its accumulation in cytosol and blood [33, 34, 38, 41, 47] which normally contain only traces of succinate. This suggests a possibility of succinate transportation from mitochondrial matrix through plasma membranes. In this connection, literature considers possibilities of increased passive permeability of membranes for succinate [57, 82]; succinate anion

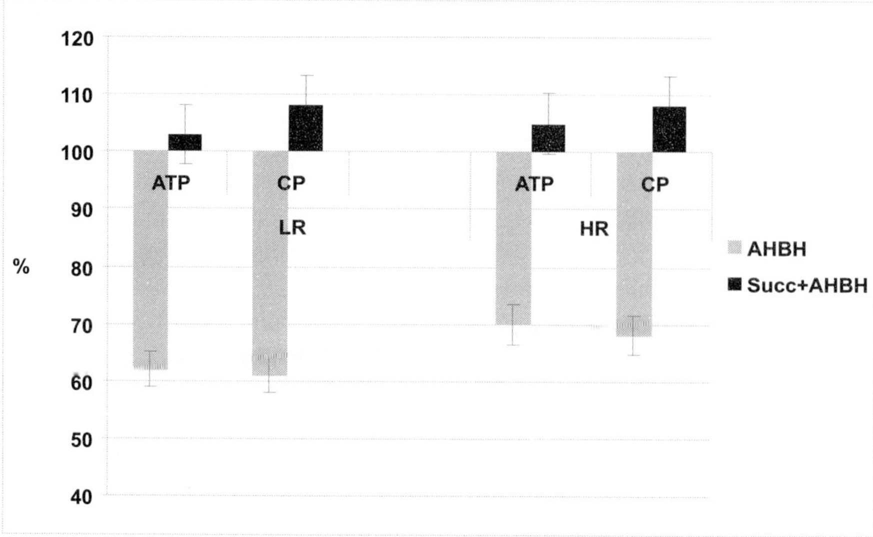

Fig. 2 Effect of the succinate-containing drug proxipin (3-hydroxypyridine, 40 mg/kg, i.p., 15 min prior to hypoxic exposure) on ATP and creatine phosphate (CP) levels (% to control) in the neocortex of rats exposed to altichamber hypoxia (acute hypobaric hypoxia, AHBH) at a simulated extreme altitude (11,500 m, 18 m/sec).

AHBH, without proxipin;

Succ+AHBH, with proxipin.

antiport with malate [48]; and succinate transport along the concentration gradient using requires carriers and organic anion transporters [26, 78].

An opposite process, exogenous succinate transportation into a cell, is also possible. As an end product of redox reactions, exogenous succinate is not metabolized in the stomach [36] but participates in oxidation reactions of energy metabolism; it becomes oxidized to CO_2 and excretes in exhaled air [58]. Isotopic analysis of succinate transformations in the body has shown that after oral administration, succinate is rapidly transported from the stomach to blood and tissues. In 30 minutes, up to 16% of the drug accumulates in different organs (skeletal muscles > liver > blood plasma > kidneys). Hypoxia potentiates this process.

The accumulation of succinate in tissues during hypoxia and the reprogramming of respiratory enzyme performance develop fast and at different levels of oxygenation. We observed changes in kinetic properties of the enzymes as early as thirty minutes after the first hypoxic exposure in different preconditioning regimen. However indirect evidence shows that this process begins much earlier, virtually within the first few minutes of a one-hour hypoxic exposure, and reaches maximum intensity in about two hours after its completion.

Thereby the reversible reprogramming of the respiratory chain performance during hypoxia (switching oxidation of NAD-related substrates to succinate oxidation) is an obligatory, evolutionary developed, urgent signaling compensatory mechanism. Due to this mechanism, the energy synthesizing function of respiratory chain is preserved in disturbed oxygen homeostasis.

2. Succinate and intracellular signaling (initiation of urgent and long-term adaptive responses to hypoxia)

Significance of succinate for formation of protective responses in hypoxia is not limited to its role as a mitochondrial energy substrate for the tricarbonic acid cycle, which possesses an advantage of oxidation under the conditions of hypoxia. Multiple facts show that succinate contributes to the formation of urgent adaptive mechanisms.

For instance, injections of succinate prior to hypoxic exposure provide a sharp increase in the body resistance as compared to the effect of hypoxia or succinate alone (fig.3).

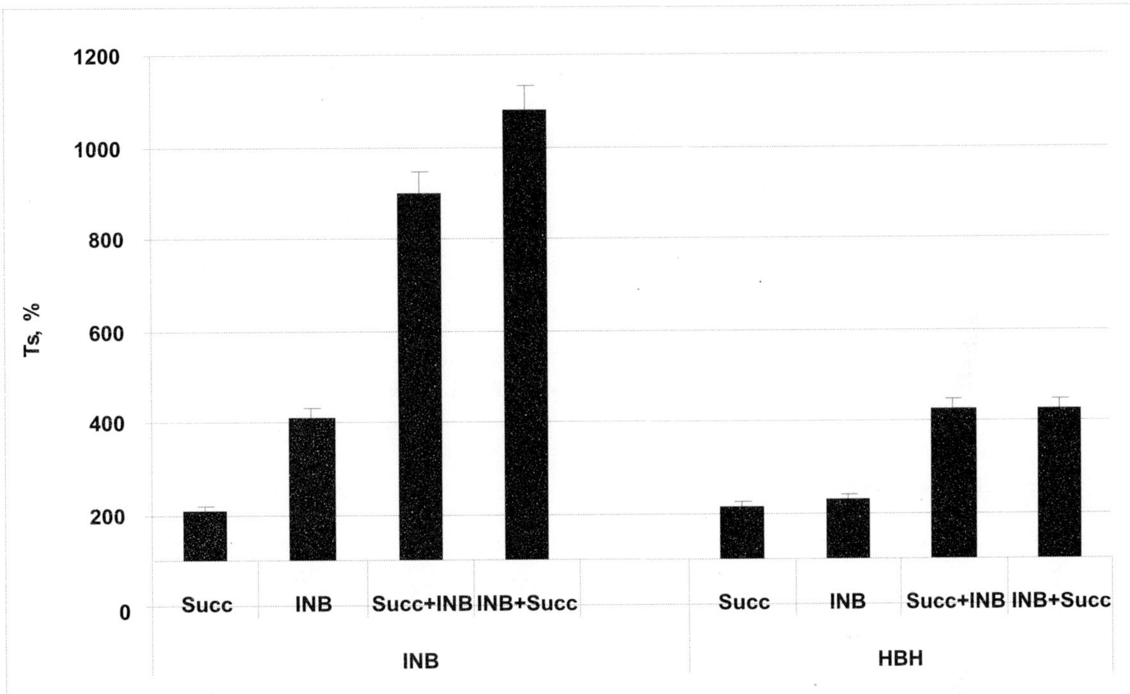

Fig. 3. Effect of succinate-containing drug proxipin on the rat resistance (time survival - Ts, % to control) in conditions of interval normobaric hypoxia (INH) and hypobaric hypoxia (HBH).

INH, interval normobaric hypoxia training consisting of alternating periods of passive breathing of a hypoxic gas mixture at constant pressure containing 10% O_2 (5 min) and breathing (3 min) atmospheric air (21% O_2). These sessions were repeated 6-7 times daily (a training cycle) and lasted for about one hour in total).

Succ-INH, proxipin (40 mg/kg, i.p., 15 minutes prior to the INH session)

INH-Succ, proxipin (40 mg/kg, i.p., immediately after a INH session). Resistance test (measuring Ts) was performed 15 minutes after INH.

HBH, hypobaric hypoxia, 5000m (10% O2, 1h) .

Succ+HBH, proxipin (40 mg/kg, i.p., 15 minutes prior to a HBH session.

HBH+Succ, proxipin (40 mg/kg, i.p., immediately after a HBH session. Resistance test (measuring Ts) was performed 15 minutes after HBH.

Resistance of rats was evaluated by time of survival to the second agonal breath (Ts) at the critical altitude (11,500 m).

Moreover, succinate accelerates the development of resistance to hypoxia. For example, a single hypobaric exposure in the regime of preconditioning results in a significant increase in the resistance of rats after 30 minutes of exposure. However, prior addition of the succinate compounds increases the development of resistance in 5 minutes (fig.4).

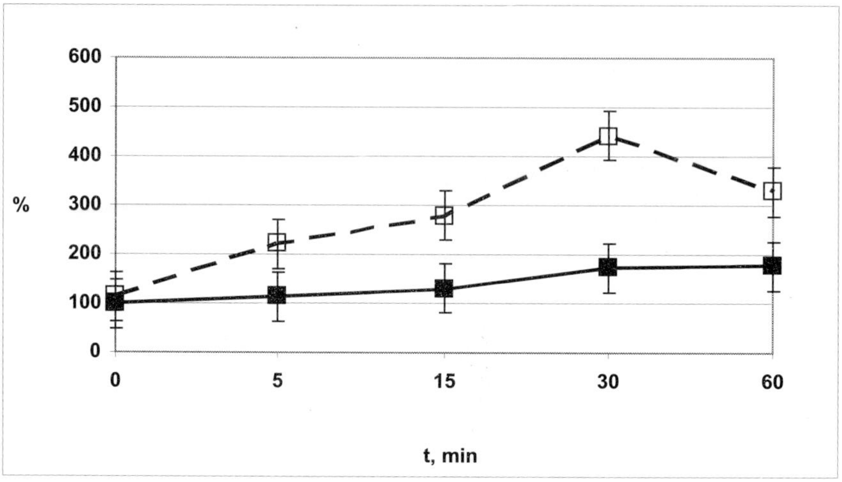

Fig. 4. Effects of succinate-containing drug proxipin in rats at the resistance (% to control) after the preconditioning in the regime of hypobaric hypoxia (HBH, 5000m, 10% O2).
Time of the preconditioning - 5 min, 15 min, 30 min and 60 min.
HBH, without proxipin;
Succ+HBH, with proxipin

All this suggests that succinate participates in the development of resistance of the body to hypoxia.

Besides, evidence is available for a special role of succinate in transcriptional activity and the development of mechanisms for long-term adaptation to hypoxia.

It is known that the process of adaptation to low oxygen level is controlled mostly by a specific transcription protein factor induced in all tissues under hypoxia, the hypoxia-inducible factor HIF-1 [80, 85]. This factor is a heterodimer, composed of 2 subunits: the α and β. The regulation of HIF-1 activity concerns mostly the α subunit. In normoxia the regulatory oxygen-dependent process of prolyl-hydroxylation and proteasomic HIF -1α degradation occurring in cytosol of cells is coupled with utilization of NAD- related substrate for the tricarbonic acid cycle, α-ketoglutarate (fig.5A). Succinate is an inhibitor of prolylhydroxylases. The increased succinate synthesis in hypoxia should provide depression of prolyl-hydroxylase activity, accumulation of HIF -1α in cytosol, and enhancement of HIF-1α activity responsible for the transcriptional activity of many genes contributing to the formation of adaptive features (fig. 5, B) [40, 35, 79, 89]. Instantaneous cell response to changes in environmental oxygen is realized within minutes or even seconds, and is related with the accumulation of HIF-1α.

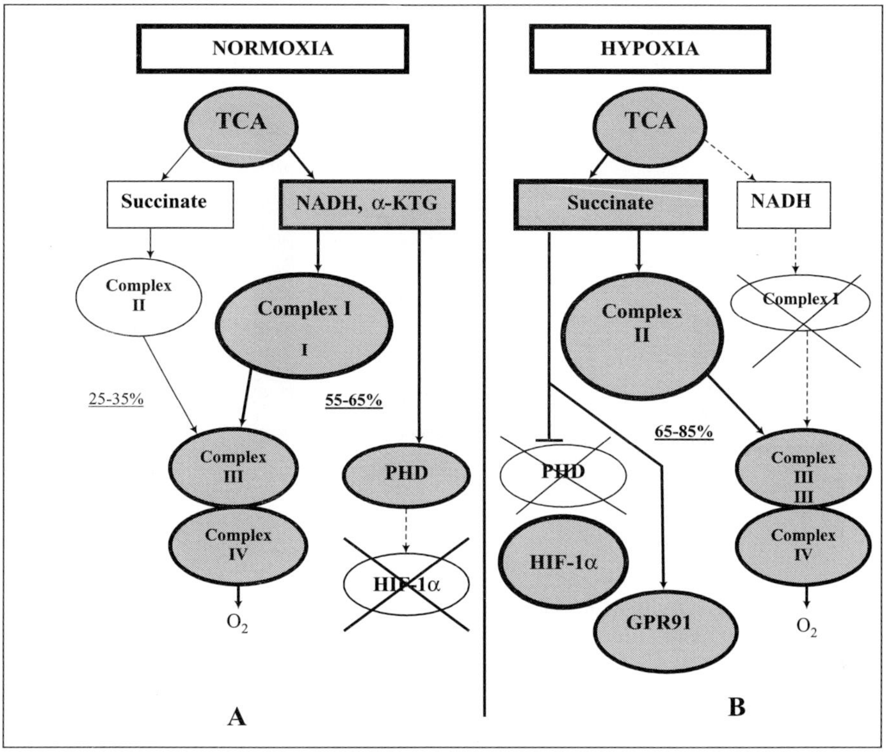

Fig. 5. Reprogramming of respiratory chain function and switching from oxidation of NAD- related substrates (complex I) to succinate oxidation (complex II) in hypoxia. (Regulatory roles of the respiratory chain complex II in the development of important and long-term mechanisms of adaptation to hypoxia).

A - normoxia

B - hypoxia

TAC, tricarbonic acid cycle

At present, conclusive evidence is available for a direct correlation between inactivation of complex I and activation of complex II and formation of HIF-1α [1, 2, 12, 89]. We also showed that activation of HIF-1α synthesis after a single hypoxic exposure in the regimen of preconditioning, correlates with increased efficiency of complex II performance and was associated with a multiple increase in the resistance of animals to acute hypoxia. This process was accompanied by a sharp increase in HSP-70 level, indicative of activated reparation processes (fig.6). In the absence of activated succinate oxidation, for example, in the neocortex of HR rats, HIF-1α was undetectable, and neither the resistance to hypoxia nor the tissue level of HSP-70 was increased.

Therefore the trigger role of succinate in initiating important adaptive processes related with HIF-1α was evident.

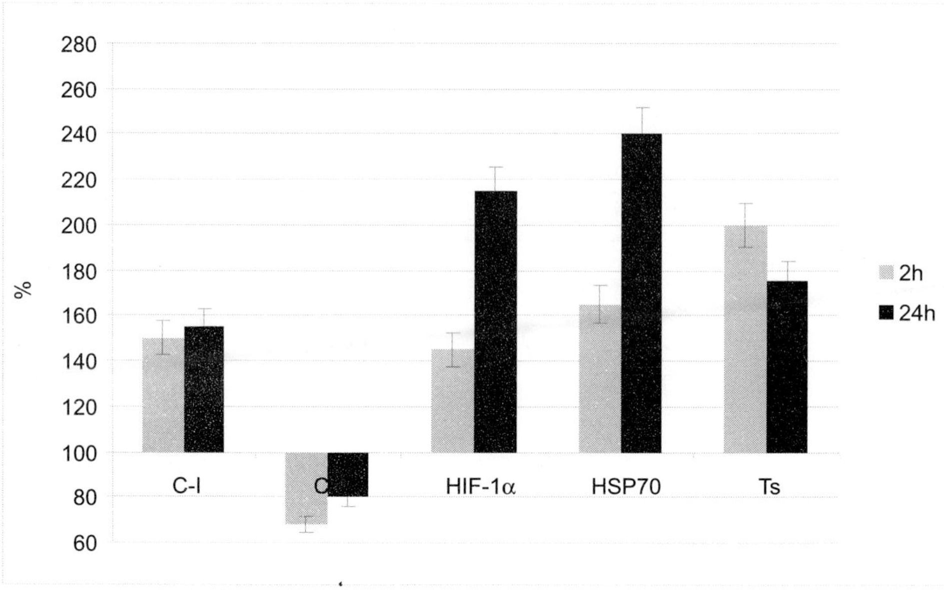

Fig. 6. Effects of INH on the activity of neocortex mitochondrial complex I (C-I) and complex II (C-II), HIF-1α and HSP70 content and time survival (Ts) of LR rats (% of control; 2 and 24 hours after Preconditioning).

However, succinate oxidase oxidation begins to lose gradually its significance in the setting up of long-term adaptation to hypoxia [53, 56]. For instance, during a course of INH, the maximum changes in kinetic properties of the complex II (reduced Km values) occur only on the first day and the normalization of them takes place on subsequent days (fig.7). Contrary to this, the KM values for complex I are increasing for several days to a maximum and remain at this level for 10-15 days, and then diminish, although not returning to the baseline values (Fig.7). As we have demonstrated earlier [53, 56], this process is associated with the appearance of new isoforms of its major enzyme with new kinetic properties, which make the possible recovery of the electron transport function in the NAD-related oxidation pathway under the conditions of high reduction of the pyridine nucleotide pool, typical for hypoxia. Therefore the process of long-term adaptation is associated with transcriptional remodeling of properties in the main enzyme complex I [51, 53, 56]. Since the succinate-related oxidation pathway is an alternative to the NAD-related oxidation pathway, preservation of high activity of complex II would hamper this process in such conditions. This is why the succinate-related oxidation pathway gradually decreases its activity in the course of adaptation to hypoxia, and kinetic parameters of the enzyme returned to baseline values.

Reducing the role of succinate oxidase oxidation pathway in the course of long-term adaptation to hypoxia may have an additional and important regulatory significance. It is related with the need to restrict the transcriptional function of HIF-1α, which may eventually result in uncontrolled enhancement of cell proliferation activity and tumor growth [80, 86].

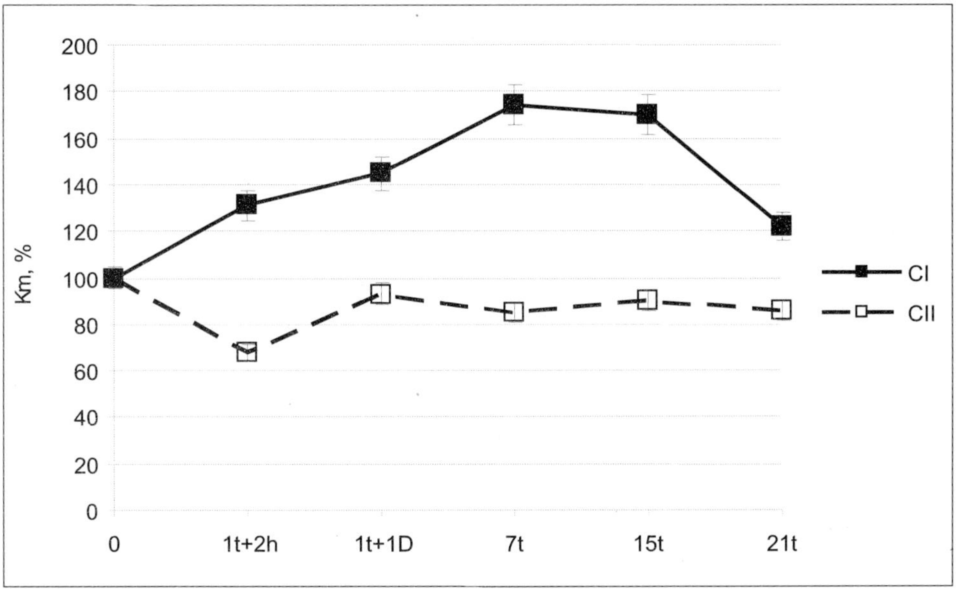

Fig. 7. Effects of interval normobaric hypoxia training (INH) on dynamics of Km for NADH- ubiquinon reductase (complex I - CI) and succinate-cytochrome c reductase (complex II -CII) in neocortex of low-resistance rats.

Effect of a INH course on Km values for complex I and complex II in LR neocortex.

A daily training cycle consisted of passive breathing of a hypoxic gas mixture at constant pressure containing 10% O_2 (5 min) and breathing (3 min) atmospheric air (21% O_2) repeated 6-7 times per hour.

Enzyme Km values were measured 2 hours after the first training cycle (1t+2h); 24 hours after the first training cycle (1t+1D); 24 hours after 7, 15 and 21 training cycles.

Therefore it is obvious that there are basic differences in the formation of the response of the respiratory chain substrate region to a single hypoxic exposure and long-term adaptation to hypoxia.

Urgent adaptation to hypoxia develops in the first few minutes and hours of a single hypoxic exposure. The major mechanism of adaptation is transient, reversible, and compensatory activation of respiratory chain complex II necessary for 1) succinate-related energy synthesis under conditions of oxygen deficiency and formation of urgent resistance in the body; 2) succinate-related stabilization of HIF-1α and initiation of its transcriptional activity.

Long-term adaptation to hypoxia develops after hours and days as a result of multiple, lengthy application of hypoxic exposures. Such adaptation is based on the HIF-1α related formation of adaptive genes and the generation of a new enzyme spectrum (including mitochondrial enzymes) capable to ensure vital activity and the energy supply under conditions of high reduction and low oxygen concentration in cells. This process is associated with the formation of a set of stable adaptive features responsible for the long-standing increase in body resistance to hypoxia.

Mandatory components of this process include: 1) transcriptional remodeling of mitochondrial complex I kinetic properties, which allow restoration of its electron transport and coupling function in the setting of hypoxia; 2) alternative limitation of the complex II role in the processes of mitochondrial oxidation; and 3) changes in kinetic properties of the respiratory chain cytochrome region, which allows transition to a more efficient level of energy production [18-22, 5-, 51,53-56]. Earlier we have shown that this process is associated with simultaneous development of a new population of small mitochondria [51, 54].

Therefore a hypoxic exposure initiates the most sophisticated, multi-step rearrangement of respiratory chain performance triggered by succinate-related processes. Furthermore, mitochondria and cytosol establish a very close crosstalk where succinate plays the role of a signaling molecule.

3. Succinate and free-radical activity in hypoxia

At present, the concept of a signal role for free radicals in the mechanisms of urgent and long-term adaptive reactions is very important [5, 8, 9, 14, 32, 74, 76, 77, 90]. According to this concept, free radicals generated in hypoxia or during the posthypoxic reoxygenation period, are a primarily cause for inactivation of prolyl hydroxylase reactions responsible for proteasomal degradation of HIF-1α and hampering its accumulation. Therefore, presumably, free radical processes may act as a trigger mechanism, that initiates transcriptional and adaptogenic HIF-1α function.

At the same time, controversial data are available, which show that a single hypoxic exposure in the preconditioning mode suppresses free radical activity and decreases redox oxidative changes [23, 24, 63, 81, 86]. In other words, hypoxia may protect the cell from oxidative injuries under certain conditions.

Such opposite points of view may be due to methodological reasons.

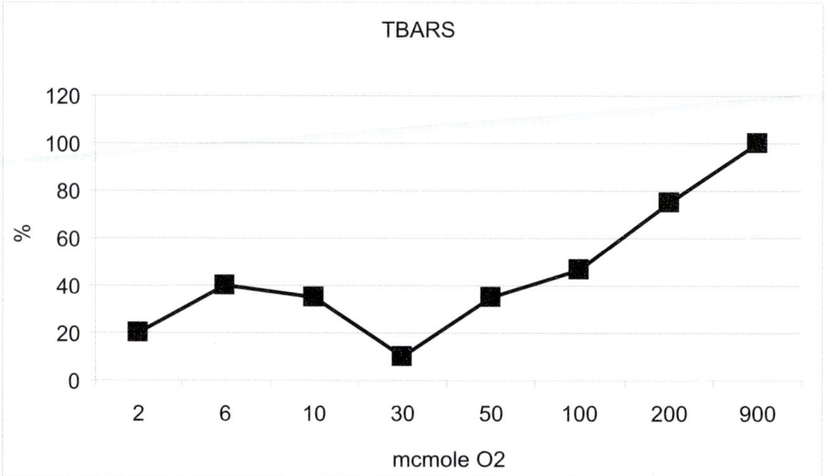

Fig. 8. Dependence of LP intensity (estimated by TBARS level) in slices of rat neocortex on oxygen concetration in the incubation medium.

Indeed, according to our early data, decreased oxygen levels result in the gradual decline of lipid peroxidation (LP) in tissue. However at very low oxygen tension (1-2% of normoxic level), secondary intensification of free-radical activity can be recorded (fig.8). Similar dynamics were observed for superoxide anion also (not shown). The increased free radical activity in the range of low oxygen concentrations results from proteolytic conversion of the cytosolic enzyme xanthine dehydrogenase into xanthine oxidase. The latter becomes activated in this range due to its kinetic properties (K_M (O_2) = 10 μM) [22, 54]. Therefore activation or suppression of ROS depends on the extent of hypoxic exposure. This may explain contradictory data of literature on this issue.

Supporters of the concept on the free-radical mechanism for inactivation of HIF-1α transcriptional function are generally based on either results of experiments performed in cell models using very lengthy, many-hour exposures to non-physiological, low levels of oxygen (1% O_2), or on a long-term, many-day exposure to acute hypoxia. In all these cases the post-hypoxic period is associated with oxidative stress with concomitant reactive oxygen species generation, increase in free radical production, nitric oxide level and LP.

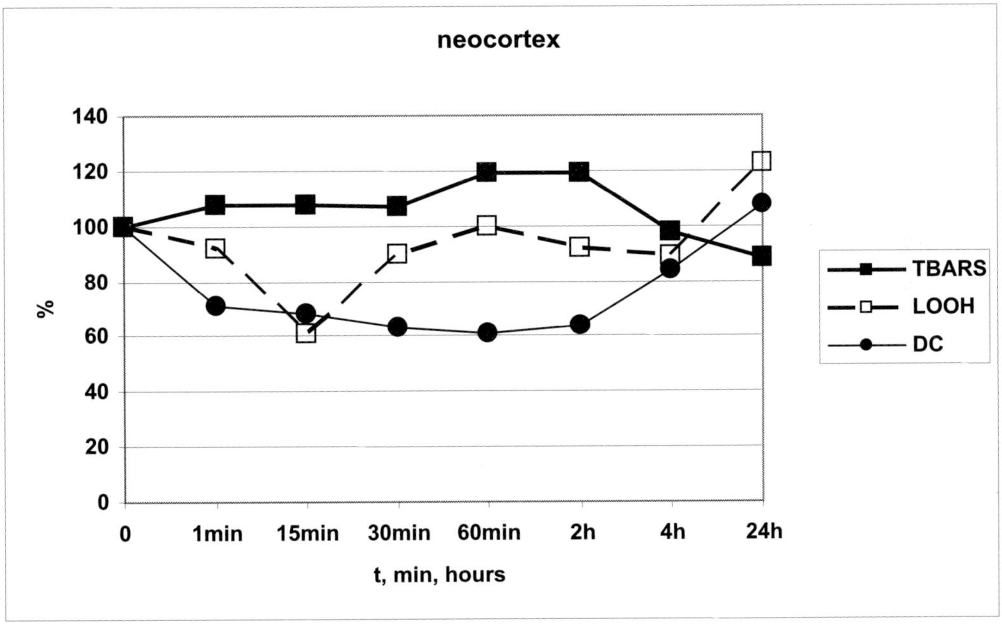

Fig. 9. Effects of a single exposure of hypobaric hypoxia (HBH) in the precondintioning regimen (5000 m, one hour) on lipid hydroperoxydation parameters in the early posthypoxic period (the neocortex of low resistant rats).

TBARS, thiobarbituric acid reactive substances; LOOH, lipid hydroperoxides; DC, conjugated dienes.

Different results can be obtained using milder hypoxic conditions. Our experiments showed that a single exposure to hypobaric hypoxia in the preconditioning regimen (HBH, 5000 m, one hour) associated with instantly increased resistance of animals to acute hypoxia, activation of complex II and accumulation of HIF-1α [56] either did not significantly influence LP or even suppressed this process within the first day of exposure (fig.9). The level of hydroperoxides (LOOH) and especially conjugated dienes (CD), the primary LP products, decreased to the greatest extent. Maximal changes were observed already in 15 minutes after the hypoxic exposure. Significant changes in malonic dialdehyde

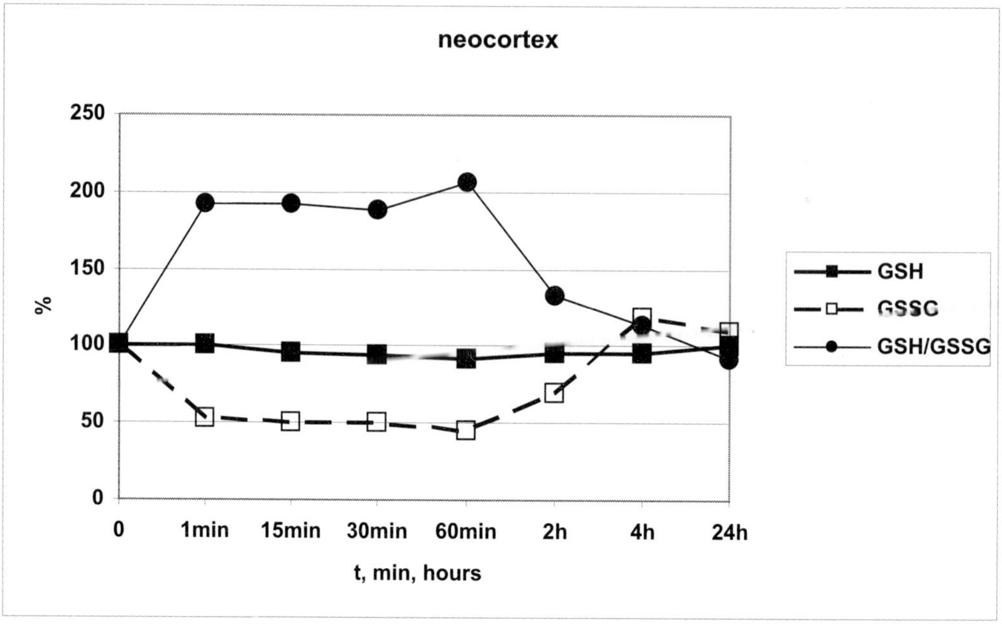

Fig. 10. Effects of single exposure of hypobaric hypoxia (HBH) in the preconditioning regimen (5000 m, one hour) on glutathion pool parameters in the early posthypoxic period (neocortex of low resistant rats). GSH, reduced glutathion; GSSG, oxidized glutathion.

(measured as thiobarbituric acid reactive substances - TBARS), a secondary LP product, were absent. The content of oxidized glutathione (GSSG), a marker for cell redox state, sharply falls during the same period (fig.10). Therefore formation of important mechanisms for adaptation to hypoxia occurs during this period in the absence of pronounced oxidative stress and increased LP, but with activated functioning of complex II. Hence it is just succinate, the inhibitor of prolyl hydroxylase reactions, rather than free radical processes that can claim to the role of an initiator of transcriptional processes.

Signs of activated free radical processes appeared only after an exposure to acute hypobaric hypoxia (AHBH): levels of TBARS and GSSG increased immediately after elevation of the animal to a critical, life-incompatible simulated altitude in an altitude chamber (11,5000m) (fig.11,A). Therefore, the extent of posthypoxic activation of free radical processes really depends on the severity of hypoxia.

Literature has some rather controversial interpretations of succinate effects on the free radical activity. In normoxia, succinate oxidation in the respiratory chain is accompanied by reverse electron transport with concomitant reduction of complex 1 respiratory carriers stimulating the formation of hydrogen peroxide [30, 66, 69, 74, 91]. This might suggest that the mechanism of succinate effect in hypoxic environments is based on stimulation of free-radical activity. At the same time, according to other data, succinate exhibits antioxidant properties in hypoxic conditions [42, 58].

All this has prompted us to perform a special study on the effect of succinate on tissue free-radical activity and the glutathione pool in hypoxia of different severity.

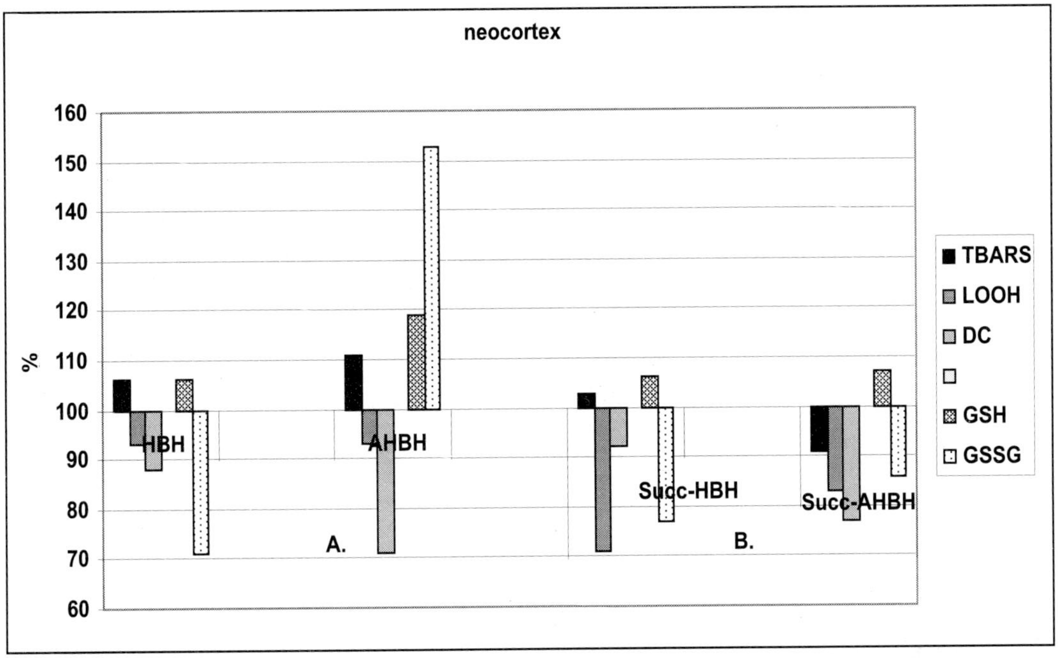

Fig. 11. Effect of succinate on free radical processes and gutathione pool in the neocortex of rats exposed to preconditioning in the regimen of hypobaric hypoxia or to acute hypobaric hypoxia. (The first minute of postypoxic period-% of control).

A – without succinate; B – with succinate.
Succ, succinate-containing drug proxipoin (40 mg/kg, i.p.).
HBH, hypobaric hypoxia (5000 m, one hour); wihtout succinate.
AHBH, acute hypobaric hypoxia (11,500 m); without succinate
Suce + HBH or Succ + AHBH – with succinate.
TBARS, the content of thiobarbuturic acid reacting products;
LOOH, lipid hydroperoxides; DC, conjugated diens; GSH and GSSG, reduced and oxidized glutathione, respectively.

We found that in normoxia, succinate did not exert any significant effect on TBARS, CD or the pool of reduced and total glutathione, but inhibited the formation of LOOH and increased the level of oxidized glutathione (GSSG). These effects may reflect formation of hydrogen peroxide during succinate oxidation (not shown) [30].

However after succinate administration, hypoxic exposure (either HBH or AHBH) resulted in complete disappearance of all signs of oxidative stress (decreasing the level of GSSG, and LOOH) (Fig. 11B) and an increase in resistance to within the first minute of posthypoxic period. Therefore in early hypoxia, succinate displayed not only antihypoxic and adaptogenic but also antioxidant properties. Moreover it follows from the above mentioned data that the mechanism for initiating the HIF-1α transcriptional activity by succinate is not related with potentiation of free-radical activity in hypoxia.

Experiments with the classic antihypoxant ionol have also documented the fact that formation of mechanisms for urgent adaptation does not depend on free radical processes.

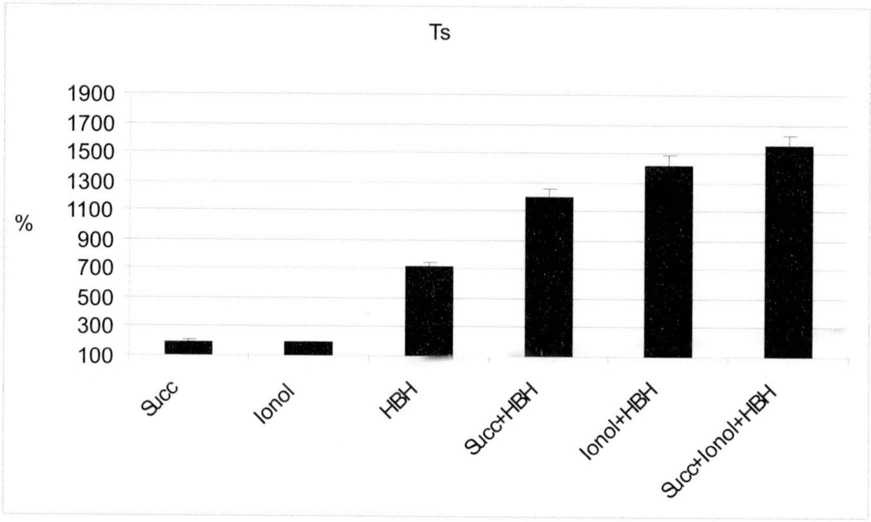

Fig. 12. Effect of ionol and succinate on tolerance of hypobaric hypoxia (HBH) (by Ts).

Ionol administered to animals prior to hypobaric hypoxic preconditioning not only exerted pronounced antioxidant activity, depressed ROS and decreased GSSG level (not shown), but also increased the resistance of animals to hypoxia. The effect of ionol was similar to the effect of succinate in the same conditions. However the combination (ionol + succinate) increased the animal resistance even more, i.e., they exerted an additive effect suggesting different targets of antihypoxic effects of these drugs (Fig. 12).

Therefore 1) formation of urgent adaptation to hypoxia can occur in the absence of oxidative stress; 2) protective effect of succinate during hypoxia is not related with amplification of free radical processes, and it increases when these processes are suppressed.

4. Role of succinate in cell-to-cell interactions

The increased level of succinate in blood and cytoplasm, which allows considering succinate a marker for hypoxic states [41], suggests that it may participate in cell-to-cell interactions and regulatory, functional and metabolic processes at the systemic level. For instance, it has long been known that succinate contributes to regulation of blood pressure by increasing renin formation in renal tubules and thereby exerting a vasodilatory effect in hypoxia [4].

Succinate possesses anti-acidic properties and exerts a normalizing effect on buffer properties of blood by enhancing hemoglobin affinity to oxygen, i.e., it facilitates oxygen release in hypoxic conditions. This improves venous outflow, tissue oxygenation, cardiovascular performance both in the central and peripheral regions, and involves underoxidized lactate and other metabolites in gluconeogenesis [58].

Literature data show that a succinate-controlled electron bypass exists between peripheral cells and lungs. In hypoxia, succinate is transported with blood from peripheral hypoxic tissues to lungs, where succinate synthesis is hampered by high oxygenation, to be used there as an energy substrate needed for pulmonary vasoconstriction. In this process, succinate is oxidized to fumarate which is again transported with blood to peripheral cells and again forms succinate through the fumarate oxidase reaction [10, 64].

In recent two decades, data on receptor function of succinate have appeared. Maria Kondrashova of Russia has demonstrated a regulatory connection of the sympathetic system with succinate and the parasympathetic system with α-ketoglutarate [45]. It appeared that stimulation of physiological functions with epinephrine involves selective activation of succinate dehydrogenase while succinate, which acts in micromolar concentrations, is a signal molecule stimulating adrenalin and noradrenalin release [45].

Studies have shown that complex II contributes to gene expression of glutamate- and dopaminergic signaling systems [61].

A sensational discovery was a study performed by a research group in San Francisco in 2004, which demonstrated that succinate was a specific ligand of GPR91 [33]. This receptor belongs to a group of guanine nucleotide - binding protein-coupled receptors which form large and diverse gene families translating extracellular signals into intracellular ones, but the sequence similarity is minimal among the most distant GPCRs [67].

It has been presently established that GPR91 is localized in plasma membranes. It has been found in more than 20 tissue types, and the array of its localizations is continuously expanding. Kidneys, liver, spleen, small intestine, and bladder are characterized with the best capability for this receptor expression. Only succinate activates it in a receptor-specific way and also provides its internalization [33].

Succinate-dependent receptor activation induces :

- accumulation of inositol phosphate and increases $[Ca^{2+}]_i$;
- extracellular signal-regulated kinase (Erk);
- mitogen-activated protein kinase, p38 and cyclooxygenase;
- synthesis and release of prostaglandin E2 and NO.

Taken together, these results suggest that GPR91 activation by succinate couples to at least several signaling pathways [33, 84]. However the succinate signal performance is apparently tissue-specific, and these pathways cannot work in all tissues.

In kidneys, the succinate-induced expression of GPR91 accompanied by its internalization involves the renin-angiotensin system and precipitates renovascular hypertension, a disease closely related to atherosclerosis, diabetes mellitus and renal insufficiency [33, 72]. This process is absent in GPR91-deficient rats [33]. It was demonstrated recently that this activation is a part of kidney-specific paracrine signal pathway initiated by high glucose levels. The signaling cascade includes local accumulation of succinate as well as expression and internalization of GPR91 receptors in endothelial cells of renal tubules. This GPR91 signal cascade can modulate renal function and provide elimination of different metabolic products through renal hyperfiltration. The GPR91 cascade is apparently a connecting link to pathologic conditions such as diabetes mellitus, syndrome of renin-angiotensin system hyperactivity, systemic hypertension, and organ injury. This succinate receptor signaling cascade can modulate renal function and provide elimination of different metabolic products through renal hyperfiltration [84].

In hepatic ischemia associated with intensive succinate release, the GPR91 receptor is expressed only in hepatic stellate cells. However the fact that succinate does not increase hepatic perfusion pressure suggests that actions of succinate in the liver do not contribute to the hypertensive effects of this metabolite. GPR91 receptor transforms the signal related with increased extracellular levels of succinate into an intracellular signal providing activation of stellate cells in response to organ injury. Therefore the effect of succinate should be regarded as a paracrine signal in this instance as well.

Succinate-dependent expression of GPR91 in ischemia was observed in dendritic cells; it results in production of various angiogenic factors including VEGF. In these cells, succinate triggers intracellular calcium mobilization, induces migratory responses and acts in synergy with Toll-like receptor ligands to produce pro-inflammatory cytokines [71].

Succinate also enhances antigen-specific activation of human and mouse helper T cells. GPR91-deficient mice have less migration of Langerhans cells to draining lymph nodes and impaired tetanus toxoid-specific recall T cell responses [81].

Data are available that succinate accumulates in the hypoxic retina of rodents and, via GPR91, acts as a potent mediator of vessel growth in the settings of both normal retina development and proliferative ischemic retinopathy. The effects of GPR91 are mediated by retinal ganglion neurons, which, in response to increased succinate levels, regulate the production of numerous angiogenic factors including VEGF. Accordingly, succinate do not have pro-angiogenic effects in RGC-deficient rats [73].

The GPR91 ligand succinate can regulate lipolysis in white adipose tissue, suggesting that signaling by this citric acid cycle intermediate may regulate energy homeostasis [68].

The receptor function and participation of succinate in B cell-to-cell signaling and systemic regulation can explain many fast effects of this metabolite including the role of succinate in development of urgent adaptation. Data on the interaction of GPR91 with other receptors allow our understanding of the regulatory participation of catecholamines and acetylcholine in the process of succinate dehydrogenase activation.

Therefore the body response to hypoxia, which initially occurs at the subcellular, mitochondrial level through compensatory potentiation of enzyme complex II, is coupled to activation of a specific succinate-dependent receptor. This receptor initiates cytoplasmic signaling pathways which coordinate and facilitate formation of adaptive processes, as well as cell-to-cell interactions at the systemic level.

5. Use of succinate-containing drugs as energotropic medicines in medical practice

Prevalence of pathology bearing a hypoxic component defines the exceptional importance and social significance of protecting the body from oxygen shortage and related energy deficiency. The above-described materials also show that understanding the role of signaling in maintaining the respiratory chain electron transport function in hypoxia is an essential practical task that has defined formation of tactics and strategy for the antihypoxic defense and development of antihypoxants with energotropic action.

The leading place among them belongs to succinate-containing drugs which have been successfully used in Russia as effective antihypoxic medicines [42, 50, 52, 56]. These drugs are used in clinical practice in the early period of acute disorders related with oxygen shortage.

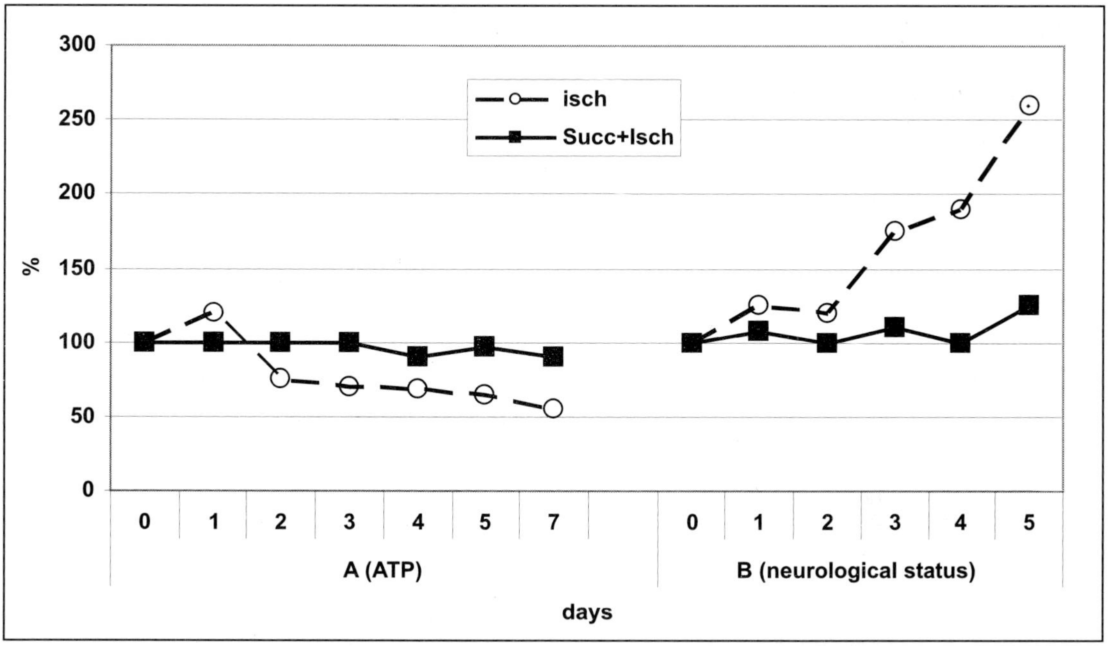

Fig. 13. Effects of three-day injetion of succinate-containing drug proxipin at the neocortex ATP content (A) and disturbances of neurological status (B) of rats during chronic brain global ischemia. Proxipin (Succ) injected once a day (40 mg/kg, i.p) during the first 3 days of global cerebral ischemia.

For example, earlier we have shown that succinate-containing drug therapy used in the phase of formation of important compensatory adaptive mechanisms, i.e., in the first 1-3 days of global cerebral ischemia, stroke, myocardial infarction, acute heart failure, traumatic shock, resuscitation after heart arrest, early postoperative period, after anesthesia, etc., exerts a pronounced protective, antihypoxic effect and an increase of ATP in tissues (fig.13) [52, 53, 55, 56].

However a longer use of succinate-containing drugs loses its protective effects. Figure 14 shows that succinate administered, prior to the first hypoxic exposure (HBH 3,000 m or 5,000 m), exerts a pronounced antihypoxic effect and increases the resistance of animals to hypoxia. However a course administration (7 days daily prior to each hypoxic session) was unfavorable for the resistance of rats (fig.14) because it began to hamper the formation of long-term mechanisms of adaptation (see Section 3).

Timely use of this drug decreases mortality and provides more fast and complete regression of general and focal cerebral symptoms in the majority of patients.

One of the first synthetic succinate-containing drugs, mexidol (3-oxypyridine derivative), is very essential in Russian medical practice. A mexidol analogue, proxipin, possesses even more pronounced antihypoxic and energotropic properties.

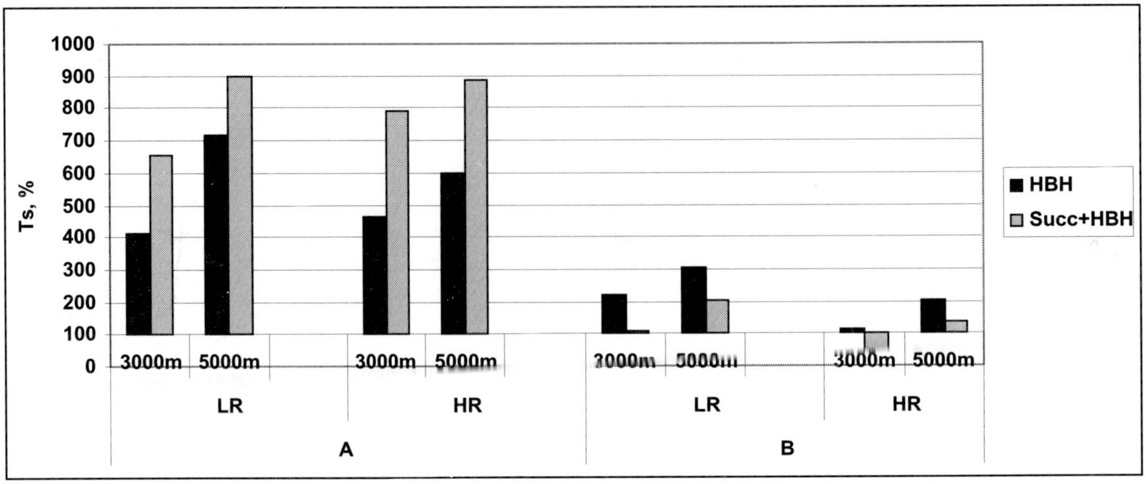

Fig. 14. Effects of succinate-containing drug proxipin (Succ) in the rats at the resistance (Ts, % to control), exposed to preconditioning in the regimen of hypobaric hypoxia (HBH) (3000m/1 h and 5000m/1 h).

A – single injection of proxipin (40 mg/kg, i.p.) 15 minutes prior to the first hypoxic exposure (HBH 3,000 m or 5,000 m),

B – course administration of proxipin (40 mg/kg, i.p.) 15 minutes prior to consecutive HBH exposure (7 HBH trainings during 7 days)

LR – low-resistance rats; HR-high-resistance rats;

Furthermore, it has been demonstrated that the energotropic effect of succinate-containing drugs is due to oxidation of succinate included into the structure of 3-oxypyridine, in the respiratory chain by the classical for succinate pathway.

All succinate-containing drugs are very rapidly absorbed in the setting of different types of hypoxia/ischemia. They exert heir stabilizing or restoring effect on intracellular ATP as soon as 15 minutes after administration.

Energotropic and antihypoxic effects of succinate-containing compounds are associated with 1) pronounced antioxidant properties; 2) modification and resynthesis of phospholipids; 3) normalizing effect on calcium metabolism; 4) catecholamine- mimetic, antiteratogenic, antitoxic, hepatoprotective, antiketogenic, and anticholesterolemic effects; 5) removal of excessive acetyl-Co-A, associated with decreases in excessive lipids and their metabolites; 6) reduction and normalization of pH and elimination of metabolic acidosis [58].

Therefore, a single preventive or therapeutic administration of succinate-containing drugs in the setting of hypoxic therapy will provide enhanced development of important resistance in the body and, therefore, it may be used in practice for activation of urgent adaptive mechanisms.

CONCLUSION

Mitochondria long have been recognized for their role as power house of the cell. The above-described materials show that this role remains also in the conditions of hypoxia. A hypoxic exposure initiates a sophisticated, multistep regulatory process of mitochondria-to-cell-to-system interactions aimed at the formation of urgent and long-term mechanisms of adaptation.

Succinate, an intermediate of tricarbonic acid cycle, is a signal molecule which becomes involved in this process in different forms of hypoxia including the preconditioning phase to provide development of important mechanisms for adaptation to oxygen shortage. In hypoxia, signal effects of succinate are realized at three levels, intramitochondrial, intracellular and intercellular.

At the level of mitochondria, reduced oxygen delivery to cells is associated with rapidly developing, reversible reprogramming of the mitochondrion respiratory chain performance. The associated suppression of mitochondrial enzyme complex I functions and alternative activation of mitochondrial enzyme complex II do not result from dysfunction of the respiratory chain. These two reactions represent the initial urgent (compensatory) stage of an evolutionary requirement developed for the signaling regulatory process responsible for maintaining the cell energy status in the hypoxic environment and participating in the formation of urgent resistance in the body. Therefore, the increased resistance to hypoxia resulting from hypoxic exposures is associated with a regulatory switch of electron flow in the respiratory chain focused on activation of the succinate oxidase pathway, a more efficient pathway in the setting of hypoxia. If such switch is absent, urgent mechanisms of adaptation cannot be formed.

The increase in intracellular succinate associated with this process provides succinate-dependent stabilization of HIF-1α and initiation of its transcriptional activity. Both these processes occur in cytoplasm and nucleus and initiate the formation of long-term adaptive mechanisms. Therefore long-term adaptation to hypoxia is a succinate-dependent process, during which the HIF -1 dependent formation of adaptive genes and the generation of a new enzyme (including mitochondrial enzymes) spectrum are capable to ensure vital activity and energy supply to cells in conditions of high reduction and low oxygen concentration. Thus, this period is characterized by remodeling of the main enzyme of complex I (changing kinetic properties of complex I to expand its functional capabilities in the setting of high reduction in the cell) as well as by restoration of its electron transport function.

Another, not less important signaling mechanism in hypoxia is the succinate activation of a succinate-specific GPR91 receptor. The receptor signaling function provides initiation and coordination of a broad spectrum of other signaling pathways which not only may facilitate formation of adaptive processes but also provide cell-to-cell interactions at the systemic level.

The succinate signaling function is not related with the signaling function of free-radical processes in early hypoxia. Suppression of free radical processes in early hypoxia potentiates the effects of succinate and facilitates the formation of urgent adaptation to hypoxia.

Understanding of biological mechanisms for adaptation to hypoxia has substantiated the development of a new strategy for antihypoxic defense. A single preventive or therapeutic administration of succinate-containing drugs in the setting of hypoxic therapy enhances the development of urgent resistance in the body and, therefore, it may be used in the practice for activation of urgent adaptive mechanisms.

References

1. Agani FH, Pichiule P, Chavez JC *et al.* The Role of Mitochondria in the Regulation of Hypoxia-inducible Factor 1 Expression during Hypoxia. JBC. 2000, 275(46): 35863-35867.
2. Agani FH, Pichule P, Chavez C, LaManna JC. Inhibitors of mitochondrial complex I attenuate the accumulation of hypoxia-inducible factor-1 during hypoxia in Hep3B cells. Compar. Biochem. Physiol. 2002, 132: 107-109.

3. Aithal HN, Ramasarma T. Activation of liver succinate dehydrogenase in rats exposed to hypobaric conditions. Biochem J. 1969, 115(1):77-83.
4. Baumbach L, Leyssac PP, Skinner SL. Studies on rennin release from isolated superfused glomeruli: effects of temperature, urea, oubain and ethacrynic acid. J. Physiol. 1976, 258:243-56.
5. Bell EL, Klimova TA, Eisenbart J, Shumacker CT, Chandel NS. Mitochondrial reactive oxygen species trigger hypoxia-inducible factor-dependent extension of the replicative life span during hypoxia. Mol Cell Biol. 2007, 27: 5737-5745.
6. Bruick RK. O_2 sensing in the hypoxic response pathway: regulation of the hypoxia-inducible transcription factor. Genes & Development. 2003, 17: 2614-2623.
7. Butow RA, Avadhani NG. Mitochondria signaling. The retrograde response. Molecular Cell. 2004, 14(1): 1-15.
8. Chandel NS, Maltepe E, Goldwasser E, Mathieu CE, Simon MC, Schumacker PT. Mitochondrial reactive oxygen species trigger hypoxia-induced transcription. Proc Natl Acad Sci U S A. 1998, 95: 11715-11720.
9. Chandel NS, Schumacker PT. Cellular oxygen sensing by mitochondria: old questions, new insight. J Am Physiol, 2000, 88: 1880-1889.
10. Cascarano J, Ades IZ, O'Conner JD. Hypoxia: a succinate-fumerate electron shuttle between peripheral cells and lung. J Exp Zool. 1976, 1976(2):149-153.
11. Caseda R, Gamboa JI, Boero JA, Monge-CC, Arregui A. Energetic metabolism in mouse cerebral cortex during chronic hypoxia. Neurosc Lett. 2001, 301(3): 171-174.
12. Chavez JC, Agani FH, Pichule P, LaManna JC. Expression of hypoxia-inducible factor - 1α in the brain of rats during chronic hypoxia. J Appl Physiol. 2000, 89(5): 1937-1942.
13. Correa PR, Kruglov EA, Thompson M, Leite MF, Dranoff JA,Nathanson MH. Succinate is a paracrine signal for liver damage. Journal of Hepatology . 2007, 47: 262-269.
14. Da Silva MM, Sartori A, Belisle E, Kowaltowsky AJ. Ischemic preconditioning inhibits mitochondrial respiration, increase H2O2 release, and enhances K+ transport. Am J Physiol Heart Circ Physiol. 2003, 285: 154-162.
15. Das J. The role of mitochondrial respiration in physiological and evolutionary adaptation. Bioessays. 2006 , 28(9): 890-901.
16. Devin A, Rigoulet M. Mechanisms of mitochondrial response to variations in energy demand in eukaryotic cells. Am J. Physiol Cell Physiol, 2007, 292(1): 52-58.
17. Duchen MR. Roles of Mitochondria in Health and Disease. Diabetes. 2004, 53: 96-102.
18. Dudchenko AM, Belousova VV, Lukyanova LD. Effects of adaptation to periodic hypoxia on the brain respiratory chain kinetic parameters. Bull. Exp. Biol. Med. 1996, 121-123:
19. Dudchenko AM, Chernobaeva GN, Belousova VV, Romanova VE, Lukyanova LD. Bioenergetic parameters of rat brain with different resistance to hypoxia. Bull. Exp. Biol. Med.1993, 115(3): 251-254.
20. Dudchenko AM, Lukyanova LD. Effects of adaptation to periodic hypoxia on the cytochrome content in rat brain and liver. Bull. Exp. Biol. Med. 1996, 121(3): 232-235.

21. Dudchenko AM, Lukyanova LD. The parameters of adenylate pool as predictors of energetic metabolism disturbances in hepatocytes during hypoxia. Bull. Exper. Biol. Med. 2003, 136 (7): 41-44.
22. Dudchenko AM, Lukyanova LD. Triggering role of energetic metabolism in the cascade of functional and metabolic disturbances during hypoxia. In: Hypoxia problems: molecular, physiological and clinical aspects. Moscow. Eds: Lukyanova LD, Ushakov IB. 2004, 32-54 "In Russian".
23. Fan J, Cai H, Yang I, Tan W. Comparison between the effects of normoxia and hypoxia on antioxidant enzymes and glutathione redox state in ex vivo culture of CD34(+) cells. Comp Biochem Physiol B Bioch Mol Biol. 2008, 151(2): 153-158.
24. Fandrey J, Frede S, Jelkman W. Role of hydrogen peroxide in hypoxia-induced erytropoetin production Bioch J. 1994, 303;507-510.
25. Feldkamp T., Kribben A., Roeser NF, Senter RA, Kemner S, Venkatachalam MA, Nissim I, Weinberg JM. Preservation of complex I function during hypoxia-reoxygenation-induced mitochondrial injury in proximal tubules. Am J Renal Physiol. 2004, 286(4):749-759.
26. Fiermonte G. Organization and sequence of the gene for the human mitochondrial dicarboxylate carrier: evolution of the carrier family. Bioch.J. 1999, 344: 953-960.
27. Genova MI, C. Casteluccio R, Fato G, Parenti-Castelli M, Merlo-Pich G, Formiggini C, Bovina M, Marchetti G, Lenaz. Major changes in Complex I activity in mitochondrian from aged rats may not be detected by direct assay of NADH-coenzymeQ reductase . Biochem. J. 1995, 311: 105-109.
28. Goldberg ND, Passonneau JV, Lowry OH. Effects of Changes in Brain Acid Cycle Intermediates. J Biol Chem. 1966, 241(17):3997-4003.
29. Grai MW, Burger G, Lang BF. Mitochondrion evolution. Science. 1999, 283: 1467-1481.
30. Grivennikova VG, Vinogradov AD. Generation of superoxide by the mitochondrial Complex I. Biochimica et Biophysica Acta (BBA) - Bioenergetics. 2006, 1757(5-6): 553-561.
31. Gutman M. Modulation of mitochondrial succinate dehydrogenese activity, mechanism and function. Mol. Cell Biochem. 1978, 20(1):41-60.
32. Guzy RD, Hoyos B, Robin E, Chen H, Liu L, Mansfield KD. Mitochondrial complex III is required for hypoxia-induced ROS production and cellular oxygen sensing. Cell Metab. 2005, 1: 401-408.
33. He W, Miao FJ, Lin DC. Citric acid cycle intermediates as ligands for orphan-G-protein-coupled receptors. Nature. 2004, 429: 188-193.
34. Hems DA, Brosnan JT. Effects of ischaemia on content of metabolites in rat liver and kidney in vivo. Biochem. J. 1970; 120: 105-111.
35. Hewitson KS, Lienard BM, McDonough MA, Clifton IJ., Butler D., Soares AS, Oldham NJ, McNeill LA, Schofield CJ. Structural and mechanistic studies on the inhibition of the hypoxia-inducible transcription factor hydroxylases by tricarbonic acid cycle intermediates. J Biol Chem. 2007, 282(5):3293-3230.
36. Hochachka PW, Somero GN. Biochemical Adaptation-Mechanism and Process in Physiological Evolution. New York: Oxford University Press, 2001.

37. Hohl C, Oestreich R, Rösen P, Wiesner R, Grieshaber M. Evidence for succinate production by reduction of fumarate during hypoxia inisolat adult rat heart cells. Arch Biochem Biophys.1987; 259(2): 527-535.
38. Hughes GM, Johnston IA. Some responses of the electric ray (Torpedo marmorata) to low ambient oxygen tensions. J Exp Biol. 1978, 73(1): 107-117.
39. Kinnula VL. Rat liver mitochondrial enzyme activities in hypoxia. Acta Physiol Scand. 1975, 95(1):54-59.
40. Kolvunen P, Hirsila M, Remes AM, Hassinen IE, Kivirikko KI, Myllyharju J. Inhibition of hypoxia-inducible factor (HIF) hydrolases by citric acid cycle intermediates: possible links between cell metabolism and stabilization of HIF. J Biol Chem. 2007, 282(7): 4524-4532.
41. Komaromy-Hiller G, Sundquist PD, Jacobsen LJ, Nuttall KL. Serum succinate by capillary zone electrophoresis: marker candidate for hypoxia. Ann Clin Lab Sci. 1997, 27(2):163-168.
42. Kondrashova MN. Therapeutic effects of succinate. 1976. Puschino. 283 p. "In Russian".
43. Kondrashova MN. The interaction of transamination and oxidation of carboxylic acids in different functional states of tissues. Biokhimiya.1991, 56: 243-310.
44. Kondrashova MN. The formation and utilization of succinate in mitochondria as a control mechanism of energization and energy state of tissue. In: Chance B. (Ed.). Biological and Biochemical Oscillators. Academic Press. New York, NY. 1993. pp. 373-397.
45. Kondrashova MN, Doliba NM. Polarografiphic observation of substrate-level phosphorylation and its stimulation by acetylcholine. FEBS Lett. 1989, 243: 153-155.
46. Kushnir MM, Komaromy-Hiller G, Shushan G, Urry FM, Roberts WL. Analysis of dicarboxylic acids by tandem mass spectrometry. Clin Chem. 2001. 47: 1993-2002.
48. La Noue K, Nicklas WJ, Williamson JR. Control of citric acid cycle activity in rat heart mitochondria. J Biol Chem. 1970; 245:102-111.
49. Lukyanova LD. Limiting steps of energy metabolism in brain in hypoxia. Neurochem. Intern. 1988. 13(I):146-147.
50. Lukyanova L. Bioenergetic hypoxia: Definitions and Methods of Correction. Exp. Biol. Med. 1997. 124 (9): 244-254.
51. Lukyanova LD. Molecular, metabolic and functional mechanisms of individual resistance to hypoxia. In: Adaptation Biology and Medicine. Sharma B.K., Takeda N., *et al.* (eds), , Narosa Publishing House New Dehli, India. 1997. I: 261-272.
52. Lukyanova LD. ?ellular mechanism responsible for beneficial effects of hypoxic therapy. In: Adaptation Biology and Medicine. Moravec, *et al.* (eds), Narosa Publishing House New Dehli, India. 2002a. 3: 290-303.
53. Lukyanova LD. Novel approaches to the understanding of molecular mechanisms of adaptation. In: Adaptation Biology and Medicine. Hargens A., Takeda N., Singal P.K.(eds). 2004, 4:11- 22.
54. Lukyanova LD, Dudchenko AM. Regulatory role of the adenylate pool in the formation of hepatocyte resistance to hypoxia. In: Adaptation Biology and Medicine. Pandolf, N. Takeda KB, Singal PK. (eds.), Narosa Publishing House New Dehli, India. 1999, 2:139-150.

55. Lukyanova LD, Dudchenko AV, Tsybina TA, Germanova EL., Tkatchuk EN. Mitochondrial signaling in adaptation to hypoxia. Adaptation. Biol. Med. Lukyanova L., Singal P. *et al.*(Eds.). Narosa Publishing House New Dehli. 2008, 5: 245-260.

56. Lukyanova LD, Dudchenko AM, Germanova EL, Tsybina TA, Kapaladse RA, Ehrenbourg IV, Tkatchouk EN. Mitochondria signaling in formation of body resistance to hypoxia. Intermitten Hypoxia: from molecular mechanisms to clinical applications. Lei Xi & Serebrovskaya T . (Eds). Nova Science Publishers. USA. Ink. 2009. 423-450.

57. Lukyanova LD, Germanova EL, Tsybina TA, Tchernobaeva GN. Energotropic effect of succinate-containing drugs. Bull. Exp. Biol. Med. 2008. Bull. Exp. Biol. Med.

58. Maevsky EI, Rosenfeld AC, Grishina EV, Kondrashova MN. Correction of metabolic acidosis by maintaining the mitochondrial function. Pushchino; 2001, 155 P. "In Russian".

59. Maklashinas E, Sher E, Zhou H-Z, Gray M. Effect of anoxia/ reperfusion on the reversible active/de-active transition of complex I in rat hear. BBA, Bioenergetics. 2002; 1556(1): 6-12.

60. MacDonald MJ, Fahien LA, Brown LJ, Hasan NM, Buss JD, Kendrick MA. Perspective emerging evidence for signaling roles of mitochondrial anaplerotic products insulin secretion. Am J Physiol Endocrinol Metab. 2005, 288: E1-E15.

61. Napolitano M, Centoze D, Gubellini P, Rossi S, Spiezia S, Bernardi G, Gulino A, Calabresi P. Inhibition of mitochondrial complex II alters strial expression of genes involved in glutamatergi signaling: possible implications for Huginton s diease. Neurobiol Dis. 2004, 15(2): 407-414.

62. Nowak G, Clifton GL, Bakajsova D. Succinate ameliorates energy deficits and prevents dysfunction of complex I in injured renal proximal tubular cells. J Pharmacol Exp Ther; 2008, 324(3):1155-1162.

63. Ohkuwa T, Itoh H. Yamamoto T, Minami C, Yamazaki Y, Kimoto S, Yoshida R. Effects of hypoxia and hypoxic training on 8-hydroxydeoxyguanosine and glutathione levels in the liver. Metabolism. 2004; 53(6): 716-719.

64. Paddenberg R., Ishak B., Goldenberg A., Faulhammer P., Rose F., Weissmann N., Braun - Dullaeus R., Kummer W. Essential role of complex II of the respiraitory chain in hypoxia-induced ROS generation in pulmonary vasculature. Am J Physiol Lung Cell Mol Physiol. 2003; 284: 1710-1719.

65. Penney D.G., Cascarno J. Anaerobic rat heart. Effect of glucose and tricarbolic acid-cycle metabolites on metabolisme and physiological performance. Biochem J. 1970; 118:221-227.

66. Pitkänen S., B.H. Robinson. Mitochondrial complex I deficiency leads to increased production of superoxide radicals and induction of superoxide dismutase. J. Clin. Invest. 1996; 98: 345-351.

67. Pitkänen S., Merante F, McLeod DR, Robinson BH. Familial cardiomyopathy with cataracts and lactic acidosis: A defect in complexes I of the mitochondrial respiratory chain. Pediatr. Res. 1996, 39:513-52.1

68. Regard JB, Sato IT. Coughlin SR. Anatomical profiling of G protein-coupled receptor expression. Cell. 2008, 135(3):561-71.

69. Robinson BH. Human Comlex I deficiency: Clinical spectrum and involvement of oxygen free radicals in the pathogenicity of the defect. Bioch. Biophys. Acta. 1998, 1364: 271-286.

70. Rouslin W, Millard RW. Canine myocardial ischemia: defect in mitochondrial electron transfer complex I. J. Mol. Cell. Cardiol. 1980, 12: 639-645.
71. Rubic T, Lametschwandtner G, Hinteregger S, Kund J. Triggering the succinate receptor GPR91 on dendritic cells enhances immunity. Nat Immunol. 2008, 9(11): 1261-9.
72. Sadagopan N, Roberds SL, Major T, Preston GM, Yu Y, Tones MA. Circulating succinate is elevated in rodent models of hypertension and metabolic disease. Am.J Hypertens. 2007, 20(11):1209-15.
73. Sapieha P, Sirinyan M, Hamel D, Zaniolo K, Joyal JS, Honore JC, Kermorvant-Duchemin E, Varma DR. The succinate receptor GPR91 in neurons has a major role in retinal angiogenesis. Nature Medicine. 2008, 14: 1067-1076.
74. Sadek HA, Sweda PA, Sweda LI. Modulation of mitochondrial complex I activity by reversible Ca2+ and NADH mediated superoxide anion dependent inhibition. Biochemistry. 2004, 43(26): 8494-502.
75. Sanborn T, Gavin W, Berkowitz S, Perille T, Lesch M. Augmented conversion of aspartate and glutamate to succinate during anoxia in rabbit heart. Am J Physiol. 1979, 237: 535-541.
76. Schumacker PT. Hypoxia-inducible factor-1 (HIF-1). Crit Care Med. 2005, 33: S423-S425.
77. Schroedel C, McClintock DS, Budinger S, Chandel NS. Hypoxic but not anoxic stabilization of HIF-1a requires mitochondrial reactive oxygen species. Am.J.Physiol.Lung Cell Mol Physiol. 2002, 283: 922-931.
78. Sekine T., Miyazaki H., Endou H. Molekular physiology of renal organic anion transporters. Am.J.Renal Physiol. 2006; 290:F251-F261.
79. Selak MA, Armour SM, McKenzie ED. Succinate links TCA cycle dysfunction to oncogenesis by inhibiting HIF-prolyl hydroxylase. Cancer Cell. 2005, 7: 77-85.
80. Semenza GL. Expression of hypoxia- inducible factor 1: mechanisms and consequences. Bioch. Pharmacol. 2000, 59: 47-53.
81. Serviddio G, Di Venosa N, Agostino D, Rollo T, Priggigalo F, Altomare E, Fiore T, Vendemiale G. Brief hypoxia before normoxic reperfusion (postconditioning) protects the heart against ischemia-reperfusion injury by preventing mitochondria peroxide production and glutathion depletion. FASEB J. 2005, 19(3):354-361.
82. Susheela L, Ramasarma T. Nature of the activation of succinate dehydrogenase by various effectors and in hypobaria and hypoxia. Biochimica et Biophysica Acta. Bioenergetics. 1973, 292(1): 50-63.
83. Taegmeyer H. Metabolic responses to cardiac hypoxia. increased production of succinate by rabbit papillary muscles. Circ Res.1978, 43:808-815.
84. Toma I, Kang JJ, Sipos A, Vargas S, Bansal E, Hanner F, Meer E, Peti-Peterdi J. Succinate receptor GPR91 provides a direct link between high glucose levels and renin release in murine and rabbit kidney. J Clin Invest. 2008, 118(7): 2526-2534.
85. Wang G, Semenza GL. Characterization of hypoxia-inducible factor I and regulation of DNA binding activity by hypoxia. J. Biol Chem. 1993, 26821513-21518.

86. Wartenberg M, Ling FC, Muschen M, Kltin F, Acker H, Gassmann M, Petrat K , Putz V, Herscheler J, Sauer H. Regulation of the multidrug resistance transporter P-glycoprotein in multicellular tumor spheroids by hypoxia-inducible factor HIF-1 and reactive oxygen species FASEB J. 2003, 17: 503-505.

87. Weinberg JM, Venkatachalam MA, Roeser NF, Saikumar P, Dong Z, Senter RA, Nissim I. Anaerobic and aerobic pathways for salvage of proximal tubules from hypoxia-induced mitochondrial injury. Am J Physiol Renal Physiol. 2000, 279: F927-F943.

88. Weinberg JM, Venkatachalam MA, Roeser NF, Senter RA, Nissim I. Energetic determinants of tyrosine phosphorylation of proximal tubule focal adhesion proteins. Am J Pathol. 2001, 158: 2153-2164.

89. Vaux EC, Metzen E, Yeates KM, Ratcliffe PJ. Regulation of hypoxia-inducible factor is preserved in the absence of a functioning respiratory chain Blood. 2001, 98: 296-302.

90. Shu X, Zuo l, Cardounel AJ, Zweiger JH, He G. Characterization of in vivo tissue redox status, oxygenation, and formation of reactive oxygen Species in postischemic myocardium. Antioxid Redox Signal. 2007, 9(4):447-455.

91. Zoccarato F, Cavallini L, Bortolami S, and Alexandre A. Succinate modulation of H2O2 release at NADH:ubiquinone oxidoreductase (Complex I) in brain mitochondria. Biochem J. 2007, 406(1): 125-129.

Adaptation Biology and Medicine (Volume 6: Cell Adaptations and Challenges)
Editors: P. Wang, C.-H. Kuo, N. Takeda and P.K. Singal
Copyright © 2011 Narosa Publishing House Pvt. Ltd., New Delhi, India

20. Role of mRNA Expression of Hypoxia Inducible Factor Subunits in Adaptation to Hypoxia

Iryna M. Mankovska*, Tetyana I. Drevytska, Victor E. Dosenko

Bogomoletz Institute of Physiology, National Academy of Sciences, Kyiv, Ukraine

Abstract

This mini review focuses on the regulatory mechanism of mRNA expression of hypoxia inducible factor (HIF) subunits in different tissues under adaptation to hypoxia. Data about regulation of mRNA HIF subunits expression including structure of HIFs promoter, factors that influence on transcriptional activity of these genes, and the results concerning changes in mRNA of all HIF subunits level are presented. Studies of the HIF transcriptional system activity under normoxia, acute hypoxia, hypoxic preconditioning, and intermitting hypoxia are reviewed.

Keywords: Hypoxia, adaptation, HIF subunits expression.

Introduction

Activation of gene expression is a crucial component of adaptive responses to hypoxia that are mediated by hypoxia inducible transcription factors (HIFs). HIFs are members of the basic helix-loop-helix Per/ARNT/Sim (bHLH-PAS) protein superfamily and consist of one of three alternative oxygen-regulated alpha-chains (HIF-1α, HIF-2α or HIF-3α) and a constitutive oxygen-independent beta-chain (HIF-1ß, aryl hydrocarbon receptor translocator, ARNT) [43, 50].

Recently molecular mechanisms of cellular adaptation to hypoxia are mainly explained by function of protein systems that regulate formation and degradation of HIF α-subunits - master regulators of oxygen homeostasis [54, 67]. This process occurs at post-translational level and leads to deceleration of HIF degradation and its transportation to the nucleus with the formation of dimer containing α-subunit and ß-subunit of HIF which in specific manner activates transcription of target genes in hypoxic conditions. Protein products of these genes supply adaptive reconstruction of cellular and tissue metabolism [2, 15, 59, 68, 70]. But question about the regulation of HIF subunits expression is practically unexplored; factors that regulate transcription of these genes are poorly understood, and

*Address for Correspondence: Dr. Iryna Mankovska Professor and Head Department of Hypoxia Bogomoletz Institute of Physiology National Academy of Sciences of Ukraine Bogomoletz st. 4 Kiev, 01024, Ukraine
Email: mankovsk@biph.kiev.ua

practically little is known about the structure of HIF promoter and about the role of changes in HIFs subunits mRNA expression under adaptation to hypoxia.

Regulation of mRNA HIF Subunits Expression

Structure of HIF-1α Promoter

Wenger *et al.* [68, 69, 71] have previously cloned and characterized the gene encoding mouse HIF-1α (*Hif1a*) and showed that it contained two different first exons (designated as I.1 and I.2). Exon I.2 is localized on 6 kb distal from the exon I.1 and contains nonmethylated CpG-island that is reached by G+C nucleotides. This kind of sequences is usually associated with promoters of house-keeping genes. The two alternative first exons give rise to two distinct mRNA isoforms differing in the composition of their 5′-untranslated regions (UTRs). The difference particularly concerns the protein products of these exons: exon I.1 encodes protein that is shorter on 12 amino acids than exon I.2. The analysis of the exon I.1 promoter of HIF-1α gene showed that its activity could not provide valid gene transcription and needed additional cis-regulated elements. Moreover, it was shown that exon I.2 is identical on 75% to 3UTR of the human HIF-1α gene. It was suggested that this human gene is mouse homolog. Human homologues of exon I.1 are still unknown.

The study of the promoter activity of both exons with use of genetic construction that included corresponding promoter and luciferase gene without promoter has shown that transcription was significantly higher in case of transfection of the cells with vector, including exon I.2 (from 6 to 9 times depending on cell type). Restriction of exon I.2 significantly decreases its promoter activity in normal conditions at shortening to 232 base pairs. At hypoxic conditions (1% O_2 in cell culture), activity of luciferase gene expression decreases even when it was insignificant shortening. The HIF-1α expression was decreased at modeling of hypoxia in mouse liver (7.5 % O_2), that elucidate the role of alternative exon I.2 in the regulation of this gene expression in mouse and probably in human [69].

Influence of Proto-oncogenes and other Factors on mRNA HIF Expression

Transcriptional activity of HIF-1α genes in human and animals is stimulated by different products of proto-oncogenes (src, ras, Her-2) and inhibited by tumor suppressor proteins (VHL, p53). It suggests that HIF promoter contains elements providing response to these factors action. Signal transduction cascade from cAMP also induces expression of HIF mRNA. Microarrays and real-time PCR show that expression of this gene during the application of forskolin (cAMP increasing agent) increases in 2.8 – 3.0 times in cell specific manner [7]. Transfection of the melanoma B16 cells and fibroblasts NIH-3T3 with fragment of HIF-1α promoter and luciferase gene leads to different changes in response to forskolin application and in modeling of hypoxia by cobalt salts. In melanoma B16 cells, the activity of luciferase expression significantly increases from 5 to 14 times at forskolin and cobalt salts combination. But in fibroblasts this activity is not changed. Authors explain these differences by specific melanocyte-specific transcription factor, or microphthalmia-associated transcription factor (MITF), that can in dose-dependent manner increase HIF-1α promoter activity. This action is provided by two specific sites of recognition on -1202 and -260 positions that contain sequence CACGTG described earlier as MITF recognizing site in another genes [7].

Probably, factors that can stimulate HIF gene expression have some tissue specificity. For example, vasoactive hormones such as angiotensin II and thrombin play a significant role in smooth muscle cells. The diacylglycerol-sensitive kinase C (PKC) and ROS-dependent activation of phosphatidilinositol-3-kinase way (PI3K) mediate the action of these hormones on increasing the HIF-1α mRNA expression [2]. Hypoxia of VSMC does not lead to activation of the promoter region of this gene. Lozartan, specific angiotensin II receptor blocker, and PKC and PI3K inhibitors eliminate this action of angiotensin II. The involvement of PI3K is necessary for stimulating the effects of different cytokines (e.g. heregulin and other EGF family members) on HIF-1α gene expression. But heregulin effect is dependent on activation of AKT and FRAP. Laughner *et al.* [36] show that deletion of 32 bases in promoter eliminates the stimulating effect of heregulin.

Some viruses can stimulate the expression of HIF gene that possibly improves effectiveness of their parasitism. Kaposi sarcoma-associated with herpes viruses (KSHV) induce HIF-1α and HIF-2α mRNA expression following the induction of other genes containing hypoxia-responsive element (HRE) [11]. It is interesting, that these elements are contained in KSHV genome. At modeling of hypoxia on infected endothelial cells, the synergic increasing of mRNA expression and HIF-1α and HIF-2α proteins content is observed. The application of SIC kinase family inhibitors prevent this effect. It shows that this family of kinases mediates the action of this virus [11].

Recently, it was determined that serine/threonine kinase - homeodomain-interacting protein kinase-2 (HIPK2) - negatively regulates the mRNA HIF-1α and its target VEGF in tumor cells, because of HIPK2 knockdown by RNA-interference induced by significant up-regulation of these genes [47]. Moreover, Calzado *et al.* [9] indicate the inducible autoregulatory loop between HIPK2 and Siah2 (ubiquitin E3 ligase). During hypoxia HIPK2 is eliminated due to increased HIPK2/Siah2 interaction, polyubiquitylation and proteasomal degradation. As HIPK2 has an important role as a negative regulator of gene expression, its elimination from promoter-associated repressor complexes allows the induction of a substantial fraction of hypoxia-induced genes. Consequently authors of this article suggest that this interaction is the "apex" of the hypoxic response.

New Classes of Transcripts (Antisense and microRNAs) - Novel Regulatory Factors of HIF and Target Gene Expression

Besides transcription factors, the regulation of HIF-1α gene expression is provided by so called natural antisence transcript (aHIF). The sequence of aHIF is complemented to 3'UTR of HIF-1α mRNA [66]. The discovery of this factor is related with study of renal nonpapillary carcinomas with mutant gene VHL. In these malignancies the permanent overexpression of aHIF is observed. Later, Rossignol *et al.* [52] showed that aHIF expresses everywhere in human tissues and its overexpression correlates with negative prognosis of breast cancer. Promoter of aHIF gene has typical HRE sequence that allows ranking gene aHIF to genes-targets of HIF-1α, i.e. products of these genes provide antihypoxic defense. Mechanisms of aHIF influence on HIF-1α mRNA stability and translation are studied insufficiently but there are some data that this antisense can overlap nonameric AU rich elements in UTR HIF-1α gene that are responsible for mRNA destabilization [56]. High conservativeness of aHIF (89% homologues between rodent and human) is additional proof of an important role of aHIF in cellular physiology and regulation of oxygen homeostasis. aHIF molecule in human consists of 1795 nucleotides and overlap 3'UTR HIF-1α gene on 1027 (other data - 882) bp, in rat this molecule is shorter and complemented to HIF-1α mRNA to 423 bp and in mouse – 484 bp [66]. At modeling of hypoxia, level of aHIF and HIF-1α mRNA is significantly increased (by 80% and 60%, respectively, at moderate hypoxia). But at modeling of severe hypoxia, only the level of aHIF mRNA is increased (more than three times compared to control) and the level of HIF-1α expression is not changed. There are some tissue peculiarities of these two factors expression: in rats under chronic hypobaric hypoxia in Lapaz (altitude – 3,700 m, 450-490-500 mmHg) the level of aHIF expression is increased in heart and gastrocnemius muscle and is not changed in brain; HIF-1α expression is increased in the heart and brain tissue but is not different from the control in the muscle tissue. Modeling of hypoxia with C_OCI_2 in different cell lines of rat (NB104, PC12) leads to significant increase of aHIF mRNA expression with simultaneous decreasing of HIF-1α mRNA [66]. Thus, interrelations (antagonistic and complementary) between HIF-1α and aHIF are dependent on different factors and especially the type of hypoxia. But the question about aHIF action on HIF-1α mRNA stability is still unsolved.

The discovery and characterization of microRNAs (miRs) in the last decade has created a new approach to our understanding of gene regulation. This new paradigm has been described extensively in recent reviews [2, 64]. MicroRNAs are evolutionarily conserved family of small non-coding RNAs of 20-23 nucleotides which negatively regulate the expression of at least a third of all translated genes [19]. The classic view of miRs action was that in mammalian cells miRs act predominantly, if not exclusively, by blocking the translation of mRNA targets. However, more recently, it has become apparent that the action is more complex, with mRNA degradation being quite frequent, in a similar fashion to siRNAs [1].

It is now estimated that the majority of translated genes are regulated by miRNAs, since each miRNA potentially regulates hundreds of different target mRNAs. Several recent studies have established a link between hypoxia and specific miRs [28, 14]. For example, miR-210 was identified as hypoxia inducible in all the cell types. Its hypoxic regulation is mediated by hypoxia-inducible factor (HIF) [10, 18, 33]. It was established that siRNA-mediated suppression of HIF-1α or HIF-2α in MCF7 cells led to abrogation of miR-210 induction under hypoxia. In primary endothelial cells, the knock-down of HIF-2α, the main HIF species in this cell type, does not affect miR-210, while siRNAs to HIF1α effectively prevents miR-210 induction, suggesting a potential HIF isoform specificity, at least in certain cell systems [33]. Moreover, the dynamic of miR-210 induction exhibits a notable difference compared to HIF, as it continues to increase beyond the peak of HIF protein level.

For human miR-210, it was found to have a highly complex spectrum of candidate targets, including genes involved in proliferation, DNA repair, chromatin remodeling, metabolism and cell migration [33]. So, the dual aspect of the hypoxic response was suggested: in addition to the well-documented hypoxia-inducible genes that are direct transcriptional targets of HIF, and the response to low oxygen triggers expression of select miRs, which in turn down-regulate select genes.

Expression of HIF Genes in Different Tissues Under Normoxia and Hypoxia

It is known that HIF-1α and ARNT (HIF-1β) mRNA are expressed in majority of human and rodent tissues. It was established on rather high level in skin, thymus, bladder, lymphatic nodes, thyroid gland, liver, kidneys, adrenal glands, pancreas, heart, lungs, placenta, prostate, macrophages, etc. Other tissues have not significant expression of these genes. HIF-2α subunit expression which is also known as endothelial PAS domain protein 1 (EPAS-1) was found in macrophages, lungs, liver, lymphatic nodes, spleen, and brain. Normal Kupffer cells also showed this subunit expression [38]. Expression of a factor similar to HIF-1α (HIF-2α) was found in vascular endothelium cells, embryonic lungs and catecholamine-producing cells [20, 55, 57]. It allows making a conclusion that these data do not coincide with data described above, but the latter show that HIF-1α plays a more general role in hypoxic signals transduction in most of cells whenever the HIF-2α and HIF-3α role is restricted or has some specific action in maintaining the O_2 homeostasis. This hypothesis is supported by comparing the analysis of knock-out by HIF-1α and HIF-2α in mouse [35, 58, 65, 72].

HIF-1α and HIF-2α have also differences in range of induced genes [6]. Micro-arrays analysis [27] shows that the amount of genes that are regulated by HIF-2α is more than for HIF-1α. But genes of glycolytic pathway are not regulated by HIF-2α when HIF-1α stimulates their expression. As it was shown by Chavez *et al.* [13], erythropoietin gene expression is linked by the activation of HIF-2α to a greater extent. Moreover, HIF-2α does not need nuclear translocation and always exists in the nucleus [38]. Two regions involved in the nuclear localization are proposed: one of them NLS (nuclear localization site) is situated in the N-terminal end of the bHLH domain and the second one is localized on the C-terminal, between N-terminal transactivation and N-terminal transactivation domains. Some point mutations in HIF-2α gene were included to support these data and 4 mutant forms of it were obtained (HIF-2αR14A, HIF-2αR14G, HIF-2αK709T, HIF-2αK738T). Thus, HIF-2α is responsible for the constitutional HIF-dependent gene expression.

Wenger *et al.* [70] have shown that up-regulation of HIF-1α on transcription level is not observed and moreover prolonged hypoxia in mouse leads to the decreasing of this gene expression. But later Heidbreder *et al.* [26] obtained other results. Quantitative analysis of all HIF mRNA subunits expression in different tissues (brain cortex, hippocampus, lung, heart, liver, and kidney) has shown that the level of HIF-3α mRNA was increased in all organs except the liver. The highest increase of expression was shown in the heart (12 times), hippocampus (7 times) and kidney (6 times). Such increase was less in the lung (5 times) and brain cortex (4 times). Changes in expression of other HIF subunits at modeling of hypoxia (decreasing of pO_2 in the air from 130 to 65 mm Hg for 2 hours) were not observed. Thus, this work became the first one that showed not only predictable and previously described increasing of HIF-1α, HIF-2α and HIF-3α proteins during hypoxia but also changes in the HIF-3α mRNA expression. The role of these changes in the HIF subunits expression as a mechanism of organism's adaptation to hypoxia remains unclear.

In our recent work [16], the mRNA expression of all HIF subunits at normoxia was detected in all rat tissues analyzed (heart, lung, kidney, muscle). To our mind, it is understandable because of HIF proteins cleavage at normal pO_2 and needs formation of new proteins that are necessary for support of organ functions in normoxia as well for effective response to possible hypoxia. It is known that activation of HIF transcriptional factors are not only with hypoxia but also with a diversity of regulatory systems (NO, $CoCI_2$, iron chelators) [2, 24]. It was also observed the highest relative level of all HIF subunits expression is in lung and kidney. It coincides with the data from some researchers that have shown the highest level of HIF-1α expression in the lung and brain.

In similar work, it has been shown that after two hours of acute hypoxia (inhalation of gas mixture with 12% O_2), the level of HIF-1α and HIF-1β mRNA expression was not changed in any rat organ when compared to control.

No significant hypoxic induction of HIF-2α subunit expression in lung and heart was observed except the highest one in kidney. It means that tendencies to enhance HIF-2α mRNA existed but was not significant and statistically not reliable. But HIF-3α gene expression was significantly increased. The levels of HIF-3α mRNA expression were increased in kidney (267±37%), heart (210±29%), and not very significant in lung and muscle (20%).

The use of real time PCR method with TaqMan system and specific primers and probe for HIF-3α gene helped us to perform quantitative analysis of changes in HIF-3α mRNA expression in acute hypoxia. It was established that the level of HIF-3α mRNA expression averaged to 12±0.55 (heart), 4.81±0.19 (lung), 1.57±0.29 (muscle) and 13.16±0.0045 (kidney) times in acute hypoxia compared to control [16]. Thus, the data obtained in general coincide with results of Heidbreder *et al.* [26] at analogical method of hypoxia modeling and its severity. But our genetic method had a higher sensitivity and that has shown a more significant increase in subunits expression registered in our work.

So, in contrast to HIF-1α and HIF-2α, moderate hypoxia that was used in our study is enough for HIF-3α gene activation in all tissues except muscle. The report [24] also shows an increase in this subunit expression in brain and kidney at moderate hypoxia. Functional role of this subunit expression increases as a response in hypoxia and is still disputable. It was shown that this subunit can interact and form dimer with HIF-1β [22, 32], and this way it stimulates target gene expression. The increase of this subunit expression level at normoxic conditions is observed in perivenous parts of the liver. It could be explained by pO_2 gradient formation due to unilateral blood stream [49]. On the other hand, there are some data concerning HIF-3α subunit as a natural antagonist of other HIF α-subunits.

In general, the question about the role of a HIF-3α cell response to hypoxia cannot be solved unambiguously as well as the question of the role of different HIF subunits. Some data show that HIF-3α protein has an opposite activity compared to the other subunits. Maynard *et al.* [44] showed that human HIF-3α can produce numerous splice variants such as HIF-3α1, HIF-3α2 (also called IPAS), HIF-3α3, HIF-3α4, HIF-3α5 and HIF-3α6. It was also shown that HIF-3α1-α3 as well as HIF-1α and HIF-2α are targets for E3 ubiquitin-ligase complex in vivo and in vitro due to the common ODD presence in its structure. Endogenous VHL significantly decreases the level of HIF-3α1 protein. Hypoxia that was modeled with $CoCl_2$ induces HIF-3α1 protein accumulation in cells with functional VHL deficiency, and hypoxia mimetic factors did not have any influence on this accumulation. It was later shown that the alternative splice-variant of human HIF-3α gene - HIF-3α4 - attenuates HIF-1 ability to bind with HRE in HIF target-gene promoters. HIF-3α4 transfection in osteosarcoma cells U2OS has decreased the hypoxia induced endogenous GLUT-1 expression. Overexpression of HIF-3α4 inhibits transcriptional activity of HIF-1 and its knockdown significantly increased the hypoxia inducible genes transcription: GLUT-1 and VEGF. Detection of endogenous HIF-3α4 expression in CC-RCC cells have invariant overexpression of hypoxia inducible genes, and there was observed a significant decrease of this splice variant level from 10 to 1000 times compared to the intact kidney cells. Based on these data as well as on comparing the HIF-3α and IPAS sequence, it was proposed that HIF-3α is human ortolog of IPAS that is also a negative regulator of HIF-1. The only difference between the mechanisms of their action is that IPAS cooperates only with HIF-1α, but HIF-3α cooperates also with HIF-1β.

It was previously shown by Hara *et al.* [25] that HIF-3α suppresses hypoxia-inducible HIF-mediated gene expression. At transfection of HIF-3α expression vector into COS-7 cells together with HIF-1α and HIF-2α vector, it was established that the transfection of HIF-3a suppresses both HIF-1α and HIF-2α-mediated gene expressions. Increase of expression of HRE containing genes was naturally detected in cells transfected both by HIF-1α and HIF-2α, but HIF-3α suppresses endogenous HRE driven genes. C-terminal deletion of HIF-3α (1-386) leads to C-terminal domain loosing but saves PAS domain and suppresses HIF-mediated gene expression. Authors explain this effect by competition between HIF-3α and HIF-1α or HIF-2α for involving their general partner - ARNT during activation of transcription [25].

Thus, it could not be excluded that HIF-3α has bimodal action. The increase of its expression at acute hypoxia that was shown in our work [16] and by Heidbreder *et al.* [26] is not only an adaptive response to hypoxia but also leads to destructive changes in cells and tissues due to depression of HIF-1 target-genes. It also could not be excluded that this suppressive effect is peculiar to some of splice-variants of HIF-3α.

Interactions Between HIF Subunits

Recently, it has been shown that the interaction of HIF subunits is not simple. For example, Murphy *et al.* [45] show that perhaps one of HIF-dimers (HIF-2α or HIF-3α) with HIF-lβ can serve as HIF-lα inhibitors. Biochemical comparison of HIF-lα and HIF-2α shows that these proteins have very similar biochemical properties but every subunit controls different biological functions. For example, during embryogenesis HIF-lα has a relation to vascularization control and HIF-2α – to catecholamines production. There are data that in endothelial cells of human lungs the level of HIF-lα and HIF-2α was increased in equal manner at hypoxia (4 hours of 0.5% O_2 exposure). At the same time, these subunits have different reactions to sustained hypoxia (12 hours, 0.5% O_2): synthesis of HIF-lα protein was decreased due to its mRNA stability decrease when HIF-2α protein level was stable and high. It can be an evidence of HIF-lα and HIF-2α specificity in relation to target genes. It was suggested that transcriptional complex with HIF-2α may have selective activity for target genes [48].

Hypoxic Preconditiong (HPC) and HIF Subunits Expression

In 1992, Shizukuda *et al.* [62, 63] demonstrated on dog hearts equipotent protective effects of hypoxia pretreatment and ischemic preconditioning and introduced the term hypoxic preconditioning (HPC) [62]. Later, protection mediated by hypoxic pretreatment has been adopted and reported for a number of cell types, tissues, and organs [4]. Many authors have shown a significant role of HIF in the protection of brain, heart, and kidney. Hypoxic preconditioning has been shown to induce HIF-1 in the brain with subsequent upregulation of HIF target genes [3]. It was demonstrated that HIF-1α and HIF-target genes are upregulated in the penumbra of brain infarcts [61]. HIF-1a and PHD2 protein were shown to be upregulated within the 24-h reoxygenation period after 3 h of hypoxia [30]. Different time periods of HPC were shown to reduce the infarct zone in the brain of mice, which was associated with HIF-1α and target gene activation [4].

In the retinal photoreceptors, HPC was able to upregulate the HIF-1α, VEGF, and EPO, and protect the neurons from light-induced damage [21]. The same authors revealed that HIF-1α activation by hypoxia (6% O_2 for 6 h followed by 4 h of reoxygenation) and systemic EPO administration (5000 IU intraperitoneally) were equipotent in mediating retinal protection.

Myocardial protection by HPC was shown to be connected with HIF system, where the latter plays an important role in restricting ischemic myocardial damage and adaptive remodeling [31]. Mechanisms of myocardial protection and nephroprotection by HPC have been described in recent review [5]. Experiments with heterozygous HIF-1α knockout mice in which the protective effect of a preconditioning protocol suggests that HIF plays an important role in this adaptation. Wild-type (wt) and heterozygous HIF-1α knockout mice (HIF-1$\alpha^{+/-}$) were exposed to 5 cycles of 6 min of hypoxia (6% O_2) followed by 6 min of reperfusion. After 24 h of recovery (21% O_2), the hearts were isolated (Langendorff preparation) and subjected to 30 min of ischemia followed by 2 h of reperfusion. Preconditioned hearts of wt mice were substantially protected as indicated by a significantly reduced infarct size and higher left ventricular diastolic pressure, whereas no difference was observed between hearts from HIF-1$\alpha^{+/-}$ mice and hearts from animals without hypoxic preconditioning. While the lack of protection was not associated with different levels of the HIF target genes VEGF, Glut-1, or NOS2, the heterozygous animals lost the capability to transactivate EPO in response to hypoxia [8].

In several studies, hypoxia was used for preconditioning in order to protect the kidney from acute injury via HIF activation. HIF-1α signals were predominant in the proximal tubule, while HIF-2α was observed in interstitial and glomerular cells under hypoxic conditions [4, 51], reflecting a clear cell selectivity of the expression pattern. Increases of haem-oxygenase-1 and EPO mRNA levels confirmed the transcriptional activation of HIF-1α and HIF-2α target genes, respectively.

Protection mediated by hypoxic preconditioning has also been demonstrated in a number of other organs, including the lung [73], the intestine [12], and the liver [34]; however, the potentially protective role of HIF in these organs is not well understood yet.

Intermittent Hypoxia and HIF Subunits Expression

Intermittent hypoxia (IH) is broadly defined as repeated episodes of hypoxia interspersed with episodes of normoxia. The actual protocols used experimentally vary greatly in cycle length, the number of hypoxic episodes per day, and the number of exposure days. Intermittent hypoxia is divided into chronic IH (hypoxic episodes lasted years or months), short-term IH (hypoxic episodes lasted days or hours), and very short-term IH (hypoxic episodes lasted minutes or seconds).There are few data about the genetic background of IH, and particularly of HIF gene expression. These data are summarized in the recent reviews of Semenza and Prabhakar (2009) and Nanduri et al (2008) [46, 53]. Studies on rodent cell cultures have shown that chronic IH activates a variety of transcription factors including the hypoxia-inducible factor-1 (HIF-1), c-fos (immediate early gene), nuclear factor of activated T-cells (NFAT), and nuclear factor κβ (NF-κβ) [46]. The signaling pathways associated with transcriptional activation connected with IH differ from continuous hypoxia (CH). When compared to same duration and intensity of CH, IH is more potent in activating HIF-1 and c-fos and results in long-lasting accumulation of HIF-1α and c-fos mRNA, a phenomenon that was not seen with CH. IH-evokes transcriptional activation by HIF-1, c-fos, as well as the resulting activator protein-1 (AP-1) and requires reactive oxygen species (ROS)-mediated signaling which involves complex feed forward interactions between HIF-1 and ROS.

In the recent work of Lin *et al.* (2008), HPC was induced by maintaining rats in a chamber constantly flushed with nitrogen to reduce the oxygen content to 10% for 8 h per day and then returning to room air after hypoxic induction. The induction time was 6 weeks [37]. It was shown that repeated reduction of tissue oxygenation during HPC stabilizes the a-subunit of the transcription factor HIF-1. As a result, a number of hypoxia-responsive genes are induced, and two important targets in this process are EPO and VEGF. A study in mice that lack an allele at the HIF-1α locus showed that EPO mRNA expression in the kidneys is not induced by intermittent hypoxia, and that this mutation abrogates the consequent cardioprotection [8]. No study to date has clearly shown whether intracardial EPO mRNA affects EPO protein synthesis or cardiac function, although bone marrow-expressed EPO appears to be involved in autocrine or paracrine regulation of erythropoiesis [17]. Partovian *et al.* (1998) have shown that there is a 5-fold increase in VEGF mRNA in left ventricle after rats were exposed to 11% oxygen for 1 day, but a prolonged exposure returned its level to baseline [49]. Exposure of rats to moderate hypoxia leads to an increase in progenitor cells in the heart and circulation [37]. This effect is dependent on EPO, which induces cell homing by increased SDF-1/CXCR4 and reduces the heart susceptibly to ischemia-reperfusion injury.

Intermittent hypoxic training (IHT) is one of the modern methods of adaptation to hypoxia and its effectiveness is proved by numerous studies [39, 40, 41, 42, 60]. Effect of IHT is based on the constructive action of O_2 lack at different levels beginning at activation of respiratory and cardio-vascular systems to adaptation on tissue, cellular, and subcellular levels [41]. After 5 sessions of IHT, the erythropoietin level in blood serum was significantly increased (45-50%) [23].

To elucidate changes in HIF-1α, HIF-1β, HIF-2α, HIF-3α mRNA expression in rat heart, lung, kidney and gastrocnemius muscle at IHT, rats underwent the IHT sessions: breathing with hypoxic gaseous mixture containing 12 % O_2 for 15 min with 15-min rest intervals, 5 times daily, for 14 days [16]. It was established that normobaric intermittent hypoxia did not lead to significant changes in the expression of any HIF subunits. Moreover, trained rats did not show HIF-3α gene expression increasing after two hours of acute hypoxia that was registered in all tissues (except muscle) of untrained rats after acute hypoxia. HIF-3α expression in muscle at combined action of IHT and acute hypoxia was highest and acute hypoxia leads to the smallest increase of HIF-3α mRNA level compared to control.

Conclusion

A role of changes in HIF genes expression at hypoxia consists in the increase of HIF genes transcription followed by HIF proteins increasing that probably provides more pronounced reaction of organisms in case of hypoxia prolongation. The data reviewed have demonstrated the peculiarities of HIF-1a promoter structure, the factors that can regulate HIF subunits on transcriptional level, such as proto-oncogenes, MITF, protein kinases HIPK2 and Siah2, antisense transcripts, microRNA and others. There are some data about the opposite action of HIF-3α or its splice-variants to other α-subunits at acute hypoxia. It is obvious that the question of adaptation to hypoxia on transcriptional level needs additional studies. Performing of experiments on the animals with knock-down of different HIFs subunits and/or of different splice-variants of HIF-3α is very important. Intermittent hypoxia in different regimen could explain a role of changes in HIF genes expression at hypoxia and the role of different HIF subunits in hypoxic resistance formation.

References

1. Bartel DP. MicroRNAs: Genomics, Biogenesis, Mechanism, and Function. *Cell.* 2004; 116; 281-297.
2. Bell EL, Klimova TA, Eisenbart J, Schumacker PT and Chandel NS. Mitochondrial reactive oxygen species trigger hypoxia-inducible factor-dependent extension of the replicative life span during hypoxia. *Mol Cell Biol.* 2007; 27: 5737-5745.
3. Bergeron M, Yu AY, Solway KE, Semenza GL and Sharp FR. Induction of hypoxia-inducible factor-1 (HIF-1) and its target genes following focal ischaemia in rat brain. *Eur. J. Neurosci.* 1999; 11: 4159-4170.
4. Bernaudin M, Tang Y, Reilly M, Petit E, and Sharp FR. Brain genomic response following hypoxia and re-oxygenation in the neonatal rat. Identification of genes that might contribute to hypoxia-induced ischemic tolerance. *J. Biol. Chem.* 2002; 277: 39728-39738.
5. Bernhardt WM, Campean V, Kany S, Jurgensen JS, Weidemann A, Warnecke C, Arend M, Klaus S, Gunzler V, Amann K, Willam C and Wiesener MS Preconditional activation of hypoxia-inducible factors ameliorates ischemic acute renal failure. *J. Am. Soc. Nephrol.* 2006; 17: 1970-1978.
6. Bracken CP, Fedele AO, Linke S, Balrak W, Lisy K, Whitelaw ML and Peet DJ Cell-specific regulation of hypoxia-inducible factor (HIF)-1α and HIF-2α stabilization and transactivation in a graded oxygen environment. *J Biol Chem.* 2006; 281: 22575-22585.
7. Buscà R, Berra E and Gaggioli C. Hypoxia-inducible factor 1a is a new target of microphthalmia-associated transcription factor (MITF) in melanoma cells. *J Cell Biol.* 2005; 170: 49-59.
8. Cai Z, Manalo DJ, Wei G, Rodriguez ER, Fox-Talbot K, Lu H, Zweier JL and Semenza GL. Hearts from rodents exposed to intermittent hypoxia or erythropoietin are protected against ischemia-reperfusion injury. *Circulation*, 2003; 108; 79-85.
9. Calzado MA, Vega L, Möller A, Bowtell DD and Schmitz ML. An inducible autoregulatory loop between HIPK2 and Siah2 at the apex of the hypoxic response. *Nat Cell Biol.* 2009; 11: 85-91.
10. Camps C, Buffa FM, Colella S, Moore J, Sotiriou C, Sheldon H, Harris AL, Gleadle JM and Ragoussis J. hsa-miR-210 Is induced by hypoxia and is an independent prognostic factor in breast cancer. *Clin Cancer Res.* 2008; 14: 1340-8.
11. Carroll PA, Kenerson HL, Yeung RS and Lagunoff M. Latent Kaposi's sarcoma-associated herpesvirus infection of endothelial cells activates hypoxia-induced factors. *J Virology.* 2006; 80: 10802-10812.
12. Ceylan H, Yuncu M, Gurel A, Armutcu F, Gergerlioglu HS, Bagci C and Demiryurek AT. Effects of whole-body hypoxic preconditioning on hypoxia/reoxygenation-induced intestinal injury in newborn rats. *Eur J Pediatr Surg.* 2005; 15: 325-332.
13. Chavez JC, Baranova O, Lin J and Pichiule P. The transcriptional activator hypoxia inducible factor 2 (HIF2/EPAS-1) regulates the oxygen-dependent expression of erythropoietin in cortical astrocytes. *J Neuroscience.* 2006; 26: 9471-9481.

14. Corn PG. Hypoxic regulation of miR-210. *Cancer Biol Ther*. 2008; 7: 265-267.
15. D'Angelo G, Duplan E, Boyer N, Vigne P and Frelin C. Hypoxia up-regulates prolyl hydroxylase activity: a feedback mechansim that limits HIF-1 responses during reoxygenation. *J Biol Chem*. 2003; 278: 38183-38187.
16. Drevytska T, Dosenko V, Nagibin V and Mankovska I. HIF-1α, HIF2α, HIF3α and HIF1β mRNA expression changes in different tissues under intermittent training. In: Ley Xi, Serebrovskaya TV, eds., *Intermittent Hypoxia*, Nova Science Publishers, Inc, N Y, USE, 2009: 451-469.
17. Fandrey J. Oxygen-dependent and tissue-specific regulation of erythropoietin gene expression. *Am J Physiol Regul Integr Comp Physiol*. 2004; 286(6): R977-88.
18. Fasanaro P, D'Alessandra Y, Di Stefano V, Melchionna R, Romani S, Pompilio G, Capogrossi MC and Martelli F. MicroRNA-210 modulates endothelial cell response to hypoxia and inhibits the receptor tyrosine-kinase ligand Ephrin-A3. *J Biol Chem*. 2008; 283: 15878-83.
19. Fish JE and Srivastava D. MicroRNAs: Opening a New Vein in Angiogenesis Research. Sci Signal. 2009; 52: 1-7.
20. Forsythe JA, Jiang BH, Iyer NV, Agani F, Leung SW, Koos RD and Semenza GL. Activation of vascular endothelial growth factor gene transcription by hypoxia-inducible factor 1. *Mol Cell Biol*. 1996; 16: 4604-4613.
21. Grimm C, Wenzel A, Groszer M, Mayser H, Seeliger M, Samardzija M, Bauer C, Gassmann M and Reme CE. HIF-1-induced erythropoietin in the hypoxic retina protects against light-induced retinal degeneration. *Nat. Med*. 2002; 8: 718-724.
22. Gu YZ, Moran SM, Hogenesch JB, Wartman L and Bradfield CA. Molecular characterization and chromosomal localization of a third alpha-class hypoxia inducible factor subunit, HIF3α. *Gene Expr*. 1998; 7: 205-213.
23. Gulyaeva NV and Tkatchouk EN. Effects of normobaric hypoxic training on immunoreactive erythropoietin and transferin levels in blood serum of healthy volunteers. *Hyp Med J*. 1998; 6: 13-17.
24. Haase VH. Hypoxia-inducible factors in the kidney. *Am J Physiol Renal Physiol*. 2006; 291: F271-F281.
25. Hara S, Hamada J, Kobayashi C, Kondo Y and Imura N. Expression and characterization of hypoxia-inducible factor (HIF)-3α in human kidney: suppression of HIF-mediated gene expression by HIF-3α. *Bioch Bioph Res Comm*. 2001; 287: 808-813.
26. Heidbreder M, Fröhlich F, Jöhren O, Dendorfer A, Qadri F and Dominiak P. Hypoxia rapidly activates HIF-3alpha mRNA expression. *FASEB J*. 2003; 17: 1541-1543.
27. Hu CJ, Wang LY, Chodosh LA, Keith B and Simon MC. Differential roles of hypoxia-inducible factor 1alpha (HIF-1alpha) and HIF-2alpha in hypoxic gene regulation. *Mol Cell Biol*. 2003; 23: 9361-9374.
28. Ivan M, Adrian L. Harris, Martelli F and Kulshreshtha R. Hypoxia response and microRNAs: no longer two separate worlds. *J. Cell. Mol. Med*. 2008; 12: 1426-1431.
29. Jones NM and Bergeron M. Hypoxic preconditioning induces changes in HIF-1 target genes in neonatal rat brain. *J. Cereb. Blood Flow Metab*. 2001; 21: 1105-1114.

30. Jones NM, Lee EM, Brown TG, Jarrott B and Beart PM. Hypoxic preconditioning produces differential expression of hypoxia-inducible factor-1alpha (HIF-1a) and its regulatory enzyme HIF prolyl hydroxylase 2 in neonatal rat brain. *Neurosci. Lett.* 2006; 404; 72-77.

31. Jurgensen JS, Rosenberger C, Wiesener MS, Warnecke C, Horstrup JH., Grafe M, Philipp S, Griethe W, Maxwell PH, Frei U, Bachmann S and Willenbrock R. Persistent induction of HIF-1α and -2α in cardiomyocytes and stromal cells of ischemic myocardium. *FASEB J.* 2004; 18: 1415-1417.

32. Kietzmann T, Cornesse Y, Brechtel K, Modaressi S and Jungermann K. Perivenous expression of the mRNA of the three hypoxia-inducible factor α-subunits, HIF-1α, HIF-2α and HIF-3α, in rat liver. Biochem J. 2001; 354: 531-537.

33. Kulshreshtha R, Ferracin M, Wojcik SE, Garzon R, Alder H, Agosto-Perez FJ, Davuluri R, Liu CG, Croce CM, Negrini M, Calin GA and Ivan M. A microRNA signature of hypoxia. *Mol Cell Biol.* 2007; 27: 1859-67.

34. Lai IR, Ma MC, Chen CF and Chang KJ. The protective role of heme oxygenase-1 on the liver after hypoxic preconditioning in rats. *Transplantation.* 2004; 77: 1004-1008.

35. Lando D, Pongratz I, Poellinger L and Murray WA. Redox mechanism controls differential DNA binding activities of hypoxia-inducible factor (HIF) 1a and the HIF-like factor. *J Biol Chem.* 2000; 275: 4618-4627.

36. Laughner E, Taghavi P, Chiles K, Mahon PC and Semenza GL. HER2 (neu) signaling increases the rate of hypoxia-inducible factor 1alpha (HIF-1alpha) synthesis: novel mechanism for HIF-1-mediated vascular endothelial growth factor expression. *Mol Cell Biol.* 2001; 21: 3995-4004.

37. Lin JS, Chen YS, Chiang HS and Ma MC. Hypoxic preconditioning protects rat hearts against ischaemia-reperfusion injury: role of erythropoietin on progenitor cell mobilization. *J Physiol.* 2008; 586: 5757-5769.

38. Luo JC and Shibuya M. A variant of nuclear localization signal of bipartite-type is required for the nuclear translocation of hypoxia inducible factors (1alpha, 2alpha and 3alpha). *Oncogene.* 2001; 20: 1435-1444.

39. Mankovska I, Serebrovskaya T, Vavilova G, Nosar V and Kharlamova O. Intermittent hypoxic training enhances myocardial metabolic and membrane defense against severe hypoxia. *Pathophysiology.* 1998; 5: 31.

40. Mankovska I, Vavilova G, Kharlamova O, Nosar V and Serebrovskaya T. Activity of the cell membrane marker enzyme in rats under adaptation to hypoxia. *Ukr Biochem J.* 1997; 69: 79-87.

41. Mankovska IN, Gavenauskas BL, Gonchar OA, Nosar VI and Rozova EV. Structurally are metabolic bases of adaptation of muscular tissue to the different types of hypoxia. *Clin Exp Pathol.* 2004; 2: 76-78.

42. Mankovskaya I, Drevitskaya T, Dosenko V, Gavenauskas B and Moiseenko E. Expression of transcriptional factor HIF subunits in rat tissues under acute and intermittent hypoxia. *Hypoxia Med. J.* 2006; 1-2: 35.

43. Maxwell PH. Hypoxia-inducible factor as a physiological regulator. *Exp. Physiol.* 2005; 90: 791-797.
44. Maynard MA, Evans AJ, Hosomi T, Hara S, Jewett MA and Ohh M. Human HIF-3_4 is a dominant-negative regulator of HIF-1 and is down-regulated in renal cell carcinoma. *FASEB J.* 2005; 19: 1396-1406.
45. Murphy BJ. Regulation of malignant progression by the hypoxia-sensitive transcription factors HIF-1alpha and MTF-1. *Comp Biochem Physiol B Biochem Mol Biol.* 2004; 139: 495-507.
46. Nanduri J, Yuan G, Kumar GK, Semenza GL and Prabhakar NR Transcriptional responses to intermittent hypoxia. *Respir Physiol Neurobiol.* 2008; 164: 277-81.
47. Nardinocchi L, Puca R, Guidolin D, Belloni AS, Bossi G, Michiels C, Sacchi A, Onisto M and D'Orazi G. Transcriptional regulation of hypoxia-inducible factor 1alpha by HIPK2 suggests a novel mechanism to restrain tumor growth. *Biochim Biophys Acta.* 2009; 1793: 368-77.
48. Pagé EL, Robitaille GA, Pouysségur J and Richard D.E. Induction of hypoxia-inducible factor-1alpha by transcriptional and translational mechanisms. *J Biol Chem.* 2002; 277: 48403-48409.
49. Partovian C, Adnot S, Eddahibi S, Teiger E, Levame M, Dreyfus P, Raffestin B and Frelin C. Heart and lung VEGF mRNA expression in rats with monocrotaline- or hypoxia-induced pulmonary hypertension. *Am J Physiol Heart Circ Physiol.* 1998; 275: H1948-H1956.
50. Ratcliffe PJ. Understanding hypoxia signalling in cells-a new therapeutic opportunity? *Clin. Med.* 2006; 6: 573-578.
51. Rosenberger C, Mandriota S, Jurgensen JS, Wiesener MS, Horstrup JH, Frei U, Ratcliffe PJ, Maxwell PH, Bachmann S and Eckardt KU. Expression of hypoxia-inducible factor-1alpha and -2alpha in hypoxic and ischemic rat kidneys. *J. Am. Soc. Nephrol.* 2002; 13: 1721-1732.
52. Rossignol F, Laplanchec E, Mouniera R, Bonnefontd J, Cayree A, Godinotb C, Simonnetc H and Clottesa E. Natural antisense transcripts of HIF-1α are conserved in rodents. *Gene.* 2004; 339: 121-130.
53. Semenza GL and Prabhakar NR. HIF-1-dependent respiratory, cardiovascular, and redox responses to chronic intermittent hypoxia. *Antioxid Redox Signal.* 2007; 9: 1391-6.
54. Semenza GL and Wang GL. A nuclear factor induced by hypoxia via de novo protein synthesis binds to the human erythropoietin gene enhancer at a site required for transcriptional activation. *Mol Cell Biol.* 1992; 12: 5447-5454.
55. Semenza GL, Roth PH, Fang HM and Wang GL. Transcriptional regulation of genes encoding glycolytic enzymes by hypoxia-inducible factor 1. *J Biol Chem.* 1994; 269: 23757-23763.
56. Semenza GL, Rue EA, Iyer NV, Pang MG and Kearns WG. Assignment of the hypoxia-inducible factor 1-alpha gene to a region of conserved synteny on mouse chromosome 12 and human chromosome 14q. *Genomics.* 1996; 34: 437-439.
57. Semenza GL. HIF-1: mediator of physiological and pathophysiological responses to hypoxia. *J Appl Physiol.* 2000; 88: 1474-1480.
58. Semenza GL. Hydroxylation of HIF-1: oxygen sensing at the molecular level *Physiology.* 2004; 176-182.

59. Semenza GL. Perspectives on oxygen sensing. *Cell*. 1999; 88: 281-284.
60. Serebrovskaya TV, Mankovskaya IM, Moiseenko EV and Drevytska TI. Role of genetic factors in the adaptation to intermittent hypoxia. *Pneumologia I Alergologia Polska*. 2006; 74: 339.
61. Sharp FR, Bergeron M and Bernaudin M. Hypoxia-inducible factor in brain. *Adv. Exp. Med. Biol*. 2001; 502: 273-291.
62. Shizukuda Y, Iwamoto T, Mallet RT and Downey HF. Hypoxic preconditioning attenuates stunning caused by repeated coronary artery occlusions in dog heart. *Cardiovasc. Res*. 1993; 27; 559-564.
63. Shizukuda Y, Mallet RT, Lee SC and Downey HF. Hypoxic preconditioning of ischaemic canine myocardium. *Cardiovasc. Res*. 1992; 26: 534-542.
64. Siomi H and Siomi M. On the road to reading the RNA-interference code. *Nature*. 2009; 45722: 396-404.
65. Talks KL, Turley H, Gatter KC, Maxwell PH, Pugh CW, Ratcliffe PJ and Harris AL. The expression and distribution of the hypoxia-inducible factors HIF-1α and HIF-2α in normal human tissues, cancers, and tumor-associated macrophages. *Am J Pathol*. 2000; 157: 411-421.
66. Thrash-Bingham CA and Tartof KD. aHIF: a natural antisense transcript overexpressed in human renal cancer and during hypoxia. *J National Cancer Inst*. 1999; 91: 143-151.
67. Vogt M, Puntschart A, Geiser J, Zuleger C, Billeter R and Hoppeler H. Molecular adaptations in human skeletal muscle to endurance training under simulated hypoxic conditions. *J Appl Physiol*. 2001; 91: 173-182.
68. Wenger RH and Gassmann M. Oxygen(es) and the hypoxia-inducible factor-1. *Biol Chem*. 1997a; 378: 609-616.
69. Wenger RH, Rolfs A, Kvietikova I, Spielmann P, Zimmermann DR and Gassmann M. The mouse gene for hypoxia-inducible factor-1alpha-genomic organization, expression and characterization of an alternative first exon and 5′ flanking sequence. *Eur J Biochem*. 1997b; 246: 155-165.
70. Wenger RH, Rolfs A, Marti HH, Guénet JL and Gassmann M. Nucleotide sequence, chromosomal assignment and mRNA expression of mouse hypoxia-inducible factor-1 alpha. *Biochem Biophys Res Commun*. 1996; 223: 54-59.
71. Wenger RH, Rolfs A, Spielmann P, Zimmermann DR and Gassmann M. Mouse hypoxia-inducible factor-1a is encoded by two different mRNA isoforms: expression from a tissue-specific and a housekeeping-type promoter. *Blood*. 1998; 91: 3471-3480.
72. Wiener CM, Booth G and Semenza GL. In vivo expression of mRNAs encoding hypoxia-inducible factor 1. *Biochem Biophys Res Commun*. 1996; 225: 485-488.
73. Zhang SX, Miller JJ, Gozal D and Wang Y. Whole-body hypoxic preconditioning protects mice against acute hypoxia by improving lung function. *J. Appl. Physiol*. 2004; 96: 392-397.
74. Zolk O, Solbach TF, Eschenhagen T, Weidemann A and Fromm MF. Activation of negative regulators of the hypoxia-inducible factor (HIF) pathway in human end-stage heart failure. *Biochem Biophys Res Commun*. 2008; 376: 315-320.

Adaptation Biology and Medicine (Volume 6: Cell Adaptations and Challenges)
Editors: P. Wang, C.-H. Kuo, N. Takeda and P.K. Singal
Copyright © 2011 Narosa Publishing House Pvt. Ltd., New Delhi, India

21. Role of Nitric Oxide (NO) Stores in Adaptive Defense Against NO Overproduction

E.B. Manukhina[1,2], H.F. Downey[2], R.T. Mallet[2], A.V. Goryacheva[1], I. Yu. Malyshev[1,3], A.F. Vanin[4]

[1]Institute of General Pathology and Pathophysiology, Moscow, Russian Federation; [2]University of North Texas Health Science Center, Fort Worth, TX, USA; [3]Moscow State University of Medicine and Dentistry, Moscow, Russian Federation; [4]Institute of Chemical Physics, Moscow, Russian Federation

Summary

Formation of NO stores is an effective mechanism for protection of the cardiovascular system against NO overproduction. Moreover, this mechanism ensures compensatory changes in the ability of blood vessels to store NO in accordance with NO concentrations in blood plasma. Focused modulation of this process may have clinical significance, allowing NO storage to play a greater role in treatment of disorders caused by either NO overproduction or NO deficiency. The subject matter has been discussed in greater detail.
Keywords: Nitric oxide, paracrine, paramagnetic resonance, cysteine and heart

Introduction

Nitric oxide (NO) is a universal regulatory factor that participates in myriad physiological and biochemical processes. NO is a free radical, possessing an unpaired electron, and its half life is brief, just 6-10 s. Despite its intrinsic instability, NO possesses both autocrine and paracrine effects; i.e., once synthesized it can exert effects not only in its cell of origin but also in neighboring and even distant cells [Lancaster, 1994; Sato & Murota, 1995]. However, before reaching distant targets, NO may be scavenged by endogenous traps, such as hemoglobin, or oxidized by O_2 and superoxide ($\cdot O_2^-$). To exert its paracrine and endocrine effects, NO must therefore be protected from oxidation and inactivation. One means of protecting NO is to incorporate it into more stable nitroso compounds, which can then release NO under appropriate conditions in target tissues or cells (Figure 1). These nitroso compounds form so-called NO stores. S-Nitrosothiols (RS-NO) and dinitrosyl iron complexes (DNIC; $[(RS^-)Fe^+(NO^+)_2]^+$) are the principal physiological forms of NO storage and transport [Vanin *et al.*, 2007; Yang & Loscalzo, 2007].

*Address for Correspondence: Dr. Eugeina B. Manukhina, Laboratory of Regulatory Mechanisms of Adaptation, Institute of General Pathology and Pathophysiology, Russian Academy of Medical Sciences, Baltijskaya Street 8, Moscow 125315, Russia
Email: manukh@netbynet.ru

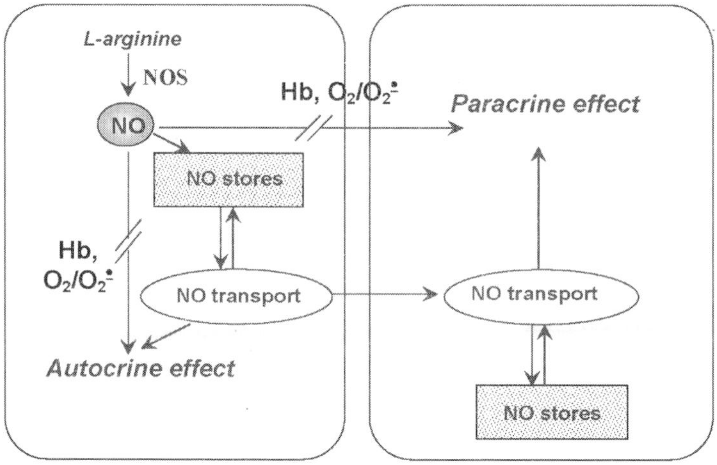

Fig. 1. Transport and storage of nitric oxide. The two rectangles represent cells. Hemoglobin, O_2 and O_2 degrade or condense with NO, limiting its biological activity.

Both RS-NO and DNIC exist in two forms: high molecular weight, in which NO is bound to thiol groups in proteins, and low molecular weight, where NO is attached to crystalloid thiols such as cysteine or glutathione (GSH). The high molecular weight complexes are much more stable than the low molecular weight ones and are considered to be the primary intracellular NO stores. For example, protein-bound DNIC can stabilize NO with $t_{1/2}$ = 4.5-8 hours [Turella *et al.*, 2003]. Increased local concentrations of low molecular weight thiols shift the balance to favor the low molecular weight DNIC. Under such conditions, the stability of the protein-bound DNIC decreases, and these stores slowly release NO [Stoclet *et al.*, 1999; Vanin & van Faassen, 2007]. These biologically active NO stores represent a dynamic system with constant interconversion of free and bound NO forms of various stabilities [Mulsch, 1994; Vanin & van Faassen, 2007].

Detection and Measurement of NO Stores

Studies to detect NO stores began in 1961, even before the discovery of endothelium-dependent relaxation of blood vessels and characterization of endothelium-derived relaxing factor. At first Furchgott et al. [1961] and later Venturini et al. [1993] and Megson et al. [1995] demonstrated that pre-contracted blood vessels relaxed after exposure to light, and they proposed that this photorelaxation was mediated by NO release from a store in smooth muscle cells. This store was exhaustible but could be replenished by NO donors or by incubation of the vessel with intact endothelium in the dark. NOS inhibitors abolished the recovery of NO stores by incubation in the dark. The aorta relaxation induced by UV light closely corresponded to the release of NO from NO stores, detected with a porphyrin microsensor [Kubaszewski *et al.*, 1993].

A current method for detecting DNIC-bound NO stores is electron paramagnetic resonance (EPR), which detects characteristic signals of NO bound to a specific trap. The signal from low-molecular weight DNIC produces a narrow, isotropic peak with semi-width 0.7 mT and g-factor 2.03. Protein-bound DNICs produce a wider, anisotropic peak (~ 4 mT) and g-factor values g = 2.04, g = 2.014 [Mülsch *et al.* 1995; Fichtlscherer & Mülsch, 2000; Vanin *et al.* 2007]. Figure 2A shows EPR signals from NO stores accumulated in a porcine coronary artery after incubation with the NO donor, S-nitrosoglutathione (GSNO), and subsequent washout [Alencar *et al.*, 2003]. The values in Figure 2B reflect quantitative measurements of EPR signals. After incubation with the NO donor, the amplitude of EPR signal increased strikingly. Selective destruction of S-nitrosothiols by $HgCl_2$ sharply lowered the EPR signals. NO released from NO stores formed from S-nitrosothiols can substantially depress contractile responses of the coronary artery (Figure 2C).

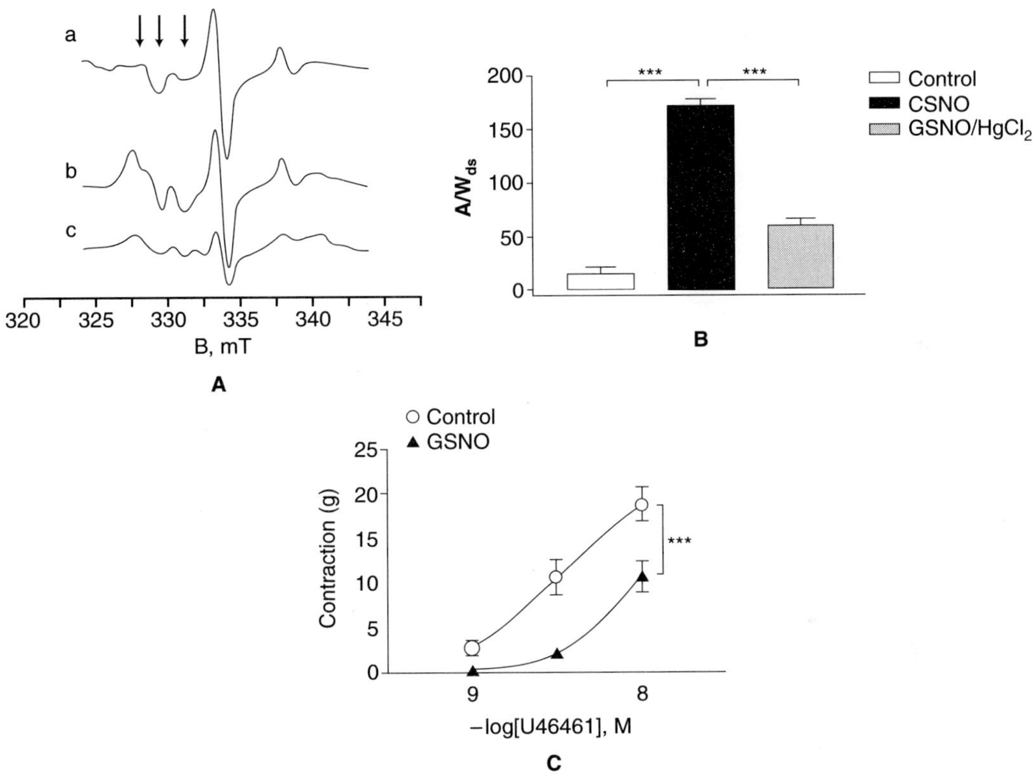

Fig. 2 Electron paramagnetic resonance (EPR) of NO. A: Typical EPR spectra from porcine coronary artery segments after incubation with [Fe(II)(DETC)$_2$]. a, control segments; b, segments pre-exposed to GSNO (100 μM followed by several washouts;c, segments treated with 1 μM $HgCl_2$ after exposure to GSNO. Arrows indicate the three hyperfine components of the[Fe(IINO(DETC)$_2$] EPR signal. B: Quantification of the amplitude of the [Fe(II)NO(DET)$_2$] Signal (A/Wds in relative units per mg, where A is amplitude and Wds is weight of the dried sample) in rat aorta segments. Values are means ± SEM from three experiments. ***P < 0.001 (ANOVA). C:Contractile effect of U46619 in porcine coronary artery segments pre-exposed to 1 μM GSNO (tringles) or vehicle (circles). Means ± SEM from 5-8 experiments. Error bars are not shown when the size of the symbol exceeds the value of the S.E.M. ***P < 0.001 (MANOVA)[Alencar et al., 2003]

Histochemical studies permitted visualization of NO stores [Flitney *et al.*, 1992]. Staining of cells for Fe2+ clusters demonstrated easy penetration and accumulation of cluster-type NO donors within the endothelial cells. At the same time, the clusters seldom entered smooth muscle cells, and they were rarely observed even in smooth muscle cells adjacent to the endothelium.

NO stores can also be visualized after incubation with NO donors, e.g. GSNO, using immunolabeling of S-nitrosated cysteine with antibodies to S-NO moieties [Alencar *et al.*, 2003]. Immunostaining was not observed in isolated vessel rings treated with $HgCl_2$ after exposure to GSNO.

NO stores are currently most often detected and measured using physiological methods. These methods are based on evaluation of vascular relaxation in response to NO released from NO stores in vascular walls using diethyldithiocarbamate (DETC) or N-acetylcysteine (NAC) (Figure 3), agents which react with DNIC and S-nitrosothiols and release NO from these stores.

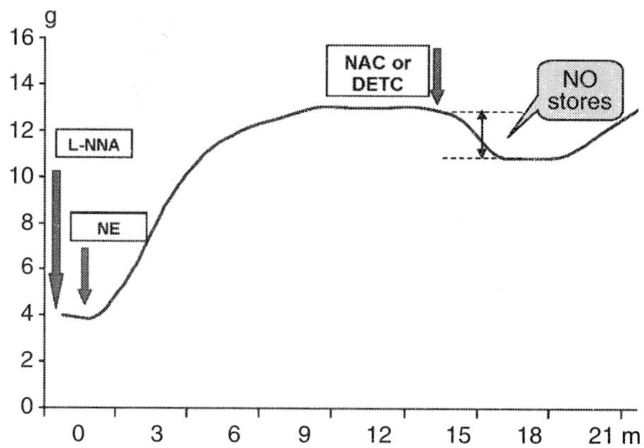

Fig. 3. Detection of available NO stores in rat aorta. Abscissa: time (min); ordinate: contraction tension of aortic segments (g). NE: norpinephrine; L-NNA: N^{w}-nitro-L-arginine; DETC: diethyldithiocarbamate; DNIC: dinitrosy1 iron complex. After 1h stabilization, the aortic segment was precontracted with NE (5×10^{-7} M), then incubated wiht NOS inhibitor L-NNA (10^{-4} M) for 20 min. No stores were detected and evaluated by relaxation of the preparation in response to DETC (3×10^{-4} M) against the background of NOS inhibition to exclude the contribution of de nove synthesized NO in the vascular response. The relaxation value was calculated as % of total contraction incuced by NE and L-NNA.

DETC reacts with DNIC to form a mononitrosyl iron complex, and this reaction releases a nitrosonium ion, NO^+ [Butler *et al.*, 1985; Stamler *et al.*, 1992]. NO^+ nitrosylates thiols to form RS-NO [Vedernikov *et al.*, 1992]. Under the influence of DETC, RS-NO also dissociates, liberating NO [Arnell *et al.*, 1997].

A low molecular weight, cell-penetrating thiol, NAC reacts with intracellular NO stores to form low-molecular weight RS-NO and DNICs, which in turn release free NO [Muller *et al.*, 1996, 2000; Vlasova *et al.*, 2003]. Low molecular weight thiols can displace NO from S-nitrosoproteins and $Fe(NO)_2$ from protein-bound DNIC. These reactions result in the formation of NO-donating agents or in transnitrosation of proteins, or in transfer of $Fe(NO)_2$ groups between thiols in proteins [Muller *et al.*, 2002]. The NAC-induced vasorelaxation is blunted by NO traps, oxyhemoglobin, the NO scavenger 2-(4-carboxyphenyl)-4,4,5,5-tetramethyl-imidazoline-L-oxyl-3-oxide (carboxy-PTIO), or by $HgCl_2$ [Alencar *et al.*, 2001].

EPR demonstrated that high concentrations of NAC induced partial conversion of protein-bound DNIC to low molecular weight DNIC [Muller *et al.*, 1996]. After iNOS induction by lipipolysaccharides and L-arginine, NO stores in the form of protein-bound DNICs were detectable in the isolated aorta by EPR. No DNICs were formed in the presence of NOS inhibitors [Muller *et al.*, 1996; Kleschyov *et al.*, 2000]. Thus, it has been hypothesized that protein-bound DNIC represent a NO store and that the low molecular weight DNIC formed in the presence of NAC either releases NO, transfers $Fe(NO)_2$ groups, or S-nitrosates another target protein [Muller *et al.*, 2002]. Relaxation induced by NAC or DETC differs from photorelaxation. The latter is mediated by release of the free radical NO from stores generated from endothelium-derived NO, while NAC- or DETC-induced relaxation is generally prominent under conditions associated with iNOS induction and the resulting NO overproduction [Muller *et al.*, 2002].

Comparison of the NO-releasing effects of DETC and NAC, alone and in combination, revealed that NO stores consist of two pools releasable by DETC or NAC [Vlasova *et al.*, 2003]. Apparently NAC can deplete both pools, since the vasorelaxation induced by DETC is smaller than the vasorelaxation induced by NAC, the sum of DETC-induced vasorelaxation and subsequent NAC-induced vasorelaxation is equal to the relaxation response to NAC alone, and DETC added after NAC does not induce any additional relaxation of isolated vessels. When DETC and NAC were tested separately, inhibition of guanylate cyclase with methylene blue completely abolished the response to DETC and reduced the response to NAC by the increment equal to DETC-induced relaxation. The authors concluded that the use of NAC is preferable to DETC, because NAC mobilizes stored NO more effectively and also is less toxic than DETC.

Aortas isolated from rats after stimulation of endogenous NO synthesis or pretreatment with an NO donor responded to DETC by relaxing. The magnitude of this vasorelaxation reflected the size of NO stores. Control preparations of rat aorta [Smirin *et al.*, 1999], rat mesenteric artery [Muller *et al.*, 1996; Alencar *et al.*, 2003] and porcine coronary artery [Alencar *et al.*, 2003] did not respond to DETC, indicating that NO stores were absent or below the detection limit of this method. However, recently when NAC was used to mobilize NO stores in isolated rat coronary and basilar arteries under basal conditions, relaxation was detected with very sensitive myography [Manukhina *et al.*, 08].

Efficiency of NO storage or the potential capacity to store NO can be assessed by the maximum amount of NO stored in the wall of an isolated blood vessel after incubation with excess NO donor for 15-20 minutes. After washout of the NO donor, NO store is assessed by the magnitude of the relaxation response upon release of stored NO by NAC or DETC (Figure 4) [Vlasova *et al.*, 2003, 2006]. A modification of this method allowed evaluation of NO stores in the whole body of conscious rats [Mashina *et al.*, 2003]. In these experiments, NO stores were formed after imposing heat stress or injecting an NO donor. NO release from stores was detected by the decrease in blood pressure to DETC administration in the presence of the NO synthase (NOS) inhibitor, L-NNA (Figures 5A, 5B). The release of stored NO was confirmed by the characteristic EPR signal from rat aorta, liver (Figure 5C), kidneys, heart and brain.

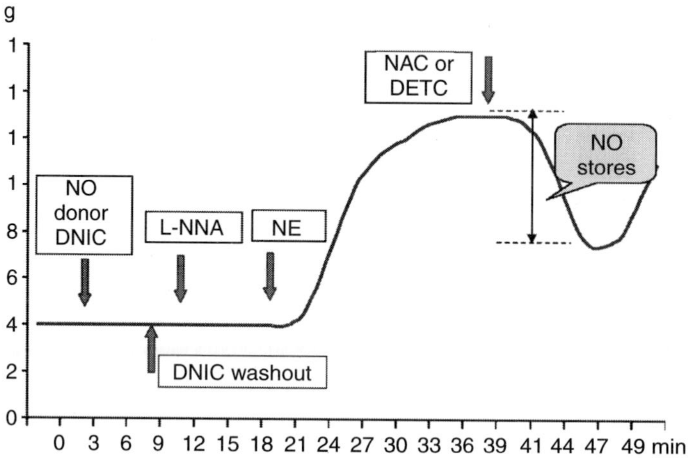

Fig. 4. Evaluation of NO storage capacity in rat aorta. Efficiency of NO storage was assessed as the potential capacity of NO stores, i.e. the maximum amount of NO which can be potentially stored in the wall of isolated vessels after incubation with an excess of NO donor. To evaluate the potential capacity of NO stores, rat aortic segment was incubated with DNIC (10^{-5} M) for 30 min after the NE-induced contractile response plateaued. The incubation was performed until vascular tone returned to the initial level after DNIC-induced relaxation, indicating complete degradation of DNIC in the solution. After washout of DNIC, NO stores were evaluated by the relaxation response induced by DETC.

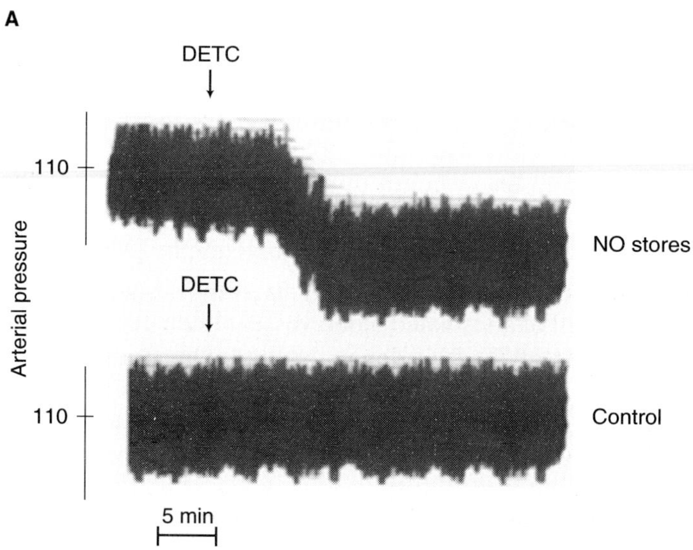

Fig. Contd..

Fig. Contd..

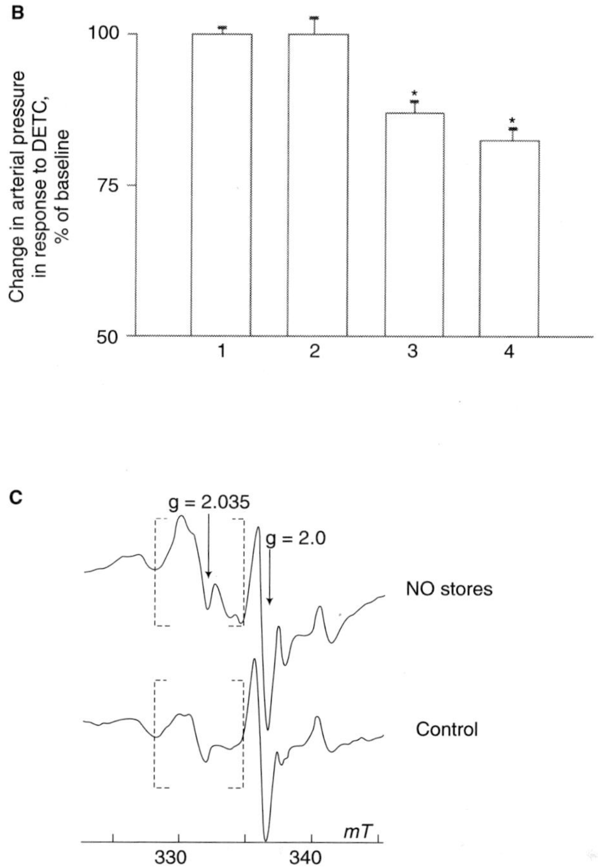

Fig. 5. In vivo assessment of NO stores formed after administration of NO donor dinitrosyl-iron complexes (DNIC) or heat shock in rats. A: Arterial pressure in conscious rats; arrow indicates administration of diethyldithiocarbamate (DETC). B: Change in arterial pressure (mm Hg) after administration of DETC superimposed on NO synthase inhibition. 1: Baseline arterial pressure before administration of DETC; 2: Arterial pressure response to DETC in untreated rats; 3: Arterial pressure after DETC mobilization of NO stores in rats pre-injected with the NO donor, DNIC; 4: Arterial pressure after DETC mobilization of NO stores in rats pre-exposed to heat shock. Values are means ± SEM from 9 experiments. *P<0.05 compared to control animals injected with DETC (condition 2). C: EPR signals of mononitrosyl-iron-diethyldithiocarbamate complexes in rat liver after the formation of NO stores by DNIC or heat shock (upper curve); control (lower curve). EPR was performed under the conditions of NO synthase inhibition. Dotted frame: component of the hyperfine triplet structure of EPR signals reflecting NO content, as in Figure 2A [Mashina *et al.,* 2003].

Mechanisms of NO Store Formation

Studies in cell cultures and intact animals suggested that NO stores are formed after any increase in NO, irrespective of the source. Indeed, myriad interventions such as administration of NO donor

DNIC, pharmacological NOS activation, exposure to heat shock, or gradual adaptation to hypoxia, physical load, or stress are accompanied by formation of NO stores in the vascular wall. These stores can then be detected by DETC-induced relaxation of the isolated vessels [reviews: Manukhina & Malyshev, 2002, 2005; Manukhina *et al*., 2006a].

The increased NO production during adaptation to various environmental factors is not generally associated with signs of NO toxicity. Apparently, development of nitrosative stress requires high rates of NO generation rather than a large cumulative NO production. For example, NO donors with long half-lives protect cells from apoptotic cell death, but the same concentrations of less stable NO donors induce apoptosis [Kim *et al*., 1997], indicating that the rate of NO production is a critical determinant of NO-mediated cell death [Kim *et al*., 2000].

Many studies of NO stores have used adaptation to intermittent hypobaric hypoxia to stimulate endogenous NO production. Indeed, one of the strongest physiological stimuli of NO synthesis is adaptation to hypobaric or normobaric hypoxia [Manukhina & Malyshev, 2002, 2005; Manukhina *et al*., 2006]. In one study, rats were exposed to a simulated altitude of 4000 m in a hypobaric chamber. The duration of hypoxia was gradually increased from 10 min to 5 h over 40 daily sessions. NO production was evaluated from the combined plasma concentrations of the stable NO metabolites NO2- plus NO3-, i.e. NOx. Hypoxic adaptation induced a gradual increase in plasma NOx, and by the end of the 40 day program, NOx was double the control level. NO stores, which were absent in non-adapted rats, appeared and considerably expanded over the course of adaptation. After hypoxic adaptation, the size of NO stores, as judged from the extent of DETC-induced vasorelaxation, significantly correlated with plasma NOx concentrations (Figure 6) [Manukhina *et al*., 1999].

Figure 7 presents tentative mechanisms for formation of NO stores during intermittent hypoxia. Each hypoxic session ends with return to normal oxygen supply. This repeated reoxygenation results in intermittent activation of processes that generate reactive oxygen species (ROS) which are essential for the cadioprotection that is conferred by adaptation to hypoxia [Kolár *et al*., 2007; Mallet *et al*., 2008]. Treatment with antioxidants during hypoxic sessions interferes with the development of cardioprotection in rats [Kolár *et al*., 2007] and dogs [Mallet *et al*., 2008]. A possible mechanism for expanding NO stores during adaptation to hypoxia is hypoxia-stimulated increase of free iron to the body [Simpson, 1992], which augments NO storage in the form of DNIC [Vanin *et al*., 1997]. Free iron increases, since ·O2- formed during each reoxygenation releases Fe2+ from the iron-containing protein, ferritin [Paul, 2000]. In addition, synthesized NO also can mobilize iron from ferritin [Richardson, Lok, 2008]. This increase in Fe2+, in turn, promotes formation of additional DNIC [Lipinski & Drapier, 1997]. Fe2+ not only facilitates NO storage but can also transiently activate NO synthesis via an L-arginine-dependent pathway, possibly by generating free radicals which damage cell membranes, allowing entry of Ca2+ which activates constitutive NOS [Kubrina *et al*., 1993].

The role of iron in the formation of NO stores and in the defense against NO toxicity afforded by NO stores was demonstrated in cultured cells [Kim *et al*., 2000]. Elevation of non-heme iron in cells significantly delayed cell death by apoptosis associated with an inhibition of NO-induced caspase-3 activation, initially, and then by an increase in S-nitrosylation of caspase-3-like enzymes. The inhibition of caspase activity and delay in apoptosis correlated with the formation of DNIC in the cells. DNIC was able to inhibit caspase activity and caspase-3-mediated DNA fragmentation in the in vitro reconstitution system. Therefore, the level of non-heme iron was an important factor in determining

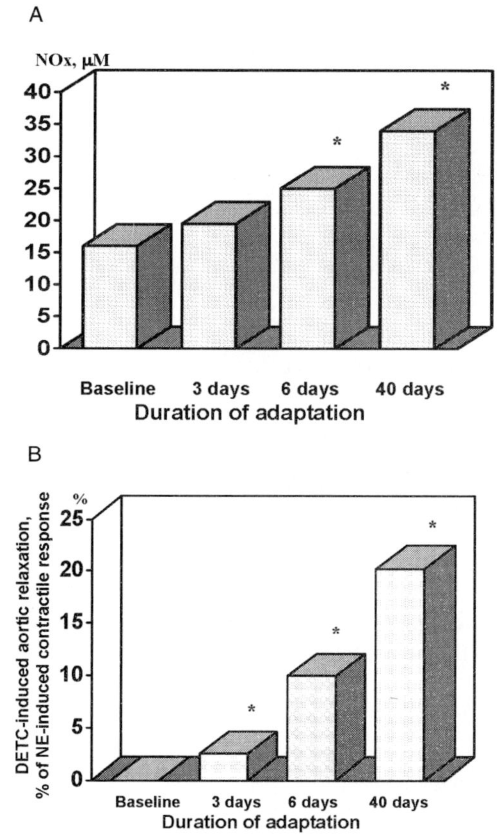

Fig. 6. Time course of plasma NOx concentration (A) and formation of NO stores in vascular walls (B) during adaptation to hypoxia. A: Plasma NO_2^- + NO_3^- concentration (M). *P<0.05 vs. baseline. B: Relaxation of NE-precontracted rat aorta induced by DETC, expressed as percent of the contractile response to NE. The magnitude of DETC-induced relaxation reflects the size of NO stores. *P<0.05 vs. baseline. Mean values from 12 experiments.

the consequence of NO exposure on cell viability. Non-heme iron effectively removed excessive NO by formation of DNIC, while DNIC regulated caspase activity in cells by facilitating S-nitrosylation of the caspase enzymes and thereby diverted NO into nontoxic or protective pathways.

The intensity of NO production is genetically predetermined, so it is conceivable that the ability to bind NO to NO stores may also be inherited. To address this possibility, the relationship between NO production and storage was studied in rats of different genetic strains: normotensive Wistar and Wistar-Kyoto (WKY) rats, August rats with higher intensity of NO synthesis [Mikoyan *et al.*, 1996], and stroke-prone spontaneously hypertensive rats (SHRSP) with lower intensity of NO synthesis or lower NO availability [Serebrovskaya *et al.*, 2008]. Formation of NO stores was induced by incubation of isolated aortas with DNIC. A direct relationship between NO storage efficiency and inherited levels of NO production in different rat strains emerged. Both NO storage efficiency and plasma NOx levels were highest in August rats, lowest in SHRSP, and intermediate in Wistar and WKY rats (Figure 8).

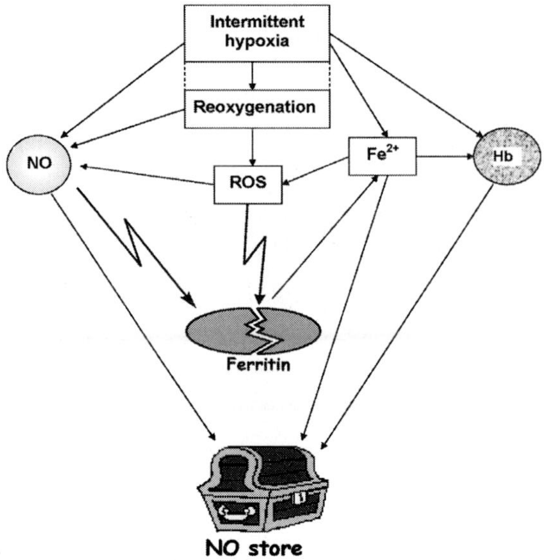

Fig. 7. Possible mechanisms of NO store formation in adaptation to hypoxia. See text for details. ROS: reactive oxygen species; Hb: hemoglobin.

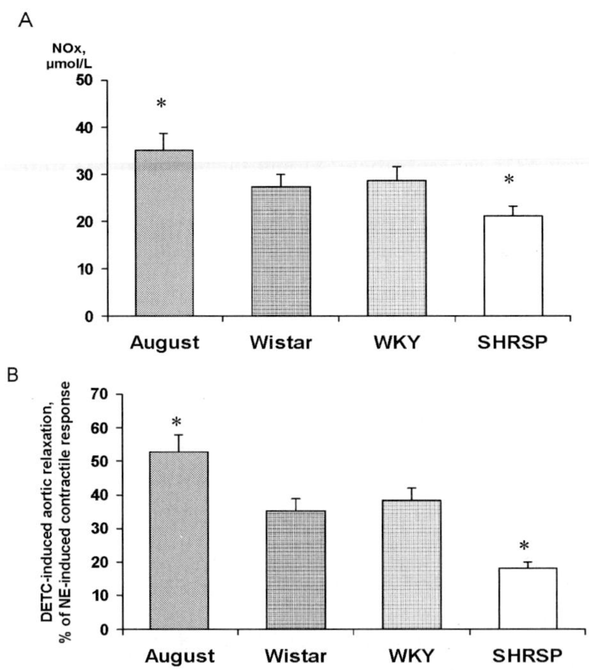

Fig. 8. Plasma NOx and NO store capacity in aorta of different rat strains. A: Plasma concentration (µM) of $NO_2^- + NO_3^-$. B: Potential capacity of NO stores (relaxation of aortic rings in response to DETC, % of NE-induced contraction). *P<0.05, August vs. Wistar and SHRSP vs. WKY. Abbreviations are defined in text. Means ± SEM from >8 experiments [Vlasova *et al.,* 2006].

If the ability to store NO is indeed genetically predetermined, blood vessels of August rats should bind NO in stores more effectively than vessels of rats with lower NO production. Adaptation to hypoxia consistently increased plasma NOx concentrations in both August and Wistar rats (Figure 9A) [Pshennikova *et al.*, 2000; Manukhina *et al.*, 2002]. However, this increase developed more slowly in August rats; consequently, plasma NOx concentrations were lower in August than Wistar rats at 40 d adaptation, despite the higher baseline levels in August rats. NO stores began to appear within the first three hypoxia sessions, and their capacity steadily increased until the end of the hypoxia program. NO stores formed more rapidly in August than in Wistar rats, so by the end of adaptation NO stores in August rats were double those of Wistar rats (Fig. 9B) [Pshennikova *et al.*, 2000; Manukhina *et al.*, 2002]. Therefore, August rats, which have a higher basal NO production, are also able to accumulate NO more efficiently than Wistar rats when NO synthesis is stimulated by adaptation to hypoxia.

Physiological and Cardioprotective Functions of NO Stores

The physiological significance of increased NO production and storage efficiency was evaluated by comparing the extent of stress-induced stomach ulceration in Wistar and August rats. Stress-induced

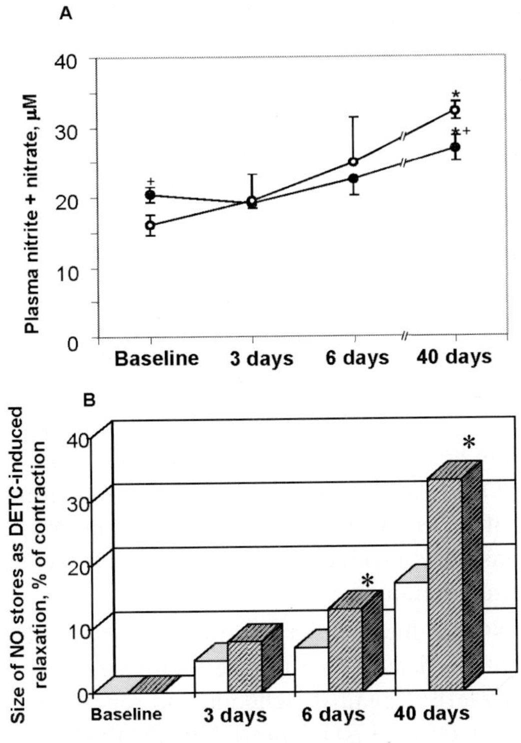

Fig. 9 Dynamics of plasma NOx (A) and NO store accumulation (B) in Wistar and August rats during adaptation to hypoxia. A: Plasma $NO_2^- + NO_3^-$ concentration (μM) over 40 d hypoxia conditioning. Open circles: Wistar rats; filled circles: August rats. *P<0.05 vs. baseline; +P<0.05 vs. Wistar rats. B: NO store capacity over 40 d hypoxia conditioning, determined from DETC-induced vasorelaxation as in Figure 4. Open bars: Wistar rats; hatched bars: August rats. *P < 0.05 vs. Wistar rats. Adapted from [Manukhina *et al.*, 2002].

stomach ulcers result from adrenergically-mediated vasoconstriction in gastric mucosa, and NO protects against this injury [Elliot &, Wallace, 1998]. Naive August rats, which have higher NO production, were considerably more resistant to stress-induced gastric ulceration than naive Wistar rats. Stomach ulcers developed three times more frequently in Wistar rats, and their total area was significantly larger than in August rats. After six days of hypoxic conditioning, a protective 'anti-ulcerogenic' effect was observed in rats of both strains, although ulcers were still larger in Wistar rats. Interestingly, a 40-day hypoxia program, as distinct from a 6-day program, not only failed to protect the rat stomach against stress ulceration but paradoxically potentiated this damage. After 40 days of hypoxia adaptation, the total area of stress ulcers was 3-10 times larger than in non-adapted rats. The potentiating effect of 40-day adaptation on stomach ulceration was more pronounced in August than in Wistar rats (ulcer area increased by 85% and 23%, respectively) [Pshennikova *et al.*, 2000; Manukhina *et al.*, 2002]. The authors suggested that the intense storage of NO in August rats could have produced a local deficiency of free NO, thus predisposing these animals to stress-induced vasoconstriction and stomach ulceration. These findings are the first indication that excessive storage of NO under certain conditions can be unfavorable for protection against stress-induced damage [Pshennikova *et al.*, 2000].

Formation of NO stores is considered to be part of the body's adaptive defenses. A key function of NO stores is protection against NO toxicity during NO overproduction, and the efficiency of NO store formation determines the effectiveness of this defense [Manukhina et al., 2008]. The contribution of NO stores to protection against NO overproduction has been studied in rats [Wong *et al.*, 1995; Smirin *et al.*, 2000; Manukhina & Malyshev, 2005]. Heat shock was applied to induce NO overproduction, which resulted in acute hypotension and 50% mortality [Manukhina *et al.*, 1996]. When animals were pre-adapted to mild restraint stress or pretreated with DNIC to produce an antecedent increase in NO stores, subsequent NO overproduction and systemic hypotension were prevented [Smirin *et al.*, 2000]. Prevention of NO overproduction was associated with formation of NO stores in the vascular walls during the course of adaptation. Indeed, small doses of NOS inhibitors in the course of adaptation prevented both vascular NO storage and the associated vasoprotection. Conversely, repeated injections of DNIC augmented NO stores and recapitulated the salutary effects of adaptation. These protective effects cannot be ascribed to a direct action of DNIC, since DNIC injected into rats is completely cleared within 3-5 h [Manukhina *et al.*, 1998; Vanin, Manukhina, 2007], whilst its enhancement of survival from heat shock and of endogenous NO synthesis persisted for at least 24 h [Manukhina *et al.*, 1997]. NO stores may participate in this protection by feedback suppression of the activity and/or expression of NOS [Assreuy *et al.*, 1993; Colasanti, Suzuki, 2000] or by buffering excess NO. These compensatory mechanisms may prevent the toxic effects of increased NO generated in the process of adaptation.

A temporary increase in NO produced either by administering an NO donor or by activating NOS in rats produces protection that persists even after the NO concentration returns to its baseline. For instance, a delayed protective effect against ischemia and reperfusion injury was observed in hearts isolated from rats 24 h after injection of NO donor DNIC [Malyshev *et al.*, 1998]. NO synthesized by iNOS following its induction also has a long-term cardioprotective effect [Gyorgy *et al.*, 1999]. The delayed NO-dependent cardioprotection can be related either to NO-induced cardiac synthesis of secondary protective factors such as antioxidants [Fukai *et al.*, 2000; Dhakshinamoorthy & Porter, 2004] and HSPs [Malyshev *et al.*, 1995, 1998] or to formation of releasable NO stores in blood vessels that provide an additional source of endogenous NO when needed [Muller *et al.*, 2000].

Some studies have demonstrated that delayed cardioprotection resulted from upregulation of inducible NOS (iNOS). Ding *et al.* [Ding *et al.*, 2005] reported improved post-ischemic function of hearts from rats adapted to intermittent hypoxia, compared to non-adapted control hearts. The pre-ischemic contents of iNOS mRNA and protein, as well as NOx concentration in myocardium of adapted rats were higher than those of non-adapted rats. Compared to the respective baselines, these variables increased in non-adapted hearts and fell in adapted hearts after reperfusion. In these experiments, the iNOS-selective inhibitor, aminoguanidine, suppressed cardioprotective effects of intermittent hypoxia. Thus, prior induction of iNOS and increased NO production stimulated by intermittent hypoxia conditioning may have prevented a burst of iNOS activity and cytotoxic NO overproduction during the initial phase of reperfusion due to the negative feedback mechanism, induction of secondary protective factors, such as, heat shock proteins or antioxidants, and/or formation of NO stores buffering high level of NO.

A 20-day intermittent hypoxia conditioning program afforded remarkable cardioprotection from ischemia-reperfusion injury in dogs [Zong *et al.*, 2004]. This protection was associated with restriction of NO overproduction during reperfusion without compromising reactive hyperemia [Ryou *et al.*, 2008; Mallet *et al.*, 2008, 2009]. Adaptation to hypoxia induced formation of NO stores in coronary vessels of the same dogs and, moreover, significantly increased the ability of blood vessels to store NO, i.e., expanded the NO store capacity (Figure 10A). When NAC was added to the diet consumed each morning 2 hours before a hypoxia session, the cardioprotection was completely abolished [Mallet *et al.*, 2008]. Measuring the size and potential capacity of NO stores showed that chronic depletion of NO stores by NAC halved NO stores and significantly impaired the efficiency of NO storage (Figure 10B). These data provide indirect evidence that formation of NO stores may be important for the cardioprotective effects of adaptation to hypoxia.

Further evidence for the protective role of NO stores against NO overproduction was obtained in a rat model of Alzheimer's disease induced by injection of a toxic-amyloid fragment (A) into basal magnocellular nuclei [Mashina *et al.*, 2006]. Development of experimental Alzheimer's disease was associated with impaired memory, loss of neurons in parieto-temporal cortex [Manukhina *et al.*, 2008, 2009a,b], NO overproduction and formation of toxic peroxynitrite in brain tissue. The latter findings were confirmed by measurements of NO_2^- and NO_3^- [Mashina *et al.*, 2006] (Figure 11A) and 3-nitrotyrosine (Figure 11B), a marker of peroxynitrite formation, respectively. Adaptation to intermittent hypoxia afforded considerable neuroprotection against experimental Alzheimer's disease, manifest as improvement of cognitive function, prevention of neuronal death in cerebral cortex and restriction of NO overproduction and peroxynitrite formation in brain [Manukhina *et al.*, 2009a,b] (Figure 11). Since the size of NO stores correlated significantly with NO production [Manukhina *et al.*, 1999], increased NO stores may be regarded in these experiments as an additional, indirect marker of NO overproduction. NO overproduction was confirmed by detection of elevated NOx in brain tissue. NO stores were absent in cerebral blood vessels of control rats. In rats pretreated with DNIC, cerebral blood flow increased, indicative of NO release from NO stores. In both hypoxia-adapted and A -treated rats, the vasodilatory response to NAC administration also revealed significant NO stores. In rats injected with A after adaptation, NO stores were increased vs. non-adapted rats treated with A (Figure 12). The authors proposed that adaptation increased the NO storage capacity of blood vessels [Mashina *et al.*, 2006].

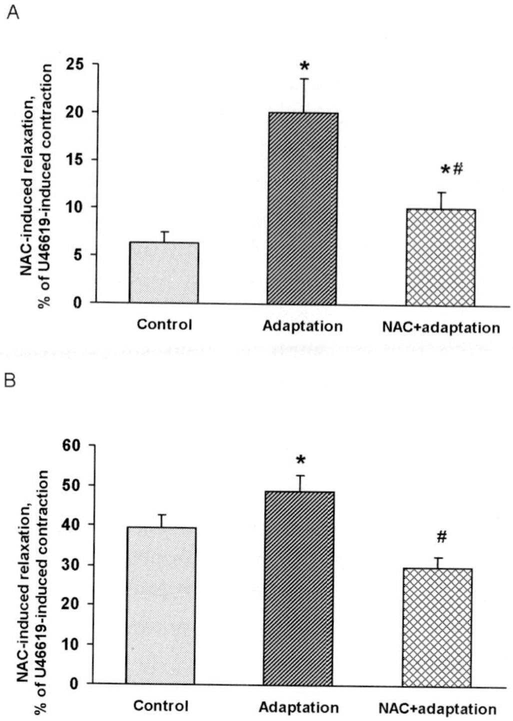

Fig. 10. Effect of N-acetylcysteine on formation of NO stores (A) and NO store capacity (B) in canine coronary vessels during adaptation to hypoxia. Mean values ± SEM for NAC-induced vasorelaxation in > 7 experiments, expressed as % of contraction induced by U46619. A: Detection of available NO-stores. B: Detection of NO stores after incubation with the NO donor DNIC and washout.

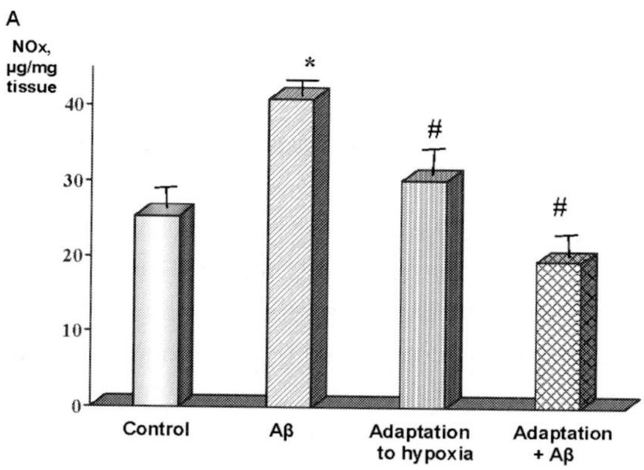

Fig. Contd..

Fig. Contd..

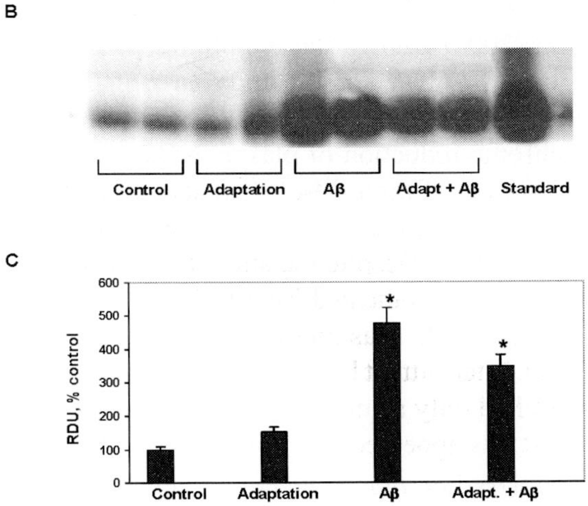

Fig. 11. Hypoxia adaptation prevents cerebral NO overproduction in β-amyloid 25-35 (Aβ)-induced Alzheimer's disease. A: NO production in rat cerebral cortex, from 8-10 experiments per group. B:Representative immunoblots of cerebrocortical homogenates probed with anti-nitrotyrosine antibody. C: Relative densitometric units (RDU) (means ± SEM from 5 experiments) of nitrotyrosine immunoblotting. Mean control value = 100%. *P<0.05 vs. control; #P<0.05 vs. Aβ.

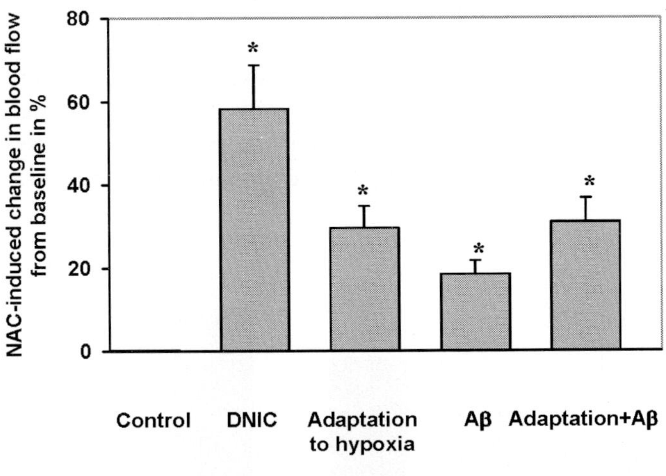

Fig. 12. Impact of hypoxia adaptation and experimental β-amyloid 25-35 (Aβ)-induced Alzheimer's disease on cerebrovascular NO stores in rats. Increase in cerebral blood flow in response to N-acetylcysteine (% of basal flow) from groups of 7-10 experiments. *P<0.05 vs. control [Mashina *et al.*, 2006].

No Stores in Hypertensive States and Metabolic Syndrome

Vlasova *et al.*, [Vlasova *et al.*, 2006] studied the possibility of targeted modulation of NO storage capacity by long-term changes in NO levels in the body. In Wistar rats, chronic elevation of plasma NOx caused by either adaptation to intermittent hypoxia or by DNIC injections increased the efficiency of NO storage. Conversely, chronic reduction of plasma NOx caused by administration of the NOS inhibitor L-NNA reduced NO storage (Figure 13). When adaptation to hypoxia was used to stimulate NO synthesis and formation of NO stores, urinary NOx of both SHRSP and normotensive WKY rats were similarly increased (Figure 14A). Despite the similar increase in plasma NOx, NO stores were twice as large in WKY as in SHRSP, indicated by 19.6% vs. 9.4% relaxation response to DETC, respectively [Manukhina *et al.*, 2000; Mashina *et al.*, 2001; Manukhina *et al.*, 2006; Vanin & Manukhina, 2007] (Figure 14B). Intermittent hypoxia significantly increased the capacity of NO stores in normotensive WKY rats but had only a small effect in SHRSP blood vessels (Figure 14C). At the same time, NO stores in August rats appeared less sensitive to chronic inhibition of NO synthesis.

Nitric oxide overproduction rather than NO deficiency is often observed in spontaneously hypertensive rats (SHR) [Wu, Yen, 1997] in contrast to the deficiency observed in SHRSP rats [Manukhina *et al.*, 2000; Mashina *et al.*, 2001]. However, NO is not sufficiently bioactive to stimulate the formation of cyclic GMP and to maintain an adequate NO-dependent vasodilatory tone in SHR. The NO production is apparently increased in blood vessels by a mechanism involving induction of endothelial NOS (eNOS) [Nava *et al.*, 1998] and/or iNOS [Racasan *et al.*, 2005]. Excessive and sustained generation of NO may contribute to oxidant-mediated endothelial dysfunction. In early hypertension, iNOS induction may be sufficient to restrict BP elevation. Later, however, excessive NO inhibits eNOS and directly damages vascular cells by suppressing mitochondrial respiration and DNA synthesis. These effects eventually lead to endothelial dysfunction and increased blood pressure, which, in turn, damage endothelium-dependent relaxation even further, creating a vicious cycle [Wu & Yen, 1997; Puppu *et al.*, 2000]. The lack of compensatory increase in NO storage capacity in hypertensive rats may exacerbate endothelial injury and dysfunction due to NO overproduction.

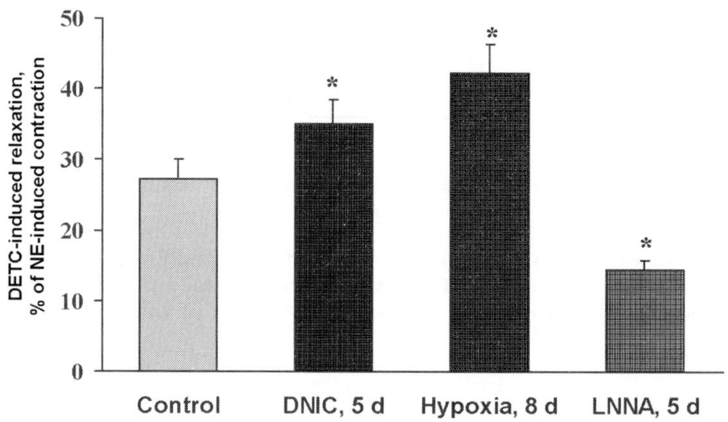

Fig. 13. Changes in capacity of NO stores in aorta of Wistar rats during chronic changes in NO formation. Capacity of NO stores assessed as DETC-induced relaxation, expressed as % NE-induced contraction. Mean values ± SEM from > 8 experiments. *P<0.05 vs. control [Vlasova *et al.*, 2006].

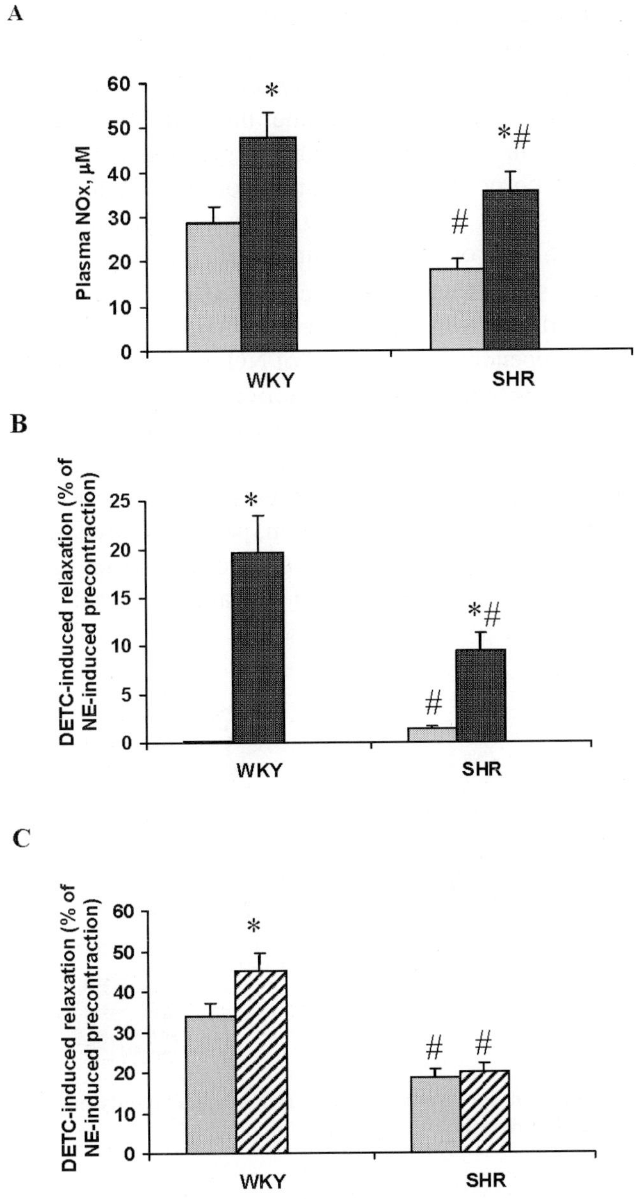

Fig. 14. Up-regulation of NO synthesis (A), enhancement of NO storage efficiency (B), and NO store formation in aortic wall (C) in normotensive WKY rats and in SHRSP rats after 8-day intermittent hypoxia. Solid bars, control; dashed bars, adaptation to hypoxia. A: plasma concentration of nitrite + nitrate, M. B, C: size of available NO stores in aorta (B) and potential capacity of NO stores (C), assessed from relaxation of aortic segments in response to DETC (% of NE-induced contraction). Means ± SEM from > 8 experiments. *P<0.05 vs. control; #P<0.05 vs. WKY [Vlasova *et al.*, 2006]

A similar situation was observed in metabolic syndrome. Metabolic syndrome is associated with NO overproduction, which is detrimental for insulin-dependent metabolic processes [Zahedi *et al.*, 2008] and endothelial function [Kagota *et al.*, 2007]. The capacity of NO stores was considerably reduced in coronary arteries from dogs with metabolic syndrome; therefore, their ability to bind excessive NO was diminished. Furthermore, in dogs with metabolic syndrome, incubation with an NO donor produced smaller NO stores in coronary vessels than in control. Collectively, these results indicated failure of the adaptive mechanism for expanding NO stores during chronic exposure to high NO levels in a canine model of metabolic syndrome [Manukhina *et al.*, 2009c].

The major finding of these studies was that efficiency of NO storage increases during long-term elevation of NO production and, vice versa, decreases if NO production is reduced. The compensatory changes in NO store capacity depending on the level of NO in the body represent a new mechanism of cardiovascular adaptation to long-term excess or deficiency of NO. Enhanced efficiency of NO storage is an adaptive mechanism which may protect blood vessels against both the cytotoxic and the excessive vasodilatory effects of NO overproduction. On the other hand, reduced capacity of NO stores restricts binding of available NO to storage complexes in vascular walls, and more NO remains free. Thus, under conditions of reduced NO production, paradoxically larger amounts of free NO are available to maintain NO-dependent physiological processes,such as endothelium-dependent vasodilation, prevention of thrombosis, antibacterial and antitumor defense, and neurotransmission (Figure 15).

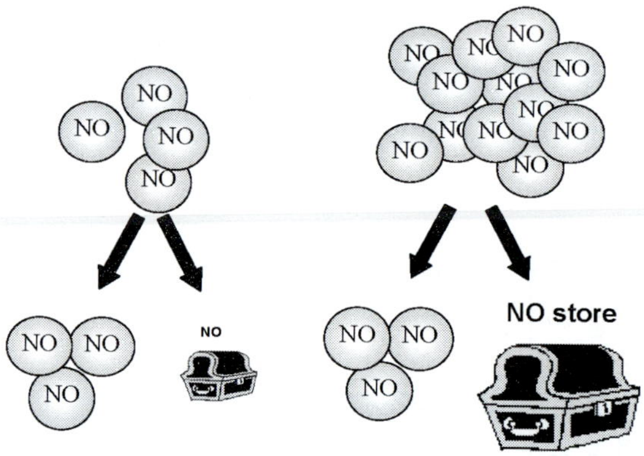

Fig. 15. Possible adaptive mechanism for modulation of NO store capacity in NO deficiency (left) and NO excess (right). See text for details.

Acknowledgements

This work was supported by the Russian Foundation for Basic Research (grants 07-04-00650 and 10-04-00980).

References

1. Alencar JL, Lobysheva I, Geffard M, Sarr M, Schott C, Schini-Kerth VB, Nepveu O, Stoclet JC, Muller B. Role of S-nitrosation of cysteine residues in long-lasting inhibitory effect of nitric oxide on arterial tone. Mol Pharmacol 2003;63:1148-58.

2. Alencar JL, Stoclet JC, Muller B. Formation of releasable NO stores on tissue thiols in blood vessels: differential effect of NO donors. Fund Clin Pharmacol 2001; 15:P085.
3. Arnell DA, Da BJ, Stamler JS. Diethyl dithiocarbamate-induced decomposition of S-nitrosothiols. Nitric Oxide 1997;1:56-64.
4. Assreuy J, Cunha FQ, Liew FY, Moncada S. Feedback inhibition of nitric oxide synthase activity by nitric oxide. Br J Pharmacol 1993;108:833-7.
5. Butler A.R., Glidewell C., Hyde A.R., Walton J.C. Formation of mononuclear iron nitrosyl complexes from diamagnetic di- and tetranuclear iron-sulphur nitrosyls: characterization by EPR spectroscopy and study of thiolate and nitrosyl ligand exchange reactions. Polyhedron 1985;4:797-809.
6. Colasanti M., Suzuki H. The dual personality of NO. Trends Pharmacol Sci 2000;21: 249-52.
7. Dhakshinamoorthy S, Porter AG. Nitric oxide-induced transcriptional up-regulation of protective genes by Nrf2 via the antioxidant response element counteracts apoptosis of neuroblastoma cells. J Biol Chem 2004;279:20096-107.
8. Ding HL, Zhu HF, Dong JW, Zhu WZ, Yang WW, Yang HT, Zhou ZH. Inducible nitric oxide synthase contributes to intermittent hypoxia against ischemia/reperfusion injury. Acta Pharmacol Sin 2005;26:315-22.
9. Elliott SN, Wallace JL. Nitric oxide: a regulator of mucosal defense and injury. J Gastroenterol 1998;33:792-803.
10. Fichtlscherer B, Mülsch A. MR imaging of nitrosyl-iron complexes: experimental study in rats. Radiology 2000;216:225-31.
11. Flitney FW, Megson IL, Flitney DE, Butler AR. Iron-sulphur cluster nitrosyls, a novel class of nitric oxide generator: mechanism of vasodilator action on rat isolated tail artery. Br J Pharmacol 1992;107:842-8.
12. Fukai T, Siegfried MR, Ushio-Fukai M, Cheng Y, Kojda G, Harrison DG. Regulation of the vascular extracellular superoxide dismutase by nitric oxide and exercise training. J Clin Invest 2000;105:1631-9.1
13. Furchgott RE, Ehrreich SJ, Greenblatt E. The photoactivated relaxation of smooth muscle of rabbit aorta. J Gen Physiol. 1961;44:499-519.
14. Gyorgy K, Muller B, Vegh A, Kleschyov AL, Stoclet J-C. Triggering role of nitric oxide in the delayed protective effect of monophosphoryl lipid A in rat heart. Br J Pharmacol. 1999;127:1892-8.
15. Kagota S, Tada Y, Kubota Y, Nejime N, Yamaguchi Y, Nakamura K, Kunimoto M, Shinozuka K. Peroxynitrite is involved in the dysfunction of vasorelaxation in SHR/NDmcr-cp rats, spontaneously hypertensive obese rats. J Cardiovasc Pharmacol. 2007;50:677-85.
16. Kim YM, de Vera ME, Watkins SC, Billiar TR. Nitric oxide protects cultured rat hepatocytes from tumor necrosis factor-alpha-induced apoptosis by inducing heat shock protein 70 expression. J Biol Chem. 1997;272:1402-11.
17. Kim YM, Chung HT, Simmons RL, Billiar TR. Cellular non-heme iron content is a determinant of nitric oxide-mediated apoptosis, necrosis, and caspase inhibition. J Biol Chem. 2000;275:10954-61.

18. Kleschyov AL, Muller B, Keravis T, Stoeckel M-E, Stoclet J-C. Adventitia-derived nitric oxide in rat aortas exposed to endotoxin: cell origin and functional consequences. Am J Physiol Heart Circ Physiol 2000;279:H2743-51.

19. Kolá? F, Jezková J, Balková P, B?eh J, Necká? J, Novák F, Nováková O, Tomášová H, Srbová M, Oštádal B, Wilhelm J, Herget J. Role of oxidative stress in PKC- upregulation and cardioprotection induced by chronic intermittent hypoxia. Am J Physiol Heart Circ Physiol 2007;292:H224-30.

20. Kubaszewski E, Peters A, McClain S, Bohr D, Malinski T. Light-activated release of nitric oxide from vascular smooth muscle of normotensive and hypertensive rats. Biochem Biophys Res Commun 1994;200:213-8.

21. Kubrina LN, Mikoyan VD, Mordvintcev PI, Vanin AF. Iron potentiates bacterial lipopolysaccharide-induced nitric oxide formation in animal organs. Biochim Biophys Acta 1993;1176:240-4.

22. Lancaster JR. Stimulation of the diffusion and reaction of endogenously produced nitric oxide. Proc Natl Acad Sci USA 1994;91:8137-41.

23. Lipinski P, Drapier JC. Interplay between ferritin metabolism, reactive oxygen species and nitric oxide. J Biol Inorg Chem 1997;2:559-66.

24. Mallet RT, Ryou M-G, Manukhina EB, Downey HF. Beta-adrenergic signaling and ROS: pivotal roles in intermittent, normobaric hypoxia-induced cardioprotection. In: Xi L., Serebrovskaya TV, eds. Intermittent Hypoxia: from Molecular Mechanisms to Clinical Applications. Nova Science Publishers, 2009, pp. 189-212.

25. Mallet RT, Ryou M-G, Manukhina EB, Downey HF. Intermittent hypoxic conditioning of canine myocardium: robust protection against ischemia-reperfusion injury. In: Lukyanova L, Takeda N, Singal PK, eds. Adaptation Biology and Medicine: Health Potentials. Narosa Publishers, New Delhi, 2008, Vol. 5, pp. 59-78.

26. Malyshev IY, Zenina TA, Golubeva LY, Saltykova VA, Manukhina EB, Mikoyan VD, Kubrina LN, Vanin AF. NO-dependent mechanisms of adaptation to hypoxia. Nitric Oxide 1999;3:105-13.

27. Malyshev IYu, Malenyuk EB, Manukhina EB, Mikoyan VD, Vanin AF. Long-term cardioprotective effect of nitric oxide: role of HSP70. Bull Exp Biol Med 1998;125:23-6 [Rus].

28. Malyshev IYu, Manukhina EB, Mikoyan VD, Kubrina LN, Vanin AF. Nitric oxide is involved in heat-induced HSP70 accumulation. FEBS Lett 1995;370:159-62.

29. Manukhina EB, Caffrey JL, Barlow M, Vanin AF, Mikoyan VD, Downey HF. Nitric oxide stores in coronary blood vessels of dogs with metabolic syndrome. FASEB J 2009;23:628.8.

30. Manukhina EB, Goryacheva A, Mashina S, Alexandrin V, Downey HF. Adaptation to hypoxia increases NO stores: a defense for cerebral blood vessels against NO overproduction in experimental Alzheimer's disease. FASEB J 2009;23:628.9.

31. Manukhina EB, Goryacheva AV, Barskov IV, Guseva AA, Pshennikova MG, Khomenko IP, Mashina S Yu, Pokidyshev DA, Malyshev IYu. Prevention of neurodegenerative brain injury in rats with experimental Alzheimer's disease by adaptation to hypoxia. Ross. Physiol. Zhurnal im. I.M. Secheniva 2009a; 95: 706-15 [Rus].

32. Manukhina EB, Pshennikova MG, Khomenko IP, Mashina SYu, Pokidyshev DA, Malyshev IYu. Role of nitric oxide in prevention of cognitive disorders in neurodegenerative brain injuries in rats. Bull Exp Biol Med 2008; 146: 391-395.

33. Manukhina EB, Vanin AF, Malyshev IYu, Mallet RT, Downey HF. Intermittent hypoxia-induced cardio- and vasoprotection: role of NO stores. In: Lukyanova L, Takeda N, Singal PK, eds. Adaptation Biology and Medicine: Health Potentials. Narosa Publishers, New Delhi, 2008, Vol. 5, 113-46.

34. Manukhina EB, Kalenchuk VU, Gavrilova SA, Goryacheva AV, Malyshev IYu, Koshelev VB. Organ specificity of nitric oxide storage in blood vessel walls during adaptation to hypoxia. Ross Fiziol Zhurnal im IM Sechenova 2008;94:198-205 [Rus].

35. Manukhina EB, Malyshev IY, Mikoyan VD, Kubrina LN, Vanin AF. Increase of nitric oxide production in rat organs in heat shock. Bull Exp Biol Med 1996; 21:520-3 [Rus].

36. Manukhina EB, Malyshev IYu, Smirin BV, Mashina SYu, Saltykova VA, Vanin AF. Production and storage of nitric oxide in adaptation to hypoxia. Nitric Oxide 1999;3:393-401.

37. Manukhina EB, Malyshev IYu. Role of free NO and NO stores in protective effects of adaptation to hypoxia. In: Singal PK, Takeda N, eds. Adaptation Biology and Medicine: Current Concepts. Narosa Publishers, New Delhi, 2005, vol. 4, pp. 82-94.

38. Manukhina EB, Malyshev IYu. Role of nitric oxide in protective effects of adaptation. In: Moravec J, Takeda N, Singal PK, eds. Adaptation Biology and Medicine. Narosa Publishers, New Delhi, 2002, vol. 3, pp. 312-27.

39. Manukhina EB, Pokidyshev DA, Malenyuk EB, Malyshev IY, Vanin AF. The protective effect of nitric oxide in heat shock. Izv Akad Nauk Ser Biol 1997;1:54-8 [Rus].

40. Manukhina EB, Smirin BV, Pshennikova MG, Vanin AF, Markov HM, Malyshev IYu. Production and storage of nitric oxide in adaptation to hypoxia: genetic peculiarities. Hypoxia Med J 2002;4:35-40.

41. Manukhina EB, Vanin AF, Markov KM, Malyshev IY. Formation and role of nitric oxide stores in adaptation to hypoxia. Adv Exp Med Biol 2006a;578:35-40.

42. Mashina SY, Aleksandrin VV, Goryacheva AV, Vlasova MA, Vanin AF, Malyshev IY, Manukhina EB. Adaptation to hypoxia prevents disturbances in cerebral blood flow during neurodegenerative process. Bull Exp Biol Med 2006;142:169-72.

43. Mashina SY, Vanin AF, Serezhenkov VA, Kubrina LN, Malenkova IV, Malyshev IY, Manukhina EB. Detection and evaluation of NO stores in awake rats. Bull Exp Biol Med 2003;136:26-9 [Rus].

44. Megson IL, Flitney FW, Bates J, Webster R. Repriming of vascular smooth muscle photorelaxation is dependent upon endothelium-derived nitric oxide. Endothelium 1995;3:39-46.

45. Mikoyan VD, Kubrina LN, Manukhina EB, Malysheva EV, Malyshev IYu, Vanin AF. Differences in the stimulation of NO synthesis during heat shock in genetically different rats. Bull Eksp Biol Med 1996;121:634-7 [Rus].

46. Muller B, Kleschyov AL, Alencar JL, Vanin A, Stoclet JC. Nitric oxide transport and storage in the cardiovascular system. Ann N Y Acad Sci 2002;962:131-9

47. Muller B, Kleschyov AL, Gyorgy K, Stoclet JC. Inducible NO synthase activity in blood vessels and heart: new insight into cell origin and consequences. Physiol Res 2000;49:19-26.
48. Muller B, Kleschyov AL, Stoclet JC. Evidence for N-acetylcysteine-sensitive nitric oxide storage as dinitrosyl iron complexes in lipopolysaccharide-treated rat aorta. Br J Pharmacol 1996;119:1281-5.
49. Mülsch A, Mcordvintcev P, Bassenge E, Jung F, Clement B, Busse R. In vivo spin trapping of glyceryl trinitrate-derived nitric oxide in rabbit blood vessels and organs. Circulation 1995;92:1876-82.
50. Mülsch A. Nitrogen monoxide transport mechanisms. Arzneimittelforschung 1994;44:408-11.
51. Nava E, Farrc AL, Moreno C, Casado S, Moreau P, Cosentino F, Luscher TF. Alterations to the nitric oxide pathway in the spontaneously hypertensive rat. J Hypertens 1998;16:609-15.
52. Paul T. Effect of a prolonged superoxide flux on transferrin and ferritin. Arch Biochem Biophys 2000;382:253-61.
53. Pshennikova MG, Smirin BV, Bondarenko ON, Malyshev IYu, Manukhina EB. Nitric oxide storage in rats of various genetic strains and its role in the antistressor effect of adaptation to hypoxia. Ross Fiziol Zh Im I M Sechenova 2000;86:174-81 [Rus].
54. Puddu P, Puddu GM, Zaca F, Muscari A. Endothelial dysfunction in hypertension. Acta Cardiol 2000;55:221 32.
55. Racasan S, Branko Braam B, Koomans HA, Joles JA. Programming blood pressure in adult SHR by shifting perinatal balance of NO and reactive oxygen species toward NO: the inverted Barker phenomenon. Am J Physiol Renal Physiol 2005;288:F626-36.
56. Ryou M-G, Sun J, Oguayo KN, Manukhina EB, Downey HF, Mallet RT. Hypoxic ?onditioning suppresses nitric oxide production upon myocardial reperfusion. Exp Biol Med 2008;233:766-74.
57. Sato I, Murota S. Paracrine function of endothelium-derived nitric oxide. Life Sci 1995;13:1079-87.
58. Serebrovskaya TV, Manukhina EB, Smith ML, Downey HF, Mallet RT. Intermittent hypoxia: cause of or therapy for systemic hypertension? Exp Biol Med 2008;233:627-50.
59. Simpson RJ. Effect of hypoxia on iron absorption in heterozygous hypotransferrinaemic mice. Ann Hematol 1992;65:260-4.
60. Smirin BV, Pokidyshev DA, Malyshev IYu, Vanin AF, Manukhina EB. Nitric oxide storage as a factor of adaptive defence. Ross Fiziol Zh Im IM Sechenova 2000;86:447-54 [Rus].
61. Smirin BV, Vanin AF, Malyshev IYu, Pokidyshev DA, Manukhina EB. Nitric oxide stores in blood vessels in vivo. Bull Eksp Biol Med 1999;127:629-32 [Rus].
62. Stamler JS, Simon DJ, Osborn JA, Mullins MS, Jarani O, Singel DJ, Loscalzo J. S-nitrosylation of proteins with nitric oxide; synthesis and characterization of biologically active components. Proc Natl Acad Sci USA 1992;89:444-8.
63. Stoclet JC, Muller B, György K, Andriantsiothaina R, Kleschyov AL. The inducible nitric oxide synthase in vascular and cardiac tissue. Eur J Pharmacol 1999;375:139-55.

64. Turella P, Pedersen JZ, Caccuri AM, de Maria F, Mastroberardino P, lo Bello M, Federici G, Ricci G. Glutathione transferase superfamily behaves like storage proteins for dinitrosyl-diglutathionyl-iron complex in heterogeneous systems. J Biol Chem 2003;278:42294-9.

65. Vanin AF, Malenkova IV, Serezhencov VA. Iron catalyzes both decomposition and synthesis of S-nitrosothiols: optical and electron paramagnetic resonance studies. Nitric Oxide 1997;1:191-203.

66. Vanin AF, Manukhina EB. Hypotensive, vasodilatory and anti-aggregative properties of dintrosyl-iron complexes. In: van Faassen E, Vanin AF, eds. Radicals forLife: The Various Forms of Nitric Oxide. Elsevier, Amsterdam, 2007, 75-96.

67. Vanin AF, Sanina NA, Serezhenkov VA, Burbaev DSh, Losinsky VI, Aldoshin SM. Dinitrosyl-iron complexes with thiol-containing ligands: spatial and electronic structures. Nitric Oxide 2007b;16:82-93.

68. Vanin AF, van Faassen E. Chemical equilibria between S-nitrosothiols and dinitrosyl iron complexes with thiol-containing ligands. In: van Faassen E, Vanin AF, eds. Radicals for Life: The Various Forms of Nitric Oxide. Elsevier, Amsterdam, 2007a, 223-52.

69. Vedernikov YP, Mordvintcev PI, Malenkova IV, Vanin AF. Effect of diethyldithiocarbamate on the activity of nitric oxide-releasing vasodilators. Eur J Pharmacol 1992;212:125-8.

70. Venturini CM, Palmer RM, Moncada S. Vascular smooth muscle contains a depletable store of a vasodilator which is light-activated and restored by donors of nitric oxide. J Pharmacol Exp Ther 1993;266:1497-500.

71. Vlasova MA, Smirin BV, Pokidyshev DA, Mashina SY, Vanin AF, Malyshev IY, Manukhina EB. Mechanism of adaptation of the vascular system to chronic changes in nitric oxide level in the organism. Bull Exp Biol Med 2006;142:670-4.

72. Vlasova MA, Vanin AF, Muller B, Smirin BV, Malyshev IY, Manukhina EB. Detection and description of various pools of nitric oxide store in vascular wall. Bull Exp Biol Med 2003;136:226-30.

73. Wong HR, Finder JD, Wasserloos K, Pitt BR. Expression of inducible nitric oxide synthase in cultured rat pulmonary artery smooth muscle cells is inhibited by the heat shock response. Am J Physiol Lung Cell Molec Physiol 1995;269:L843-8.

74. Wu CC, Yen MH. Nitric oxide synthase in spontaneously hypertensive rats. Biomed Sci 1997;4:249 55.

75. Yang Y, Loscalzo J. S-Nitrosated proteins: formation, metabolism and function. In: van Faassen E, Vanin AF, eds. Radicals for Life: The Various Forms of Nitric Oxide. Elsevier, Amsterdam, 2007, 201-21.

76. Zahedi Asl S, Ghasemi A, Azizi F. Serum nitric oxide metabolites in subjects with metabolic syndrome. Clin Biochem 2008;41:1342-7.

77. Zong P, Setty S, Sun W, Martinez R, Tune JD, Ehrenbourg IV, Tkatchouk EN, Mallet RT, Downey HF. Intermittent hypoxia training protects canine myocardium from infarction. Exp Biol Med 2004;229:806-12.

Adaptation Biology and Medicine (Volume 6: Cell Adaptations and Challenges)
Editors: P. Wang, C.-H. Kuo, N. Takeda and P.K. Singal
Copyright © 2011 Narosa Publishing House Pvt. Ltd., New Delhi, India

22. Biological Approach using Hyperbaric Oxygen Therapy to Improve Ligament Healing in a Rabbit Model

Yi-Sheng Chan M.D.*

Professor and Chief Division of Sports Medicine, Department of Orthopaedic Surgery and Hyperbaric Oxygen Therapy Center, Chang Gung Memorial Hospital and Chang Gung University College of Medicine, Taiwan

Abstract

Hyperbaric oxygen (HBO) enhances oxygen transport in hypoxic tissue and increases oxygen diffusion through tissue fluids. HBO treatment affected various cell activities in diverse systems; they include stimulated fibroblast proliferation, affected endothelial cell fibrinolysis, promote healing in bone, tendon, ligaments, muscle and skin, and increased collagen synthesis of tendon and ligaments. The non-invasive HBO therapy had been widely used in clinical applications. The author's study demonstrated the positive effect of hyperbaric oxygen on the healing of the medial collateral ligament (MCL) in vitro and in vivo using a rabbit model. HBO treatment decreases the Type III collagen/Type I collagen content ratio, possibly resulting in mechanically stronger collagen fibrils. The repaired ligaments had better biomechanical performance and earlier histological evidence of healing in the HBO-treated group. However, the effect was dependent on the dosage of HBO. HBO therapy may have considerable clinical potential for use in shortening the healing time of injured ligament

Keywords: Hyperbaric oxygen, Medial collateral ligament, Ligament healing

Introduction

Hyperbaric oxygen (HBO) therapy has been shown to enhance the healing of bone, muscle, skin, and wound, particularly in conditions of ischemia and low oxygen tension.[1-9] Preliminary data show that HBO may be beneficial in the treatment of muscle-tendon and ligament injury. [10-14] This has obvious implications for sports medicine personnel, given the frequency with which these injuries occur and the wide variety of treatment options available. In the United States and Europe, some athletes with partial or complete periarticular ligament disruptions have been treated with hyperbaric oxygen therapy, and anecdotal reports indicate a faster return to competition after using hyperbaric oxygen therapy.[10, 12]

*Address for Correspondence: Yi-Sheng Chan, MD Professor and Chief Department of Orthopaedic Surgery and Hyperbaric Oxygen Therapy Center Chang Gung Memorial Hospital and Chang Gung University College of Medicine 5, Fu-Hsing St. 333, Kweishan, Taoyuan, Taiwan Tel: 866-3-3281200 ext 3882 Fax: 886-3-3284564 Email: chan512@adm.cgmh.org.tw

However, there were no better controlled trials or clinical research to delineate and define the benefits of HBO therapy in ligament healing, and the effect of hyperbaric oxygen therapy on ligament healing has not been thoroughly investigated.

Ligament healing is a slow and time-consuming but highly organized process involving hemorrhage, inflammation, proliferation, and remodeling.[15] Ligaments and tendons are composed of bundles of collagen fibrils. Normal medial collateral ligament (MCL) is composed of six genetically distinct types of collagen.[16] Type I collagen is the major fibrillar collagen of the MCL and type III and V collagen are quantitatively minor components.[17,18,19] The amount of type III collagen present increases with age or after ligaments injury.[20] Previous studies have demonstrated that during early phases of ligament healing, type III collagen is highly elevated relative to type I collagen,[21] which is suspected to result in small collagen fibrils.[22] The healed ligaments with smaller collagen fibrils are mechanically weaker than those with the normal collagen fibrils.[23]

Fibroblast proliferation and collagen synthesis are closely related to oxygen availability and cannot proceed without the presence of oxygen.[24, 25] Oxygen is required for the hydroxylation of proline and lysine, a step that is essential for the synthesis of collagen by fibroblast.[26] Under hypoxic conditions, fibroblasts produce an intracellular peptide collagen precursor, but fail to release it.[27,28] Maturation and cross-linking of collagen increases linearly with an elevated ambient oxygen concentration.[28]. Collagen synthesized under hypoxic conditions is produced at a slow rate and has poor mechanical strength compared to that produced in a normoxic environment.[29]

The purpose of this study was to use biomechanical and histological analyses to assess the effect of hyperbaric oxygen on the healing of the medial collateral ligament in a rabbit model.[29,30,31]

Materials and Methods

Cell isolation and cell culture. New Zealand white rabbits, 3 kg of weight, were sacrificed by an intravenous injection of ketamine hydrochloride (Ketalar, ParkeDavis, Taiwan, ROC) and Rompum (Bayer, Leverkusen, Germany) and the knee joint was exposed by a medial parapatellar incision. MCL ligaments were harvested under aseptic conditions and collected into sterile biological safety hood. A 2-mm portion of the ligaments at the tibial and femoral insertion sites was discarded to avoid possible cartilage and bone tissue contamination. The remaining ligament was cut into pieces of approximately 1-mm size and placed in 60 mm culture dishes (Faclon) containing 4 ml of growth medium. Growth medium consisted of HSE 3:1 (DMEM: Ham's F12) supplemented with 10% (vol/vol) FBS and antibiotics (mixture of 100 units/ml of penicillin and 100 ìg/ml of streptomycin; Gibco, Grand Island, NY). Cultures were incubated in a humidified atmosphere of 5% CO_2 / 95% air. After cells reached confluence, explants were removed and cells detached by treating with 0.05% trypsin-EDTA (Gibco, Grand Island, NY). Cells were centrifuged (1000 rpm for 5 minutes), washed with PBS, and seeded in T-75 tissue culture flasks (Falcon) in 15 ml of medium HSE 3:1 supplemented with 10% FBS and antibiotics. Cultures were incubated in a humidified atmosphere of 5% CO_2 / 95% air until cell confluence.

Responses Of MCL Fibroblast Cultures to Intermittent HBO

Control cells were maintained in 5% CO_2 / 95% air (non-HBO) through the experimental protocol. All hyperoxic cells were exposed to 100% O_2 for 25 minutes, then to air for 5 minutes at 2.5 ATA (atmospheres absolute) in a hyperbaric chamber (Sigma II, Perry Baromedical Corporation, Riviera Beach, FL, USA) with a total treatment of 120 minutes per 48 hours. Cells were plated at 10^5 cells per 60-mm tissue culture dish (Falcon) in 4 ml of medium HSE 3:1 containing 2% FBS for 24 h to make the cells quiescent. At intervals of 48 hours, 96 hours, and 144 hours, cell number was determined from triplicate dishes by harvesting cells by trypsin treatment and then counting cells in a Coulter Multisizer II (Beckman Coulter, Inc., Taiwan Branch). Cell growth was measured by increase in cell number after hyperbaric or normobaric treatments.

Cell Cycle Detection by Flow Cytometry

Forty-eight hours after treatments, cells were trypsinized, pelleted (1200 revolutions per minute for 5 minute), and washed twice with cold phosphate buffered saline (PBS)-1% bovine serum albumin (BSA) (Sigma, St. Louis, MO), and fixed in ice-cold 70% ethanol. Cells were kept for at least 1 hour at –20 °C, washed once at 4 °C with PBS, and finally resuspended in 1 ml of 0.5% Triton X-100, 0.05% RNase (Sigma, St. Louis, MO) PBS solution and incubated for 1 hour at 37 °C in a temperature controlled bath. At the end of the incubation time, add 1 ml of 50ìg/ml Propidium Iodide solution (in PBS) and incubated in the dark for 30 minutes at 37 °C. Samples were analyzed in a FACS (Beckman Coulter, Inc., Taiwan Branch).

SDS-polyacrylamide Gel Electrophoresis and Immunoblotting for Type I Collagen and Type III Collagen

Forty-eight hours after each treatment, cell culture conditioned medium was collected, and the cells washed with ice-cold PBS three times. Cells were lysed by freeze-thrawing on dry ice and supernatant fraction collected by centrifugation at 3,000 g for 10 min. The collected conditioned medium was lyophilized and combined with the cell fraction. Aliquots of samples were first treated with 100 ìg/ml pepsin (Sigma) dissolved in 1 M acetic acid at 4 °C for 12 hours, dialyzed against 0.05 M ammonium bicarbonate for 48 hours, and lyophilized. Protein content was estimated using a commercially available protein assay kit (Sigma). Samples were separated by SDS-polyacrylamide gel electrophoresis (PAGE) on 10% gel according to Laemmli under reducing conditions. Gels were stained with Coomassie blue-R250, destained, and photographs taken. Prestained proteins (Bio-Rad) were employed as molecular weight markers.

For immunoblotting, proteins were separated by SDS-PAGE on 10% gel and transferred onto nitrocellulose membranes electrophoretically, using a protein transfer unit (Hoeffer, CA, U.S.A.). After blocking with 5% non-fat milk, membranes were incubated for 1.5 h at room temperature with 300-fold diluted mouse monoclonal antibodies to Type I collagen (Oncogene, MA, U.S.A.) or Type III collagen (Oncogene, MA, U.S.A.). After washing, membranes were further incubated for 2 hours with 1,000-fold diluted goat anti-mouse IgG conjugated to horseradish peroxidase. Membranes were washed and rinsed with ECL detection reagents (Amersham Pharmacia Biotech,UK). Photographs of the band images were taken using ECL Hyperfilm (Amersham Pharmacia Biotech,UK). Each image was captured by a digital camera (DCS 460; Eastman KODAK, Rochester, NY) and the density of the band was quantified by KODAK 1D image analysis software (Eastman KODAK company, Rochester, NY).

Animal Experimental Design

The Institutional Animal Care and Use committee of Taiwan approved the methods described herein. All of the studied animals were cared for in accordance with the regulations of the National Institutes of Health of the Republic of China (Taiwan), under the supervision of a licensed veterinarian.

Sixty-four skeletally mature (12 to 17 months of age) male New Zealand white rabbits, each weighing 3.5 ± 0.6 kg (mean ± SD), were randomly divided into two groups. The experimental group (n = 32) went through intermittent 2.5 atmosphere absolute (ATA) hyperbaric oxygen therapy, the control group (n = 32) did not go through hyperbaric oxygen therapy but were kept in the HBO chamber. Under sterile conditions and general anesthesia with ketamine hydrochloride (Ketalar, Parke-Davis, Taiwan, ROC) and Rompun (Bayer, Leverkusen, Germany) intravenous injection, each animal underwent a standardized complete severance of the right medial collateral ligament (MCL). The MCLs of both hindlimbs were exposed through longitudinal medial incisions in the skin and fascia. The MCL of the right leg was completely severed by passing a 3-0 braided steel suture beneath it at the level of the suprameniscal recess. The MCL in the left leg of each animal was simply exposed. Skin was approximated with 3-0 interrupted nylon sutures. All animals were intravenously injected with Keflin prophylactic antibiotics. After the surgery, the animals were permitted unrestricted activity in cages and monitored daily for food and water intake, wound and ambulation. Each group of animals was housed in a hyperbaric research chamber (Model 17-48-100, Reimers System Inc, Springfield, VA) for 2 hours daily. When they were in the chamber, the experimental group animals were exposed to 2.5 ATA of pure oxygen for 2 hours for 5 days, beginning on the day of surgery. The control group animals were exposed to 1 ATA of normal air using the same regimen.

The control and experimental animals were then separated into 4 subgroups of 6 rabbits each for study. These subgroups were killed at 2, 4, 8, and 12 weeks, respectively, for study by biomechanical methods.

Biomechanical Test

After the animals were killed, both hind limbs of each animal were then disarticulated at the hip, wrapped in saline-soaked gauze, and stored at –20°C. The night before biomechanical testing, the specimens were thawed at 4°C. For the first stage of biomechanical testing, the skin, subcutaneous tissue, and muscles were removed, taking care not to damage the ligaments or joint capsule. The specimens were kept moist with 0.9% saline solution during testing. All soft tissue was stripped from the femur and tibia, leaving only the periarticular soft tissue about the knee. A unique 3mm diameter surgical grade drill was used to drill the holes perpendicular to the sagittal plane on both distal femur and proximal tibia sides, which allowed insertion of two 3 mm diameter rods. In order to obtain the optimum relative position of specimens before testing, each rod was placed on a custom-designed symmetrical grip with two hoops that allowed the specimen adjustable on both perpendicular (linear movement and rotation) and horizontal (linear movement) axises. The remaining soft tissue about the knee, including the intraarticular ligaments, was divided sharply, leaving only the medial collateral ligament intact. To provide consistent comparisons between the specimens, all specimens were maintained on condition of 0° flexion of the knee joint and the medial collateral ligament parallel to the axis of tension by way of adjusting the grip to the optimum position prior to tensile test. After the specimens were clamped, an axial tensile force was applied at a crosshead rate of 7.2 mm/min using a materials testing machine (Bionix 858, MTS Company, Minneapolis, MN, USA). The relationship between force and displacement was recorded simultaneously at an increment of 0.05mm using MTS Teststar II software. To assess the effect of hyperbaric oxygen therapy on the healing rate of ligament at various periods, the magnitude of force at failure for each individual specimen was selected for comparison. Using the other intact limb in the same rabbit as a control, the percentage of failure force of the ruptured medial collateral ligament of the each rabbit was calculated.

Histological Analysis

Two sacrificed animals from each group were sent for histological analysis. The entire strips of MCLs of both hindlimbs were dissected from these representative specimens. The strips included areas of old ligament at both ends, connected by scar in the middle or from comparable areas of controls. After fixing in 10% buffered formalin and embedding in paraffin, the specimens were stained by standard Masson's trichrome stain,[24] hematoxylin and eosin procedure for subsequent examination. By the method of trichrome stain, collagen fibers were identified by blue color stain and muscle fibers were identified by red color stain.

Statistical Analysis

Differences between two groups were compared by two-tailed Student's t test. Values were expressed as the mean ± SD; $p < 0.05$ was considered statistically significant.

Results

Dose Dependent Effect of HBO on MCL Cell Growth

Treatment dose of HBO dependently increased MCL cell number in comparison with control cells cultured under the same medium (Fig. 1, Control v. HBO treatment one time: $2.3 \times 10^5 \pm 1.4 \times 10^4$ v.s. $2.8 \times 10^5 \pm 1.5 \times 10^4$, $p<0.05$; Control v. HBO treatment two times: $3.6 \times 10^5 \pm 3.2 \times 10^4$ v.s. $5.2 \times 10^5 \pm 2.4 \times 10^4$, $p<0.05$; Control v. HBO treatment three times: $5.2 \times 10^5 \pm 3.1 \times 10^4$ v.s. $6.6 \times 10^5 \pm 2.9 \times 10^4$, $p<0.05$; n = 6)

Cell cycle detection. Using PI staining and FACS analysis, the results for control group, and HBO group treatment on cell cycle alteration are shown in Table 1. Treatment with HBO significantly increased the percentage of cells in the S-phase and decreased percentage of cells in the G_0/G_1 phase but did not produce a significant difference in the proportion of cells in G_2/M phase. .

Table 1. Effects of treatments on cell cycle phases (n = 6)

	Control	HBO
G_0/G_1	74.1% ± 2.6%	68.5% ± 3.4%
S	4.4% ± 0.6%	8.0% ± 1.2%
G_2/M	12.6% ± 2.6%	14.3% ± 2.5%

Effects of HBO on Type I Collagen and Type III Collagen Synthesis

After quantification by immunoblotting and image analyzer, no significant changes were apparent in Type I collagen content after each treatment (Fig. 3, HBO/Control ratio: 111.2% ± 8.2%, n = 2). Type

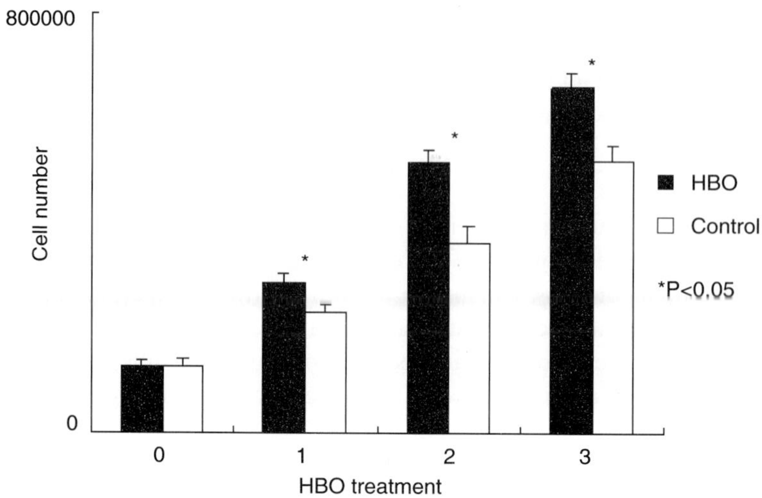

Fig. 1. Dose dependent HBO treatment increased MCL cell number in comparison with control cells cultured under the same medium. Data are expressed as mean ± SD of findings in six laboratory animals in each group; $p < 0.05$ was considered statistically significant.

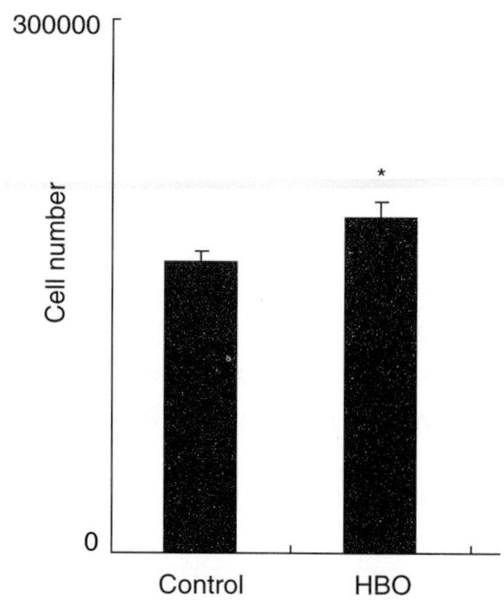

Fig. 2. Treatment with HBO significantly affected MCL cell growth. Data are expressed as mean ± SD of findings in six animals in each group; $p < 0.05$ was considered statistically significant.

III collagen content was significantly decreased after HBO treatment ($p<0.05$) (Fig. 4, HBO/Control ratio: 59.5% ± 7.6%, n = 2). Table 2 lists total data for six samples. Treatment with HBO significantly decreased Type III collagen/Type I collagen content ratio.

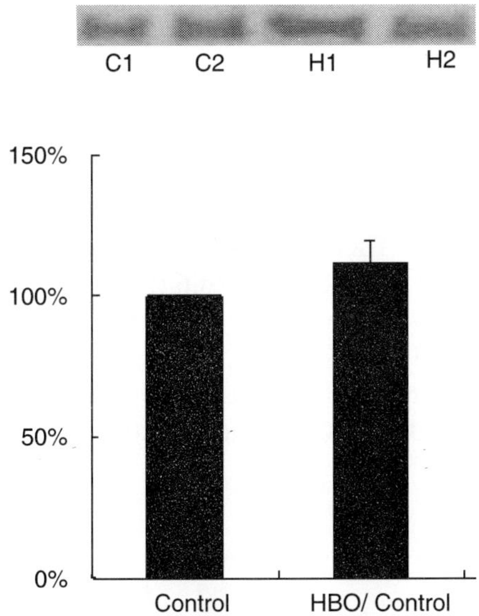

Fig. 3. No significant change in Type I collagen content after each treatment. Data are mean ± SD of findings in two animals in each group; $p < 0.05$ was considered statistically significant.

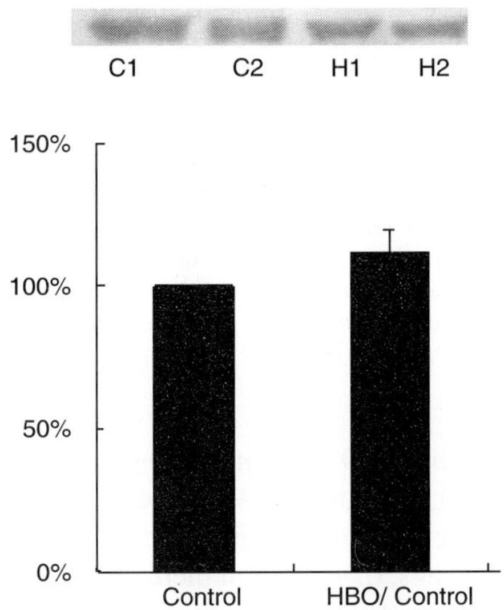

Fig. 4. Type III collagen content significantly decreased after HBO treatment ($p<0.05$) Data are expressed as mean ± SD of findings in two rabbits in each group; $p < 0.05$ was considered statistically significant.

Table 2. Type III / Type I collagen content ratio (n = 6)

	Control	HBO
Mean	9.5%	4.9%
SD	1.8%	0.7%

Biomechanical Test

Surgery was successful in all 48 rabbits. There were no infections. The mean percentage of failure force in the HBO group at 2, 4, 8, and 12 weeks were 20.6 (SD 4.8) %, 49.1 (SD 11.0) %, 63.9 (SD 9.5) %, and 76.3 (SD 10.2) %, respectively, and in the non-HBO group were 13.5 (SD 4.3) %, 25.3 (SD 6.6) %, 36.5 (SD 8.9) %, and 57.3 (SD 10.9) %, respectively (Fig 5). The HBO group had a significantly higher failure force as compared to that of the non-HBO group at all time intervals (Fig. 5, two-tailed *t* test, $p< 0.05$, $p< 0.01$, $p< 0.01$, and $p< 0.05$, respectively).

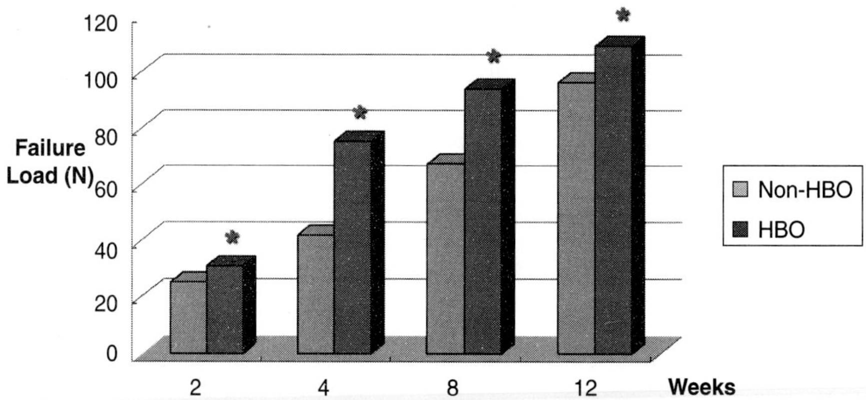

Fig. 5. The time course of the mean percent failure load in the HBO group and non-HBO group. The percent failure load = mean failure load of the right severed medial collateral ligament / mean failure load of the left intact medial collateral ligament in the same rabbit. The HBO group had a significantly higher failure force as compared to that of the non-HBO group at all time intervals. * $P < 0.05$.

Histological Analysis

Representative sections (Fig. 6) from the HBO and non-HBO groups at different intervals showed that the healing processes were identical in both groups. At 2 weeks, the defects were filled by a vascular inflammatory tissue. By 4 weeks, the reparative zone in the midsubstance areas of healing MCLs were occupied by active fibroblasts. At 8 weeks, there was an increase in the number and size of fibroblasts, and some evidence of longitudinal (along the long axis of the ligament) alignment of their nuclei. At 12 weeks, continued remodeling had occurred, with improving realignment. Although the phases of healing in the severed MCLs from both groups were identical, the amount of fibroblasts and the extent of reparative zone were predominantly higher in the HBO group. By the method of trichrome stain, the HBO group showed higher density and better alignment of the collagen fibers than the non-HBO group at 2, 4, 8, and 12 weeks. The histological findings were compatible with the biomechanical results.

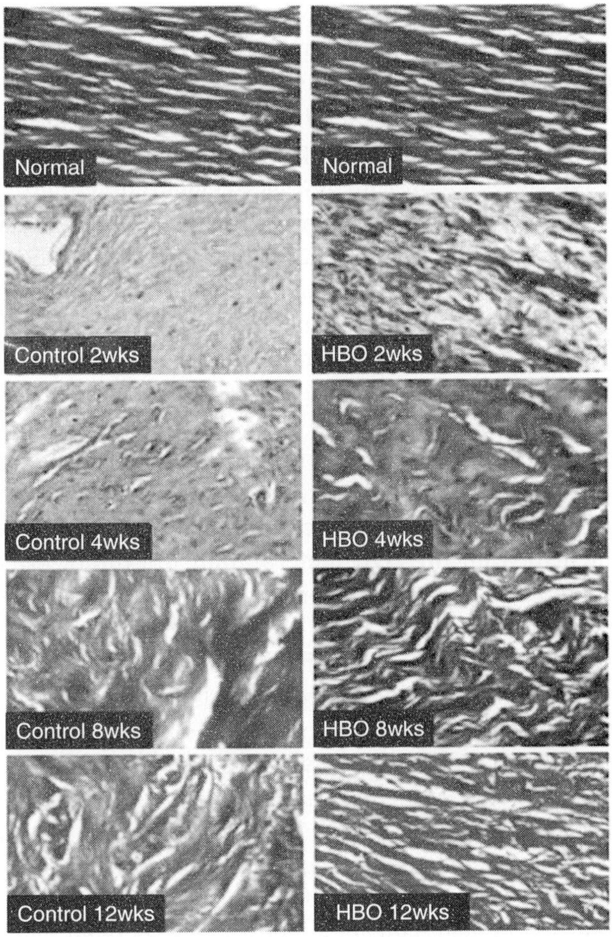

Fig. 6. Representative sections of midsubstance areas of healing MCLs at different intervals taken from non-HBO and HBO groups. Panels A, B, C, and D show the severed MCL in non-HBO group at 2, 4, 8, and 12 weeks after injury. Panels E, F, G, and H show the severed MCL in HBO group at 2, 4, 8, and 12 weeks after injury. Both groups showed progressive changes in cell number, size (metabolic activity), distribution, and orientation are noted as stages of inflammation, proliferation, and remodeling take place. However, the collagen fibers (blue color stain) in the HBO group had denser blue color (higher density) and better alignment than the non-HBO group at 2, 4, 8, and 12 weeks (Trichrome stain, x 200).

Discussion

Hyperbaric oxygen therapy has been shown to enhance the rate of healing in various problem wounds including diabetic foot ulcer, osteoradionecrosis, necrotizing soft tissue infections, refractory osteomyelitis, and ischemic tissue flaps.[32-38] Kivisaari and Niinikoski showed that hyperbaric oxygen enhanced the rate of wound healing in ischemic wounds in rats.[35] Hammarlund and Sundberg reported a significant enhancement of healing of chronic leg ulcers with the use of hyperbaric oxygen therapy in a randomized study of 16 patients.[32] Hunt et al. found that the greatest benefit of hyperbaric oxygenation was achieved in situations in which the nutritive flow and oxygen supply to the repair tissue are markedly compromised, as for instance by local injury or infection, but in which the regional blood is intact.[34] Hyperbaric oxygen has also been shown to accelerate normal fracture healing and normal skin healing in animal models.[1,2,4,9] Various mechanisms have been postulated that may explain the positive effect that hyperbaric oxygen on injured tissue including increasing fibroblasts proliferation and collagen deposition,[6,7] stimulation of vascular proliferation at the site of injury,[1,5,6] and decreased leukocyte adhesion to injured endothelium with decreased elaboration of inflammatory cytokines.[7]

Collagen is the most abundant substance in ligament[39] , and the mechanical properties of ligaments depend primarily on collagen fibers. More than 90% of collagen in normal ligaments is Type I collagen. Ligament injuries heal by scar formation rather than by actual ligament regeneration. The functional recovery of ligaments is most apparent in an examination of their tensile properties. Ultimate load and stiffness are the most common measures of the tensile properties of healing ligament. Horn *et al.*[40] demonstrated that administration of HBO increases the ultimate load in lacerated MCL despite no significant difference in the stiffness. However, no previous studies have related the effects or mechanism of HBO to fibroblasts from MCL. The oxygen required for ligament healing is the rationale for HBO therapy. This type of therapy increases tissue oxygen tension and improves collagen synthesis and angiogenesis. Furthermore, due to complex cell-to-cell interaction and the activity of soluble factors controlling ligament healing in vivo, determining the direct effects of oxygen on cells themselves and predicting optimal oxygenation conditions for the function of these cells is difficult. Tompach *et al.* investigated the effects of HBO on cells involved in wound healing.[28] Cultured endothelial cells and fibroblasts were exposed to HBO. To measure the effect of HBO on the P_{02} of the culture medium, a small-bore catheter was threaded through a port in the HBO chamber with the tip placed in the culture medium. After cultures were exposed to HBO at 2.4 ATA for 90 minutes, the partial pressure of oxygen in the medium of culture cells was over 1011 mmHg. The current study followed the same in vitro model and cell culture medium as Tompach *et al.* The rapid increase in P_{02} of the culture medium with HBO treatment could be achieved. Tolerance to HBO can be increased by intermittent exposure when HBO treatments are required for humans. This study applied a clinical HBO therapy protocol almost identical to the protocol used by the United States Navy. Treatment with HBO every 48 hours may reduce free radical damage to cells. Cells secreted cytokines, growth factors or extracellular matrix to the cell culture media. The cells and media were pooled for collagen measurements. Cells were seeded in T-75 tissue culture flasks until cell confluence was achieved, then plated at 10^5 cells per 60-mm tissue culture dish for control group or HBO group. Although extracellular collagen deposits might have remained in the T-75 flasks, the purpose of the procedure was to study the effect of HBO on cells cultured in the 60 mm tissue culture dish.

Collateral ligaments undergo changes in their complement of collagens during fetal development, postnatal development, ageing and healing. Treatment with HBO increases total collagen synthesis of tendon and ligaments, [23,41], but the complementary modulation of collagen has never been reported. Exogenous TGF-α increases Type I collagen secretion and thus decreases Type III collagen /Type I collagen in healing ligament. In the current study, these monoclonal antibodies were specific for Type I collagen and Type III collagen, respectively, and did not react with other classes. The data indicated that Type III collagen content significantly decreased after HBO treatment. Treatment with HBO increased cell growth and collagen synthesis. Because the proportion of Type I collagen was much higher than that of Type III collagen, Type I collagen content was increased after HBO treatment.

However, the increased Type I collagen content (increased/original) was not statistically significant. Treatment with HBO produces a stronger regulatory effect on collagen complement modulation and results in the suppression of Type III collagen secretion. Treatment with HBO decreases the ratio of Type III collagen/Type I collagen, which may result in mechanically stronger collagen fibrils.

The remodeling power and duration of human ligament healing is more similar to that of rabbits than rats. Ligament scars in humans and rabbits are thought to require as much as 12 months or more to complete remodeling. Previous studies found that the tensile strength of the injured ligament reached only 60% of control at one year. In this study, we used an established rabbit model of medial collateral ligament healing,[10] and biomechanical and histologic analyses of the healed segments to investigate the effect of HBO on ligament healing. Using the other intact medial collateral ligament in the same rabbit as a control, we found that the mean percent failure load of the severed medial collateral ligament of the HBO group was significantly higher than that of the non HBO group at 2, 4, 8, and 12 weeks, respectively. The mechanical strength of the two groups begins to differ significantly as early as 2 weeks and through 4, 8 and 12 weeks of ligament healing. In this animal model, hyperbaric oxygen therapy clearly enhances the mechanical properties of early ligament healing in clinical practice. Evidence of the beneficial effect of hyperbaric oxygen on ligament healing using the rabbit model has not been previously reported. The results are measured biomechanically and using conventional light microscopy demonstrate an increased rate of tensile strength development and collagen formation in the hyperbaric oxygen treated animals as compared to the controls. There are no biochemical or cell biology studies so that there is no information on mechanisms. The mechanism of their effect remains unclear. It would be interesting to have carried the experiment out over a longer period of time to determine if the trends of the treated and nontreated group eventually reach the same level or if the hyperbaric treated animals had a permanent increase in ligament strength.

The results of this study suggest that hyperbaric oxygen therapy enhances regenerative collagen formation in injured ligaments. HBO treatment decreases the Type III collagen/Type I collagen content ratio, possibly resulting in mechanically stronger collagen fibrils. The repaired ligaments had better biomechanical performance and earlier histological evidence of healing in the HBO-treated group. We recommend HBO therapy is a potentially effective treatment for MCL healing.

Acknowledgement

The authors are grateful to Chang Gung Memorial Hospital of the Republic of China, Taiwan for financially supporting this study under Contract No. CMRP 981

References

1. E. Barth, T. Sullivan and E. Berg, Animal model for evaluating bone repair with and without adjuvant hyperbaric oxygen therapy: Comparing dose schedules, *J Invest Surg.* 1990; 3, 387-392.
2. I.G. Yablon and R.L. Cruess, The effect of hyperbaric oxygen on fracture healing in rats, *J Trauma*. 1968; 8, 186-202.
3. R. Penttinen, J. Niinikoski and E. Kulonen, Hyperbaric oxygenation and fracture healing. *Acta Chir Scand.* 1972;138: 39-44.

4. S.W.N. Ueng, S.S. Lee, S.S. Lin, C.R. Wang, S.J. Liu, C.L. Tai and C.H. Shih, Bone healing of tibia lengthening is enhanced by hyperbaric oxygen therapy: a study of bone mineral density and torsion strength on rabbits. *J Trauma* 1998; 44: 676-681.
5. E.P. Kindwell, L.J. Gottlieb and D.L. Larson, Hyperbaric oxygen therapy in plastic surgery: A review article. *Plast Reconstr Surg.* 1991; 88: 898-908.
6. D.N. Roth and L.D. Weiss, Hyperbaric oxygen and wound healing. *Clin Dermatol* 1994; 12: 141-156.
7. G. Oriani, A. Marroni and F. Wattel, Physiologic Principles of Hyperbaric Oxygenation. In: G. Oriani, A. Marroni and F. Wattel, eds, *Handbook of Hyperbaric Oxygen Therapy*. Milan, Springer Verlag; 1996, pp. 35-58.
8. E. Uhl, A. Sirsjo, T. Haapaniemi, G. Nilssonand and G. Nylander, Hyperbaric oxygen improves wound healing in normal and ischemic skin tissue. *Plast Reconstr Surg.* 1994; 96: 835-841.
9. T. Kitakoji, S. Takashi, Y. Ono, T. Hattori, H. Takahashi and H. Iwata, Effect of hyperbaric oxygenation treatment on lengthened callus. *Undersea & Hyperbaric Medicine* 1999; 26: 165-168.
10. P.B. James, B. Scott, M.W. Allen, Hyperbaric oxygen therapy in sports injuries: A preliminary study. *Physiotherapy* 1993; 79: 571-572.
11. A.G. Nelson, E.G. Wolf and B. Li, Influence of delayed hyperbaric oxygenation on recovery from mechanically induced damage. *Undersea Hyperb Med.* 1994; 21: 185-191.
12. C. Potera, Healing under pressure. *Physician Sportsmed* 1995; 25: 46-47.
13. J.R. Staples, Effects of intermittent hyperbaric oxygen on pain perception and eccentric strength in a human injury model. PhD Dissertation. University of British Columbia, Vancouver, British Columbia, Canada, 1996.
14. J.R. Staples, D.B. Clement and D.C. McKenzie, The effects of intermittent hyperbaric oxygen on biochemical muscle metabolites of eccentrically exercised rats. *Can J Appl Physiol.* Suppl 1995; 20: 49.
15. C. Frank, S.L.Y. Woo, D. Amiel, F. Harwood, M. Gomez and W. Akeson, Medial collateral ligament healing: A multidisciplinary assessment in rabbits. *Am J Sports Med.* 1983; 11: 379-389.
16. Niyibiza C, Saggaria-Visconti C, Kavalkovich K, Woo SLY: Collagens in an adult bovine medial collateral ligament. Immunofluorescence localization by confocal microscopy reveals that type XIV collagen predominates at the ligament-bone junction. *Matrix Biol* 1995; 14: 743-751.
17. Amiel D, Frank C, Harwood DF, Fronek J, Akeson W: Tendons and ligaments: a morphological and biochemical comparison. *J Orthop Res* 1984;1: 257-265.
18. Niyibiza C, Saggaria-Visconti C, Kavalkovich K, Woo SLY: Collagens in an adult bovine medial collateral ligament. Immunofluorescence localization by confocal microscopy reveals that type XIV collagen predominates at the ligament-bone junction. *Matrix Biol* 1995; 14: 743-751.
19. Watanabe M, Nojuma M, Shibata T, Hamada M: Maturaiton-related biochemical changes in swine anterior cruciate ligament and tibialis posterior tendon. *J Orthop Res* 12: 1994; 12: 672-682.

20. Bland YS, Ashhurst DE: Changes in the distribution of fibrillar collagens in the collateral and cruciate ligaments of the rabbit knee joint during fetal and postnatal development. *Histochemicla J* 1996; 28: 325-334.
21. Inoue M, Woo SLY, Gomez MA, Amiel D, Ohland KJ, Hitabayashi LR: Effect of surgical treatment and immobilization on the healing of the medial collateral ligament: a long-term multidisciplinary study. *Connect Tissue Res* 1990; 25: 13-26.
22. Amiel D, Frank CB, Harwood FL, Akeson WH, Kleiner JB: Collagen alteration in medial collateral ligament healing in a rabbit model. *Connect Tissue Res* 1987; 16: 357-366.
23. Doilon C, Dunn MG, Bender E, Silver F. Collagen fiber formation in repair tissue: development of strength and toughness. *Collagen Relat Res* 1992; 5: 481-492.
24. Lepistö J, Peltonen J, Vähä-Kreula M, Ninikoski J, Laato M: Platelet-derived growth factor isoforms PDGF-AA, -AB and –BB exert specific effects on collagen gene expression and mitotic activity of cultured human wound fibroblasts. *Biochem. Biophys. Res. Commun.* 1995; 209: 393-399.
25. Niiniskoski J. Effect of oxygen on wound healing and formation of experimental granuration tissue. *Acta Physiol Scand* 1969; Suppl 334: 1-72.
26. Prockop D, Kaplan A, Udenfriend S. Oxygen-18 studies on the convension of proline to hydroxyproline. *Biochem Biophys Res Commun* 1992; 9: 162-166.
27. Juva K, Prockop DJ, Cooper GW, Lash JW. Hydroxylation of proline and the intracellular accumulation of a polypeptide precursor of collagen. *Science* 1966; 152: 92-94.
28. Tompach PC, Lew D, Stoll JL. Cell response to hyperbaric oxygen treatment. *Int. J. Oral Maxillofac. Surg* 1997; 26: 82-86.
29. Chan YS; Chen ACY; Yuan LJ; Lin SS; Yang CY, Lee MSS; Ueng SWN. Effects of hyperbaric oxygen and platelet derived growth factor on medial collateral ligament fibroblasts *Undersea & Hyperbaric Medicine* 2007; 34(3):181-190.
30. Chan YS; Chang CH; Yuan LJ; Lin YC; Lee MSS; Ueng SWN. Effects of Hyperbaric Oxygen on Rabbit Medial Collateral Ligament Healing: An in Vitro Study (Abstract). Undersea & Hyperbaric Medicine Vol 32(4): 286, 2006.
31. Chan YS ; Chang CH; Yuan LJ; Lin YC; Lee MSS; Ueng SWN. Effects of Hyperbaric Oxygen on Medial Collateral Ligament Healing: An in Vivo Study Using a Rabbit Model (Abstract). *Undersea & Hyperbaric Medicine* 2006 ; 32: 286-287.
32. C. Hammarlund and T. Sundberg, Hyperbaric oxygen reduced size of chronic leg ulcers. A randomized double-blind study. *J Plast Reconstr Surg*. 1994; 93: 829-834.
33. S.S Lee, C.Y. Chen, Y.S. Chan, C.Y. Yen, E.K. Chao and S.W.N. Ueng, Hyperbaric oxygen in the treatment of diabetic foot infection. *Chang Gung Medical Journal* 1997; 20: 17-22.
34. T.K. Hunt, J. Niinikoski and B.H.I. Zederfeldt, Oxygen in wound healing enhancement: Cellular effects of oxygen. In: Davis JC, Hunt TK, ed., *Hyperbaric Oxygen Therapy.* Bethesda, MD: Undersea and Hyperbaric Medical Society; 1977: pp. 111.

35. J. Kivisaari, and J. Niinikoski, Effects of hyperbaric oxygenation and prolonged hypoxia on the healing of the open wound. *Acta Chir Scand.* 1975; 141: 14-19.
36. E.G. Mainous, Hyperbaric oxygen in maxillofacial osteomyelitis, osteoradionecrosis and osteogenesis enhancement. In: Davis JD, Hunt TK, eds. *Hyperbaric oxygen therapy.* Bethesda, MD: Undersea Medical Society, 1977: PP. 191-203.
37. G.B. Hart and E.G. Maninous, The treatment of radiation necrosis with hyperbaric oxygen. *Cancer* 1976; 37: 2580-2586.
38. C.Y. Chen, S.S. Lee, Y.S. Chan, C.Y. Yen, E.K. Chau and S.W.N. Ueng, Chronic Refractoy Tibial Osteomyelitis Treated with Adjuvent Hyperbaric Oxygen: A Preliminary Report. *Chang Gung Medical Journal* 1998; 21: 165-171.
39. Frank C, Amiel D, Woo SL-Y, Akeson W. Normal ligament properties and ligament healing. *Clin Orthop* 1985; 196: 15-25.
40. P.C. Horn, D.A. Webster, H.M. Amin, M.F. Mascia, F.W. Werner and M.D. Fortino, The effect of hyperbaric oxygen on medial collateral ligament healing in a rat model. *Clin Orthop* 1999; 360: 238-242.
41. Yuan L.J, Ueng S.W.N., Lin S.S, Yeh WL, Yang CY, Lin PY. Attenuation of Apoptosis and Enhancement of Proteoglycan Synthesis in Rabbit Cartilage Defects by Hyperbaric Oxygen Treatment Are Related to the Suppression of Nitric Oxide Production. *J Ortho Res* 2004; 22(5):1126-34.

Adaptation Biology and Medicine (Volume 6: Cell Adaptations and Challenges)
Editors: P. Wang, C.-H. Kuo, N. Takeda and P.K. Singal
Copyright © 2011 Narosa Publishing House Pvt. Ltd., New Delhi, India

23. Mild Traumatic Brain Injury in Athletes

Douglas G. Chang* **and Alan R. Hargens**

Physical Medicine and Rehabilitation, Department of Orthopaedic Surgery University of California, San Diego, San Diego, USA

Abstract

Concussion is a pathologic clinical syndrome originating from biomechanical trauma to the brain[1]. There are several common features which define concussions. They are caused by "impulsive forces" transmitted to the head from a blow to the head, face, neck or elsewhere. They result in acute impairments of neurologic function with spontaneous resolution. The clinical impairments reflect a problem of neurocognitive function rather than a structural anatomic injury. In fact, no abnormalities are seen in concussion with today's standard neuroimaging studies. Concussion produces a spectrum of clinical symptoms which may or may not involve a loss of consciousness. The resolution of such symptoms follows a sequential course over days to weeks to even months (in a small percentage of cases). 80-90% of cases resolve within 7-10 days, but children and adolescents may require more time for resolution. Expert opinion recommends that all concussions should be taken seriously and that older concussion classifications schemes are overly simplistic given the diversity of clinical presentations.

Keywords: Brain trauma, neurocognitive function, concussion

Introduction

In the U.S., 300,000 sports-related concussions and brain injuries occur annually[2]. These injuries tend to be mild and recurrent (concussions), but serious and even catastrophic traumatic brain injuries do occur well.

Most head injuries are attributable to three sports; soccer[3], boxing, and American football[4]. Other sporting activities (e.g. ice hockey, rugby, lacrosse, skiing, pole vaulting, and cheerleading) are also implicated[5, 6]. Males have higher rates of head injuries, with a different mechanism and spectrum of injury, than females, and they appear to suffer less from the injury[7]. To date, the sports-related concussion literature describes epidemiology patterns[8,9,10], neuro-cognitive performance[11, 12,] body fluid composition changes[13, 14], postural balance[15], head motion and surface contact forces[2, 4, 16].

*Address for Correspondence:Correspondence information: Dr. Douglas G. Chang Associate Clinical Professor, Chief, Physical Medicine and Rehabilitation, Department of Orthopaedic Surgery University of California, San Diego 350 Dickinson St., San Diego, CA 92103-8894, USA Tel: +1(858) 543-2542 Email: dchang@ucsd.edu

Epidemiology

Guskiewicz *et al*[8] followed about 3000 American collegiate football players in 1999-2001. About 6% of the players were identified to have a concussion, and 6% of these players had repeat concussions in the same season. 82 hours of headache was the chief complaint. Players with a history of 3 or more previous concussions were 3 times more likely to sustain another concussion than players with no concussion history. The majority (92%) of the repeat concussions occurred within 10 days of the first injury. A history of multiple concussions was also associated with slowed recovery.

Among American high school athletes, Schulz *et al*[9] measured concussion rates in terms of an "athlete-exposure," or one athlete participating in one practice or game. Concussion rates ranged from 9 per 100,000 athlete exposures for cheerleaders to 33 per 100,000 for football players. History of previous concussion was a moderately strong risk factor for another concussion. Among the cheerleaders, their concussion rate was higher during practice than during competition, and mostly involved two-level pyramids.

In the National Football League, 887 concussions were identified from 1996-2001 and 10. 8% of these were "prolonged," lasting more than 7 days. 1.5% of all the concussions involved prolonged post concussion syndrome. The prolonged concussions occurred mainly among quarterbacks, special teams return units and defensive secondaries. The chief complaint was disorientation to time, retrograde amnesia, fatigue and cognition problems. More than a minute loss of consciousness was a predictor of prolonged concussion.

Neurocognitive Function

Psychometric tests of neurocognitive function evaluate attention, concentration, reaction time, executive planning and visual perceptual function. These higher-order brain functions are particularly sensitive to impairment after injury[17]. As an indicator of brain performance, the tests have been shown to be reliable in middle-school aged children[11] and college students [18]. Furthermore, the tests can be used sequentially to monitor recovery of function over the days to weeks after an injury. These performance tests complement clinical evaluation of symptoms (e.g. headache, dizziness, mood changes) and neuro-imaging in the evaluation of concussion.

There are three levels of neurocognitive function: baseline (pre-competition) testing, sideline screening, and traditional comprehensive examination. The baseline studies typically take about 30 minutes to complete. Computer programs allow testing of large numbers of athletes, and can assist with return-to-play decisions [19]. Some popular computer programs include the Immediate Post-Concussion Assessment, CogSport, HeadMinder Resolution Index, and Automated Neuropsychologic Assessment Metrics [15]. One popular sideline examination is the Standardized Assessment of Concussion (SAC). It is a 30 point scale that measures orientation, immediate memory, attention and retrograde memory that takes about 5 minutes to administer [20]. However, orientation tests during a competition setting have been shown unreliable [1]. Therefore, the Zurich 2009 conference emphasizes an alternative sideline test, the Sport Concussion Assessment Tool 2 (SCAT2). Last, the traditional tests of neurocognitive function, typically take hours to complete and are most comprehensive. They require a specially trained Neuropsychologist to administer and may detect subtle deficits beyond the acute concussive event. These tests evaluate domains of motor, attention, language, memory, problem solving, and executive performance [17]

Biochemical Markers

Stalnacke *et al.* showed elevations of two serum biochemical markers of CNS tissue damage (S-100B and NSE, neurone specific enolase) in elite soccer players after a match. The S-100B elevations were significantly correlated to the number of headers performed in the match. These serum markers were shown to rise after ice hockey and basketball competitions, and seem to be correlated to the number of acceleration/deceleration events experienced by the players during a match. One player who happened to sustain a concussion had serum levels elevated greater than the other players, who were not injured, in the cohort [21]. Zetterberg *et al*[14]. performed lumbar punctures on amateur boxers and controls. They measured several proteins after a boxing match, including markers of neuronal and axonal injury (neurofilament light protein, total tau) and astroglial injury (glial fibrillary acidic protein). These markers were elevated immediately after a match, and most elevated in boxers who sustained greater frequency or impact energy to the head. This suggests that amateur boxing is associated with acute neuronal and astroglial injury.

Biomechanical Measurement and Modeling

Head acceleration in 52 football players during two years of competition was measured using an instrumented helmet system with remote data transmission [2], [16]. 11,600 head impacts were studied. An instrumented dummy was subjected to head blows from Olympic boxers [4]. High speed video of the boxing and National Football League head impacts were compared. Head and neck accelerations were measured. Mathematical models showed that the greatest strain was generated in the midbrain [4].

Clinical Management of Concussions [1]

Acute concussions may result in a range of symptoms and signs. Common symptoms include cognitive (difficulty with memory, attention and concentration), emotional (irritability, apathy), physical (loss of consciousness, impaired balance), and somatic (headache, sleep disturbance) domains.

If an athlete exhibits the signs of a concussion, they must be subject to standard emergency medical evaluation by a healthcare provider, with particular emphasis on cervical spine management. A sideline screening tool, such as the SCAT2 or SAC, should be administered. The athlete should not be left alone, and should be monitored for signs of neurological deterioration over the ensuing hours. It is worth noting that concussion symptoms may take several hours before they present. For these reasons, athletes with a diagnosed concussion should be removed from play for the remainder of the day.

If further clinical evaluation is necessary, the athlete should be subject to a detailed neurological exam involving mental status, cognitive function, gait and balance. There should be consideration for advanced neuro-imaging to evaluate for more serious brain injuries. Last, the findings should be interpreted within the context of the time of injury to estimate interval progression or deterioration.

Return to Play

The return to play decision is a clinical one, meaning that a medical decision must be made on an individual case-by-case basis. However, there are clinical guidelines and recommendations. In some highly experienced adult settings, there are a few circumstances where an athlete might return to play in the same day. Such a decision is based on the availability of an exceptional level of medical resources and experience. However in most situations, an athlete who suffers a concussion must not return to play until they are symptom free. The initial recommendation is for rest, both physical and cognitive. Once there are no symptoms, the athlete should participate in a slow, graduated program of exertion. This graduated program typically occurs in 24 hour increments, such that a typical concussed athlete returns to play in about one week. The athlete progresses through light aerobic exercise, to sport-specific exercise and drills, to full contact play and competition. After a concussion, there are medications that an athlete may take to address symptoms (headache, insomnia, anxiety, depression). The decision to return to play should be made when the athlete is symptom-free and medication-free in general. Again it must be emphasized that the decision to return to play is a clinical decision, and must be made on case by case basis.

Future Concussion Investigations

Conventional neuroimaging techniques (MRI, CT) are not helpful in concussion evaluation (because no abnormalities are seen). Advanced MRI techniques, such as diffusion weighted imaging studies, have greater sensitivity but have not seen widespread clinical use or validation. At the research level, compelling studies have involved positron emission tomography (PET), magnetic resonance spectroscopy and functional MRI. In one clinical study of 41 athletes, abnormal fMRI results during the first week of injury were associated with symptom severity and predictive of clinical recovery [1,22]. At this time, such modalities are not ready for clinical use.

Intra-cranial pressure (ICP) is an essential clinical management parameter in patients with severe head injuries [23], [24]. However, no studies have examined ICP in athletes at risk for more mild head injuries during the course of actual sports participation, because conventional measurement techniques are invasive. Yet, a clear catalog of dynamic ICP responses could provide scientists data to evaluate the risks of particular sporting activities and suggest mechanisms to reduce the likelihood of injuries.

Recently, an ultrasonic device was developed to measure ICP waveforms non-invasively, using a pulsed phase lock loop (PPLL) technique. Briefly, an ultrasonic transducer, the size of a quarter, is placed superficially over the temporal bone (see figure 1). Ultrasonic waves pass through the skull and reflect off the opposite side of the inner skull surface [25], [26].

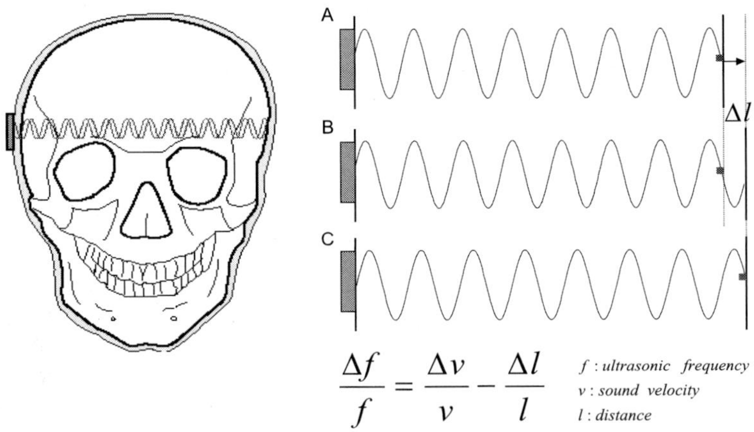

$$\frac{\Delta f}{f} = \frac{\Delta v}{v} - \frac{\Delta l}{l}$$

f : ultrasonic frequency
v : sound velocity
l : distance

Fig. 1. Principle of PPLL

The device records the slight changes in skull dimensions (on the order of micrometers), which change with fluctuations in ICP. Furthermore, the ultrasound waveforms have been shown to correlate reliably with invasively measured ICP waveforms in head trauma patients, in both the frequency and time domains [26].

Preliminary Studies

Previous studies have shown that the ultrasound system reliably measures distance (figure 2) [25], and represents invasively measured ICP (figure 3) [26].

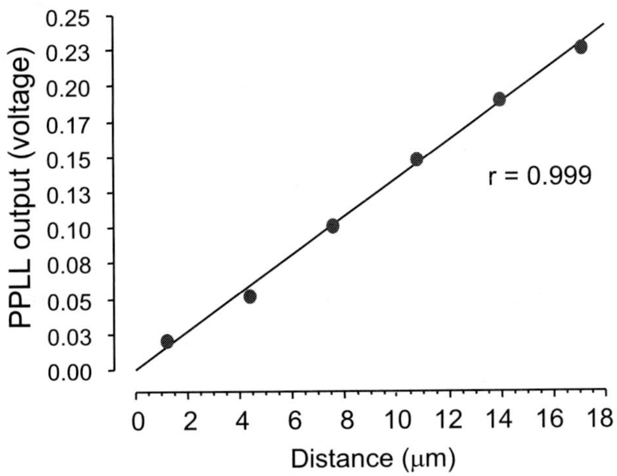

Fig. 2. Relationship between PPLL output and distance.

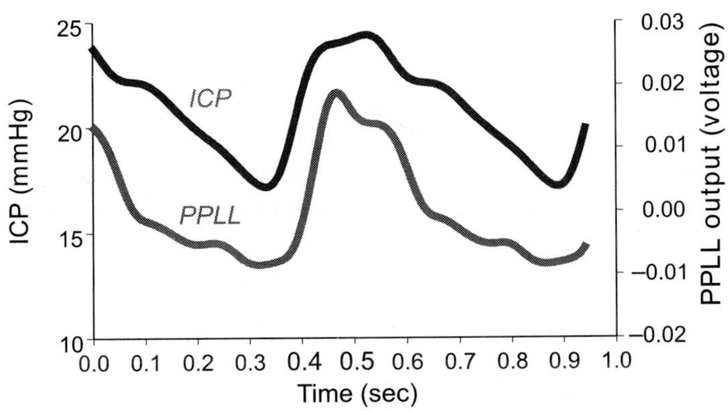

Fig. 3. Typical waveforms of ICP and PPLL output during a cardiac cycle have high correlation.

The system has sufficient sensitivity to detect ICP changes due to orthostatic changes (Figure 4) [27] and fluid shifts [28] due to prolonged bed rest.

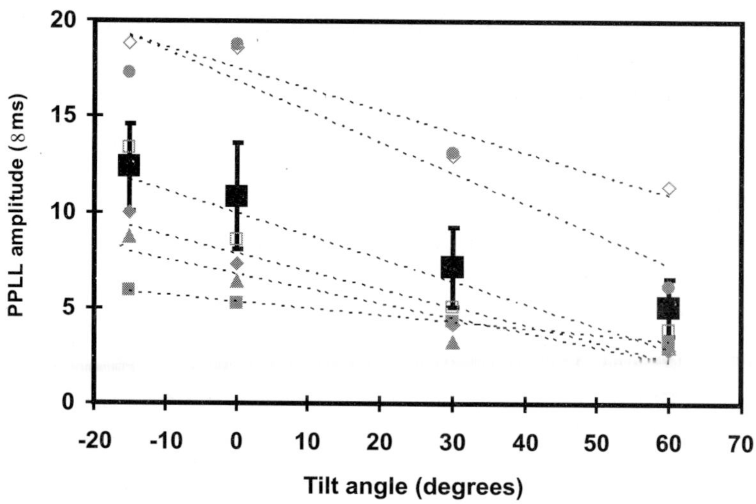

Fig. 4 **Changes in PPLL output during whole body tilting.**

Alternative Techniques to Measure Intracranial Pressure Non-invasively

Classic acoustic methods are affected by body temperature and superficial soft tissue conditions such as contusions [29]. Transcranial doppler sonography measures cerebral blood flow, but requires bulky equipment and does not account for other intracranial volume reserves [30]. Venous opthaldynamometry is based on the principle that the optic nerve is surrounded by a dural sheath filled with cerebrospinal fluid. Pressure in the retinal vein is influenced by intracranial pressure. However opthaldynamometry requires compression of the eyeball and is not optimal for sideline testing and validation among athletes [31]. Tympanic membrane displacement may not be clinically useful because of predictive limitations [32], and artifact due to ear canal obstructions such as cerumen. Tissue resonance analysis techniques have not been validated in prospective studies [33, 34].

Summary

In the future, the non-invasive measurement of ICP in an athletic setting may help improve our understanding about the mortality and morbidity associated with sports-related head injuries. The authors are involved in projects to study the effects of exercise and soccer heading on the intra-cranial pressure of an able-bodied population on the athletic playing fields. Such a device could be used also to study other sports such as boxing. In the end, the regulation of intracranial pressure and orthostatic function may be an important parameter of concussion evaluation and management. Applications of such knowledge may extend well beyond athletic settings (for instance, affecting our understanding of blast injuries to the head and return-to-duty decision making).

References

1. McCrory P, Meeuwisse W, Johnston K, *et al.* Consensus Statement on Concussion in Sport: the 3rd International Conference on Concussion in Sport held in Zurich, November 2008. Br. J. Sports Med. 2009;43:i76-i84.

2. Brolinson PG, Manoogian S, McNeely D, Goforth M, Greenwald R, Duma S. Analysis of linear head accelerations from collegiate football impacts. Curr Sports Med Rep. Feb 2006;5(1):23-28.
3. Janda DH, Bir CA, Cheney AL. An evaluation of the cumulative concussive effect of soccer heading in the youth population. Inj Control Saf Promot. Mar 2002;9(1):25-31.
4. Viano DC, Casson IR, Pellman EJ, et al. Concussion in professional football: comparison with boxing head impacts--part 10. Neurosurgery. Dec 2005;57(6):1154-1172; discussion 1154-1172.
5. McCrory P, Johnston K, Meeuwisse W, et al. Summary and agreement statement of the 2nd International Conference on Concussion in Sport, Prague 2004. Br. J. Sports Med. Apr 2005;39(4):196-204.
6. Bailes JE, Cantu RC. Head injury in athletes. Neurosurgery. Jan 2001;48(1):26-45; discussion 45-26.
7. Dvorak J, McCrory P, Kirkendall D. Head injuries in the female football player: incidence, mechanisms, risk factors and management. Br. J. Sports Med. Aug 2007;41(Suppl 1):i44-i46.
8. Guskiewicz KM, McCrea M, Marshall SW, et al. Cumulative effects associated with recurrent concussion in collegiate football players: the NCAA Concussion Study. JAMA. Nov 19 2003;290(19):2549-2555.
9. Schulz MR, Marshall SW, Mueller FO, et al. Incidence and risk factors for concussion in high school athletes, North Carolina, 1996-1999. Am. J. Epidemiol. Nov 15 2004;160(10):937-944.
10. Pellman EJ, Viano DC, Casson IR, Arfken C, Powell J. Concussion in professional football: injuries involving 7 or more days out--Part 5. Neurosurgery. Nov 2004;55(5):1100-1119.
11. Valovich McLeod TC, Barr WB, McCrea M, Guskiewicz KM. Psychometric and measurement properties of concussion assessment tools in youth sports. J Athl Train. Oct-Dec 2006;41(4):399-408.
12. Echemendia RJ, Cantu RC. Return to play following sports-related mild traumatic brain injury: the role for neuropsychology. Appl Neuropsychol. 2003;10(1):48-55.
13. Stalnacke BM, Tegner Y, Sojka P. Playing soccer increases serum concentrations of the biochemical markers of brain damage S-100B and neuron-specific enolase in elite players: a pilot study. Brain Inj. Sep 2004;18(9):899-909.
14. Zetterberg H, Hietala MA, Jonsson M, et al. Neurochemical aftermath of amateur boxing. Arch. Neurol. Sep 2006;63(9):1277-1280.
15. Cohen J, Gioia G, Atabaki S, Teach S. Sports-related concussions in pediatrics. Curr. Opin. Pediatr. 2009;21:288-293.
16. Crisco JJ, Chu JJ, Greenwald RM. An algorithm for estimating acceleration magnitude and impact location using multiple nonorthogonal single-axis accelerometers. J. Biomech. Eng. Dec 2004;126(6):849-854.
17. Randolph C. Implementation of neuropsychological testing models for the high school, collegiate, and professional sports settings. Journal of athletic training. 2001;36(3):288-296.

18. Echemendia RJ, Putukian M, Mackin RS, Julian L, Shoss N. Neuropsychological test performance prior to and following sports-related mild traumatic brain injury. Clinical journal of sport medicine : official journal of the Canadian Academy of Sport Medicine. Jan 2001;11(1):23-31.
19. Erlanger D, Feldman D, Kutner K, et al. Development and validation of a web-based neuropsychological test protocol for sports-related return-to-play decision making. Archives of Clinical Neuropsychology. 2003;18:293-316.
20. McCrea M. Standardized mental status testing on the sideline after sport-related concussion. Journal of athletic training. 2001;36(3):274-279.
21. Stalnacke BM, Tegner Y, Soika P. Playing ice hockey and basketball increased serum levels of S-100B in elite players. A pilot study. Clinical journal of sport medicine : official journal of the Canadian Academy of Sport Medicine. 2003;13(5):292-302.
22. Lovell M, Pardini J, Welling J, et al. Functional brain abnormalities are related to clinical recovery and time to return-to-play in athletes. Neurosurgery. 2007;61:352-360.
23. Adamides AA, Winter CD, Lewis PM, Cooper DJ, Kossmann T, Rosenfeld JV. Current controversies in the management of patients with severe traumatic brain injury. ANZ J Surg. Mar 2006;76(3):163-174.
24. Stocchetti N, Penny KI, Dearden M, et al. Intensive care management of head-injured patients in Europe: a survey from the European brain injury consortium. Intensive Care Med. Feb 2001;27(2):400-406.
25. Ueno T, Ballard RE, Shuer LM, Cantrell JH, Yost WT, Hargens AR. Noninvasive measurement of pulsatile intracranial pressure using ultrasound. Acta Neurochir Suppl. 1998;71:66-69.
26. Ueno T, Macias BR, Yost WT, Hargens AR. Noninvasive assessment of intracranial pressure waveforms by using pulsed phase lock loop technology. Technical note. J. Neurosurg. Aug 2005;103(2):361-367.
27. Ueno T, Ballard RE, Macias BR, Yost WT, Hargens AR. Cranial diameter pulsations measured by non-invasive ultrasound decrease with tilt. Aviat. Space Environ. Med. Aug 2003;74(8):882-885.
28. Steinbach G, Macias BR, Tanaka K, Yost WT, Hargens AR. Intracranial pressure dynamics assessed by noninvasive ultrasound during 30 days of bed rest. Aviat. Space Environ. Med. February 2005 2003;76(2):85-90.
29. Ragauskas A, Daubaris G, Ragaisis V, Petkus V. Implementation of on-invasive brain physiologic monitoring concepts. Med. Eng. Phys. 2003;25:667-678.
30. Schmidt B, Czosnyka M, Schwarze J, et al. Evaluation of a method for noninvasive intracranial pressure assessment during infusion studies in patients with hydrocephalus. J. Neurosurg. May 2000;92(5):793-800.
31. Firsching R, Schutze M, Motschmann M, Behrens-Bauman W. Venous opthalmodynamometry: a noninvasive method for assessment of intracranial pressure. J. Neurosurg. Jul 2000;93:33-36.

32. Shimbles S, Dodd C, Banister K, Mendelow A, Chambers I. Clinical comparison of tympanic membrane displacement with invasive ICP measurements. Acta Neurochir Suppl. 2005 2005;95:197-199.
33. Michaeli D, Rappaport Z. Tissue resonance analysis; a novel method for noninvasive monitoring of intracranial pressure. Technical note. J. Neurosurg. 2002;96(6):1132-1137.
34. Raksin P, Alperin N, Sivaramakrishnan A, Surapaneni S, Lichtor T. Noninvasive intracranial compliance and pressure based on dynamic magnetic resonance imaging of blood flow and cerebrospinal flow: review of principles, implementation, and other noninvasive approaches. Neurosurgery Focus. 2003;14(4):1-8.

Adaptation Biology and Medicine (Volume 6: Cell Adaptations and Challenges)
Editors: P. Wang, C.-H. Kuo, N. Takeda and P.K. Singal
Copyright © 2011 Narosa Publishing House Pvt. Ltd., New Delhi, India

24. Protective Effect of Endurance Exercise Training on Brain Antioxidant System in Old Age Rats Treated with Alcohol: A Comparison with Young Age Rats

K. Mallikarjuna[1,2], Somasekhara N. Reddy[1], K.R. Shanmugam[1], Chia-Hua Kuo[2] and Sathyavelu K. Reddy[1*]

[1]Division of Molecular Biology and Exercise Physiology, Department of Zoology Sri Venkateswara University, Tirupati - 517 502, A. P, India, [2]Laboratory of Exercise Biochemistry, Taipei Physical Education College, Taipei-11153, Taiwan, ROC

Abstract

Elder subjects are more prone to alcohol-induced oxidative damages, particularly to brain tissue. The present study was carried out to examine the impact of exercise training on alcohol-induced oxidative damage and diminished antioxidant status in the brain of aged rats comparing with young ones. The age-matched Wistar albino rats (3 months young, n=24; 18 months old, n=24) were evenly divided into four groups and treated as, normal control (NC), exercise trained (Ex), alcohol treated (At) and exercise plus alcohol treated (Ex+At). With alcohol treatment, brain antioxidant enzyme activities including superoxide dismutase (SOD), catalase (CAT), glutathione peroxidase (GPx), glutathione reductase (GR) and the content of glutathione (GSH) were significantly ($p < .001$) decreased. These changes were found to be greater in the aged rats than those of the young rats. However, in our study, reduced antioxidant status with alcohol treatment was significantly reversed by exercise training. This was shown as improved activities of antioxidant enzymes along with GSH levels by exercise combination treatment in the brain of old rats. These results suggest that exercise training is potential to reverse the deteriorated antioxidant enzyme status in the brain of elderly alcohol treated rats.

Keywords: Ethanol, Aging, Oxidative damage, Glutathione

Introduction

The free radical theory of aging proposed by Harman (1981) states the involvement of active oxygen species in ageing process. Free radicals damage to membranes of the nerve cells has been suggested to be the primary cause of aging of an organism (Zs-Nagy, 1988). The cellular antioxidant defense system

*Address for Correspondence: Dr. Sathyavelu K. Reddy Professor of Zoology Division of Molecular Biology Sri Venkateswara University Tirupati - 517502, A.P, India Telephone: +91 877-2289304 Ext: 104, + 91 9849023339 Fax: +91 877-2261801 E-mail: sathyakreddy@hotmail.com/sathyakreddy@rediffmail.com

plays a vital role to protect the organism against free radicals mediated oxidative damages. Aging has been shown to be associated with declined capacities of both enzymatic and non-enzymatic antioxidant systems in various important organs (Harman, 1986; Mallikarjuna *et al.* 2007; Pushpalatha *et al.* 2008).

Alcoholic beverages are used universally and alcohol is the world's most widely used psychoactive drug, but chronic excessive alcohol consumption leads to permanent organ damage or death. Alcohol-induced brain damage produces some of the most subtle effects of alcoholism, including cognitive deficits such as learning and memory impairment and changes in the physiological and biochemical processes of the whole organism and in the cells (Clemens and Jerrells, 2004; Oba *et al.* 2005). Alcohol treatment elevates the production of reactive oxygen species (ROS) and decreases enzymatic and non-enzymatic antioxidant levels particularly in liver and brain (Somani *et al.* 1996; Loguercio *et al.* 1996; Mallikarjuna *et al.* 2007). Existing scientific knowledge clearly emphasizing that age and alcohol are the two principal sources for ROS production that ultimately cause oxidative damages in the brain.

It has been observed that individuals undergoing regular exercise training have high levels of antioxidant enzymes and non-enzymatic antioxidants in different tissues demonstrate greater resistance to exercise-induced or imposed oxidative stress (Ji, 1995; Vina *et al.* 1996). Previous studies showed that endurance exercise training can attenuate oxidative damages concurrently with increased antioxidant status in the liver and brain (Somani *et al.* 1996; Husain and Somani, 1997; Navarro *et al.* 2004). A recent study from our laboratory demonstrated that alcohol-induced oxidative damage was significantly reversed by regular exercise training in liver of rats (Mallikarjuna *et al.* 2009).

Although, information is available on the recovery of antioxidant enzymes in young rats, little is known about the recovery of brain antioxidant enzymes in old alcohol treated rats. Hence, for a comprehensive interpretation, this study was carried out to investigate the protective effect of endurance exercise training against alcohol-induced disturbances in brain antioxidant enzyme system in old age rats as well as the young.

Materials and Methods

Animal Care and Maintenance

Wistar strain male albino rats of two different age groups, that is, young (3 months n = 24) weighing 170 ± 10 g and old (18 months n = 24) weighing 240 ± 10 g, were used in the present investigation. This study was approved by the Institutional Animal Ethics Committee (Regd. No. 438/01/a/CP CSEA/ dt.17.07.2001; resolution number 9/IAEC/SVU/ 2001/dt. 04.03.2002). The rats in the 3-month age group were considered as "younger" and the 18-month age group were considered as "older" or adult rats as per the life span of Wistar strain (Jang *et al.* 2001). The rats were housed in clean polypropylene cages, six rats per cage and maintained under hygienic conditions, in a temperature controlled room (27 ± 2°C) with a photoperiod of 12-h light and 12-h dark cycles. The rats were fed with standard laboratory chow (Hindustan Lever Ltd, Mumbai, India) and water ad libitum.

Reagents and Chemicals

All the chemicals used in the present study were Analar grade and obtained from the following scientific companies: Sigma (St. Louis, MO, USA), Fisher (Pittsburgh, PA, USA), Merck (Mumbai, India), Ranbaxy (New Delhi, India), and Qualigens (Mumbai, India).

Treatment to the Animals

Age-matched rats were divided into four groups of six in each, and treated as follows.

Group I- Normal Control (NC)

Six rats were put on a six-channel, motor-driven treadmill for 5 days/week for a period of 2 months and given 2 m/min exercise for 5 min; the rats also received normal saline (0.9%) orally via orogastric tube for equivalent handling.

Group II-Exercise Trained (Ex)

Six rats were given endurance exercise training on a six channel motor-driven treadmill for 5 days/week for a period of 2 months at the running speed of 23 m/min, 30 min at a constant gradient of 7.5%. All animals in both age groups completed the 2-month period of exercise training protocol. The running program was scheduled between 6.00 and 8.00 a.m. The treadmill was custom-built at University Scientific Instrumentation Center, Sri Venkateswara University, Tirupati, India.

Group III- Alcohol Drinking/Treated (At)

Six rats received 20% of alcohol (v/v) at the dose of 2.0 g/kg bodyweight/day via an orogastric tube for a period of 2 months. Before treatment, the rats were subjected to exercise training on a six-channel motor-driven treadmill as described in group I for equivalent handling.

Group IV - Exercise Plus Alcohol (Ex + At)

Six rats were exercised on a treadmill for 30 min as described in group II and the same rats were received 20% of alcohol as described in group III after exercise training.

Tissue Processing

The animals were sacrificed after 24hrs of the last exercise session by cervical dislocation. The brain was excised at 4°C and washed with ice cold saline. After the atria and blood vessels were trimmed, the brain was immediately immersed in liquid nitrogen and stored at -80ºC for biochemical analysis.

Biochemical Assays

Brain superoxide dismutase (SOD) activity was assayed in the tissue homogenates by the method of Misra and Fridovich (1972) at 480 nm for 4 min on a Hitachi U-2000 spectrophotometer. Activity was expressed as the amount of enzyme that inhibits the oxidation of epinephrine by 50%, which is equal to 1 U per milligram of protein. Catalase (CAT) activity was determined at room temperature by using the modified version of Aebi (1984) and absorbance of the sample was measured at 240 nm for 1 min in a UV-spectrophotometer. Activity of glutathione peroxidase (GPx) was determined by the method of Flohe and Gunzler (1984) in the presence of NADPH and absorbance was measured at 340 nm using cumene hydrogen peroxide. Brain glutathione reductase (GR) enzyme activity was determined according to the method of Carlberg and Mannervik (1985). The concentration of reduced GSH in brain homogenates was measured as described by Akerboom and Sies (1981). All of enzyme activities were expressed per mg protein and the tissue protein was estimated according to the method of Lowry *et al.* (1951) using bovine serum albumin (BSA) as a standard.

Statistical Analysis

The data has been analyzed by using SPSS (Version 13.5; SPSS Inc., Chicago, IL, USA) and M.S. Office, Excel Software for the significance of the main effects and along with their interactions. The data has been compared using two-way ANOVA and differences were considered significant at $p < .05$.

Results

In the present study, we observed a significant ($p < .001$) reduction in all antioxidant enzyme activities including SOD, CAT, GPx, GR and GSH content in the brain of rat with advancing of age (Figs. 1, 2, 3, 4 and 5). In addition, alcohol consumption further decreased all above mentioned antioxidant enzymes along with GSH content in the brain of old rats. A significant increase ($p < .001$) in the antioxidant status was observed with response to exercise training in both the age groups, compared to respective normal control rats. In this study, we found that decreased SOD, CAT, GPx, GR activities and GSH levels with alcohol consumption was significantly elevated by 2 months exercise training in young as well as in old rats. However, this exercise-induced elevation of SOD, CAT, GPx, GR activities and GSH content were more pronounced in the 18 months old rats when compared to 3 months old rats. Increased brain antioxidant status with combination treatment, indicates exercise training can protect the brain against alcohol-induced oxidative damages even at old age.

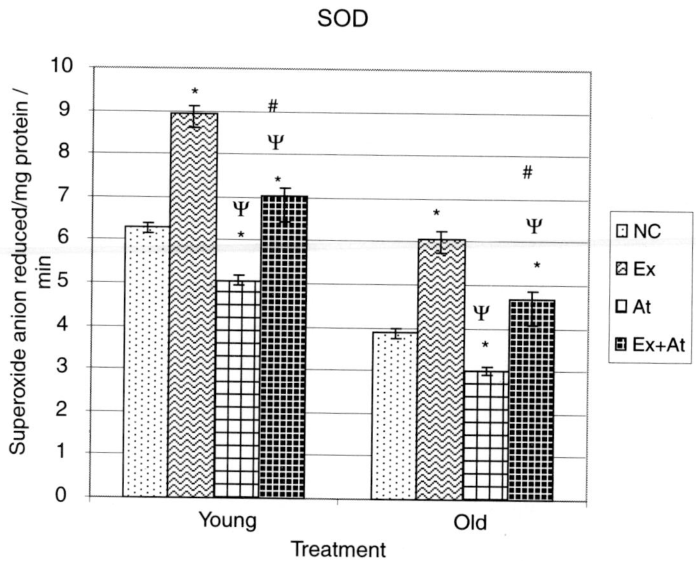

Fig. 1. Effect of exercise trained (Ex), alcohol drinking/treated (At), and combination of both (Ex + At) on SOD activity in the brain tissue of young and old male albino rats. The values are significant compared to the following: control (*P<.001), exercise (ΨP<.001), and alcohol (#P<.001).

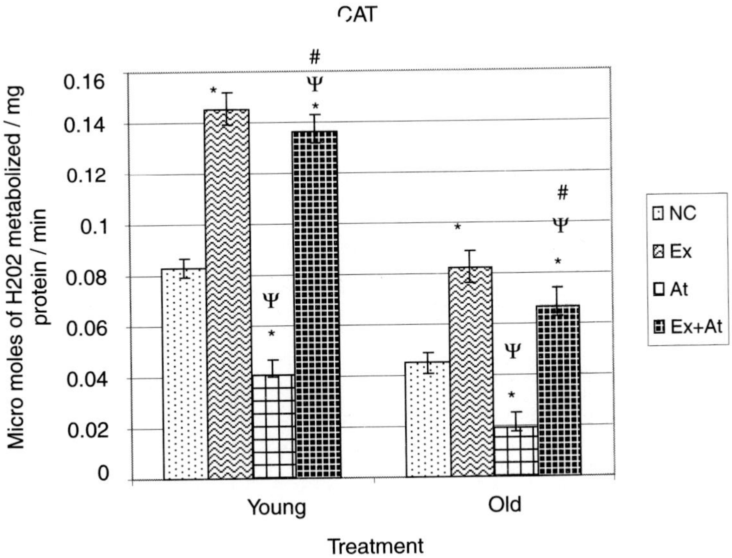

Fig. 2. **Effect of exercise trained (Ex), alcohol drinking/treated (At), and combination of both (Ex + At) on CAT activity in the brain tissue of young and old male albino rats. The values are significant compared to the following: control (*P<.001), exercise (ΨP<.001), and alcohol (#P<.001).**

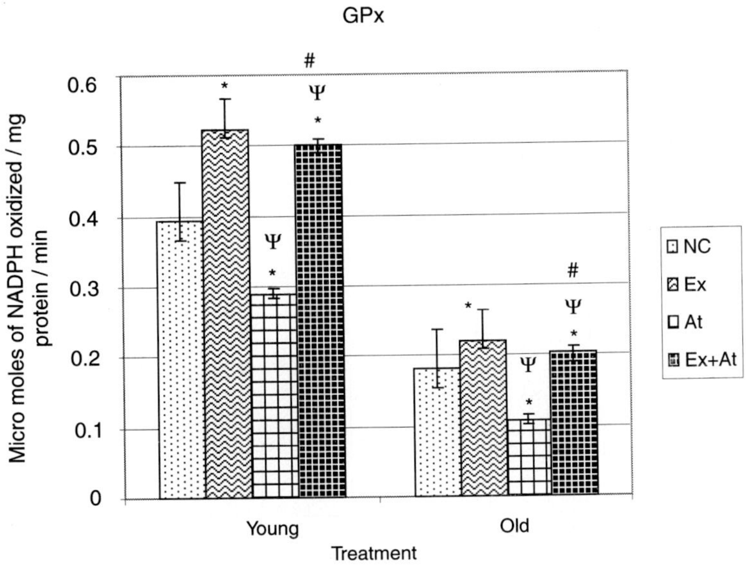

Fig. 3. **Effect of exercise trained (Ex), alcohol drinking/treated (At), and combination of both (Ex + At) on GPx activity in the brain tissue of young and old male albino rats. The values are significant compared to the following: control (*P<.001), exercise (ΨP<.001), and alcohol (#P<.001).**

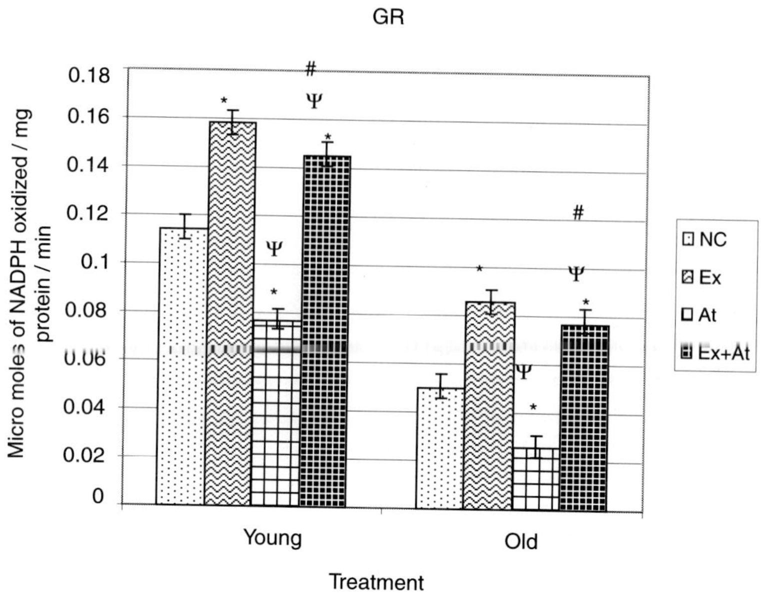

Fig. 4. **Effect of exercise trained (Ex), alcohol drinking/treated (At), and combination of both (Ex + At) on GR activity in the brain tissue of young and old male albino rats. The values are significant compared to the following: control (*P<.001), exercise (ΨP<.001), and alcohol (#P<.001).**

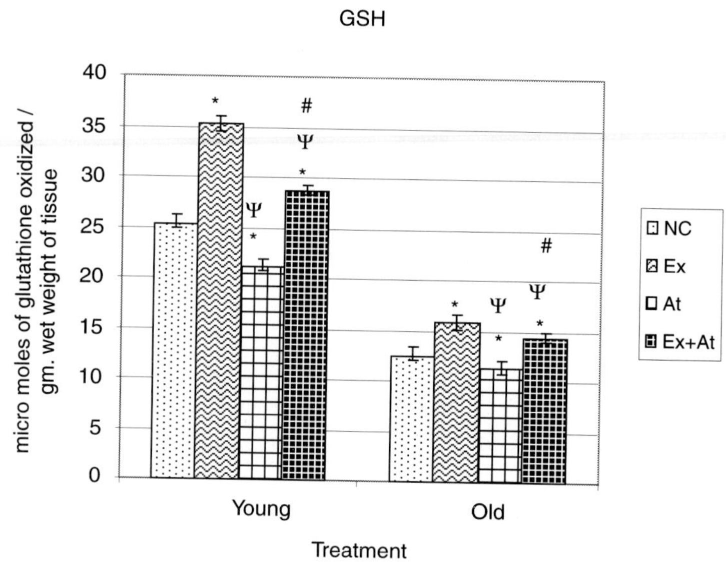

Fig. 5. **Effect of exercise trained (Ex), alcohol drinking/treated (At), and combination of both (Ex + At) on GSH level in the brain tissue of young and old male albino rats. The values are significant compared to the following: control (*P<.001), exercise (ΨP<.001), and alcohol (#P<.001).**

Discussion

In the previous investigations, it has been reported that alcohol-induced oxidative damage and decreased antioxidant status was reversed by exercise training only in young rats (Husain and Somani, 1997; Somani *et al.* 1996). In this study, we found that the alcohol-induced decrease in antioxidant status in aged rat brain was significantly reversed by exercise training.

The results of the present study clearly demonstrated that decrease in SOD activity in the brain of old rats was prevalent. Similar decrease was also reported by several investigators in experimental animals (Rao *et al.* 1990; Vohra *et al.* 2001). Similar to our findings, Mahendran and Shyamala Devi (2001) and Skrzydlewska *et al.* (2002) reported decrease in SOD activities with alcohol treatment in the liver, kidney and brain tissues. Somani *et al.* (1996) reported significant decrease in plasma and brain SOD activity with alcohol treatment. The over production of superoxide radicals due to alcohol intoxication implies the over utilization of SOD, and this may indicate its low activity under alcohol induced oxidative stress condition. However, we found significant increase in SOD activity with exercise training in alcohol treated rats in both age groups. Previously several studies reported that exercise has been shown to increase the SOD and CAT activities in various tissues including brain (Ji and Fu, 1992; Somani *et al.* 1997). Increased SOD activity with exercise training indicates effective elimination of superoxide radicals from the cell and protects the brain from possible oxidative damage.

Age related decrease in CAT activity was found in the brain of old rats. Decreased CAT activity, which was also noticed with alcohol treatment suggesting the increased tissue damage as a result of uncontrolled generation of partially reduced oxygen species (Mahendran and Shyamala devi, 2001), and this damage was further exacerbated with aging. The decreased CAT activity with alcohol treatment indicates inefficient scavenging of hydrogen peroxide due to oxidative inactivation of catalase (Pigeolet *et al.* 1990). However, in the present investigation, group IV rats that were received both exercise training plus alcohol treatment showed a significant elevation in CAT activity. The increased CAT activity with combination treatment was noticed in both age groups of rats, whereas, this increase was more in younger rats than older rats. Our findings suggest that exercise training may be helpful to develop a resistance to cope with alcohol-induced oxidative injuries in the brain of rats.

In this study an age-dependent decrease in GPx was observed in the brain of old rats. Vohra *et al.* (2001) reported a decrease in GPx activity in different brain regions of 32 months old guinea pigs. The age related decrease in brain GPx activity in the current study was also supported by earlier reports (Asha Devi and Ravi Kiran, 2004). Age related increase in free radicals especially hydrogen peroxide (H_2O_2) may be responsible for the low activity of brain GPx in older rats. The results obtained from the present study reveal that endurance exercise training enhanced the brain GPx activity in both age groups of rats. Many investigators reported that exercise training significantly elevated the activities of GPx and GR activities in different tissues (Atalay *et al.* 1997: McArdle and Jackson, 2000). Increased GPx activity indicates an efficient elimination of organic peroxides (Ohkuwa *et al.* 1997). The available reports suggest the fact that higher SOD activity may be responsible in part for higher GPx activity (Ray and Husain, 2002).

A significant decrease in specific activity of GR in brain tissue of rats during aging can be attributed to increased oxidation or decreased synthesis of GSH. The reduced availability of NADPH may also cause a decrease in GR activity. In conjunction, a decreased glucose-6-phosphate dehydrogenase (G-

6-PDH) activity associated with age results in the reduced synthesis of NADPH. Alcohol administration resulted in the lower GR activity in the brain of young and old rats. Das and Vasudevan (2005) in their studies found a significant decrease in GR activity in the hepatic tissue of rat. Somani *et al.* (1996) also reported the decreased GR activity with alcohol treatment in brain of rat. A significant decrease in GR activity indicates impaired reduction of GSSH to GSH due to depletion of reducing equivalent (NADPH) which is a co-substrate for GR activity.

With combination of exercise training plus alcohol treatment GR activity was found to increase in the brain of rats. This reveals that exercise may influence the GSH/GSSG levels at high concentration. Exercise training prevents the formation of protein adducts and increases temperature during exercise training and may rapidly metabolize the alcohol to less toxic acetate (El-Syed *et al.* 2005). Thus, exercise training plays a prominent role in preventing alcohol-induced oxidative stress by promoting the GR activity in the brain of old alcohol treated rats. Increased GPx and GR activities with combination treatment suggest that exercise training may help to develop a resistance in the brain to cope with alcohol induced oxidative injuries (Somani *et al.* 1996).

Glutathione serves as a sensitive marker of oxidative stress and it plays an important role in maintaining the integrity of the cell system (Das and Vasudevan, 2005). In the present study, the significant depletion of brain GSH content induced by alcohol was greater in aged rats. Generally GSH is synthesized from the amino acids by the particular enzymes in the liver. During aging process brain mitochondria have inadequate quantity of the enzymes in the γ-glutamyl cycle (Martensson et al. 1990) and it leads to low levels of GSH at old age. Increased acetaldehyde binding, higher efflux from liver, and inhibition of biosynthesis have been suggested to be the mechanism accounting for the GSH depletion by alcohol (Choi *et al.* 2000; Kim *et al.* 2003). Here, we demonstrated that the age-dependent deterioration against alcohol-induced ROS insult was reversed by exercise training. In our previous investigations, we reported the same trend of increased GSH content against alcohol in the liver of old rats (Mallikarjuna *et al.* 2009). Increase in GSH content after exercise may be due to the increase in the synthesis of precursors for GSH formation and increase in the γ-glutamyl-cystine-glycine enzyme (Somasekhara Reddy, 2008, Mallikarjuna *et al.* 2009). The data from the present study clearly indicates the role of GSH in detoxifying the free radical toxicity due to alcohol consumption thereby enhancing the GSH content with combined treatment.

From this study, we are providing the additional information that alcohol-induced decrease in antioxidant status was exacerbated in old rats. We assumed that the aged brain might be more susceptible to the alcohol-induced oxidative damages than younger. The major finding of the study was that prominent reduction in brain antioxidant status in old alcohol treated rats was significantly reversed by 2-month treadmill exercise training. Thus, exercise training could protect the brain from possible oxidative damages.

References

1. Aebi H. Catalase, Methods Enzymol 1984; 105:125-126.
2. Akerboom TPM and Sies H. Assay of glutathione, glutathione disulphide and glutathione mixed disulphide in biological samples. In: Jakoby WB, eds., Methods Enzymol. Academic Press, New York; 1981:Vol. 77,373-382.

3. Asha Devi S and Ravi Kiran. Regional responses in antioxidant system to exercise training and dietary vitamin E in aging rat brain, Neurobiol Aging, 2004; 25:501-508.
4. Atalay M, Marnila P, Lilius EM, Hanninnen O and Sen CK. Glutathione dependent modulation of exhausting exercise induced changes in neurtophil function of rats. Eur J Appl Phyiol Occup Physiol 1996; 74:342-347.
5. Carlberg I and Mannervik B. Glutathione reductase, Methods Enzymol 1985; 113:484-490.
6. Choi DW, Kim SY, Kim SK and Kim YC. Factors involved in hepatic glutathione depletion induced by acute ethanol administration. J Toxicol Environ Hlth Part A, 2000; 60:459-469.
7. Clemens DL and Jerrells TR. Ethanol consumption potentiates viral pancreatitis and may inhibit pancreas regeneration. Preliminary findings. Alcohol 2004; 33:183-189.
8. Das SK and Vasudevan DM. Effect of ethanol on liver antioxidant defense system: A dose dependent study. Ind J Clin Biochem 2005; 20(1):80-84.
9. El-Sayed MS, Ali N and El-Sayed AZ. Interaction between alcohol and exercise: Physiological and hematological implications. Sports Med, 2005; 35:257-269.
10. Flohe L and Gunzler WA. Glutathione peroxidase, Methods Enzymol 1984; 105:115-121.
11. Harman D. Free radical theory of aging: Role of free radicals in the origination and evolution of life, aging and disease process. In: Biology of Aging, Johnosn J, Walford JV, Harman R, Miquel J eds. New York: Liss; 1986.
12. Harman D. The ageing process. Proc Natl Acad Sci USA 1981; 78:7124-7128.
13. Hazelton GA and Cang CA. Glutathione content of tissue in the aging mouse. Biochem J 1980; 88:2f.
14. Husain K and Somani SM. Interaction of exercise and ethanol on hepatic and plasma antioxidant system in rat. Pathophysiology 1997; 4:69-74.
15. Jang I, Chae K and Cho J. Effects of age and strain on small intestinal and hepatic antioxidant defense enzymes in Wister and Fisher 344 rats. Mech Ageing Dev 2001;122(6): 561-570.
16. Ji LL and Fu R. Response of glutathione system and antioxidants enzymes to exhaustive exercise and hydro peroxides. J Appl Physiol 1992; 72:549-554.
17. Ji LL. Oxidative stress during exercise: implication of antioxidant nutrients. Free Rad Biol Med 1995; 18:1079-1086.
18. Kim YC, Kim YS and Sohn YR. Effect of age increase on metabolism and toxicity of ethanol in female rats. Life Sci 2003; 74:509-519.
19. Loguercio C, Piscopo P, Guerriero C, De Girolamo V, Diaslvo D and Del Vecchio Blanco C. Effect of alcohol abuse and glutathione administration on the circulating levels of glutathione and on antipyrine metabolism in patient with alcoholic liver cirrhosis. Scand J Clin Lab Invest 1996; 56:441-447.
20. Lowry OH, Rosebrough NC and Farr AL, Randall RJ. Protein measurements with folinphenol reagent, J Biol Chem 1951; 193:265-275.
21. Mahendran P and Shyamala Devi CS. The modulating effect of Garcinia Cambogia extract on ethanol induced peroxidative damage in rats. Ind J Pharmacol 2001; 33:87-91.

22. Mallikarjuna K, Nishanth K and Sathyavelu Reddy K. Hepatic glutathione mediated antioxidant system in ethanol treated rats: decline with age. Pathophysiology, 2007; 14:17-21.
23. Mallikarjuna K, Nishanth K, Hou CW, Kuo CH and Sathyavelu Reddy K. Effect of exercise training on ethanol-induced oxidative damage in aged rats. Alcohol, 2009; 43:59-64.
24. Martensson J, Lai JCK, and Meister A. High affinity transport of glutathione is part of multicomponent system essential for mitochondrial function. Proc Natl Acad Sci USA, 1990; 87:7185-7189.
25. McArdle A and Jackson MJ. Exercise oxidative stress and aging. J Anat 2000; 197:539-541.
26. Meister A and Anderson ME. Glutathione. Ann Rev Biochem1991; 52:711-722.
27. Misra HP and Fridovich I. The role of superoxide anion in the auto-oxidation of epinephrine and a simple assay for superoxide dismutase, J Biol Chem 1972; 247:3170-3175.
28. Navarro A, Gomez C, Lopez-Cepero JM and Boveris A. Beneficial effect of moderate exercise on mice aging: survival, behavior, oxidative stress and mitochondrial electron transfer. Am. J Physiol Regul Intergr Comp Physiol 2004; 286:505-511.
29. Oba T, Maeno Y and Ishida K. Differential contribution of clinical amounts of acetaldehyde to skeletal and cardiac muscle dysfunction in alcoholic myopathy. Curr Pharm Des 2005; 11:791-800.
30. Ohkuwa T, Sato Y and Naoi M. Glutathione status and reactive oxygen generation in tissues of young and old exercised rats. Acta Physiol Scand 1997; 159:237-244.
31. Pigeolet E, Corbisier P, Houbion A, Lambert D, Michiels C, Raes M, Zachary MD, and Ramacle J. Glutathione peroxidase, superoxide dismutase and catalase inactivation by peroxides and oxygen derived free radicals. Mech Aging Dev 1990; 51:283-297.
32. Pushpalatha K, Sathyavelu Reddy K and Leeuwenburgh Christiana. Exercise training with ageing protects against ethanol induced myocardial glutathione homeostasis. Free Rad Res, 2008; 42(5):428-434.
33. Rao G, Xia E, Nadakavukaren MJ and Richardson A. Effect of dietary restriction on the age-dependent changes in the expression of antioxidant enzymes in rat liver. J Nutr. 1990; 120: 602-609.
34. Ray G and Husain SA. Oxidants, antioxidants and carcinogenesis. Ind J Exp Biol 2002;40: 1213-1232.
35. Skrzydlewska E, Ostrowska J, Stankiewicz A and Farbiszewski R. Green tea as a potent antioxidant in alcoholic intoxication. Add Biol 2002; 7(3):307-314.
36. Somani SM, Husain K and Schlorff EC. Response of antioxidant system to physical and chemical stress. In: Baskin SI and Salem H. eds., Oxidants, Antioxidants and Free Radicals. Washington: Taylor and Francis Publications. 1997; 125-143.
37. Somani SM, Husain K, Dalz Philips L, Lanzotti DJ, Karets KR, and Trammell GL. Interaction of exercise and ethanol on antioxidant enzymes in brain regions of the rat. Alcohol 1996; 13(6):603-610.
38. Somasekhara Reddy N. Protective effect of exercise on brain antioxidant system under alcohol treatment in two different age groups of male albino rats. Ph.D. thesis, Sri Venkateswara University, Tirupati, India. 2008.

39. Vina J, Servera E, Asensi M, Sastre J, Pallardo FV, Ferrero JA, Garcia-De-La-Asuncion J, Anton V and Marin J. Exercise causes blood glutathione oxidation in chronic obstructive pulmonary disease: prevention by O_2 therapy. J Appl Physiol 1996; 81(5):2198-202.
40. Vohra BPS, Sharma SP and Kansal VK. Age dependent variations in mitochondrial and cytosolic antioxidant enzymes and lipid peroxidation in different regions of central nervous system of guinea pigs. Ind J Biochem Biophy 2001; 38: 321-326.
41. Zs-Nagy I. The theoretical background and cellular autoregulation of biological waste product formation. In: Lipofuscin-1987: State of the Art. Zs-Nagy I. eds., Akademiai Kiado, Budapest and Elsevier Science Publishers, Amsterdam, 1988; p23-50.

Adaptation Biology and Medicine (Volume 6: Cell Adaptations and Challenges)
Editors: P. Wang, C.-H. Kuo, N. Takeda and P.K. Singal
Copyright © 2011 Narosa Publishing House Pvt. Ltd., New Delhi, India

25. Assessment of Physiological Biomarkers Involved with Acute Fatigue/Physical Exhaustion

D Fogt[1*], M Contreras[1], W Cooke[2], B Falcon[1], G Gutierrez[1], T Jones[1], J Kalns[3]

[1]Exercise Biochemistry & Metabolism Laboratory, The University of Texas at San Antonio, Texas
[2]Laboratory for Applied Autonomic Neurophysiology, The University of Texas at San Antonio
[3]Hyperion Biotechnology Inc., Brooks City Base, San Antonio, Texas

Summary

Overwhelming Fatigue/physical/exaustion (FPE) induced by sleep-deprivation, dehydration, and exercise has been extensively studied. We have created a safe, but effective protocol to fatigue subjects within 24 h. The decrease we observed in cognitive performance occurred at the same time during the protocol that subjective fatigue dramatically increased. Cognitive performance, like fatigue, remained low throughout the second day and night time periods. Using subjective POMS surveys we demonstrated that at 24 h of sleep deprivation results in substantially higher negative mood states regardless of hydration status (HS) or energy deficit induced by physical activity (PA). Negative mood state remained stable albeit at higher levels compared to sleep controls from 24 to 48 h. The POMS "fatigue" index was mirrored by the additional questionnaire's "how fatigued do you feel" answer scores for each group. Our additional subjective indices of FPE also rose dramatically during the 12-18 h period and remained above initial levels for the rest of the protocol period. Additional stress of exercise and dehydration does not significantly affect these scores at 48 h. However, the exercised groups operated at a much higher state of FPE throughout the second day. Fatigue clearly increases over the course of the first day with a pronounced rise during the first night in all groups. This is not surprising in the sleep control group as their sleep was interrupted every 3 h for data collection. There was no effect of dehydration on these redundant indices over the course of the 48 h. Mental capacities of task performance and cognitive processing are compromised after only 24 h of sleep deprivation. Exercise and dehydration did not appear to affect the decrease in mental capacity in sleep-deprived subjects. Finally, our study demonstrates that a single overnight period of sleep deprivation results in a dramatic fall in cognitive performance. Thus, military commanders should be vigilant to maintaining regular sleep intervals when at all possible to avoid military action-related accidents, injury, and loss of life.

Keywords: Mental capabilities; cognitive processing; acute fatigue; physical activity

*Address for Correspondence: Corresponding author: Donovan L. Fogt, PhD Department of Health and Kinesiology The University of Texas at San Antonio San Antonio, Texas, 78249 U.S.A. Email: Donovan.Fogt@utsa.edu

Introduction

When the body's ability to respond to a stress challenge is overwhelmed, physiologic reserve deteriorates and fatigue/physical exhaustion (FPE) ensues, reflecting a failure of the regulatory pathways related to a given stress response. Intensive military training accompanied by inconsistent and inadequate nutrition and sleep deprivation can last several weeks. Such training is conducive to the onset of chronic FPE and neuroendocrine changes that are expressed by the diminished stress adaptability (Aakwaag A *et al.* 1978; Guézennec CY *et al.* 1994). Physiological fatigue may also indirectly increase perceived exertion and unwillingness to perform physical work at a desired performance level. For instance, an altered psychic or perceptual state such as inexperience, fear, performance standing (e.g., winning vs. losing), ego, motivation, embarrassment, etc. may modulate performance independent of physiological signals. Overt FPE represents a failure of appropriate physiologic regulatory mechanisms to chronic stress. Detection of these failures affords opportunity to survey and quantify complex communication pathways linking organ systems, cells, and genes.

Several early sleep deprivation studies provided insight to fatigue protocols, the circadian rhythms of stress hormones, and other indices of fatigue. The circadian rhythms affecting physiological arousal and reflecting stress or fatigue are much more pronounced during sleep deprivation compared to normal sleep cycles (Alluisi & Chiles, 1967). Alluisi and Chiles (1967) observed that 48 h sleep deprivation elicited circadian rhythms that had not been apparent before the sleepless vigil. In a study of four days of sleep deprivation, Fiorica *et al.* (1968) reported a "flattened" circadian curve for urinary catecholamine (i.e., epinephrine and norepinephrine) excretion. A 72 h sleep deprivation protocol with 3 h data collection intervals was used by Fröberg *et al.* (1975) to evaluate catecholamine excretion, shooting performance and self-ratings of fatigue in Swedish soldiers. Shooting performance was maintained during the first sleep-deprived night with normal declines in catecholamine excretion and subjective arousal. Recently, Jouanin JC *et al.* (2004) monitored 23 military cadets prior to and following a one month ranger training course culminating in a 5 day field exercise simulation with physical and psychological stress. They observed increased parasympathetic tone on heart rate modulation in conjunction with decreased plasma testosterone both suggesting failure of normal stress responses during prolonged FPE. Similar results have been reported following intensive training in athletes (Opstad PK, 1992) and following 3 weeks of a military training program (Gomez-Merino D *et al.* 2003).

Identification of FPE

Chronic FPE in a warfighter could compromise training and mission effectiveness, and therefore must be readily identifiable.

It is not currently possible to measure and monitor neuroendocrine changes in the field and a need exists to develop simple, non-invasive techniques to assist in the identification of warfighters with FPE. Current work to develop a Warfighter Physiological Status Monitoring system (WPSM) includes assessment of hydration status, thermal load, cognitive function, and ballistic impact detection. Simple sensors to detect respiratory rate, body temperature, and heart rate will likely be incorporated into some future algorithm to provide insight into combat readiness (Hoyt RW *et al.* 2002; Savell CT *et al.* 2004), but exactly how these sensors should be incorporated has not been worked out in detail. Furthermore, it is well known that blood pressure and respiratory rate provide little insight into injury severity or combat readiness on the battlefield (Brasel KJ *et al.* 2007; Orlinsky M *et al.* 2001).

Potential Use of Heart Rate Variability for Identification of FPE

The autonomic nervous system actively compensates for injury or fatigue by modulating the balance between parasympathetic and sympathetic cardiovascular control mechanisms. Understanding such autonomic compensations can provide earlier, and perhaps more sensitive feedback on physiological status. Although autonomic function cannot be measured directly for field applications, it can be estimated through analysis of heart rate variability (HRV) and applied clinically (Gang Y & Malik, 2002; Task Force 1996). For example, frequency-domain analysis of HRV via spectral analysis has been used to classify injury severity in ICU patients (Winchell & Hoyt 1996; Winchell & Hoyt, 1997). In the pre-hospital arena, frequency-domain analysis of HRV separated patients who would eventually die from those who would survive traumatic injury (Cooke WH et al, 2006a; Cooke WH *et al.* 2006b). However, it is not possible at this time to run real-time spectral analyses for practical use in the field.

A relatively new method for assessing autonomic nervous system control of heart rate - heart rate volatility - has been developed recently as a practical, field-based monitoring strategy. The heart rate volatility measure is based on the standard deviation in integer 5 second heart rate recordings computed over 5 minute intervals (HR_{SD5}). This method is highly correlated with actual HRV determination via multiple spectra and time-domain measures known to reflect ANS function (Morris JA Jr *et al.* 2004). HR_{SD5} has numerous advantages over traditional spectral waveform analyses. For example, the HR_{SD5} integer data require significantly less computation and can be roughly computed in real time, while spectral analysis typically requires extensive offline computation. Together, these findings demonstrated that measurement of cardiac interbeat interval characteristics is suitable for continuous real-time field assessment. Small, highly portable, and relatively inexpensive heart rate monitoring devices (e.g., 2-lead ECG) such as those used by athletes during training and competition can be employed to collect the heart rate data for subsequent download to a computer via telemetry. Radespiel-Tröger *et al.* (2003) validated HRV field data from a relatively inexpensive exercise heart rate monitor (Polar Electro, Kempele, Finland) against clinically-established reference HRV data.

Interbeat interval detectors such as those developed by Polar Electro will likely be incorporated into some future WPSM system, but exactly how these data should be interpreted remains unknown. My colleagues and I have embarked on a line of investigation to identify novel FPE-responsive biomarker relationships for development of strategies to optimize physical performance. Our studies will add greatly to our knowledge of how various known biomarkers are affected by interventions during various physical challenges in diverse military-relevant and civilian populations. We hypothesize that our data will allow development of a novel mathematical algorithm for individual soldier's physiological stress reactivity so that medical staff and commanders can objectively determine the physical capability of their troops in near-real time. It could impact development of new training regimes, dietary supplementation and equipment selection. Such an algorithm-derived physiological "score" would also allow smaller, low cost studies to evaluate the overall physical capability of military personnel with reasonable confidence. In addition, an easy, non-invasive point of care test based on the "score" could impact general wellness, competitive athletics, nutritional supplements and healthcare industries.

Progression of Preliminary Studies to Current FPE Proposal

In our preliminary study, asymptomatic, military recruit-aged adults were recruited as subjects in a series of experiments to validate heart rate HR_{SD5} as a tool for monitoring military training and recovery to avoid excessive fatigue that could impair the performance of soldiers. Thus far, heart rate parameters measured with the field-based Polar 810 monitor have corresponded extremely well with concurrent laboratory-based ECG recordings. We generated a baseline database of this relationship during supine and 70° tilt, representing an orthostatic challenge to which the autonomic nervous system and cardiovascular system must respond. In addition, we can qualify our findings by age, gender, menstrual phase, body weight and composition, psychological state, stress hormone levels, hydration status and fitness level.

Using our baseline database, we tested the hypothesis that resting heart rate variability (HRV), including both time- and frequency-domain assessments, would correlate with hydration status and aerobic capacity in military recruit-age men and women with varying fitness levels (Fogt *et al.* 2009). Cardiac interbeat intervals were recorded using a heart R-R monitor during 20 minutes of quiet, supine rest with paced breathing (0.25 Hz). HRV metrics included: average R-R interval (RRI_{avg}); R-R interval standard deviation (RRISD); the percentage of adjacent R-R intervals varying by ≥ 50 ms (pNN50); and integrated areas of R-R interval spectral power at the high (0.15 – 0.4 Hz; RRIHF) and low (0.04 – 0.15 Hz; RRILF) frequencies. Treadmill maximal oxygen uptake (VO_2 peak), segmental bioimpedance estimates of total body water (TBW) and urine specific gravity (USG) were also assessed.

The results of the initial study are encouraging with respect to development of a novel, field-based tool for real-time monitoring of human ANS function during any number of perturbations, be it exercise, acute stress or FPE. Our subjects demonstrated above average age group predictions for VO_2 peak, while subjects' TBW (≥ 60% and ≥ 50% body weight for males and females, respectively) and mean USG reflected adequate hydration (≤1.020). Despite the normality of these parameters, the absolute ranges of individual values were considerable. This individual variation allowed us to demonstrate correlations between HRV and TBW and VO_2 peak (Figure 1).

Results from our initial study demonstrate how fitness level and hydration status may affect cardiac function via changes in vagal activity. The ANS is fundamentally responsible for defending arterial pressure through modulation of heart rate and peripheral vascular resistance. Reductions of central blood volume, as occurs with dehydration associated with FPE, trigger vagal withdrawal and sympathetic activation demonstrated by lower HRV. Conversely, aerobic capacity is related directly to increased central blood volume (Rowell LB, 1986). Individuals with high aerobic capacities tend to have greater plasma volume and ventricular compliance allowing the hypothalamic/medullary cardiovascular control centers to maintain resting cardiac output with a greater stroke volume but lower heart rate (i.e., increased R-R interval) and greater vagal activity at rest (Rowell LB, 1986; Dixon EM *et al.* 1992). Our results are consistent with these predictions in that RRI_{avg} correlated directly with VO_2 max and TBW, and inversely (but weakly) with USG (Figure 1). Because changes of RRI are related directly to changes of vagal-cardiac control (Buchheit & Gindre, 2006; Katona & Jih, 1975), our data suggest that high aerobic capacities and TBW are associated with high relative parasympathetic tone at rest, and can be identified with simple interbeat interval analysis as opposed to HRV metrics. It is important to note that our subjects for this initial study were well-rested, fed,

Aerobic capacity (VO2peak), total body water (TBW), and urine specific gravity (USG) plotted in relation to resting supine R-Rintervals (RRI; n = 104 (Fog DL et al., 2009).

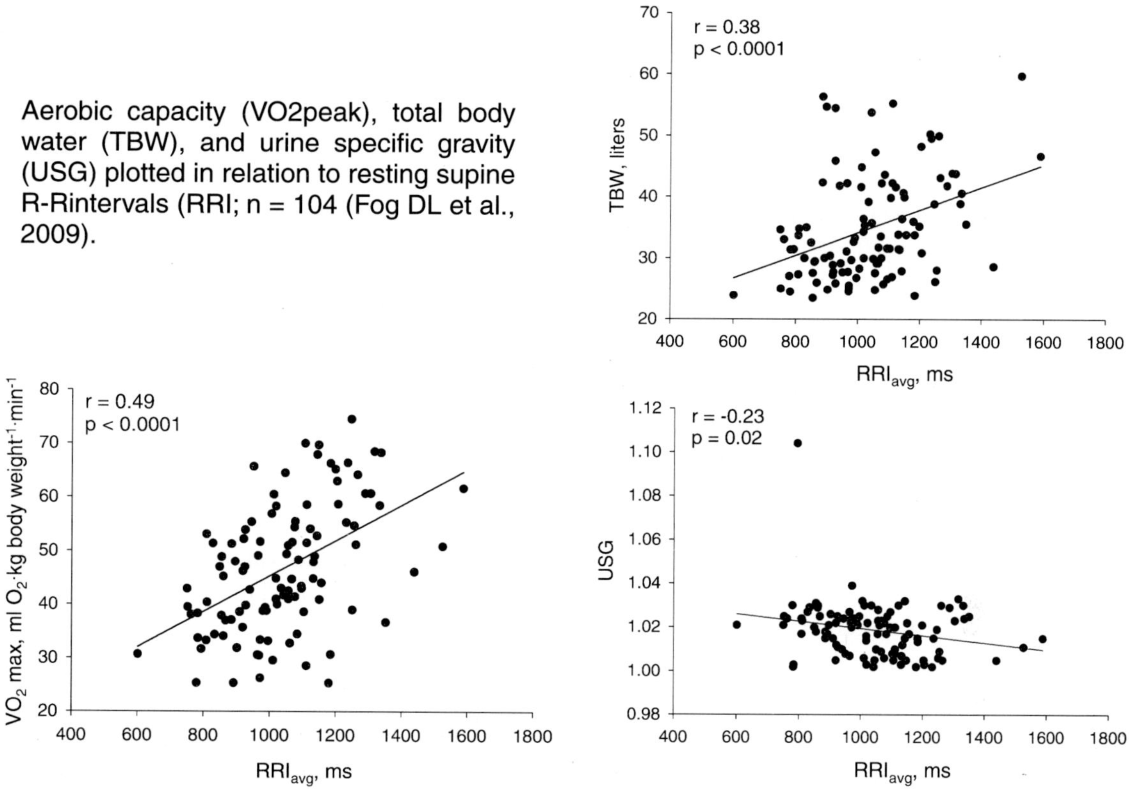

Fig. 1. Correlations for resting HRV, VO_2 peak, TBW, and USG.

hydrated and by no means fatigued. However, individual variation observed in the non-fatigued, resting state highlights the need to establish baseline relationships between physiological parameters prior to deployment of a field-based HRV monitor.

Complex frequency-domain analyses of HRV failed to provide insight into associations among hydration status, aerobic capacity, and ANS control. This is likely due to statistical limitations related extreme frequency-domain variability in our subjects (data not shown).

As previously mentioned, time-domain HR_{SD5} integer data requires significantly less computation and can be roughly computed in real time, while spectral analysis typically requires extensive offline computation to derive frequency-domain indices. Therefore, the lack of statistical significance in our frequency-domain indices and strength of basic cardiac interbeat (time-domain) data ultimately play in our favor. We are most interested in simple, pre-existing and deployable field-based technologies to eventually employ for a putative physiological algorithm. Interestingly, Cooke *et al.* (2008) have shown recently that standard time-domain measures of HRV (which reflect overall ANS cardiac influences) track reductions of blood volume as induced with lower body negative pressure as well as Fourier-derived variables. Grogon *et al.* (2005) have also concluded that simple time-domain statistics are useful for monitoring patient status, and therefore it is reasonable to propose the use of interbeat interval detectors for fast, convenient assessment of ANS function in the field without the necessity of including complicated Fourier-based algorithms in future monitoring devices.

The results of the initial study highlight the impact of individual differences in aerobic fitness and hydration status on resting HRV. Identification of altered ANS activity using field-based HRV will likely prove effective in preventing and diagnosing FPE during accelerated training or missions involving sleep deprivation, inadequate food and fluid intake and physical or mental challenges. Considering for such individual differences would be imperative when developing a field-based HRV monitor of combat readiness. It is important to note that correlations derived in the current study resulted from using a relatively homogenous group with respect to age, fitness level and hydration status. The application of the relationship between HRV and individual characteristics while controlling for the confounding effects of acute and chronic physical activity, dehydration, stress and fatigue warrant further study. However, the benefits of an inexpensive and non-invasive monitor of ANS function in military, athletic and civilian (first responders, airline pilots, etc.) applications are lucid.

Current FPE protocol development.

Our initial results were encouraging with respect to further development of a novel, field-based tool for real-time monitoring of human autonomic function during any number of perturbations, be it exercise, acute stress, or FPE. However, in order to establish this tool we need to be able to account for individual differences. The preliminary correlations between HRV and individual characteristics such as TBW and VO_2 peak will help guide subsequent examination of HRV-based measures of individuals' stress and fatigue management. Furthermore, these results provide direction as to how fitness level and hydration status may affect cardiac function via changes in autonomic tone.

We want to be able to determine the best way to predict and prevent a level of FPE that would negatively affect soldier performance. In order to do so, we need to identify the progression towards a FPE threshold where performance begins to decline. Knowing that both the rate of progression and threshold will likely vary across individuals we needed a FPE-inducing protocol which would allow assessment of individual FPE reactivity. Thus, we have designed a study designed to determine the effects of sleep deprivation with or without exercise and dehydration on several relatively basic human performance indicators. Such a protocol allows us to gradually and safely induce mental and physical fatigue while collecting data throughout the 48 h period. We assessed subject timed performances of 1) an eye-hand coordination target test; 2) a mathematical division problem task; and 3) the word, color, and word-color components of the Stroop color conflict cognitive exercise. Performance of these tests was evaluated every 3 hours over a 48 hour period. Based on our recent exploratory study, we believe that various applications of cardiac interbeat intervals, especially time-domain assessments of HRV, can be correlated with markers of fatigue, dehydration, as well as physical and mental performance using this novel protocol.

Subjects, screening and selection criteria: We recruited males of 18-35 years (military recruit age) for inclusion in the current study. Potential subjects included healthy (asymptomatic) college students, University of Texas at San Antonio ROTC cadet trainees, and recreational (non-varsity) athletes self-reporting as healthy and fit enough to enter military service, be free of disease, non-smokers, and not taking any medications or dietary supplements that would alter sympathetic or metabolic function.

Subjects meeting our eligibility criteria were asked to read and sign an informed consent document approved by the Committee for the Protection of Human Subjects in Research of the University of Texas at San Antonio. Consented subjects completed a basic medical history screening form, a standard physical examination, and a clinical graded treadmill stress test with electrocardiogram monitoring to indentify potentially dangerous or limiting pre-existing heart conditions that might compromise a subject's safe participation. Final selection of subjects was based on normal clinical results and history with written approval from an active-military physician indicating eligibility for safe participation in the study. All protocols were conducted in the Exercise Biochemistry and Metabolism Laboratory of the Department of Health and Kinesiology at The University of Texas at San Antonio.

Experimental design: All eligible subjects were randomly assigned to 1 of the 4 groups: 1) Group 1, diurnal sleep control; 2) Group 2, no sleep control; 3) Group 3, fatigue-hydrated; 4) Group 4, fatigue-dehydrated. All subjects participated in a single experimental group and each of the 4 experimental groups consisted of 14-16 subjects (Figure 2).

Baseline testing: Baseline testing was conducted 1 week prior to the start of the 48 h control or FPE protocol to allow any testing-related muscle soreness or other acute physiological response to subside by the start of the 48 h control or FPE protocol.

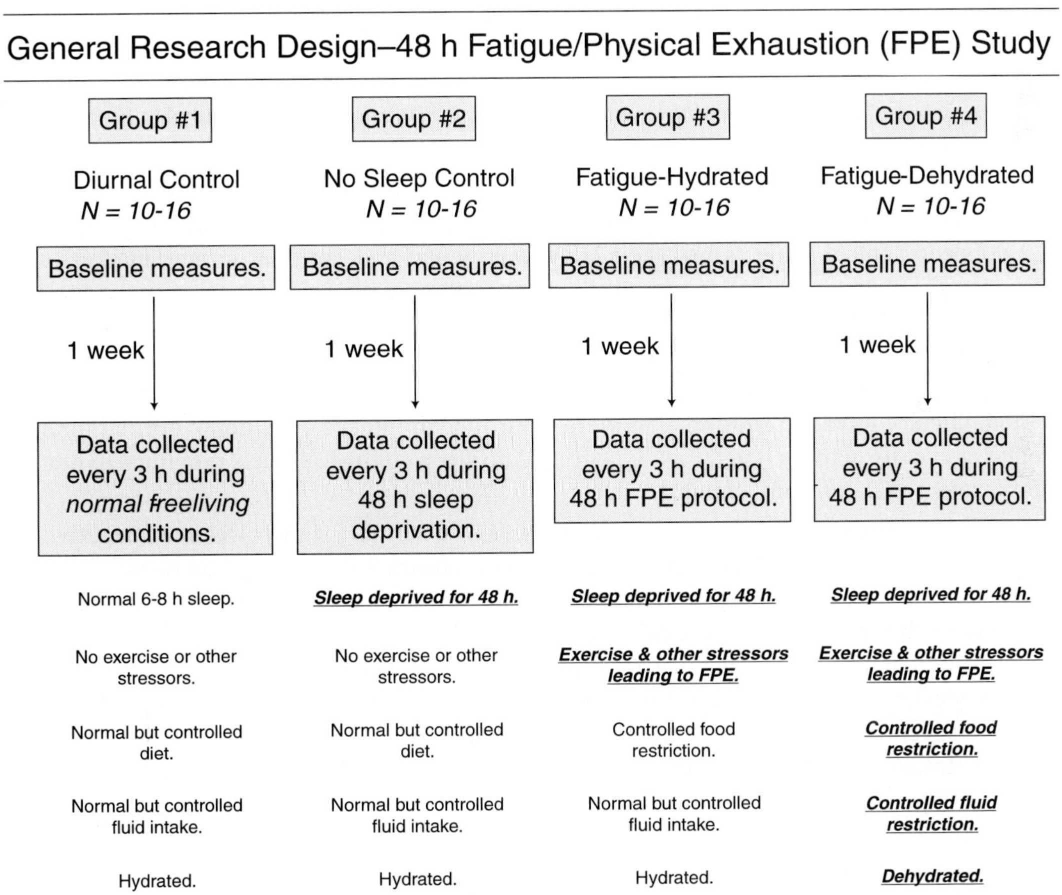

Fig. 2. General research design of current 48 h FPE protocol.

Forty-eight hour control or FPE protocol: The aim of this protocol was to gradually and safely induce mental and physical exhaustion over the course of 48 h. The 48 h sleep deprivation period has been previously shown as effective in inducing mental and physical fatigue (Åkerstedt and Fröberg, 1979; Fröberg, JE *et al.* 1975). Therefore, we employed a modified version of the sleep deprivation protocol used by Fröberg *et al.* (1975). These investigators evaluated catecholamine excretion, shooting performance and self-ratings of fatigue every 3 h during 72 h of sleep deprivation in Swedish soldiers.

The diurnal sleep control group (Group 1) was employed to permit elucidation of the affects of sleep deprivation (used in Groups 2-4) and normal sleep patterns as well as identify normal 24 h day-light cycle (i.e., circadian or diurnal) fluctuations in autonomic function and mental/physiological markers of fatigue as no single measure or quantification of FPE would be accurate without first taking into account the normal, daily fluctuations in human physiological status. Group 1 had their HRV and mental and physical fatigue markers assessed every 3 h over the course of the 48 h FPE protocol but were allowed 6 h of sleep with no daily stressors or exercise. Group 1 subjects were allowed to sleep in between each 3 h data collection point during the hours of 2300-0700 h. Group 1 subjects maintained a controlled but normal daily schedule of very light activities (e.g., studying or reading for enjoyment), eating/drinking, rest, and sleep except for the data collection periods every 3 h during the 48 h period. Dietary food and fluid intakes were controlled at intakes considered normal for the subject's age, weight, and daily activity level. Finally, Group 1 diurnal sleep control subjects remained hydrated (i.e., total body water ~60% of body weight).

The no sleep control group (Group 2) permitted quantification of the effects of sleep deprivation alone during the 48 h FPE protocol on autonomic function and mental/physiological markers of fatigue. Group 2 no sleep control subjects had their HRV and mental and physical fatigue markers assessed every 3 h over the course of the 48 h with no sleep but with no daily stressors or exercise as described below in the 48 h FPE procedure section. As with Group 1, the Group 2 subjects maintained a controlled but normal daily schedule of very light activities (e.g., studying or reading for enjoyment), eating/drinking, and non-sleep rest periods except for the data collection periods every 3 h during the 48 h period. Dietary food and fluid intakes were controlled at intakes considered normal for subject's age, weight, and daily activity level. Finally, Group 2 no sleep control subjects remained hydrated (i.e., total body water ~60% of body weight).

The fatigue-hydrated group (Group 3) was utilized to account for the effects of negative energy balance during the 48 h sleep deprivation FPE protocol. Group 3 fatigue-hydrated subjects had their HRV and mental and physical fatigue markers assessed every 3 h over the course of the 48 h with no sleep but were subjected to bouts of moderate intensity exercise every 3 h throughout the 48 h FPE protocol. Group 3 dietary food intakes were restricted to 3000 kcal/d or ~65% of normal as based on our pilot study estimations of subjects' daily energy expenditure based on the daily activities proposed and based on several military training studies (described in detail below). Subject fluid intakes were controlled at intakes considered normal for subject's age, weight, and daily total activity level. Therefore, Group 3 fatigue-hydrated subjects were kept hydrated (i.e., total body water = ~60% of body weight).

The fatigue-dehydrated group (Group 4) was implemented to determine the effects of a simulated military mission or task conducted over a progressive and controlled physically- and mentally-stressing, sleepless, food- and water-limited 48 h period. Group 4 fatigue-hydrated subjects had their

HRV and mental and physical fatigue markers assessed every 3 h over the course of the 48 h with no sleep. As with Group 3 participants, Group 4 subjects were subjected to bouts of moderate intensity exercise every 3 h throughout the 48 h FPE protocol. Group 4 dietary food intakes were restricted to 3000 kcal/d (~65% of normal) as based on our pilot study estimations of subjects' daily energy expenditure based on the daily activities proposed as well as several military training studies (described below). Finally, Group 4 fatigue-dehydrated subjects were fluid-restricted in order to gradually dehydrate subjects by approximately 3-4 kg (i.e., 6% deficit in total body water) by the end of the 48 h period.

The 48 h control or FPE protocol took place ≥1 week following the baseline testing. Subjects reported to the lab at 0500 h following an 8 h fast (including caffeine or other stimulants) the first morning of the 48 h control or FPE protocol (Figure 3). Subjects were then provided a light breakfast and evaluated for general health and ability to start with the 48 h control or FPE protocol. From this point, subjects were fed and provided water throughout the 48 h on a strict schedule detailed below. Three subjects in each group were outfitted with activity meters (Sensewear; BodyMedia Inc., Pittsburgh, PA) are worn on the upper arm and transmit information on caloric energy expenditure (approximately 3000 kcal for non-active and 5000 kcal/d for extremely active young adult males) in order to track energy expenditure/level of protocol activities in real time during the study as well as standardize FPE protocol expenditures between groups and compare expenditures between control and FRE groups.

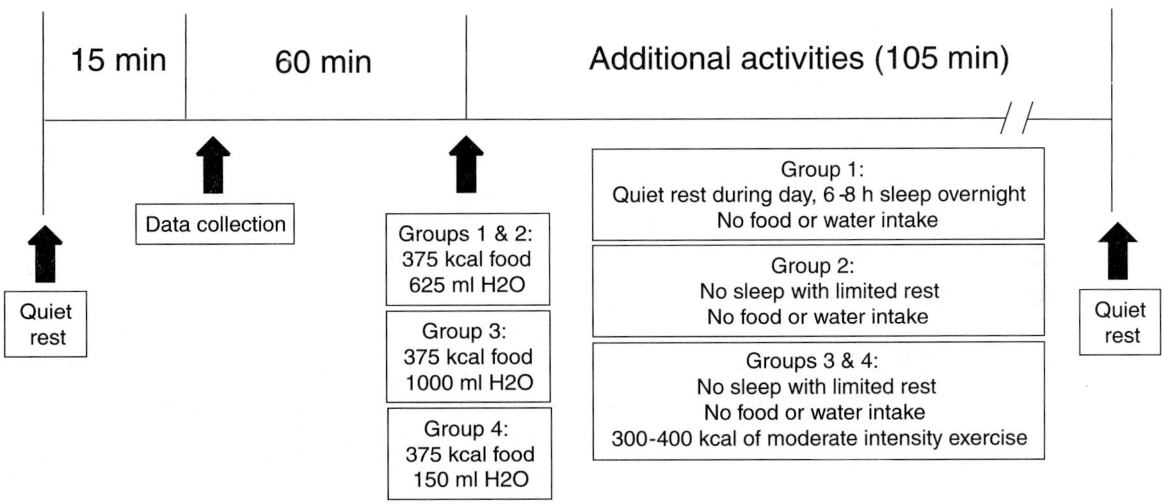

Fig. 3. Data collection schedule for current 48 h FPE protocol.

Three hour data collections: Every 3 h, subjects sat quietly for 15 min prior to having their vital signs (heart rate, breathing rate, tympanic temperature and blood pressure) measured and compared to baseline values as part of our laboratory Safety monitoring plan (described below). Additionally, at each 3 h time point, a number of tests were performed (described in the Specific testing protocol details section) according to the data collection schedule outlined below. At the conclusion of each 3 h data collection period, subjects were fed and provided water according to that scheduled for their experimental group (i.e., Groups #1-4).

After each 3 h testing period, the sleep control Group 1 and Group 2 ate a small sandwich with 0.5 cup serving of raw vegetables and 3 small cookies for a total of 8 feedings per day. Sensewear activity meter estimation of caloric expenditure for these groups was determined to be ~3000 kcal/d. Correspondingly, 3 h feedings consisted of 375 kcal (3000 total kcal/d) with 50% carbohydrate (~47 g), 30% fat (~13 g), and 20% protein (~19 g) with 625 ml of water (2.6 cups or 21 oz) for a total of 5.0 L/d for Groups 1 and 2 (euhydrated). Based on the daily energy expenditures (~5000 kcal) numbers from the Sensewear activity meters used in our pilot work, the Group 3 and 4 subjects were food restricted by approximately 40% (~2000 kcal/d deficit by providing only 3000 kcal/d total caloric intake) as they received the same meals and intake schedule as that for Groups 1 and 2. Group 3 received 600 ml (3 cups or 24 oz) of water with each 3 h meal before each exercise session and another 600 ml after each the exercise session (8 sessions/d for 10 L/d; to maintain hydration status for active adult performing moderate to heavy exercise in a hot, humid environment). At each of the 8 daily data collection/feeding points, Group 4 received 150 ml (0.7 cup or 5 oz) of water for a total of 1.2 L/d following their meal but before their exercise session. Using a predicted daily fluid need of 10 L, we attempted to generate an approximate 9 L/d daily deficit for individuals in Group 4.

Our desired caloric deficit of 2000 kcal/d (3000 total kcal/d) was based on our estimations of subjects' daily energy expenditure (5000 kcal) which, in turn, were based on the daily activities proposed as well as several military training studies (Aakwaag *et al.* 1978; Guezennec *et al.* 1994; Nindl *et al.* 1997; Opstad *et al.* 1992; Opstad *et al.* 1983). In our current protocol, we were able to control energy expenditures at ~5000 kcal/d using a normal adult male 24 h expenditure of ~3000 kcal plus the planned 300 kcal of exercise between each 3 h data collection point save the final 48 h point (300 kcal × 7 collection points = 2100 kcal).

Our 48 h FPE protocol rate of dehydration goal (≤ 4% body weight per day) corresponds with that demonstrated to decrease capacity to perform high-intensity exercise (Pinchan *et al.* 1988; Sawka and Coyle, 1999) but has also been demonstrated as relatively safe up to 5 successive days (Costill *et al.* 1975). It was our goal to gradually dehydrate the subjects over 48 h with minimal exposure to unsafe hot and humid environmental conditions.

Activities in between 3 h data collection periods: Following each 3 h meal, volunteers were sequestered in the laboratory complex and limited to seated rest except for performance of exercise in Groups 3 and 4. In an attempt to minimize the data collection during the effects of previous exercise, all moderate intensity exercise bouts took place early in each 3 h between-collection periods. We had the subjects perform the same total caloric expenditure during each between-collection period to avoid affecting circadian rhythms or longitudinal metabolic data. Therefore, subjects performed 300 kcal of exercise following the 3 h time point through 45 h time point for a daily total of 2250 kcal of moderate intensity exercise. Because we did not want to jeopardize our data with unnecessary extraneous variables, all subjects were exposed to the same activities or environments regardless of experimental group (i.e., Groups #1-4).

Specific testing protocol details

Subjective fatigue and mood questionnaires: Each 3 h data collection period was preceded by 20-30 min of quiet rest. Following this quiet rest, subjects completed questionnaires assessing psychological mood state, stress management capability and susceptibility (POMS) followed by a self-reported Likert-scaled stress/fatigue level prior to the mental challenge or eye-hand coordination testing.

Mental challenge testing: The Stroop color conflict test (Frankenhaeuser & Johansson, 1976) is a commonly used test of cognitive function challenges executive control which refers to a subset of cognitive processes involved with the intentional component of environmental interaction. It includes planning, working memory, coordination, and inhibitory control, as well as other effortful processes that must occur under conscious control to avoid making an error (Meyer DE & Kieras DE, 1997). Briefly, subjects were instructed to read aloud as many items as possible in 45 sec during each of three conditions (word, color, incongruent color-word pairs). The latter condition necessitated the greatest amount of interference control because subjects must read aloud the color of the ink and inhibit the task of reading the printed word. No performance feedback was provided to subjects for any of the tests

Statistical analyses: Subjects' data at 3 h through 48 h was collected at approximately 2 h post-prandial and all subjects received 1200 ml (40 oz) of water with breakfast after reporting to the laboratory for the fatigue protocol. The food intake pattern was identical between groups and each 3 h data collection point whereas water intake with each meal was determined by group assignment. Therefore, longitudinal FPE protocol data was compared to that collected at 3 h. Longitudinal (48 h) data were compared between experimental groups (#1-4). A multivariate mixed effects model was employed to determine whether time-dependent changes in test scores were significantly different by group. Main effects describing differences between groups at each time point as well as the entire 48 h were also evaluated. Interactions between the factors were further defined by appropriate post-hoc evaluation. Our significance level was set a priori at $p < 0.05$.

Results

Subjects' baseline characteristics. The groups' anthropometric, vital signs, and fitness levels were representative of those expected for college- or military recruit-aged, active males (ACSM Guidelines). Body mass index for the groups were considered as high "normal" (groups 1 and 2) or low "overweight" (groups 3 and 4) while body fat percentages for the groups ranged between below average and average. As a result of our random group assignments, we did observe some group baseline data differences ($p < 0.05$).

Body weight. Body weight for groups 1-3 did not change during the 48 h FPE protocol (Figure 4). However, body weight was reduced by 2.8 kg (3.2%) in group 4 over the course of the 48 h FPE protocol. While our 9 L water deficit would correspond with a 9 kg body weight loss per day, it is important to remember that our subjects' body weights would not have dropped dramatically as they were fed every 3 h during our protocol. It is unlikely that the body weight change over 48 h in group 4 was due primarily to body water loss as the time course of the protocol and the interspersed exercise bouts would not be predicted to result in measureable body fat or fat-free mass loss. Additionally, group 3 body weight did not change while subjected to the same feeding and exercise schedule as that of group 4, lending to the notion that body weight loss in group 4 can be contributed to water loss.

Hydration status. Compared to the 3 h data collection, urine specific gravity (usg) was significantly lowered ($p < 0.001$) at 48 h in groups 1 and 2 but was significantly higher ($p < 0.001$) at 48 h in group 4 indicating an increase in the solute level of the urine representing a dehydrated state (i.e., >1.020;

Figure 4). While exercise would affect the solute concentration of the urine, regardless of hydration status, group 3 h usg indicated a well-hydrated state (i.e., <1.020) and was unchanged over the 48 h FPE protocol.

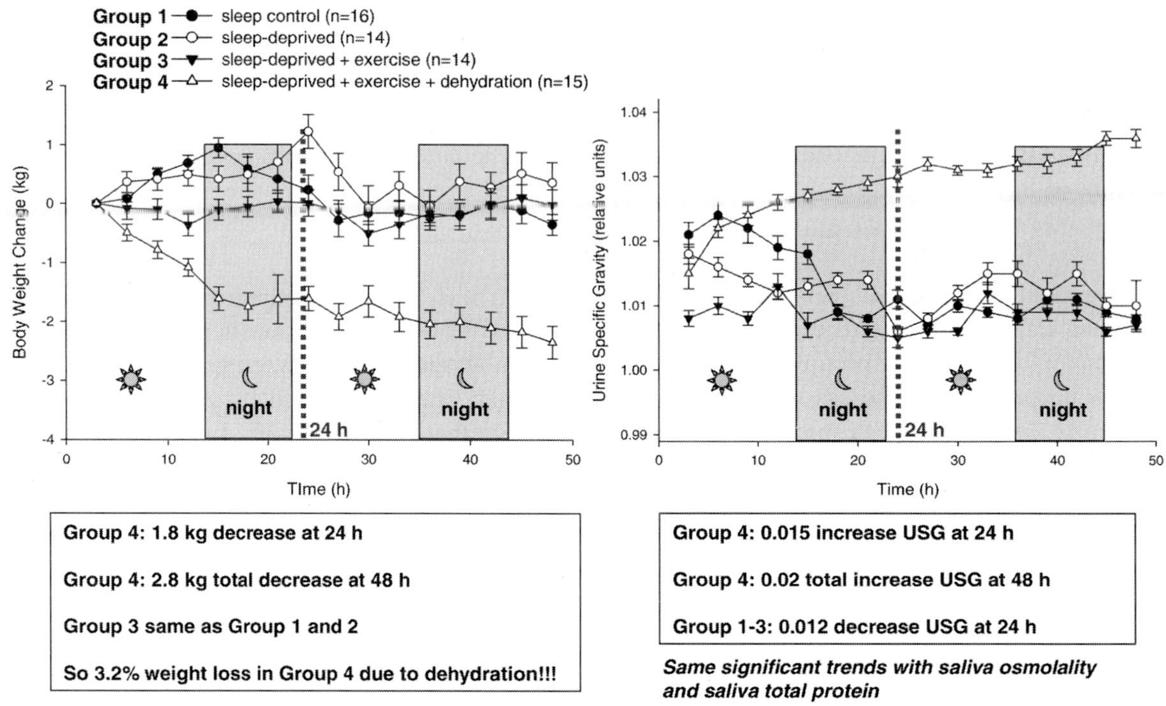

Fig. 4. Body weight and urine specific gravity (usg) assessed every 3 h over the 48 h sleep control or FPE protocol.

Subjective fatigue and mood questionnaires: Subjective indices of manifest FPE, weakness, exhaustion, irritation, and fatigue itself increased distinctively in all groups during the first 24 h. All but irritation exhibited significantly higher levels over the second day compared to controls with the highest levels observed in the exercised groups. Reported irritation followed a similar daily trend as the other subjective fatigue indices. However, the effects of dehydration was pronounced during the latter hours of day 1, remaining elevated compared to the other groups throughout the second 24 h. Thus, our findings demonstrate that 24 h of sleep deprivation significantly affects subjective indices of fatigue and remaining elevated through 48 h. Additional stress of exercise and dehydration do not significantly affect these scores at 48 h. However, the exercised groups operated at a much higher state of FPE throughout the second day.

Just 24 h of sleep deprivation resulted in substantially higher negative mood states regardless of hydration status (HS) or energy deficit induced by physical activity (PA) (Figure 6). Negative mood state remained stable albeit at higher levels compared to sleep controls from 24 to 48 h. The POMS "fatigue" index was mirrored but the additional questionnaire's "how fatigued do you feel" answer

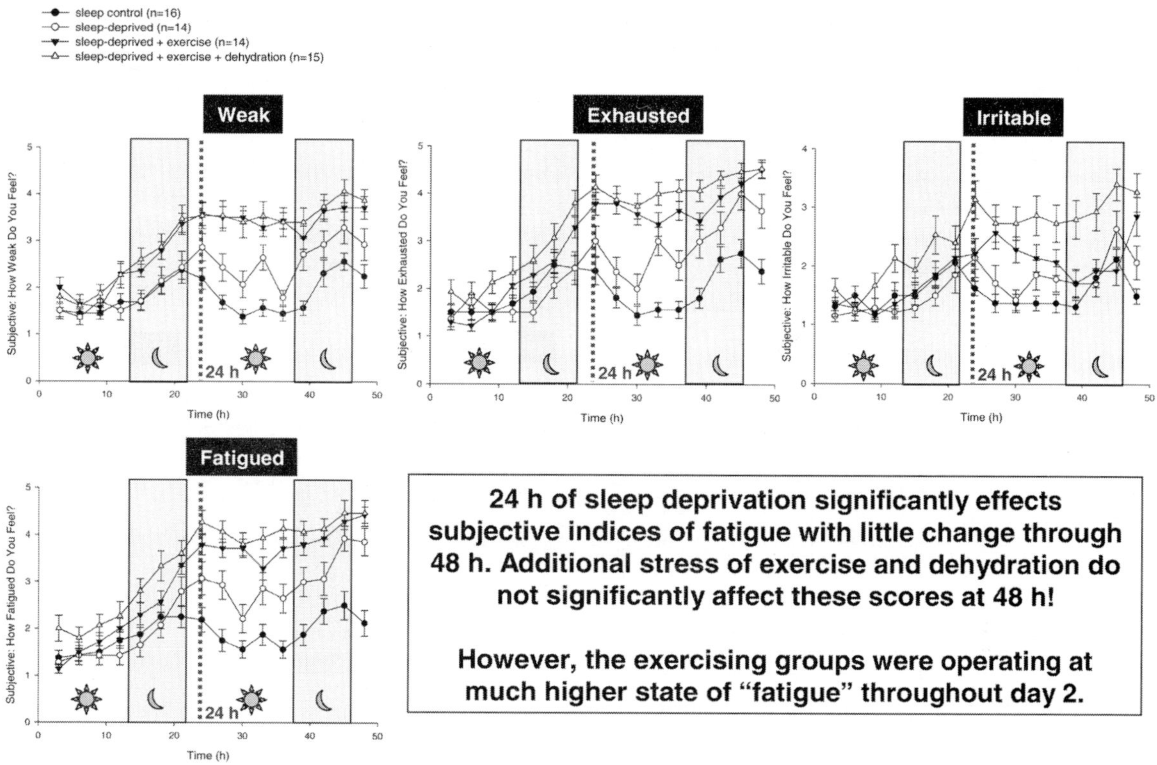

Fig. 5. Questionnaire of subjective FPE indices.

scores for each group. Fatigue clearly increases over the course of the first day with a pronounced rise during the first night in all groups. This is not surprising in the sleep control group as their sleep was interrupted every 3 h for data collection. At the end of the first night the exercised groups exhibit a higher level of fatigue than the non-exercised but sleep-deprived Group 2. Group 1 recovers to baseline levels of subjective fatigue by the start of the second night. However, Groups 2-4 remain at elevated fatigue levels with sleep-deprived plus exercise eliciting a greater fatigue than sleep-deprivation alone. All groups' fatigue increases again over the second night with Group 2 exhibiting the greatest increase such that at 48 h there is no statistically significant difference between the 3 sleep-deprived groups with no effect of exercise or dehydration.

Mental challenge testing: Performance of the Stroop test Word, Color, and Color-word tests were unchanged over 48 h in the sleep control group 1 (Figure 7). However, groups 2-4 demonstrated a decrease in performance over 48 h in all three Stroop tests ($p < 0.05$). Groups 2-4 Stroop test results over 48 h were significantly different compared to group 1 ($p < 0.001$). Word and word-color test performances were unchanged in groups 2 and 3 while color test performance was 17% lower in group 2 and 23% lower in group 3 after 48 h. Group 4 word and color test performance was unchanged while word-color test performance was significantly lower (28%; $p < 0.001$) after the FPE protocol.

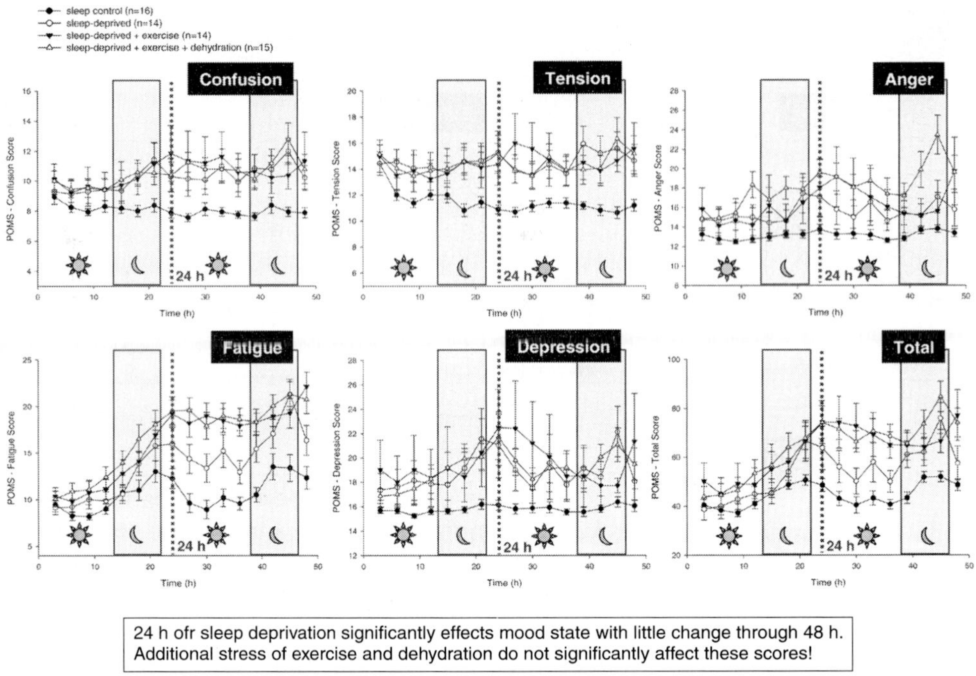

Fig. 6. POMS survey of subjective FPE indicies.

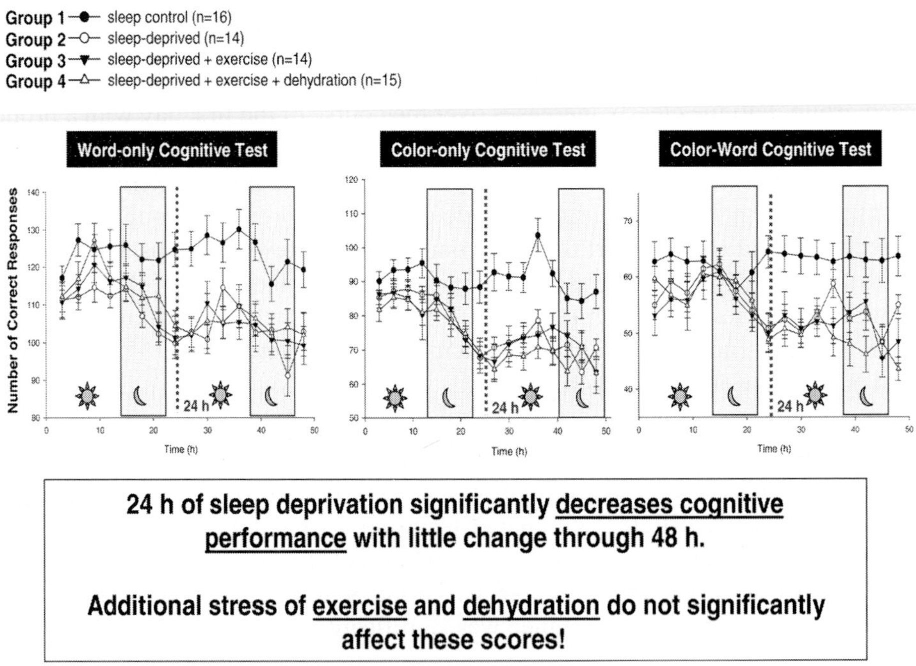

Fig. 7. Stoop 45 second tests of cognitive performance.

References

1. Aakwaag A, Sand T, Opstad PK, Fonnum F. Hormonal changes in serum of young men during prolonged physical strain. Eur J Appl Physiol. 1978;39: 283-91.
2. Åkerstedt T, Fröberg JE. Sleep and stressor exposure in relation to circadian rhythms in catecholamine excretion. Biological Psychology. 1979;8: 69-80.
3. Alluisi EA, Chiles WD. Sustained performance, work-rest scheduling, and diurnal rhythms in man. Acta Psychologica. 1967;27: 436-442.
4. Brasel KJ, Guse C, Gentiello LM, Nirula R. Heart rate: is it truly a vital sign? J Trauma. 2007;62:812-817.
5. Buchheit M, Gindre C. Cardiac parasympathetic regulation: respective associations with cardiorespiratory fitness and training load. Am J Physiol Heart Circ Physiol. 2006;291: 451-58.
6. Cooke WH, Rickards CA, Ryan KL, Convertino VA. Autonomic compensation to simulated hemorrhage monitored with heart period variability. Crit Care Med. 2008;36: 1892-99.
7. Cooke WH, Salinas J, Convertino VA, Duke JA, Hinds D, Moore FA, Holcomb JB. Heart rate variability and its association with mortality in pre-hospital trauma patients. J Trauma. 2006;60:363-370.
8. Cooke WH, Salinas J, McManus JG, Ryan KL, Rickards CA, Holcomb JB, and Convertino VA. Heart period variability in trauma patients may predict mortality and allow remote triage. Aviat Space Environ Med. 2006;77:1107-1112.
9. Costill DL, Coté R, Miller E, Miller T, Wynder S. Water and electrolyte replacement during repeated days of work in the heat. Aviat Space Environ Med. 1975; 46: 795-800.
10. Dixon EM, Kamath MV, McCartney N, Fallen EL: Neural regulation of heart rate variability in endurance athletes and sedentary controls. Cardiovasc Res. 1992;26: 713-19.
11. Fiorica V, Higgins EA, Lampeter PF, Livegola MT, Dawis AW. Physiological responses of men during sleep deprivation. J Applied Physiol. 1968;24: 167-176.
12. Fogt DL, Cooper PJ, Freeman CN, Kalns JE, Cooke WH. Heart rate variability to assess combat readiness. Mil Med. 2009;174: 491-495.
13. Frankenhaeuser M, Johansson G. Task demand as reflected in catecholamine excretion and heart rate. J Human Stress. 1976;2: 15-23.
14. Fröberg JE, Karlsson CG, Levi L, Lindberg L. Circadian rhythms of catecholamine excretion, shooting range performance and self-ratings of fatigue during sleep deprivation. Biological Psychology. 1975;2. 175-188.
15. Gang Y, Malik M. Heart rate variability in critical care medicine. Current Opinions in Critical Care. 2002;8:371-375.
16. Gomez-Merino D, Chennaoui M, Drogou C, Bonneau D, Guézennec CY. Decrease in serum leptin after prolonged activity in men. Med Sci Sports Exerc. 2003;11: 1594-1599.
17. Grogan EL, Norris PR, Speroff T, Ozdas A, France DJ, Harris PA, Jenkins JM, Stiles R, Dittus R, Morris JA. Volatility: A new vital sign identified using a novel bedside monitoring strategy. J Trauma. 2005;58: 7-14.

18. Guézennec CY, Satabin P, Legrand H, Bigard AX. Physical performance and metabolic changes induced by combined prolonged exercise and different energy intakes in humans. Eur J Appl Physiol. 1994;68: 525-530.
19. Hoyt RW, Reifman J, Coster TS, Buller MJ. Combat medical informatics: present and future. Proc AMTA Symp. 2002; 335-339.
20. Jouanin JC, Dussault C, Peres M, Satabin P, Pierard C, Guézennec CY. Analysis of heart rate variability after a ranger training course. Mil Med. 2004; 169: 583-587.
21. Katona PG, Jih F. Respiratory sinus arrhythmia: noninvasive measure of parasympathetic cardiac control. J Appl Physiol. 1975;39: 801-05.
22. Meyer DE, Kieras DE. A computational theory of executive cognitive processes and multi-task performance: part 1. Basic mechanisms. Psychol Rev. 1997;104: 3-65.
23. Morris JA Jr., Harris PA, Norris PR, Ozdas A, Stiles R. HRSD5. An efficient, continuous, immediate indicator of autonomic nervous system function in ICU patients. Crit Care Med. 2004;33: A2.
24. Nindl BC, Friedl KE, Frykman PN, Marchitelli LJ, Shippe RL, Patton JF. Physical performance and metabolic recovery among lean, healthy men following a prolonged energy deficit. Int J Sports Med. 1997;18: 317-324.
25. Opstad PK. Androgenic hormones during prolonged physical stress, sleep, and energy deficiency. J Clin Endocrinol Metab. 1992;74: 1176-1183.
26. Opstad PK, Aakvaag A. The effect of sleep deprivation on the plasma levels of hormones during prolonged physical strain and caloric deficiency. Eur J Appl Physiol. 1983;51: 97-107.
27. Orlinsky M, Shoemaker W, Reis ED, Kerstein MD. Current controversies in shock and resuscitation. Surg Clin North Am. 2001;81:1217-1262.
28. Pinchan G, Gauttam RK, Tomar OS, Babaj AC. Effects of primary hypohydration on physical work capacity. Int J Biometeorol. 1988;32: 176-180.
29. Radespiel-Tröger M, Rauh R, Mahlke C, Gottschalk T, Mück-Weymann M. Agreement of two different methods for measurement of heart rate variability. Clin Auto Res. 2003;13: 99-102.
30. Rowell LB. Human Circulation-Reulation During Physical Stress. New York: Oxford University Press. 1986.
31. Savell CT, Borosotto M, Reifman J, Hoyt RW. Life sign decision support algorithms. Stud Health Technol Inform. 2004;107: 1453-1457.
32. Sawka MN, Coyle EF. Influence of body water and blood volume on thermoregulation and exercise performance in the heat. Exerc Sport Sci Rev. 1999;27: 167-218.
33. Task Force of the European Society of Cardiology and the North American Society of Pacing and Electrophysiology. Heart rate variability: standards of measurement, physiological interpretation, and clinical use. Circulation. 1996; 93:1043-1065.
34. Winchell RJ, Hoyt DB. Spectral analysis of heart rate variability in the ICU: a measure of autonomic function. J Surg Res. 1996;63: 11-16.
35. Winchell JR, Hoyt DB. Analysis of heart-rate variability: a noninvasive predictor of death and poor outcome in patients with severe head injury. J Trauma. 1997;43: 927-33.

Adaptation Biology and Medicine (Volume 6: Cell Adaptations and Challenges)
Editors: P. Wang, C.-H. Kuo, N. Takeda and P.K. Singal
Copyright © 2011 Narosa Publishing House Pvt. Ltd., New Delhi, India

26. New Anti–Inflammatory Molecule (AISA 5203-L) Promotes Adaptive Strategies in Cell Aging

Patrizia d'Alessio, MD, PhD*

University Paris Sud-11 Inserm Unit 935 and ESTEAM Stem Cell Core Facility CHU, Paul Brousse 12, ave Paul Vaillant-Couturier, 94807 Villejuif, France

Abstract

A biological link between repeated inflammation and accelerated cell aging has been clearly established. Our group has been able to demonstrate that terpenes display striking anti-inflammatory effects. Unexpectedly, AISA exhibit anti-stress and tissue repair properties as well. Moreover, certain aspects of cell senescence are reversed, suggesting a link between inflammation, aging and stress. Traditionally, herbal remedies containing these molecules have been used for their general uplifting effect and to promote wound healing.

Keywords: Inflammation biology, chronic inflammatory disease, protein C, Il-6, TNF and anti-TNF activity; anti-inflammatory compounds; vascular endothelium; endothelial senescence; stress; adhesion molecules; actin cytoskeleton; monoterpens; *d*-Limonene

Introduction

Aging is a pleiotropic multi-factorial process, initially determined by genetic and environmental conditions. We have chosen to concentrate our attention to the measurement of the changes of endothelial cells in the course of the inflammatory process. In fact, the high concentration of pro-inflammatory cytokines will promote premature senescence of vascular lining of endothelial cells. Chronic inflammatory disease contributes to the deterioration of endothelial cell function further increasing their premature senescence. We have characterized a family of molecules able to inhibit inflammatory markers in endothelial cells and to reverse the consequences of replicative senescence. We have named these molecules "AISA" (Anti Inflammatory Senescence Actives). One of them ("AISA 5203-L") also showed a compelling anti-stress activity, as measured by a FOB (Functional Observation Battery), whereby analgesic effects were associated to enhanced motility and less irritability in non-pathological stress situations.

*Address for Correspondence: Dr. Patrizia d'Alessio University Paris Sud-11 Inserm Unit 935 ESTEAM Stem Cell Core Facility CHU Paul Brousse 12, ave Paul Vaillant-Couturier 94807 Villejuif, France
Email: patrizia.d-alessio@inserm.fr and endocell@wanadoo.fr

Moreover, an exceptional capacity to restore both the epithelia of colon and dermal layer from pro-inflammatory agonists and toxic substances were described and patented. Most probably the protective effect is due to the capacity of AISA 5203-L to inhibit circulating Tumor Necrosis Factor-α (TNF-α), as well as Interlukin-6 (IL-6) and Interlukin-1 (IL-1), that are cytokines, most relevant to the inflammatory reaction that have been also put in relationship to the aging process.

Chronic Inflammation as Promoter of Age-Related Disease

Chronic inflammation is an underlying cause of many apparently unrelated, age-related diseases. As humans grow older, systemic inflammation can inflict devastating degenerative effects throughout the body. This fact is often overlooked, yet persuasive scientific evidence exists that correcting a chronic inflammatory disorder will enable many of the infirmities of aging to be prevented or reversed.

When we envisage a link between aging and recurrent or chronic inflammation, we refer to pathological consequences of inflammation in well-documented medical literature. Regrettably, the consequences of systemic inflammation continue to be ignored. By following specific prevention protocols, the inflammatory stimulation could be significantly reduced. An important role in preventing the onset of a chronic inflammatory condition has been attributed either to the practice of a physical activity or to the prescription of a personalized diet, or both. In the frame of the EU Capacities study RISTOMED (www.ristomed.eu), AISA Therapeutics is participating with the dietary supplementation of AISA 5203-L in addition to a controlled diet to a cohort of the elderly otherwise healthy persons (65-85 years).

Inflammatory Status of Auto-Immune and Degenerative Diseases

The immune function is affected in aging. Concomitantly, aging status results in an increase of inflammatory cytokines, such as TNF-α or Il-6. Inflammation-promoting cytokines are vital for tissue defense from microorganisms, but have been shown also to contribute to the progression of many degenerative diseases. Rheumatoid arthritis is a classic autoimmune disorder in which excess levels of cytokines such as IL-6, IL-1β and/or IL-8 are known to cause or contribute to the inflammatory syndrome (1-6).

But chronic inflammation is also involved in diseases associated to the metabolic syndrome resulting in atherosclerosis, heart valve dysfunction, obesity, diabetes, congestive heart failure, and gastro-intestinal diseases. Cancer and Alzheimer's diseases have both been shown to benefit from a systemic inflammation for their progression. In aged people with multiple degenerative diseases, the inflammatory marker C-reactive protein is often elevated, indicating the presence of an underlying inflammatory condition. Moreover, when a cytokine blood profile is conducted on people in a weakened condition, an excess level of one or more of the inflammatory cytokines, e.g., TNF-α, IL-6, IL-1β, as well as IL-8, are usually found.

Relevant Markers of Aging

In 2000 the New England Journal of Medicine published several studies showing that the blood indicators of inflammation are strong predictive factors for determining susceptibility to undergo a heart attack. Many international studies have validated this first communication (7-14).

A critical inflammatory marker is C-reactive protein. This marker indicates an increased risk for destabilized atherosclerotic plaque and abnormal blood clotting, which can lead to an acute heart attack. One of these studies (7) showed that people with high levels of C-reactive protein were almost three times as likely to die from a heart attack. This also implicates that elevated C-reactive protein, IL-6 and other inflammatory cytokines indicate significantly greater risks of contracting or dying from specific diseases (heart attack, stroke, Alzheimer's disease).

Moreover, C-reactive protein and IL-6 could also predict the risk of all-cause mortality as addressed by a study conducted on a sample of 1,293 healthy elderly people (8) followed for a period of 4.6 years. Higher IL-6 levels were associated with a twofold greater risk of death. Higher C-reactive protein was also associated with a greater risk of death, but to a lesser extent than elevated IL-6. Subjects with both high C-reactive protein and IL-6 were 2.6 times more likely to die during follow up than those with low levels of both of these measurements of inflammation. Concluding, it would seem C-reactive protein and IL-6 may be useful for identification of high-risk subgroups for anti-inflammatory interventions. Indeed, in 2003, the American Heart Association and Centers for Disease Control & Prevention (CDC) jointly endorsed the C-reactive protein test to screen for coronary-artery inflammation to identify patients at risk for heart attack.

Frailty in Elderly is Associated to Enhanced Inflammatory Profile

These results were further addressed by a new study on almost 140 elderly people (9) that was comparing frail seniors to their more active counterparts for the presence of increased inflammation markers. Associated to the elevated blood inflammatory markers, these frail seniors also tended to show an enhanced clotting activity, muscle weakness, fatigue and disability when compared to the active elderly people. We are not able to document, in how far these clinical outcomes are the origin or the consequence of inflammatory status, but once we recognize that they are interdependent, we can address them by prevention and treatment.

Collectively, these studies should motivate public health policies as well as conscious individuals to monitor their inflammatory status. If C-reactive protein is elevated, then the Inflammatory Cytokine Test Panel would also be highly recommended. Secondly, all those who suffer from any type of chronic disease may also consider to access to the Inflammatory Cytokine Test Panel in order to identify the specific inflammatory mediator that is causing or contributing to their health problem. In the aforementioned RISTOMED study, all the known pro-inflammatory markers are measured to characterize the specific degree of the inflammatory reaction in the population studied, as well as the capacity by AISA compounds to lower them. In fact at this moment a clinical trial conducted in University Hospitals of Berlin, Bordeaux and Rome is administrating low doses of AISA 5203-L daily to a cohort of 65-85 old people, and reports on its anti-inflammatory activity will be available in 2011.

Strategically Addressing Inflammatory Markers

For a long time the identification and production of anti-inflammatory drugs has concentrated on symptom remission, ignoring vascular mechanisms of inflammation and their consequences in the long term, such as increased cells' replicative senescence, and promoting degenerative disease. Now we know that the presence of pro-inflammatory markers in blood may in part be responsible for degenerative disease, characteristic of aging. Through their clinical relevance, inflammatory markers can witness, via endothelial dysfunction, their implication in the occurring of disease (15-20).

In the past, monoterpenes, sesquiterpenes and diterpenes had been characterized by several authors as potential anti-cancer drugs on the basis of *in vitro* and *in vivo* studies, but their role as anti-inflammatory drugs has remained elusive.

In 2002, following our bio-guided selection, performed by means of our *in vitro* cell biology screening platform, we were able to identify four molecules out of 2000 as able to reverse inflammatory markers in senescent endothelial cells. Focusing on functional criteria, we were aiming at the identification of non-toxic molecules able to inhibit *in vitro*, the hallmarks of the inflammatory response, such as the expression of adhesive molecules (ICAM-1, VCAM 1, selectins), as well as the concomitant actin polymerization in endothelial cells. The four monoterpenes were contained in plants extracts (kindly provided by the University of Hanoi to the ICSN of the CNRS) originating from regional medicinal plants. The first patent has been filed in order to protect the therapeutic applications of "AISA" in pathologies associated with acute or chronic inflammatory stress. Two more international patents were filed protecting terpenes and in particular *d*-Limonene's clinical relevance in treating mood alterations and tissue repair. Moreover, our pre-clinical work brought up evidence for the consistency between *in vitro* and *in vivo* results concerning the anti-inflammatory effects of (4R)-1-methyl-4-isopropenylcyclohex-1-ene (*d*-Limonene), which we named "AISA 5203-L".

d-Limonene Pre-Clinical Studies

After an acute toxicity study and several dose-response studies aiming at the appreciation of the therapeutic window, pre-clincal studies were performed on a female rat TNBS induced colitis model and a murine ShK TPA model. These studies showed that the inhibition of adhesion molecules were comparable in the *in vitro/in vivo* experiments (21-23).

They also allowed us to establish that the therapeutic windows were corresponding to the *in vivo* pharmacologically active dose of 10 mg/kg given either *per os* or applied topically. Moreover, our *in vivo* data showed that plasma concentration of TNF-α is greatly reduced by the administration of *d*-Limonene and the score of post-lesional tissue regeneration enhanced and was comparable to that of ibuprofen. Unlike ibuprofen, AISA 5203-L also importantly contributes to pain tolerance and mood matching. Finally, on the quite differentiated capacity to elicit adhesion molecule expression follows TNF stimulation by different steroid and non steroid anti-inflammatory drugs (7-14).

Pharmacological Pertinence of Monoterpenes

Monoterpenes (31-37) are a class of isoprenoid molecules derived from the anabolism of acetate by the mevalonic acid branch biosynthetic pathways of plants. *d*-Limonene, a major component of orange peel oil, is formed by the cyclization of the 10-carbon isoprene intermediate geranylpyrophosphate. Interest in *d*-Limonene came from the ability of the compound to inhibit carcinogenesis in the murine benzo(*a*)pyrene-induced skin tumor model and inhibition of dibenzopyrene-induced s.c. sarcomas. The mechanisms by which *d*-Limonene and other cyclic monoterpenes inhibit tumor growth have not been firmly established. Geranylpyrophosphate, the isoprene intermediate from which these compounds are derived, is required for the synthesis of cholesterol, coenzyme Q (ubiquinone), and substrates used in the isoprenylation of several cellular proteins. Crowell *et al.* (32) found that *d*-Limonene and other monoterpenes inhibited isoprenylation of M_r 21,000–26,000 proteins, including p21^{ras} and other members of the ras family of GTP-binding proteins that are involved in signal transduction and growth regulation. The post-translational isoprenylation of these and other proteins is an essential covalent modification that is required for protein localization and function. For example, farnesylation is required for plasma membrane association and signaling function of p21^{ras}. Other intracellular proteins require isoprenylation by addition of a farnesyl (15-carbon) or geranylgeranyl (20-carbon) group to the COOH terminus for localization to a cellular compartment or for interaction with other proteins. The four molecules, identified by the AISA Therapeutics cell biology platform, for their specific anti-inflammatory activity, following an *in vitro* screening on endothelial targets associating cyto – protective and adhesion inhibiting activities, turned out to be monoterpens: geraniol, geranyl acetate, d-limonene and iso-menthone, are intimately linked by a metabolic loop.

Although data available emphasized the anti-cancer activities of geraniol and *d*-Limonene, we were tempted to find out about the *in vitro/in vivo* consistency of our data in models adapted to the study of acute and chronic inflammation. In confirmation to our *in vitro* results, the capacity of geraniol (AISA 5202-G) to inhibit the adhesion of leukocytes following TNF-α stimulation had already been established. As for *d*-Limonene (AISA 5203-L), in consideration of the efficacy of its metabolite, perillyl alcohol (POH), already tested in clinical trials in patients with refractory solid malignancies, it seems plausible that it plays the role of a precursor. In conclusion, the complex sequence of events of the inflammatory response that includes endothelial adhesive molecule expression for the vascular recruitment of leukocytes to the site of injury, it seems to be concomitant with actin polymerization and challenge of the signaling pathway of the rho GTPase family. The activation of these proteins requires a post-translational iso-prenylation. *We think that the same mechanisms of action of the anti-cancer effects reported for geraniol and d-Limonene could equally be at the origin of their anti-inflammatory properties, reported here*.

d-Limonene's Action on Stress

Important effects on mood in presence of stress situations had been documented by us previously in a rodent model thus motivating our choice to explore more in detail this unexpected effect (24-30).

As established by our Functional Observation Battery (FOB), *d*-Limonene was able to substantially contribute to pain tolerance and mood stabilization. The most intriguing result however that retained our attention, was the fact that the stressed animal (by a so-called non-pathological stress stimulating anxiety, comparable to maternal deprivation), instead of developing a freezing attitude, after oral administration of *d*-Limonene developed a "ludic" activity, starting to play with the wheel next to it. For us this was particularly interesting when compared to other mood or anxiety treating molecules, displaying substantially a hypnotic effect.

Discussion

To recapitulate links between inflammation, senescence and stress have been addressed in a fragmentary way whereas they should be considered by an integrated approach. Therefore identifying molecules able to prove anti- inflammatory/effective on replicative senescence and having a subtle but tangent effect on stress became a way to us to put this link in evidence. In this regard, I would evoke the historical and almost anecdotal properties of such molecules in food and recipes throughout the ages.

Linking Inflammation and Cell Senescence

Inflammatory diseases are numerous and systemic inflammation is a silent companion of stress and age. On the other hand, psychological stress in response to pain appears as an important customer of inflammation. The pharmacological strategies trying to inhibit inflammatory symptoms and related clinical episodes have been going far, and in case a proper indication is respected for their prescription, can be told to be successful, despite the recent side effects of several of them reported. But disease is, independently from its etio-pathology, a stressing agent by itself, able to promote inflammatory procrastination by not yet totally unraveled mechanisms. Unfortunately, a sustained anti-inflammatory treatment is inevitably associated to adverse effects, thus opening the field of the research and development of new less or not toxic and more tolerated ant-inflammatory strategies. In particular, we could provide evidence that the expression of vascular adhesion molecules is challenged by most used anti-inflammatory steroid and non-steroid drugs when compared to the effect of a triterpen contained in an edible plant used by Chinese populations since centuries to prevent rheumatoid arthritis (10).

Linking Inflammation and Stress

Recently, much attention has been given to stress as promoter of disease and syndromes implied in health decline and we have addressed this issue in a review and research article (3, 11).

Indeed, compounds found in natural environment, mostly plants, have acquired a new status of valid pharmacological candidates for the development of new drugs preventing, maintaining and curing on the basis of body integrity and substantially addressing lifestyle more than health. We think that if the aging process depends on genetic stability, metabolic control, and resistance to stress, longevity in particular seems related to this latter. If responses to stress anticipate adaptation to an unacceptable disparity between real or imagined personal experience and expectation, they include adaptive stress, anxiety, and depression. However, if stress persists, it may lead to chronic diseases, ranging from inflammation and cancer to degenerative diseases. If in the past only remarkable stress was acknowledged to induce immune and vascular alterations, such as infection or hypertension, now it is known that moderate stress independent of conventional risk factors can induce a potent alteration of health conditions and consequently shorten life quality and lifespan. Uncontrolled inflammation is a critical defense mechanism, which in chronic conditions results in inflammatory pathogenesis, and in stressful conditions induces a diffuse (systemic) pro-inflammatory status. If sub-clinical chronic inflammation is an important pathogenic factor in the development of metabolic syndrome, then a cluster of common pathologies, including cardiovascular disease, will include markers associated with endothelial activation and dysfunction. Literature concerning monoterpenes essentially concentrate on the anti-cancer activities of *d*-Limonene and geraniol, apparently due to the inhibiting activity of the isoprenoid biosynthesis and/or the inhibition of the transferases catalyzing the iso-prenylation of proteins. Our laboratory as well as others have well documented the capacity to modulate not only metastasis recruitment but also leukocyte recruitment via an inhibitory activity towards adhesive molecule expression, and thus the inflammatory reaction.

Possible Mode of Action of *d*-Limonene in Inflammation

d-Limonene has been used by us in an *in vitro* model challenging human primary endothelial cells with the pro-inflammatory cytokine TNF-α and hydrogen peroxide (H_2O_2) for its capacity to down-regulate endothelial adhesive molecules. In fact, in the course of the inflammatory response, peroxides, TNF-α, ICAM-1 expression and cytoskeleton proteins are all well-integrated elements of a complex sequence of events. In particular, ICAM-1 and cytoskeleton actin fibers interact via a signaling pathway implying the proteins of the rho GTPase family (35, 36). For their activation, these proteins require a post-translational isoprenylation, already described for *d*-Limonene anti-cancer effects. This effect displayed by monoterpenes, could also be responsible for the resulting anti-inflammatory activity shown by our *in vitro* and *in vivo* studies. In fact adhesive molecules are expressed by endothelial cells, responding to a pro-inflammatory environment, which is always present in cancer at least at its dissemination stade. ICAM-1, VCAM-1 and P- and E-selectins are indeed able to recruit circulating metastasis, but are primarily devoted to the regulation of leukocyte diapedesis during immune survey and inflammatory reaction.

Possible Mode of Action of d-Limonene in Stress

The interesting results obtained on mood and pain modulation by *d*-Limonene supplementation, may rely on an effect displayed at the cytoskeleton level, though not at vascular endothelial but at neuronal level. Recently, the relevance of the significant contribution of actin cytoskeletal regulation to basal synaptic transmission and to various forms of synaptic plasticity has shown the great importance for functional responses (37). They imply molecular signaling machineries, able to coordinate real-time modifications in synapse activity and actin-dependent dynamics in pre- and post-synaptic terminals. Possibly, *d*-Limonene, which passes the BBB easily because of its reduced MW, contributes to the re-initialization of the functional and morphological plasticity of individual synapses (possibly within the limbic area), thus influencing neighboring synapses to affect the overall neuronal circuit. In fact in our view the quenching of endogenous opioid neuro-transmitters is blocked during inflammation because of stress fiber formation at the synaptic level, that would be de-polymerized by the monoterpenes action, thus permitting their liberation into the circulation, associated with a mood modulating effect.

Addressing Inflammation and Stress in Chorus

d-Limonene has a tolerated pharmacokinetic profile, is rapidly metabolized into perillic and di-hydro-perillic acid or perillyl alcohol, monoterpenes themselves are more active than *d*-Limonene itself, as reported by studies showing an interest of phase 1 pharmaco-kinetics of perillyl acid. With a biodisponibility of 43%, a plasmatic concentration of 11.3 mg/ml, after 58 min. and a bi-phasic T/2 corresponding to 34 and then 337 min., *d*-Limonene would thus play the role of a pro-drug. In consideration of the relationship between mood alterations and inflammation, as seen in stress contributing to the development of a systemic inflammation as well as hypertension and atherosclerosis, we think that the mood stabilizing properties of AISA 5203-L permits us to extrapolate this drug as a candidate for anti-inflammatory applications.

Perspectives

It would be strategically interesting to confirm the anti-inflammatory mechanism of action of AISA molecules, by their impact on isoprenylation of proteins, measuring surrogate markers of the type ras farnesylation, as shown in anti-cancer studies (32). This type of study should also integrate the metabolites of AISA molecules, that could be more efficacious, as this has been shown for R-(+)-

Limonene and by our pre-clinical results on PA. Coming towards the end, the reader should be alerted about the few things that can be confirmed (38-42).

Today our study contributes to enhance the evidence for the relevance of a specific class of molecules contained in substances which may have been used either in the domesticated fruit and vegetable environment as food, such as the oleocantal ibuprofen – like molecule contained in olive oil (39), or as ritual substances, such as the incense and myrrh (42) containing anti-inflammatory and mood modulating terpenes. We presume that for centuries these raw materials were integrated in lean and sacred recipes devoted to the maintenance of health and the prevention of aging, because of their content in biologically active molecules, displaying their curing properties, either as anti-inflammatory remedies or inducing mood modulation allowing an enhanced perception of life.

Identifying a molecular target, this approach would result in a rational concept of original molecules ready to be patented as such, permitting to optimize their pharmacological efficacy. This vision would also contribute to better support the rational of the drug discovery project and to position the molecules as candidates in other therapeutic applications.

References

1. Toussaint O., Royer V., Salmon M., Remacle J., Stress-induced premature senescence and tissue ageing. *Biochem Pharmacol*. 2002; 64, 1007-9.
2. Ostan R. *et al.* Immunosenescence and immunogenetics of human longevity. *Neuroimmunomodulation* 2008;15, 224-40.
3. d'Alessio P., Aging and the endothelium. *Journal of Experimental Gerontology* 2004; 39, 165-171.
4. AISA Patent Family n°1, "Composition for treating or preventing cell degeneration using at least one molecule capable of maintaining adhesion molecule expression reversibility and inhibit vascular endothelium actin fiber polymerization" (FR 2 869 230, WO 2005/105074, EP1748771, US-2009-0012162-A1).
5. Maier A.B. and Westendorp R.G., Relation between replicative senescence of human fibroblasts and life history characteristics. *Ageing Res Rev*. 2009; 8, 237-43.
6. Holmes C. *et al.* Systemic inflammation and disease progression in Alzheimer disease. *Neurology* 2009; 73, 768-74.
7. Ridker PM, *et al.* Inflammation, aspirin, and the risk of cardiovascular disease in apparently healthy men. *N Engl J Med.* 1997; 336(14), 973-9.
8. Harris TB, *et al.* Associations of elevated interleukin-6 and C-reactive protein levels with mortality in the elderly. *Am J Med.* 1999 ; 506-12.
9. Walston J, *et al.* Frailty and activation of the inflammation and coagulation systems with and without clinical comorbidities: results from the Cardiovascular Health Study. Cardiovascular Health Study. *Arch Intern Med.* 2002;162(20), 2333-41.
10. Zhang D.H. *et al.* Tripterine inhibits the expression of adhesion molecules in activated endothelial cells, *Journ Leuko Biol*, 2006; 80, 309-319.
11. Bisson J.-F., Menut C., and d'Alessio P., New pharmaceutical interventions in aging. *Rejuvenation Research* 2008; 11, 399-407.

12. Ziccardi P. *et al.* Reduction of inflammatory cytokine concentrations after weight loss over one year. *Circulation* 2002;105, 804-9.
13. Clément K. *et al.* Weight loss regulates inflammation-related genes in white adipose tissue of obese subjects. *FASEB J.* 2004;18, 1657-69.
14. Hundsberger H. *et al.* TNF: a moonlighting protein at the interface between cancer and infection. *Front Biosci.* 2008; 13, 5374-86.
15. Edelman G., The golden age for adhesion. *Cell Adhes Commun.* 1993; 1, 1-7.
16. Vanier B., Intercellular adhesion molecule-1 (ICAM-1) in ulcerative colitis. In : *Digestive Diseases Sciences* 2005; 54, 313-327.
17. Farhadi A., Banan A. and Keshavarzian A. Role of cytoskeletal structure in modulation of intestinal permeability. In : *Archives Iranian Medicine* 2003; 6, 49-53.
18. García-Cardeña G., Gimbrone M.A. Jr Biomechanical modulation of endothelial phenotype: implications for health and disease. *Handb Exp Pharmacol* 2006; 2, 79-95.
19. Ingber D.E., Mechanobiology and diseases of mechanotransduction. *Ann Med.* 2003; 35, 564-77.
20. Marconi, A., Darquenne S., Boulmerka A., Mosnier M., d'Alessio P., Naftidrofuryl-driven regulation of endothelial ICAM-1 involves Nitric Oxide. *Free Rad. Biol. Med.* 2003; 34, 616-625.
21. Yamada Y., Marshall S., Specian R.D., Grisham M.B., A comparative analysis of two models of colitis in rats. *Gastroenterology*, 1992 ; 102,1524-34.
22. Medeiros R., Otuki M.F., Avellar M.C., Calixto J.B. Mechanisms underlying the inhibitory actions of the pentacyclic triterpene alpha-amyrin in the mouse skin inflammation induced by phorbol ester 12-O- tetradecanoylphorbol-13-acetate. *Eur J Pharmacol.* 2007; 559, 227-35.
23. AISA Patent Family n°3 « Use of a monoterpene to increase tissue repair » (EP2042167, WO 2009/040420).
24. AISA Patent Family n°2 « Use of a monoterpen to treat or prevent stress » (EP1990047, WO 2008/138905).
25. MacPhail R.C., Observational batteries and motor activity. *Zentralbl Bakteriol Mikrobiol Hyg B.* 1987 ;185, 21-7.
26. Shibeshi W.A., Young-Xu Y., Blatt C.M., Anxiety worsens prognosis in patients with coronary artery disease. *Journal of American College of Cardiology* 2007; 49, 2021–2027.
27. Esler M. *et al.* Chronic mental stress is a cause of essential hypertension: presence of biological markers of stress. *Clinical and Experimental Pharmacology and Physiology* 2008; 35, 498-502.
28. Queré N., Noël E., Lieutaud A., d'Alessio P., Fasciatherapy combined with Pulsology induces changes in blood turbulence potentially beneficial for the Endothelium. *J Bodyw Mov Ther.* 2009; 3, 239-45.
29. May L. *et al.* Adverse environmental conditions influence age-related innate immune responsiveness. *Immun Ageing* 2009; 30, 6-7.

30. Chandola T. *et al.* Work stress and coronary heart disease: what are the mechanisms? *European Heart Journal* 2008; 29, 640-8.
31. Vigushin D.M. *et al.* Phase I and pharmacokinetic study of *d*-limonene in patients with advanced cancer. Cancer Research Campaign Phase I/II Clinical Trials Committee. *Cancer Chemother Pharmacol*, 1998; 42, 111–117.
32. Crowell P.L., Lin S., Vedejs E., Gould M.N., Identification of metabolites of the anti-tumor agent *d*-Limonene capable of inhibiting protein isoprenylation and cell growth. *Cancer Chemother Pharmacol* 1992; 31, 205–212.
33. Hardcastle I.R. *et al.* Inhibition of protein prenylation by metabolites of limonene. *Biochem Pharmacol*, 1999; 57, 801- 809.
34. Xu Y. *et al.* Immunomodulatory drugs reorganize cytoskeleton by modulating Rho GTPases. *Blood* 2009; 114, 338-45.
35. Burridge K. and Wennerberg K., Rho and Rac take center stage. *Cell* 2004; 116, 167–179.
36. Millan J. and Ridley A.J., Rho GTPases and leucocyte-induced endothelial remodelling. *Biochem J* 2005; 385, 329–337.
37. Dillon C. and Goda Y., The actin cytoskeleton: integrating form and function at the synapse. *Annu Rev Neurosci*. 2005; 28, 25-55.
38. Galeno (Galen) *La dieta dimagrante* (The thinning diet), 45-47 (Paravia ed., 1973, Torino.
39. Beauchamp G.K. *et al.* Phytochemistry: ibuprofen-like activity in extra-virgin olive oil. *Nature* 2005; 437, 45-6.
40. Issuree P.D., Pushparaj P.N., Pervaiz S., and Melendez A.J., Resveratrol attenuates C5a-induced inflammatory responses *in vitro* and *in vivo* by inhibiting phospholipase D and sphingosine kinase activities *FASEB J.* 2009; 23, 2412-2424.
41. Atzei A.D., « Le piante nella tradizione popolare della Sardegna » (Plants of the popular tradition in Sardinia). *Carlo Delfino ed.,* 2004; Sassari.
42. Nomicos E.Y., Myrrh: medical marvel or myth of the Magi? *Holist Nurs Pract.* 2007; 21, 308-23.

Adaptation Biology and Medicine (Volume 6: Cell Adaptations and Challenges)
Editors: P. Wang, C.-H. Kuo, N. Takeda and P.K. Singal
Copyright © 2011 Narosa Publishing House Pvt. Ltd., New Delhi, India

27. Adaptation to Low-Dose Ionizing Radiation: Usefulness in Human Health

C. M. Pathak*, G. Kaushik, K. L. Khanduja and P. K. Singal[1]

Department of Biophysics, Post Graduate Institute of Medical Education and Research, Chandigarh-160012, India.

[1]Institute of Cardiovascular Sciences, St. Boniface General Hospital Research Centre, University of Manitoba, Winnipeg, Canada.

Abstract

Ionizing radiations are integral part of our life. All living organisms exist in a sea of ionizing radiations showered from cosmic, telluric and internal sources. It was initially thought that very low radiation dose, even to the extent of a single hit, was harmful to the biological system (the so-called linear no-threshold (LNT) hypothesis). However, epidemiological studies show prevalence of low cancer incidents in inhabitants residing in high radiation background areas and radiation occupational workers. Furthermore, the results from various animal experiments cast doubt on this LNT theory especially at doses below 50mSv. Interestingly, the recent genomic, cellular, animal and human data have shown that exposure to low-dose whole body irradiation (LD-WBI) appears to stimulate positive cellular and genomic adaptations like activation of immune functions, antioxidant defense system, DNA repair mechanisms, induction of apoptosis etc. These adaptive responses are considered to be beneficial in the amelioration of various reactive oxygen spices (ROS) and immune deficiency related diseases and also useful in effective cancer prevention. Repeated exposure to LD-WBI with gamma-radiation has been in use in the clinical treatment of lymphomatous malignancies like chronic lymphocytic leukemia (CLL) and non-Hodgkin's lymphoma (NHL) in patients with long term remissions in majority of the cases. The use of LD-WBI thus appears to be an effective new treatment modality in prevention and cure of certain diseases. This review brings in focus the current epidemiological and experimental data related to the clinical aspects of adaptation to low-dose radiation.

Keywords: Low-dose radiation, adaptation, gamma-rays, immunity, anti-oxidant defense, human health

Introduction

Ionizing radiation has been an integral part in the life of all living organisms that exist on earth. These radiations arise from the natural sources such as cosmic rays, radio nuclides present in the earth's crust (telluric) and in humans (internal origin) artificial/man made medical and industrial radiation sources, nuclear exposures, industrial accidents etc. On an average the annual effective dose (AED) from natural sources to an individual is about 3.0 mSv (2.0 mSv from radon and 1 mSv from cosmic,

*Address for Correspondence: Prof. C. M. Pathak Department of Biophysics, Post Graduate Institute of Medical Education and Research, Chandigarh-160012, India. Phone +91-172-2755248, Fax +91-172-2744401
Email: chander_pathak@rediffmail.com

terrestrial and internal origin) (Eisenbud and Gessell, 1997). Currently, the use of radiation in medical diagnosis in high technology societies is expanding and in the United States medical exposure now makes up over 50% of the effective dose for an average person (Mettler *et al.* 2008).

The deleterious effects due to low radiation dose are theoretically extrapolated from the high-dose by the "Linear No Threshold" (LNT) model (Figure-1). The LNT model is based on the biophysical concept that the passage of even a single charged particle or photon through the cell could damage the DNA which could result in cancer induction.

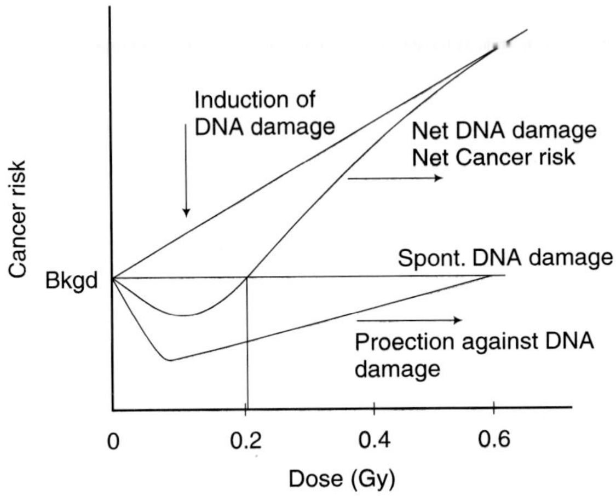

Fig. 1. The dual effect of single low-dose irradiation is schematically analysed according to a simplified model. The net cancer risk derives from the difference between cancer induction through DNA damage and its prevention at the various dose levels (Feinendegen 2005)

According to LNT Theory

1. The effects of low doses of ionizing radiation can be estimated by linear extrapolation from the effects observed at high doses.
2. There is not any 'safe dose' because even very low doses of ionizing radiation produce some biological effects.

In 1959, the International Commission of Radiation Protection (ICRP) adopted the LNT theory. However, the scientific data of many investigators including our own do not support the LNT theory.

A recent epidemiological evidence from Japanese A-bomb survivors show that there exists a linear relationship between risk of cancer and organ dose between 100 mSv (milli-Sievert) and 2.5 Sv, but for doses outside this range the relationship to organ dose is uncertain (Hall 2009, Martin *et al.* 2009). Although there is no cut off value to define low-dose range but a value < 50 mSv is generally considered as low as no harmful effects are usually observed at this dose in the DNA responsible for cancer generation. United Nations Scientific Committee on the Effects of Atomic Radiation (UNSCEAR) in 1986, defined low-doses and low-dose rates of ionizing radiation as those which are below 0.1-0.2 Gy and below 0.05-0.1 mGy/min respectively. Whereas, The French Academy of Sciences report quantified low doses as <100 mSv and very low doses as < 10 mSv (Tubiana *et al.* 2005).

Interestingly, over the past decade, some scientists have reported that low-dose ionizing radiation is not only harmless but often has a hormetic (adaptive) effect. Thus low-level ionizing radiation may be considered as essential trace energy for life, analogous to essential trace elements like iodine, zinc, selenium etc. According to an interesting hypothesis by Luckey 1999, one third of all cancer deaths are premature and preventable by low-level ionizing radiation. The combined results led to the conclusion that a sub-clinical deficiency of ionizing radiation is prevalent in most human populations. Although seemingly heretical in the face of current radiation policy dogma, the evidence indicates that external and/or dietary radiation supplementation with 0.4cGy (centi-Gray) /month to an annual maximum dose of 5 cGy would reduce cancer rates in humans and could be beneficial to health and longevity. To achieve this level of radiation a dose of 100mg of ^{40}KCl per week is needed to be ingested to maintain a reasonable level in tissue (Luckey 1999).

The health effects of low-dose radiation exposure are of public concern. Although molecular events in the cellular response to high-dose-rate radiation exposure have been investigated to a large extent, but the effects of long-term or single exposure to low-dose radiation remain unclear. Despite the fact that high doses of ionizing radiation are detrimental, substantial data from both humans and experimental animals show that a few biological functions get stimulated by exposure to low-dose radiation. The physiological benefits from low LET (linear energy transfer) radiation like X- or γ-rays usually become appreciable after exposure of the animals/organism in the range of 1-50 cGy. Based on the scientific data during the last few decades, a single or chronic exposure to low-dose whole body irradiation (LD-WBI) the following biological responses have been observed (Table 1).

Table 1. Biological responses to low-dose ionizing radiations (Luckey 1999)

Increased	Decreased
Mean life span	Total mortality rate
Growth rate	Cancer mortality rate
Development rate	Cardiovascular death rate
Neurological activity	Respiratory death rate
Immune competence	Infections
Fecundity	Female sterility
Radio résistance	Leukemia death rate

The recent advancements in molecular biology reveal that the living organisms do not respond to ionizing radiation in a linear manner at a dose below 0.50 Gy. Rather this low-dose restores the homeostasis both *in vitro* and *in vivo* by the normal processes such as, normal cellular and DNA repair processes, immune reactions and antioxidant defense system (Calabrese and Baldwin 2002a, 2002b, Pollycove 2004, Avti *et al.* 2005, Pathak *et al.* 2007), adaptive responses (Wang *et al.* 2004), activation of immune functions (Safwat 2000a, Ina and Sakai 2005, Liu 2007), stimulation of growth and reduction in death rate due to cancer (Luckey 1982, 1997), increase in the life span of mice (Tiku and Kale 2004, Ina and Sakai 2004) etc. The LNT model thus ignores these fundamental cellular processes and little is known about the molecular mechanisms involved in these events. However, some of the epidemiological studies show reduced cancer incidence in those who are exposed to high back ground radiation.

Epidemiological Studies on Radiation Adaptation

Although radiation adaptation data on humans is still very limited, extensive epidemiological studies establish that the cancer incidence has been low in those who are exposed to either single or chronic low-dose ionizing radiations. A brief review on this irrefutable evidence is as follows:

Japanese Survivors Studies

Follow up of the Japanese atomic bomb victims by the Atomic Bomb Casualty Commission and its successor, the Radiation Effect Research Foundation, in Hiroshima and Nagasaki since 1948 is generally considered to provide the most reliable index for the effect of actual external radiation in humans. These studies include analysis of birth defects, chromosomal aberrations, alteration of plasma and erythrocyte proteins as well as epidemiologic data on mortality (any cause) and cancer incidence. There is thus so far, no indication of genetic effects in the off springs of the survivors. Interestingly, those who were exposed to LD-TBI had a lower cancer death rate than the controls. Rather for every ten thousand persons exposed to 1-1.9 cGy there were 3 fewer leukemia deaths and 50 fewer solid cancer deaths than in age and sex matched controls (Shimizu *et al* 1992).

Although leukemia mortality rates increase dramatically with doses exceeding the threshold, atomic bomb survivors exposed to 5-10 cSv had 5.5 fewer leukemia deaths per 10000 persons than controls. Later studies confirm radiation hormesis in cancer mortality, leukemia mortality and the average life span of Japanese atomic bomb survivors (Kondo 1993). Such results negate the concept that all radiation is harmful.

According to United Nations Scientific Committee on Effects of Atomic Radiation (UNSCEAR) report (1994), among A-bomb survivors who received doses lower than 1.2 cGy, there was no increase in the number of total cancer deaths. Mortality caused by leukemia was even lower in this population at doses below 10 cSv than age-matched control cohorts (Shimizu *et al.* 1992).

The summary of the cancer deaths due to leukemia and other cancers in Japanese survivors after atomic bomb blasts is shown in Table 2.

Table 2. Cancer deaths in Japanese survivors

			Dose, cGy*		
	Control	0.7-1.2	1.3-3.2	3.3-6.6	6.7-13
Persons	45,148	7,430	9,235	6,439	5,316
Leukemia	81	11	14	8	11
Leukemia per 10^5	179	148	152	124	207
Change per 10^5	0	-31	-27	-55	28
Cancer	3,246	498	717	516	400
Cancer per 10^4	719	670	776	801	752
Change per 10^4	0	−49	57	82	33

*Aprox. exposures not counting radiation in fall out and ingestion; (Shimizu 1992)

Background Radiation Studies

A number of epidemiological studies have been conducted during the last 3 decades to evaluate the effect of elevated natural background radiation (HNBR). These include descriptive studies in HNBR areas of Brazil, India, China and Iran, as well as analytical studies, mainly case-control studies of lung cancer risk associated with radiation exposure. Epidemiological surveys consistently find an inverse relationship between radiation dose rate and cancer mortality rates. Many generations exposed to relatively high levels of HNBR show improved health (Lucky 1997). This includes over 70,000 exposed (3.3 mGy/y) and 70,000 control (1.07 mGy/y) Chinese peasants, several villagers in Brazil (20-35 mGy/y), the people living in mountains of Kerala (10 mGy/y), villagers on the coasts of Kerala (4-13 mGy/y), and Ramsar, Iran (7-480 mGy/y) (Luckey 1999). The preliminary cytogenic studies, focusing on chromosomal aberrations in peripheral blood lymphocytes (PBLs) performed on 21 subjects from HNBR area of Ramsar, Iran did not show any difference in lymphocyte aberrations when compared with control (Ghiassi-Nejad *et al* 2002). There is not even a single report, which demonstrates that the cancer is caused by low-dose radiation.

Nuclear Workers Studies

Exposed workers in nuclear plants provide the best data for the chronic exposure to an increased dose rate of ionizing radiations. Life time exposures of 2-25 cSv obtained during 2-30 years of work represent an increase of about 1cSv/year above the background radiation of 3mSv. Luckey (1999) has reviewed various studies on the occupational workers from different atomic energy establishments around the globe which have been summarized in Table 3.

Table 3. Cancer Mortality in Nuclear Workers

Group	cGy	Persons	Mortality/1,000 Con/Exp	Exp/Con %
LANL weapons	2	11,145/2,935	17.7/20.1	88
Nuclear shipyard	≥1	40,774/70,983	19.4/29.8	65
US bomb observers	1.3	11,057/20,958	22.0/26.5	83
Canada military	1.3	1,908/954	33.4/38.1	88
US weapons	3.2	20,619/15,314	20.8/34.8	60
British weapons	3.4	58,945/36,272	2.30/9.9	23
Canada energy	2	4,717/4,260	20.3/23.7	86
Total weighted average				51

Abbreviations are as follows: Con, Control; Exp, exposed workers; LANL, Los Alamos National Laboratory (Luckey 1999)

The above data clearly show decrease in cancer deaths in occupational workers as compared to matched control normal population.

Molecular Mechanisms Involved in LD-WBI

Biological effects of low-dose radiation are dependant upon many confounding factors like: nature of the radiation (radiation weighting factor), dose-rate, effective dose, radiation sensitivity of the biological system (animal species, organ, sex, age etc), time interval between the fractionated radiation doses, environmental factors, dietary conditions, end point assessment of the indicators etc. However, the effects of low-dose are still controversial because of the variability of the results on account of the complexity of the confounding factors stated earlier. Animal studies under controlled conditions are therefore important in order to compensate for the uncertainties of the epidemiological studies. In order to generalize animal data to humans, a study of mechanism of radiation effect is required. As mentioned earlier a number of recent studies furnish extensive documentation of low-dose stimulation of many cellular functions which include anti-oxidant prevention (Pathak *et al.* 2007), immune system (Liu 2007, Pollycove 2007), DNA repair rate (Ghiassi-Najed *et al.* 2002), apoptosis (Yu *et al.* 2007). Figure-2 describes some of the defense mechanisms which have been reported to be involved in the process of radiation adaptation.

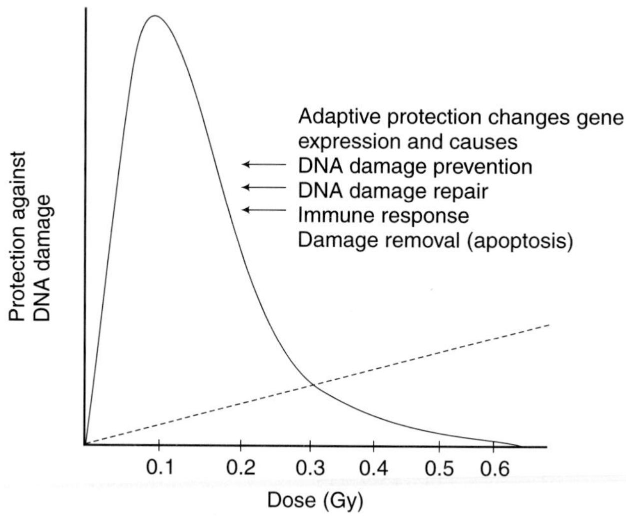

Fig. 2. Single or fractionated low-dose induced adaptive responses have a protective function through various mechanisms. Note that mechanisms of DNA damage prevention and repair and the immune stimulation decrease after a dose range between 0.1Gy and 0.2Gy, in contrast to apoptosis (Feinendegen 2005)

The data from other workers and the recent studies from the author's laboratory (Kojima *et al.* 1997, Avti *et al.* 2005, Pathak *et al.* 2007) suggest the stimulation of antioxidant defense system (reduced glutathione, superoxide dismutase, catalase and glutathione peroxidase) in different organs (liver, lungs and kidneys) of the animals when exposed to LD-WBI from 10-50 cGy (Table 4). These antioxidant defense system enzymes remained activated in these organs to variable extents between 4 and 24 hrs after LD-WBI. This differential response might be due to variable sensitivities in the activation of related genes in these organs towards low-doses of radiation. Similar results have also been observed in various *in vitro* cell culture studies by different authors.

Table 4. Antioxidant defense system stimulation by LD-WBI in various organs of Balb/c mice (Avti et al. 2005; Pathak *et al.* 2007)

Parameters	Dose (cGy)	Time (Hr)	% Change w.r.t control		
			Liver	Kidneys	Lungs
GSH	10	12 & 24	NC	21% & NL	NC & 28%
	25	4, 12 & 24	10%, 50% & NL	12%, 44% & 50%	NC, 22% & 32%
	50	12 & 24	12%, NL	21% & 17%	NC
SOD	10	12 & 24	NC	47% & 32%	NC
	25	12 & 24	8% & NL	35.6% & NL	NC
	50	12 & 24	34% & 28%	37% & NL	NC
CAT	10	12 & 24	NC	18.5% & 12.2%	NC
	25	12 & 24	NC	NC & 10%	NC
	50	4, 12 & 24	13% 12.9% & NL	NC, 11.6% & NL	NC
GPx	10	12 & 24	NC	NC	NC
	25	12 & 24	NC	NC	NC
	50	4, 12 & 24	22%, 14% & NL	30%, 14% & NL	NC
GR	10	12 & 24	NC	NC	NC
	25	12 & 24	10% & NL	10.2% & NL	NC
	50	12 & 24	10% & NL	12.2% & NL	NC

Abbreviations:] GSH: Reduced glutathione; SOD: Superoxide Dismutase; CAT: Catalase; GPx: Glutathione Peroxidase; GR: Glutathione Reductase; NL: Normalized to control values; NC: No change and cGy: Centi-Gray. Methods are given in Pathak *et al.* 2007.

Stimulation of immune system by LD-WBI is manifested in various aspects. Table 5 shows molecular mechanisms involved in various physiological and cellular pathways modulated by LD-WBI. The immunologic parameters studied in various laboratories demonstrating up-regulation of immune

Table 5. Proposed molecular mechanisms involved in various physiological and cellular pathways modulated by LD-WBI

Pathway	Radiation dose	Model	Mechanisms involved	Molecules involved	References
Immune System	500 mGy	Mice	NK cell activity enhancement	IL-2 and GSH upregulation, TNF-alpha, Perforin & Granzyme B mediated	Fan and Liu, 1989
	75 mGy	Mice	CTL activity enhancement	Perforin & Granzyme B mediated	Zhang and Liu, 1996
	75 mGy	Mice	ADCC enhancement	Cytokines and interleukins mediated e.g. IFN-γ, IL-2, IL-12	Liu, 1996
	77-106 mGy	Mice	T Cell proliferation, clonal expansion of T cells	PKC & [Ca2+] up regulation	Su *et al.* 1997, Liu *et al.* 2000.
	75 mGy	Mice	Erythrocyte immune function homeostasis maintenance	Hormesis of RBC-C3bR & RBC-C3bRR activity	Yu *et al.* 2004
	0.2 Gy	Mice	Increased number of circulating lymphocytes	Interleukins, CD markers and cytokines	Feinendegen, 2005

Contd..

Contd..

Cell signaling	75 mGy	Mice	Transcriptional up-regulation of cytokines	*IL-2, IL-12, IL-18, IL-1β* & *GM-CSF*	Liu, 2007
	0.01 Gy	Mice	Activation & self renewal leading to proliferation & differentiation of stem & progenitor population	Interleukins, colony stimulating factors & cytokines	Gridley *et al.* 2009
	0.01 Gy	Mice	Down regulate Inha-mediated cell proliferation	Inhibition of IFN-γ, IL-12 production via Th1 & DC cells. Up-regulation of IL-4 & IL-10 by Th2 cells	Gridley *et al.* 2009, Segerer *et al.* 2008
	1-50 cGy	Mice	Cell growth and proliferation	Stimulated expression of PLAU, IGFBP3	Sanzari *et al.* 2009
	0.1Gy	Mice	CD4+ T cells activation	Up regulation of IL-1R, IL-27, Tbox 21, Transcription factor CP-21	Gridley *et al.* 2009
	0.1Gy	Mice	Down regulation of Inflammation	Socs-5 down regulation	Gridley *et al.* 2009
	0.01 Gy	Mice	TH1 differentiation, Activation of T-Lymphocytes	CD2, CD3 & CD28 up-regulation	Liu, 2003a, Liu 2003b, Liu *et al.* 2001
DNA repair	0.01 cGy	ML-1 cells	Arrest cells for DNA repair	p53-regulated gene e.g. CDKN1A, GADD45A, MDM2	Amundson *et al.* 2003
	25 & 50 cGy	Human	DNA double strand break repair	*RAD50, RAD52, RAD1L1, XAB2* up regulation	Fachin *et al.* 2007 & 2009
	50 cGy	Human	DNA cross link repair	DCLREIC (DNA cross link repair protein) up regulation	Fachin *et al.* 2007
	50 cGy	Human	DNA repair/Replication	REV3L up regulation	Fachin *et al.* 2007
	1 mGy/yr	Frog	Resistance to chromosomal breakage from high doses	DNA repair mechanism	Mitchel, 2006
Antioxidants	25 & 50 cGy	Mice	Homeostatic Redox balance maintenance	Glutathione (GSH), GR, GPx, γ-GCS activation	Kojima *et al.* 2004
	10-50 cGy	Mice	Free radical quenching	SOD, GSH & CAT activation	Avti *et al.* 2005
	10 - 20 cGy	Mice	ROS detoxification, upregulation of antioxidant defense system	SOD, GSH & CAT activation	Pathak *et al.* 2007

Abbreviations:

ADCC: Antibody dependent cell mediated cytotoxicity; CAT: Catalase; CD: Cluster of differentiation; CDKNIA: Cyclin dependent kinase inhibitor 1; CTL: Cytotoxic T lymphocyte; DC cell: Dendritic cell; GADD45A: Growth arrest and DNA damage inducible protein alpha; GM-CSF: Granulocyte macrophage colony stimulating factor; GPx: Glutathione peroxidase; GR: Glutathione reductase; GSH: Reduced glutathione; IGFBP3: Insulin-like growth factor binding protein 3; IL: Interleukin; MDM2: Murine double minute 2; NK cell: Natural killer cell; PKC: Protein kinase C; PLAU: Urokinase-type plasminogen activator; RAD50: Radix 50; RBC-C3Rb: Red blood cell C3b receptor; RBC-C3RRb: RBC-C3b receptor yeast rosette rate; REV3L: Gene encoding enzyme DNA polymerase zeta catalytic subunit; Socs-5: Suppressor of cytokine signaling-5; SOD: Superoxide dismutase; Tbox 21: Tbox 21: T box protein 21; Th1: T helper cell type 1; Th2: T helper cell type 2; TNF-α: Tumor necrosis factor alpha; XAB2: Xeroderma pigmentosum group A (XPA)–binding protein 2; γ-GCS: Gamma glutamylcysteine synthetase functions by the release of a variety of cytokines from immune cells of the animals exposed to low-dose single as well as chronic irradiation are given in Table 5 (Hashimoto *et al.* 1999, Sun and Liu 2000, Safwat 2000a, Liu *et al.* 2001a, Liu *et al.* 2001b).

Recently, it has been reported that exposure of mice or rats to low-doses (<0.2 Gy) of X- or γ-rays markedly stimulated the proliferation and/or cytotoxic activity of NK (Natural Killer) cells and macrophages obtained from the irradiated animals which suppressed tumor metastases and inhibited the tumor growth (Kojima *et al.* 2004, Cheda *et al.* 2004a, 2004b, Ina *et al.* 2005, Nowosielska *et al.* 2006). In another study, low-dose exposure of dendritic cells increased the T cell activation capacity through the cytokine production of IL-2, IL-12 and IFN-γ (Shigematsu *et al.* 2007). Nakatsukasa and his co-workers (2008) have reported suppression of pro-inflammatory cytokines like TNF-α, IFN-γ and IL-6 in the splenocytes obtained from the collagen-induced arthritis in DBA/IJ mice after LD-WBI with γ-rays (0.5 Gy per wk for 5 wks). Recently, Tsukimoto and his colleagues (2009) reported that exposure to 0.5 Gy of γ-ray radiation caused up-regulation of MKP-1 (Mitogen-activated Protein Kinase Phosphatase-1), leading to inactivation of p38 MAPK and suppression of TNF-α production, in cells of mouse macrophages cell line. On the other hand, no change in the expression of mRNA of IL-10 was seen in the splenocytes obtained from the animals exposed to LD-WBI at a dose of 0.2 Gy (Hashimoto *et al.* 1999).

Determination of global gene expression pattern using cDNA or oligonucleotide microarrays is an effective approach to understand the complex mechanism(s) of radiation effects. Recently, in order to decipher the organ specific responses to low-dose-rate radiation, a few studies have been reported related to DNA metabolism, DNA damage, DNA replication, immunologic system, apoptosis and anti-oxidant defense system (Amundson *et al.* 2003, Yu *et al.* 2007, Roudkenar *et al.* 2008, Taki *et al.* 2009, Gridley *et al.* 2009). The limited data show that single whole body exposure or protracted low-dose exposure has dose- and time-dependant effects on various immune cells and there is specific modulation of gene expression in different organs.

Clinical Aspects of LD-WBI

Low-dose single or fractionated total body irradiation has been used in the past for the treatment or prevention of a wide variety of diseases. Nearly 150 different diseases were reported to be treated using LD-WBI through the use of radioactive water or external irradiations. Calabrese and Baldwin in 2000 have given a brief historical review of clinical literature relating to the treatment of various diseases (fruncles, carbuncles and other pyogenic infections, pneumonia, trachoma, gas-bacillus, peritonitis etc) using single or fractionated radiation doses in the range of 10-100 rads. In one of the very important study, Kelly and Dowell in 1941 reviewed the low-dose radiation treatment of 364 cases of gas gangrene infection from 1928 to 1940 and found several doses of X-rays (50 or 75 rad) applied locally reduced mortality from approximately 50% with amputations to 5% without amputation.

Recently, a number of experimentally induced diseases in different animal models have been reported to be suppressed/prevented/delayed employing LD-WBI (<0.5 Gy). These studies include carbon tetrachloride-induced liver damage in mice (Yamaoka *et al.* 2004), 1-methyl-4-phenyl-1,2,3,6-tetrahydropyrine- or Fe-NTA-induced brain damage in mice (Kojima *et al.* 1999, Yamaoka *et al.* 2002), collagen-induced arthritis in MRL-lpr/lpr mice (Nakatsukasa *et al.* 2008), amelioration of autoimmune diseases involving splenomegaly, lymphadenopathy and proteinuria in immune deficient MRL-lpr/lpr mice (Tanaka *et al.* 2005), prevention of Type I diabetes in NOD (non obese diabetes) mice (Takahashi *et al.* 2000, Wang *et al.* 2008), suppression of atopic dermatitis and tumor metastasis induced by Radon-222 in NC/Nga mice (Takahashi and Kojima 2006), delay in tumor appearance after inoculation of tumor cell lines in mice (Ito *et al.* 2008), and decrease in tumor volume in mice (Kojima *et al.* 2004).

Clinical applications of the LD-WBI in the treatment of cancer has shown enlightening results. When patients with non-Hodgkins lymphoma under chemotherapy were given whole body or half body (to the thorax) low dose radiation in a regimen of 0.1 Gy three times a week or 0.15 Gy two times a week for 5 consecutive weeks with a total dose of 1.5 Gy, the survival rate in 9 years rose to 84% from 50% of the control with chemotherapy alone ($p<0.01$). Half body LDR of women given 15-30r by 75-150 fluoroscopic lung examinations similarly decreased breast cancer mortality. Breast cancer mortality decreased by 34% when the female patients were exposed to doses between 10-20r (Cuttler and Pollycove 2003). The LD-WBI treated patients also showed enhanced anti-cancer immunity (Sakamoto *et al.* 1997, Safwat 2000b). Similar encouraging results on humans have also been reported earlier from two Harvard University clinical trials on LD-WBI as reviewed by Pollycove and Feinendegen (2008).

Recently, LD-WBI has also been reported to enhance the therapeutic efficacy of chemotherapeutic agents like mitomycin-C and cyclophosphamide against tumor growth in mice when administered either before or after irradiation (Zhang *et al.* 1999, Yu *et al.* 2007). It suggests that the use of such low-doses of radiation with a chemotherapeutic agent is a novel approach to achieve significant chemo-potentiation. If this strategy is successful, further investigations into the benefits of this approach in clinical setting could follow.

Conclusions

The stimulation of immunity (enhanced antibody formation, NK, CTL, macrophage, ADCC activities, secretion of interferon and other cytokines), activation of antioxidant defense system (GSH, SOD, GPx, GPR, CAT, Rhodanese), enhancement of DNA repair mechanism (poly (ADP-ribose) polymerase (PARP), apurinic/apyrimidinic (AP)-endonuclease) and induction of apoptosis (cytochrome c release, caspase-3 activity) as well as other cellular changes by LD-WBI concerns most anticancer parameters. Animal studies have revealed that LD-WBI retards tumor growth, decrease cancer metastasis, and suppresses experimentally induced diseases. Optimal protocols should be established by determining the optimum magnitude and duration of these responses coupled with the minimal marrow suppression achieved by reduced irradiation of body volume. Clinical trials are also indicated to determine the effectiveness of LD-WBI in patients with early HIV and other infectious diseases. LD-WBI of patients appears to be asymptomatic with minimal side effects, a rational and promising way of using anti-mutagenic system to control cancer and infection. Current rapid increase of published genetic, molecular biology, and animal data that confirm the stimulatory effects of low-dose ionizing radiation will overcome the resistance of the radiation regulatory authorities. However, more research is warranted before LD-WBI could be employed as another therapeutic modality for the amelioration of certain diseases.

Acknowledgements

Prof. C.M. Pathak was a Visiting Scientist in the Cell Pathophysiology Laboratory, Institute of Cardiovascular Sciences, Winnipeg. Dr. Pawan K. Singal holds the Naranjan Dhalla Chair in Cardiovascular Research supported by the St. Boniface Hospital and Research Foundation.

References

1. Amundson SA, Lee RA, Koch-Paiz CA, Bittner ML, Meltzer P, Trent JM, Fornace AJ.. Differential responses of stress genes to low dose-rate γ-irradiation. Mol. Can. Res. 1: 445-452 (2003).
2. Avti PK, Pathak CM, Kumar S , Kaushik G , Kaushik T , Farooque A , Khanduja KL , Sharma SC. Low dose gamma-irradiation differentially modulates antioxidant defense in liver and lungs of Balb/c mice. Int. J. Raidat. Biol. 81: 901-910 (2005).
3. Calabrese EJ, Baldwin L A. Defining hormesis. Hum. Exp. Toxicol. 21: 91-97 (2002a)
4. Calabrese EJ, Baldwin LA. Radiation hormesis: its historical foundations as a biological hypothesis. Hum. Exp. Toxicol. 19:41-75 (2000)
5. Calabrese EJ, Baldwin LA. Radiation hormesis and cancer. Hum. Ecol. Risk Assess. 8: 327-353 (2002b).
6. Cheda A, Wrembel-Wargocka J, Lisiak E, Nowosielska EM, Marciniak M and Janiak MK. Single low doses of X-rays inhibit the development of experimental tumor metastases and trigger the activities of NK cells in mice. Radiat. Res. 161: 335-340 (2004a).
7. Cheda A, Wrembel-Wargocka J, Lisika E, Marciniak M, Nowosielska EM, Janiak MK. Inhibition of the development of pulmonary tumor nodules and stimulation of the activity of NK cells and macrophages in mice by single low dose of low-LET radiation. Int. J. Low Radiat. 1:171-179 (2004b).
8. Cuttler JM, Pollycove M. Can cancer be treated with low doses of radiation? J. Am. Phy. Surg. 8: 108-111 (2003).
9. Eisenbud M, Gesell T. Environmental Radioactivity from Natural, Industrial, and Military Sources (San Deigo, CA: Academic) Chapter 6 (Natural Radioactivity) (1997).
10. Fachin AL, Mello SS, Garcia PS, Junta CM, Donadi EA, Passos GAS, Sakamoto-Hojo ET. Gene Expression Profiles in Human Lymphocytes Irradiated In Vitro with Low Doses of Gamma Rays. Radiat. Res. 168:650-665 (2007).
11. Fachin AL, Mello SS, Garcia PS, Junta CM, Ghilardi-Netto T, Donadi EA, Passosa ADS, Sakamoto-Hojo ET. Gene Expression Profiles in radiation workers occupationally exposed to ionizing radiation. J. Radiat. Res. 50:61-71 (2009).
12. Fan XH, Liu SZ. Radiation Effect on NK Activity in Mouse Spleen. J. N. Bethune. Univ. Med. Sci. 15:551-553 (1989).
13. Feinendegen LE. Evidence for beneficial low level radiation effects and radiation hormesis. Br. J. Radiol. 78: 3-7 (2005).
14. Ghiassi-nejad M, Mortazavi SMJ, Cameron JR, Niroomand-rad A, Karam P A. Very High Background Radiation Areas of Ramsar, Iran: Preliminary Biological Studies. Health Phys. 82: 87-93 (2002).

15. Gridley S, Rizvi A, Luo-Owen, Makinde AY, Pecaut MJ. Low dose, low dose rate photon radiation modifies leukocyte distribution and gene expression in CD4+ T cells. J. Radiat. Res. 50: 139-150 (2009).
16. Hall E. Is there a place for quantitative risk assessment? J. Radiol. Prot. 29: A171 (2009).
17. Hashimoto S, Shirato H, Hosokawa M, Nishioka T, Kuramitsu Y, Matushita K, Kobayashi M, Miyasaka K. The suppression of metastases and the change in the host immune response after low-dose total-body irradiation in tumor-bearing rats. Radiat. Res.151:717-724 (1999).
18. Ina Y, Sakai K. Activation of immunological network by chronic low-dose-rate irradiation in wild-type mouse strains; Analysis of immune cell populations surface molecaules. Int. J. Radiat. Biol. 81: 721-729 (2005).
19. Ina Y, Sakai K. Prolongation of Life Span Associated with Immunological Modification by Chronic Low-Dose-Rate Irradiation in MRL-lpr/lpr Mice. Radiat. Res. 161:168-173. (2004).
20. Ina Y, Tanooka H, Yamada T, Sakai K. Suppression of thymic lymphoma induction by life-long low-dose rate irradiation accompanied by immune activation in C57BL/6 mice. Radiat. Res. 163:153-158 (2005).
21. Ito M, Shibamoto Y, Ayakawa S, Tomita N, Sugie C, Ogino H. Effect of Low-Dose Total-Body Irradiation on Transplantability of Tumor Cells in Syngeneic Mice. J. Radiat. Res. 49: 197-201 (2008).
22. Kelley MJ, Dowell DA. Twelve-year review of X-ray therapy of gas gangrene. Radiology 37: 421-439 (1941).
23. Kojima S, Matsuki O, Kinoshita I, Gonzalez TV, Shimura N, Kubodera A. Does small dose gamma-ray radiation induced endogenous antioxidant status in vivo? Biol. Pharm. Bull. 20: 601-604 (1997).
24. Kojima S, Matsuki O, Nomura T, Yamaoka K, Takahashi M, Niki E, Elevation of antioxidant potency in the brain of mice by low-dose γ-ray irradiation and its effect on 1-methyl-4-phenyl-1,2,3,6-tetrahydropyridine (MPTP)-induced brain damage. Free Rad. Biol. Med. 26: 388-395 (1999).
25. Kojima S, Nakayama K, Ishida H. Low dose gamma-rays activate immune functions via induction of glutathione and delay tumor growth. J. Radiat. Res. 45: 33-39 (2004).
26. Kondo, S. Health Effects of Low-Level Radiation, Kinki University Press, Osaka, (1993).
27. Liu SZ, Bai O, Chen D, Ye F. Genes and protein molecules involved in the cellular activation induced by low dose radiation. J. Radiat. Res. Radiat. Proc. 18: 175-186 (2000).
28. Liu SZ, Jin SZ, Liu XD, Sun YM. Role of CD28/B7 Co-stimulation and IL-12/IL-10 Interaction in the Radiation-Induced Immune Changes. BMC Immunology 2:8 (2001).
29. Liu SZ. Cancer Control Related To Stimulation Of Immunity By Low-Dose Radiation. Dose-Response. 5:39-47 (2007).
30. Liu SZ. Nonlinear Dose-Response Relationship in the Immune System Following Exposure to Ionizing Radiation: Mechanisms and Implications. Nonlinearity in Biol. Toxicol. Med. 1:71-92 (2003a).

31. Liu SZ. On Radiation Hormesis Expressed in the Immune System. Crit. Rev. Toxicol. 33:431-441 (2003b).
32. Liu SZ. Radiation Hormesis with Low Level Exposures. 250-254. Beijing: Scientific Press (1996).
33. Liu XD, Liu SZ, Ma SM, Liu Y. Expression of IL-10 in mouse spleen mRNA and protein level after whole-body X-irradiation. Chin. J. Radiol. Med. Prot. 22: 10-12 (2001a).
34. Liu XD, Liu SZ, Ma SM, Liu Y. Opposite changes of IL-10 and IL-12 expression in mice after low-dose whole-body X-irradiation. J. Radiat. Res. Radiat. Proc. 19:253-258 (2001b).
35. Luckey T.D. Nurture with Ionizing Radiation: A Provocative Hypothesis. Nutr. Can. 34: 1-11 (1999).
36. Luckey TD. Physiological Benefits from Low Levels of Ionizing Radiation. Health Phys. 43: 771-789 (1982).
37. Luckey, T.D. Low-Dose Irradiation Reduces Cancer Deaths Radiat. Protect. Manag. 14:58-64 (1997).
38. Martin CJ, Sutton DG, West CM, Wright EG. The radiobiology/radiation protection interface in health care. J. Radiol. Prot. A1-A20 (2009).
39. Mettler FA Jr, Thomadsen BR, Bhargvan M, Gilley DB, Gray JE, Lipoti JA, McCrohan J, Yosizumi TT, Mahesh REJ. Medical Radiation exposure in US in 2006: preliminary results. Health Phys. 95: 502-507 (2008).
40. Mitchel REJ. Low doses of radiation reduce risk in vivo. Dose-Response. 5:1-10 (2006).
41. Nakatsukasa H, Tsukimoto M, Ohshima Y, Tago F, Masada A, Kojima S. Supressing effect of low-dose gamma-ray irradiation on collagen-induced arthritis. J. Radiat. Res. 49: 381-389 (2008).
42. Nowosielska EM, Wrembel-Wargocka J, Cheda A, Lisiak E, Janiak MK. Enhanced cytotoxicity activity of macrophages and suppressed tumor metastases in mice irradiated with low doses of X-rays. J. Radiat. Res. 47: 229-236 (2006).
43. Pathak CM, Avti PK, Kumar S, Khanduja KL, Sharma SC. Whole body exposure to low-dose gamma radiation promotes kidney antioxidant status in Balb/c mice. J. Radiat. Res. 48: 113-120 (2007).
44. Pollycove M, Feinendegen LE. Low-dose radioimmuno-therapy of cancer. Hum. Exp. Toxicol. 27: 169-175 (2008).
45. Pollycove M. Radiobiological basis of low dose irradiation in prevention and therapy of cancer. 14th Pacific Basin Nuclear Conference, Honolulu, Hawaii, March 21-25, pp 647-653 (2004).
46. Pollycove, M. Radiobiological basis of low-dose irradiation in prevention and therapy of cancer. Dos. Res. 5:26-38 (2007).
47. Roudkenar MH, Li L, Baba T, Kuwahara Y, Nagakawa H, Wang L, Kasaoka S, Ohokubo Y, Ono K, Fukumoto M. Gene expression profiles in mouse liver cells after exposures to different types of radiation. J. Radiat. Res.49: 29-40 (2008).
48. Safwat A. The immunobiology of low-dose total-body irradiation: More questions than answers. Radiat. Res.153: 599-604 (2000a).

49. Safwat A. The role of low-dose total body irradiation in treatment of non-Hodgkin's lymphoma: a new look at an old method. Radioth. Oncol. 56: 1-8 (2000b).
50. Sakamoto K, Myogin Y, Hosoi Y. Fundamental and clinical studies on cancre control with total or upper half-body irradiation. J. Jpn. Ther. Radiol. Oncol. 9:161-175 (1997).
51. Sanzari JK, Nuth M, Kennedy AR. Induction of Cytokine Gene Expression in Human Thyroid Epithelial Cells Irradiated with HZE Particles (Iron Ions). Radiat. Res. 172: 437-443 (2009)
52. Segerer SE, Müller N, Brandt J, Kapp M, Dietl J, Reichardt HM, Rieger L, Kämmerer U. The glycoprotein-hormones activin A and inhibin A interfere with dendritic cell maturation. Reprod. Biol. Endocrinol. 6: 17-27 (2008).
53. Shigematsu A, Adachi Y, Koike-Kiriyama N, Suzuki Y, Iwasaki M, Koide Y, Nakano K, Mukaide H, Imamura M, Ikehara S. Effects of low-dose irradiation on enhancement of immunity by dendritic cells. J. Radiat. Res. 48: 51-55 (2007).
54. Shimizu, Y, Kato, H, Schull, W.J. and Mabuchi, K, "Dose-Response Analysis Among Atomic Bomb Survivors Exposed to Low-Dose Irradiation," pp.71-74 in Sugahara, T, Sagan, L. and Aoyama, T, (eds), Low-Dose Irradiation and Biological Defense Mechanisms. Excerpta Medica, Amsterdam (1992).
55. Su X, Liu SZ, Ohyama H. Influence of Dose Rate on Radiation Hormesis with Low Dose Exposure. J. N. Bethune. Univ. Med. Sci. 23:582-584 (1997).
56. Sun YM, Liu SZ. Changes in TNF-α expression in mouse peritoneal macrophages after whole-body X-irradiation. J. Radiat. Res. Radiat. Proc. 18: 235-239 (2000).
57. Takahashi M, Kojima S, Yamaoka K, Niki E. Prevention of type I diabetes by low-dose gamma irradiation in NOD mice. Radiat. Res. 154: 680-685 (2000).
58. Takahashi M, Kojima S. Suppression of atopic dermatitis and tumor metastasis in mice by small amounts of radon. Radiat. Res. 165: 337-342 (2006).
59. Taki K, Wang B, Nakajima T, Wu J, Ono T, Uehara Y, Matsumoto T, Oghiso Y, Tanaka K, Ichinohe K, Nakamura S, Tanaka S, Magae J, Kakimoto A, Nenoi M. Microarray analysis of differentially expressed genes in the kidneys and testes of mice after long-term irradiation with low -dose- rate γ-rays. J. Radiat. Res. 50: 241-52 (2009).
60. Tanaka T, Tago F, Fang S.P, Shimura N, Kojima S. Repeated 0.5-Gy γ-ray irradiation attenuates autoimmune manifestations in MRL-lpr/lpr mice. Int. J. Radiat. Biol. 81: 731-740 (2005).
61. Tiku AB, Kale RK. Adaptive response and split-dose effect of radiation on the survival of mice. J. Biosci. 29:111-117 (2004).
62. Tsukimoto M, Homma T, Mutou Y, Kojima S. 0.5 Gy gamma radiation suppresses production of TNF-α through up-regulation of MKP-1 in mouse macrophage RAW264.7 cells. Radiat. Res. 171: 219-224 (2009).
63. Tubiana M, Aurengo A, Averbeck D, Bonnin D, Leu-Guen B, Masse R, Monier R, Valleron AJ, de Vathaire F. Dose-Effect Relationships and estimation of the carcinogenic effects of low doses of ionizing. Radiation Academic National de Medecine (Paris: Institut de France-Academie des Sciences) (2005).

64. United Nations Scientic Committee on Effects of Atomic Radiation : Adaptive responses to radiation in cells and organisms. New York : United Nations General Assembly, (Publ A/AC.82/ R.542), (1994).

65. United Nations Scientific Committee on Effects of Atomic Radiation: Genetic and Somatic Effects of Ionizing Radiation. In: Report to the General Assembly, with Annexes. p170, (1986).

66. Wang B, Ohyama H, Shang Y, Tanaka K, Aizawa S, Yukawa O, Hayata I. Adaptive Response in Embryogenesis: V. Existence of Two Efficient Dose-Rate Ranges for 0.3 Gy of Priming Irradiation to Adapt Mouse Fetuses. Radiat. Res. 161: 264-272 (2004).

67. Wang GJ, Li XK, Sakai K, Cai L. Low-dose radiation and its clinical implications: diabetes. Hum. Exp. Toxicol. 27:135-142 (2008).

68. Yamaoka K, Kataoka T, Nomura T, Taguchi T, Wang D H, Mori S, Hanamoto K, Kira S. Inhibitory Effects of Prior Low-dose X-ray Irradiation on Carbon Tetrachloride-induced Hepatopathy in Acatalasemic Mice. J. Radiat. Res. 45: 89-95 (2004).

69. Yamaoka K, Mori S, Nomura T, Taguchi T, Ito T, Hanamoto K, Kojima S. Elevation of antioxidant potency in mice brain by low-dose X-ray irradiation and its effect on Fe-NTA-induced brain damage. Physiol. Chem. Phys. Med. NMR 34: 119-132 (2002).

70. Yu HS, Song AG, Lu YA, Qiu WS, Shen FZ. Effects of low-dose radiation on tumor growth, erythrocyte immune function and SOD activity in tumor-bearing mice. Chin. Med. J. 117:1036-1039 (2004).

71. Yu HS, Xue HW, Guo CB, Song AQ, Shen FZ, Liang J, Deng C. Low dose radiation increased the therapeutic efficacy of cyclophosphamide on S180 sarcoma bearing mice. J Radiat. Res. 48: 281-288 (2007).

72. Zhang Y, Liu SH. Effect of Low Dose Radiation on Immune Functions of Tumor-Bearing Mice. Chin. J. Radiol. Health 5:235-237 (1996).

73. Zhang, Y., Lu, Z., Li, X.Y. Effect of combined whole-body low dose irradiation and chemotherapy on growth, metastasis and immune functions in tumor bearing mice. Radiat. Prot. 19: 127-131 (1999).

Adaptation Biology and Medicine (Volume 6: Cell Adaptations and Challenges)
Editors: P. Wang, C.-H. Kuo, N. Takeda and P.K. Singal
Copyright © 2011 Narosa Publishing House Pvt. Ltd., New Delhi, India

28. The Dehydroepiandrosterone Sulfate and Adaptation Against Environmental Challenges

Shu-Man Chen[1], Li-Chen Lee[1], Chia-Hua Kuo[2*]

[1]Committee for General Education, Shih Hsin University and [2]Laboratory of Exercise Biochemistry, Taipei Physical Education College, Taipei, Taiwan

Summary

Failure in coping with daily environmental fluctuation can be regarded as the fundamental cause for onset of diseases and eventual death. Human body continuously copes with alterations in physical activity, ambient temperature, oxygen availability and psychological challenges. Successful adaptation against those daily fluctuations or challenges ensures survival and wellbeing. Loss in adaptive capability is commonly observed during aging. Circulating dehydroepiandrosterone sulfate (DHEA-S) concentration has been found to closely associate with human longevity and fitness, and is known to decline with advancing aging. Based on this view, we performed a series of human experiments to test the hypothesis that DHEA-S deficiency is linked with inadequate adaptation against external challenges. This review paper describes our recent findings, which might help to elucidate the role of DHEA-S in the adaptation against the changing environment.

Keywords: longevity, aging, biomarker, coping, DHEA, DHEA-S, life span

Background

Dehydroepiandrosterone (DHEA) and its sulfate derivative (DHEA-S) are produced in the adrenal gland and brain[1-4], which serve as primary precursors in the biosynthesis of many steroid hormones, such as androgens and estrogens. Due to its enrichment in brain and its role in cognitive function[5], it has been considered as a neurosteroid by many researchers. The protective and neurotrophic effects of the steroids have been extensively studied in the brain, spinal cord, and cognitive function after traumatic challenge[6-8].

Circulating DHEA-S level in animal is proportional to their lifespan across several different species. For example, it is not detectable in rats which have an average lifespan of only 3-4 years[9]. In a short-lived primate - the grey mouse lemur - with lifespan of 10-13 years, has 20-200 ng/mL of circulating DHEA-S[10], whereas the range is from 30-300 ng/mL in other non-human primates[11]. Human has very high DHEA-S level (500-5000 ng/mL) compared to all other reported species[12, 13]. Administration of

*Address for Correspondence: Correspondence should be addressed to: Chia-Hua Kuo, Ph.D. Laboratory of Exercise Biochemistry Taipei Physical Education College 101 Jhongcheng. Rd, Sec. 2 Taipei, Taiwan 111 Phone: + 886-2-28716470 ext. 5101 Fax: +886-2-28753383 E-mail: kuochiahua@gmail.com

DHEA directly increases DHEA-S level[14], leads to a wide array of health benefit in rats, whereas human experiment receives less positive results. Thus, human subjects would be more realistic model to elucidate the biological role of DHEA-S in human health.

A progressive decrease in circulating DHEA-S levels with age has long been documented in humans, with peak levels occurring between the ages of 20 and 25 and decreasing progressively thereafter by 90% after the age of 85[15]. This age-dependent decline has been linked to the gradually increasing prevalence of reduced muscular fitness, atherosclerosis, obesity, and diabetes in the elderly[16-20]. There is great individual variation in circulating DHEA-S levels among humans[13, 21]. Combined results from NIA Primate Aging Study and Baltimore Longitudinal Study of Aging provide strong evidence that DHEA-S is a biomarker for human longevity[22].

Since daily coping (adaptation) with the environment changes is directly linked with survival and wellbeing[23], we hypothesized that DHEA-S could be a biomaterial consumed during recovery from external stress. In this paper, we summarize evidence from our previous published works to elucidate the role of DHEA-S in physiological and psychological adaptations against external challenges[14, 21, 24-27].

Impaired Exercise Training Adaptation for the Elderly with Low DHEA-S[21]

Although exercise is a challenge to human body, the beneficial effects of exercise training on health and life expectancy are well documented[28]. Based on this evidence, exercise training is generally recommended for elderly to maintain health status. However, it was also reported that the training effect on improving glycemic control for elderly is not as effective as for young individuals[29]. DHEA-S declines with age for both males and females. We have previously investigated the effect of a 4-month exercise training program on glucose tolerance and insulin sensitivity in a group of elderly greater than 80 years, in relation to their baseline DHEA-S levels. Our result shows that the chronic exercise training effect on improving insulin sensitivity was absent in the low DHEA-S subjects. This study was the first report demonstrating that the elderly subjects with low DHEA-S level were poor responders to exercise training intervention.

The degree of insulin resistance is increasing with age, and has been suggested as an origin of many age-onset metabolic syndromes, including hypertension, vascular complications, type 2 diabetes, and cancers[30]. The related metabolic parameters such as, blood pressure and fasting glucose, have also been shown as significant predictors for longevity[13]. Reduction in diastolic BP by exercise training was evident in the elderly subjects with high DHEA-S, but not for the low DHEA-S halves[21]. High BP is a known risk factor leading to stroke and heart failure. According to our data[14, 21], individual variations in DHEA-S level can be one possibility that accounts for the discrepancy among many earlier studies presenting inconsistent results in BP-lowering effect of exercise training.

The elderly subjects with greater DHEA-S levels also exhibited greater enhancement in motor performance by training. This result could be related to the improvements in neuromuscular components secondary to the improvement in insulin sensitivity. Increasing insulin action could result in better capability to store glycogen and reduce the rate of muscle protein degradation. Additionally, the sugar for DNA synthesis during cell turnover is primarily from intracellular glycogen and glucose transported from circulation. Thus high insulin sensitivity is essential for preserving greater anaerobic fuel, normal contractile property of skeletal muscle, and replacement of damaged muscle cell after acute physical challenge. Elderly are usually faced with the problem of poor insulin sensitivity and high risk of type 2 diabetes that can have major impact on nutrient delivery to peripheral motor neuron due to microvascular complication. DHEA-S has been found to exert a neuroprotective effect on motor neuron, evidenced by the fact that treating DHEA in diet for more than 5 weeks prevents the diabetes-induced development of neural dysfunction[31].

Improvement in carbohydrate metabolism by exercise training may be functionally relevant to survival and longevity[13]. Environmental stress is persistently occurring during the entire lifetime and threatening human survival. Age-dependent reduction in insulin-stimulated glycogen storage may be linked with insufficient adaptation against daily stress and explains the decrease in cumulative survival in human. Under stress condition, demand for ATP is increasing within a short period of time. Carbohydrate has the advantage of fast degradation rate in muscle cell and can occur without oxygen supply for rapid ATP resynthesis. With advanced aging, the metabolism of human body is characterized by less reliance on carbohydrates[32]. In young individuals, exercise is a stress condition that leads to rapid consumption of muscle glycogen. Afterward, the whole-body glucose tolerance and the rate of muscle glycogen storage increase during recovery, resulting in glycogen supercompensation. Exercise adaptation in improving insulin sensitivity ensures human body reserve of more carbohydrate fuel for coping with the recurrence of similar challenge.

Muscle-damaging Exercise Reduces Circulating DHEA-S[27]

As we have mentioned earlier, circulating dehydroepiandrosterone sulfate (DHEA-S) concentration is closely associated with the whole-body insulin sensitivity[13, 22, 33], and both insulin (hyperinsulinemia indicated insulin resistance) and DHEA-S levels have been suggested as biomarkers of human longevity[34]. DHEA-S has been reported to have an anti-glucocorticoid effect[35]. Glucocorticoids are known to have a potent inhibitory effect on insulin action[36]. DHEA supplementation elevates erythrocyte insulin receptor binding in dexamethasone-treated patient[37], indicating a direct effect of DHEA-S on insulin signaling pathway. Therefore, low DHEA-S level could unmask the cortisol action, which may in turn reduce insulin sensitivity without a noticeable change in cortisol level.

To determine the direct effect of DHEA-S on insulin sensitivity, we increased circulating DHEA-S level (+ 3 folds) in middle-aged subjects with oral DHEA administration. A significant improvement in the whole-body insulin sensitivity was observed[14]. Although resistance exercise is known to enhance insulin sensitivity in humans, it does generate considerable muscle damage which may or may not induce a protracted repair process. To determine the effect of resistance exercise on DHEA-S concentration during a 72-h post-exercise recovery in relation to insulin sensitivity, blood samples were collected from 19 male volunteers (aged 21.1±0.4 years) 24 h before resistance exercise and 24 h, 48 h, and 72 h following exercise. We found a persistent reduction in DHEA-S level with concurrent improvement in insulin sensitivity during the recovery period[27].

DHEA-S decline phenomenon could be due to increased demand for muscle tissue repair and recovery. This steroid has been documented to have a buffering action during various stress[38, 39], and was found to reduce in trauma and disease[8]. Thus we speculate that the increased demand on DHEA-S following resistance exercise could account for the depression or consumption of this steroid during recovery, based on the evidence that exogenous DHEA administration following various conditions causing tissue damage may enhance functional recovery in human and animal[40, 41].

High Altitude Challenge Lowers DHEA-S[26]

High altitude hiking represents a major stress condition to human body since oxygen demand is increasing during activity while oxygen availability is reduced at higher altitude. The role of DHEA-S on altitude adaptation was evaluated in 12 young males participated in a 25-day mountaineering activity, who were evenly divided into lower and upper halves according to their sea-level DHEA-S levels: low DHEA-S (N = 6, 1480 ± 247 ng/ml) and high DHEA-S groups (N = 6, 3588 ± 590 ng/ml). We found that serum DHEA-S level was significantly reduced on Day 3 of the altitude activity, only in the young subject with high DHEA-S level[42], while DHEA-S level in the low DHEA-S subjects remained unchanged. To further determine whether the DHEA-S reduction in the upper halves was due to decreased adrenal function in response to this altitude stress, we also measured the levels of another adrenal hormones, cortisol. Our data did not provide a strong support for the possibility of reduced adrenal function in the altitude activity in the high DHEA-S subjects, since serum cortisol was increased above sea-level in higher altitude for those DHEA-S declined subjects. Therefore, the DHEA-S depletion phenomenon is more likely due to an increased consumption for altitude adaptation. This hypothesis is strengthened by the fact that only the high DHEA-S group exhibited significant adaptive responses in improved glucose tolerance and increased erythropoiesis compared to the low DHEA counterparts.

Increase in red blood cell (RBC) concentration is a classic adaptive response in altitude exposure. We found that only the subjects with greater DHEA-S level have the normal altitude response in erythropoiesis. Intriguingly, the low DHEA-S group without increase in RBC exhibited about 2-fold greater serum EPO concentration, at both sea-level and altitude, compared to the high DHEA-S subjects. This result suggests that EPO sensitivity for stimulating erythropoiesis in bone marrow was lower in the subjects with low DHEA-S level. The increased RBC concentration in the high DHEA-S group is not likely to be explained by reduction in plasma volume in response to altitude exposure, since white blood cell (WBC) concentration in both the low DHEA-S and the high DHEA-S groups was not significantly altered during the 25-day altitude activity period.

It appears that whether subjects can successfully induce an improvement in glucose tolerance and insulin sensitivity by the altitude activity is associated with their serum DHEA-S concentration. Glucose is a major anaerobic fuel for most of the tissue in human body, which is particularly essential for maintaining normal ATP resynthesis as the fatty acid oxidation is hindered in hypoxic environment. Therefore, improvement in glucose tolerance could be functionally meaningful in ensuring better survival at higher/different altitudes.

Improvement in maximal oxygen consumption by altitude exercise training has been frequently documented[48], partly related to erythropoiesis induced by EPO[49]. This adaptation is particularly important when the rate of oxygen utilization is higher than the rate of oxygen supply during prolonged altitude activity. Despite variations in increasing hemoglobin, haematocrit, and RBC by the altitude activity is related to their basal DHEA-S level, the low DHEA-S group exhibits a comparable improvement in maximal oxygen consumption without improvement in hemoglobin, haematocrit, and RBC as those high DHEA-S subjects. Since the sea-level maximal oxygen consumption of the low DHEA-S group was lower than those of the high DHEA-S group at the beginning, the relative training intensity during altitude hiking activity for the low DHEA-S subjects was greater than those for the high DHEA-S counterpart, since both groups hiked at the same pace. Greater relative exercise intensity may result in greater adaptation in arterial-venous extraction, secondary to the increase in capillary density.

Increasing anaerobic fuel utilization and enhancing oxygen carrying capability are the two fundamental physiologic strategies for survival during prolonged altitude activity. In this study, we observed that the circulating DHEA-S level was significantly reduced during the altitude activity in the subjects exhibiting greater baseline DHEA-S concentration. Furthermore, we demonstrated that the magnitude of altitude activity effects on improving glucose tolerance, insulin sensitivity, and increasing RBC concentration were varied with the baseline DHEA-S level.

Hot Spring Bath can Cause Insulin Resistance in The Low DHEA-S Individuals[24]

Coping with heat stress is one of the most important survival mechanisms for human body, since increasing body temperature can cause permanent change in protein conformation that leads to death. In our living environment, ambient temperature is continuously changing and this factor has been reported to affect human mortality[50]. For human, sweat evaporation from skin surface is the most important mechanism to dissipate excessive body heat. Although hot spring bath has been frequently used to alleviate stress feeling, this model exerts greater magnitude of heat stress compared to dry hot weather, since skin was surrounded by hot water which blocks sweat dissipation. A state of insulin resistance, evidenced by increased area under curve of glucose and HOMA-IR values (insulin resistance indicator), was induced in 16 healthy males with a 30-min hot spring immersion at 41°C. A wide range of individual variation for the heat-induced stress response was identified. All subjects were then evenly divided into lower and upper halves according to their baseline DHEA-S concentrations. In this study, changes in insulin resistance measures in relation to DHEA-S level were observed. We found that hot spring immersion-induced insulin resistance state was restricted only to the low DHEA-S group. This result, together with the evidence from a number of studies from our group and others investigating various types of acute stress[42, 51-53], implicates that the DHEA-S plays a role for physiologic coping during acute stress.

To maintain the set-point of body temperature, human body must increase skin blood flow, and thus result in an acute reduction in diastolic blood pressure. Under this condition, heart rate must increase to prevent an acute reduction of blood supplies (oxygen and substrates) to the rest of peripheral tissues other than skin. In the study we found that the low DHEA-S group exhibited slower heart rate recovery after hot spring immersion, suggesting a lower coping capability with the heat stress.

During hot spring immersion, circulating growth hormone and cortisol are increasing, together with the onset of insulin resistance[54]. Cortisol and growth hormone are potent antagonists in insulin action[55, 56]. Increases in growth hormone might not contribute to the insulin resistance induced after hot spring immersion, since the levels of growth hormone in both groups exhibit similar changes. Thus, the result of this heat-stress study suggests that the insulin resistance induced by hot spring immersion in the low DHEA-S group was associated with increased cortisol. This hormone functions to increase glucose output from liver and reduce glucose uptake by skeletal muscle[57, 58], which could elevate plasma glucose availability for brain aimed to prevent hypoglycemia during heat stress.

In the heat-stress study, DHEA-S was also reduced after hot spring immersion. Together with our previous altitude study, this result points to a possibility that DHEA-S is associated with universal stress coping but convergent to lack of oxygen[42]. In fact, heat stress has been demonstrated to cause a hypoxia-mimetic effect[59-62]. Hypoxia-induced factor-1α (HIF-1α), a marker of hypoxia stress, has been reported to be induced by mild heat environment[60, 63]. It is possible that increasing blood distribution to skin can cause a short-term reduction in oxygen delivery to the rest of tissues in the human body. Oxygen is the final acceptor of electron transported from reducing substrates for ATP production in virtually all cells and tissues. Mismatch in oxygen demand and supply to tissues represents coping failure against acute heat stress. Evidence from this study suggests a possibility that DHEA-S is an essential coping resource for regulating physiological homeostasis against environmental challenge leading to oxygen redistribution among tissues.

Our observation on lower baseline growth hormone in the low DHEA-S subjects suggests that DHEA-S may affect the normal production of growth hormone. Growth hormone functions to increase lipolysis[64] and protein synthesis[65]. Its biological action under heat stress could be important to reverse protein denature and degradation at the cost of fat consumption. Increasing DHEA-S level by DHEA administration[66] has been reported to increase the levels of growth hormone and insulin-like growth factor-1 (IGF-1)[67] and lower the growth hormone dose requirement for those of the hypopituitary patients[68]. Conversely, growth hormone treatment has no effect on DHEA-S levels[69]. This evidence highlights the causal relationship between DHEA-S and growth hormone production.

As we have mentioned earlier, the lower of DHEA-S and cortisol levels may be the indication of adrenal hypofunction (mild hypoadrenocorticism) or low hypothalamic-pituitary-adrenal (HPA) function[70]. Furthermore, previous study has shown that age-associated mortality is concomitantly associated with the high insulin (an indicator of insulin resistance) and low DHEA-S levels[34]. This concurrent age-dependent change leads to the hypothesis that adrenal hypofunction can be involved with the development of insulin resistance and age-associated metabolic disorders. Again, in the study, the low DHEA-S subjects had significantly greater cortisol response. Since both steroids are produced in adrenal grand, the possibility of hypoadrenocorticism or low HPA function is minimal.

Psychological Challenge Lowers DHEA-S[25]

Changes in mood state have been suggested to play an adaptive role in mobilizing personal resource to cope with the demands of a task[71-75]. We previously hypothesized that the internal coping resource for psychological adjustment against negative athletic competition outcome could be the neurosteroid DHEA-S. Therefore, a study was conducted in 14 elite golfers participating in a major national golf tournament, who were divided into two groups according to their competition outcomes: made the cut ($n = 8$) and failed to make the cut ($n = 6$). Mood state with Profile of Mood States (POMS) inventory and plasma concentrations of DHEA-S were measured at baseline and during 5 days after the players' final competition. We found a prolonged decrease in circulating DHEA-S for 5 days in the golfers who received negative competition outcome, whereas the overall mood state was well-maintained. In contrast, the total mood disturbance scores and DHEA-S were not changed for the made the cut group throughout the entire observation period. Mood has been proposed to be less intense but more prolonged psychological experience that may play a role for adaptation in mobilizing personal resource[73, 74]. Thus the prolonged decline in DHEA-S might be related to increased consumption of the steroid for psychological adaptation based on previous clinical studies have shown decreases in DHEA-S concentration with concurrent improvement in mood state and functioning[76] and increasing DHEA-S level that could reduce depressive symptoms in patients[77-80].

In this study, depression subscale of POMS was elevated on day 1, reflecting a situational response on the negative psychological impact. Since mood is a general feeling of state combined with multiple emotions[74], small changes in depression were masked as the overall mood score is also influenced by other dimensions such as tension, confusion, vigor, fatigue, and anger. Depressed mood has been suggested a catalyst for other mood state changes and have shown evidence that this is the case in numerous studies[81]. Nevertheless, according to our current observation, the depression level on golfers recovered quickly demonstrating that the psychological readjustment was successful along with the DHEA-S decline in the elite golfers. In the present study, the possibility that muscle damage with golf competition contributes to DHEA-S decline is precluded, since the DHEA-S decline was concurrent with a comparable decrease in post-competition muscle creatine kinase in circulation for both failed and made the cut groups. Therefore, decrease in DHEA-S was more linked with the adaptation (coping) for psychological stress.

This study provides new knowledge on how athletes coping with psychological challenge at the cost of endogenous resource. Coping is defined as efforts to manage environmental demands and internal conflicts, which cost a person's resources[74]. Psychological coping, generally related to the situations not amenable to alter, involves managing the distress mood evoked by the adverse situation. Successful coping strategies are reported to relate with the maintenance of psychological wellbeing[82, 83]. This is important in the view that our environment is changing variably, and an individual experiences a specific event as challenges are largely depending on the effectiveness of the internal coping mechanism. This appears to be linked with the availability of endogenous coping resource like DHEA-S. Successful coping, reflected by fast recovery, could minimize the negative influence of challenge that disrupts the psychological integrity. Thus the psychological homeostasis would be preserved for a challenger. The involvement of DHEA-S in the psychological coping mechanism against stress provides an additional support for the link of DHEA-S to human wellbeing[22, 84].

Conclusion

Our evidence suggests that DHEA-S is the common coping resource with both physiological and psychological challenges[42, 76, 85]. We firstly found that low DHEA-S elderly (aged above 80 years old) received less benefits from exercise training in metabolic improvements, whereas the high DHEA-S counterparts display normal training adaptation. In the young subjects (aged ~30 years old) with low DHEA-S, normal altitude adaptation in erythropoiesis is blunt and erythropoietin level is double compared to high DHEA-S in the elderly. Under heat exposure, the baseline stress hormone levels were raised to greater extent and cardiovascular adaptations were less efficient for those low DHEA-S young individuals compared to high DHEA-S age-matched individals. Furthermore, muscle damage induced by resistance exercise results in a protracted drop of serum DHEA-S level in young subjects. The DHEA-S lowering effect is not confined to physical challenge. Psychological distress can also lower the steroid in young athletes against negative competition outcome, while mood can be well-maintained at this cost. DHEA supplementation is one way to enhance the circulating DHEA-S level, which can significantly suppress the marker of post-exercise muscle damage (unpublished data). Since survival and wellbeing are directly linked with successful adaptation against the changing environment, our results suggest that (1) DHEA-S is consumed after external stress, which involves with recovery from external challenges; (2) Human with low initial DHEA-S has low adaptability due to poor recovery from challenges, which implicates the lower longevity in the low DHEA-S cohorts reported previously.

References

1. Krug AW, Ziegler CG, Bornstein SR. DHEA and DHEA-S, and their functions in the brain and adrenal medulla. Neuroactive Steroids in Brain Function, Behavior and Neuropsychiatric Disorders; 2008:227-239.
2. Zwain IH, Yen SSC. Neurosteroidogenesis in astrocytes, oligodendrocytes, and neurons of cerebral cortex of rat brain. Endocrinology. 1999;140:3843-3852.
3. Zwain IH, Yen SSC. Dehydroepiandrosterone: biosynthesis and betabolism in the brain. Endocrinology. 1999;140:880-887.
4. Baulieu E-E, Robel P. Dehydroepiandrosterone (DHEA) and dehydroepiandrosterone sulfate (DHEAS) as neuroactive neurosteroids. Proc Natl Acad Sci U S A. 1998;95:4089-4091.
5. Paul R, Etienne-Emile B. Dehydroepiandrosterone (DHEA) Is a neuroactive neurosteroid. Ann N Y Acad Sci. 1995;774:82-110.
6. Milman A, Zohar O, Maayan R, Weizman R, Pick CG. DHEAS repeated treatment improves cognitive and behavioral deficits after mild traumatic brain injury. Eur Neuropsychopharmacol. 2008;18:181-187.
7. Herbert J. Neurosteroids, brain damage, and mental illness. Exp Gerontol. 1998;33:713-727.
8. Eftal G, Kagan O, Brian C, Krzysztof S, Earl B, Maria S. Dehydroepiandrosterone as an enhancer of functional recovery following crush injury to rat sciatic nerve. Microsurgery. 2002;22:234-241.
9. Cutler GBJ, Glenn M, Bush M, Hodgen GD, Graham CE, Loriaux DL. Adrenarche: A survey of rodents, domestic animals, and primates. Endocrinology. 1978;103:2112-2118.
10. Perret M, Aujard F. Aging and season affect plasma dehydroepiandrosterone sulfate (DHEA-S) levels in a primate. Exp Gerontol. 2005;40:582-587.
11. Pelletier G, Labrie C, Martel C, Labrie F. Chronic administration of dehydroepiandrosterone (DHEA) to female monkey and rat has no effect on mammary gland histology. J Steroid Biochem Mol Biol. 2008;108:102-108.
12. Villareal DT, Holloszy JO. DHEA enhances effects of weight training on muscle mass and strength in elderly women and men. Am J Physiol. 2006;291:E1003-1008.
13. Mika E, Hisashi A, Ako F, Kumiko F, Akira S, Maki O, Shun-ichi K, Yasuki N, Yoshiyuki S, Tsutomu I. Serum dehydroepiandrosterone sulfate levels predict longevity in men: 27-year follow-up study in a community-based cohort (Tanushimaru Study). J Am Geriatr Soc. 2008;56:994-998.
14. Yang SC, Chen CY, Liao YH, Lin FC, Lee WC, Cho YM, Chen MT, Chou CH, Kuo CH. Interactive effect of an acute bout of resistance exercise and dehydroepiandrosterone administration on glucose tolerance and serum lipids in middle-aged women. Chin J Physiol. 2005;48:23-29.
15. Lamberts SWJ, van den Beld AW, van der Lely AJ. The endocrinology of aging. Science. 1997;278:419-424.
16. Wellman M, Shane-McWhorter L, Jennings JP. The role of dehydroepiandrosterone in diabetes mellitus. Pharmacotherapy. 1999;19:582-591.

17. Barrett-Conner E, Khaw K, Yen S. A prospective study of dehydroepi-androsterone sulfate, mortality and cardiovascular disease. N Engl J Med. 1986;315:1519-1524.
18. Abbasi A, Duthie EH, Sheldahl L, Wilson C, Sasse E, Rudman I, Mattson DE. Association of dehydroepiandrosterone sulfate, body composition, and physical fitness in independent community dwelling older men and women. J Am Geriatr Soc. 1998;46:263-273.
19. Herranz L, Megia A, Grande C, Gonzalez-Gancedo P, Pallardo F. Dehydroepiandrosterone sulfate, body fat distribution and insulin in obese men. Int J Obes. 1995;19:57-60.
20. Valenti G, Denti L, Maggio M, Ceda G, Volpato S, Bandinelli S, Ceresini G, Cappola A, Guralnik JM, Ferrucci L. Effect of DHEAS on skeletal muscle over the life span: The InCHIANTI Study. J Gerontol A Biol Sci Med Sci. 2004;59:M466-472.
21. Huang Y, Chen M, Fang C, Lee W, Yang S, Kuo C. A possible link between exercise-training adaptation and dehydroepiandrosterone sulfate- an oldest-old female study. Int J Med Sci. 2006;3:141-147.
22. Roth GS, Lane MA, Ingram DK, Mattison JA, Elahi D, Tobin JD, Muller D, Metter EJ. Biomarkers of caloric restriction may predict longevity in humans. Science. 2002;297:811.
23. Kirkwood TB. Time of our lives. What controls the length of life? EMBO Reports. 2005;6:S4-S8.
24. Wang JS, Chen SM, Lee SP, Lee SD, Huang CY, Hsieh CC, Kuo CH. Dehydroepiandrosterone sulfate linked to physiologic response against hot spring immersion. Steroids. 2009;74:945-949.
25. Wang HT, Chen SM, Lee SD, Hsu MC, Chen KN, Liou YF, Kuo CH. The role of DHEA-S in the mood adjustment against negative competition outcome in golfers. J Sports Sci. 2009;27:291 - 297.
26. Lee WC, Chen SM, Wu MC, Hou CW, Lai YC, Laio YH, Lin CH, Kuo CH. The role of dehydroepiandrosterone levels on physiologic acclimatization to chronic mountaineering activity. High Alt Med Biol. 2006;7:228-236.
27. Tsai YM, Chou SW, Lin YC, Hou CW, Hung KC, Kung HW, Lin TW, Chen SM, Lin CY, Kuo CH. Effect of resistance exercise on dehydroepiandrosterone sulfate concentrations during a 72-h recovery: Relation to glucose tolerance and insulin response. Life Sciences. 2006;79:1281-1286.
28. Byberg L, Melhus H, Gedeborg R, Sundstrom J, Ahlbom A, Zethelius B, Berglund LG, Wolk A, Michaelsson K. Total mortality after changes in leisure time physical activity in 50 year old men: 35 year follow-up of population based cohort. BMJ. 2009;338:b688.
29. Short KR, Vittone JL, Bigelow ML. Impact of aerobic exercise training on age-related changes in insulin sensitivity and muscle oxidative capacity. Diabetes. 2003;52:1888-1896.
30. Facchini FS, Hua N, Abbasi F, Reaven GM. Insulin resistance as a predictor of age-related diseases. J Clin Endocrinol Metab. 2001;86:3574-3578.
31. Yorek MA, Coppey LJ, Gellett JS, Davidson EP, Bing X, Lund DD, Dillon JS. Effect of treatment of diabetic rats with dehydroepiandrosterone on vascular and neural function. Am J Physiol. 2002;283:E1067-E1075.

32. Rizzo MR, Mari D, Barbieri M, Ragno E, Grella R, Provenzano R, Villa I, Esposito K, Giugliano D, Paolisso G. Resting metabolic mate and respiratory quotient in human longevity. J Clin Endocrinol Metab. 2005;90:409-413.
33. Villareal DT, Holloszy JO. Effect of DHEA on abdominal fat and insulin action in elderly women and men: a randomized controlled trial. JAMA. 2004;10:2243-2248.
34. Roth GS, Lane MA, Ingram DK, Mattison JA, Elahi D, Tobin JD, Muller D, Metter EJ. Biomarkers of caloric restriction may predict longevity in humans. Science. 2002;297:811.
35. Kalimi M, Shafagoj Y, Loria R, Padgett D, Regelson W. Anti-glucocorticoid effects of dehydroepiandrosterone (DHEA). Mol Cell Biochem. 1994;131:99-104.
36. Rizza R, Mandarino LJ, Gerich JE. Cortisol-induced insulin resistance in man: impaired suppression of glucose production and stimulation of glucose utilization due to a postreceptor defect of insulin action. J Clin Endocrinol Metab. 1982;54:131-138.
37. Buffington CK, Pourmotabbed G, Kitabchi AE. Case report: amelioration of insulin resistance in diabetes with dehydroepiandrosterone. Am J Med Sci. 1993;306:320-324.
38. Cruess DG, Antoni MH, Kumar M, Schneiderman N. Reductions in salivary cortisol are associated with mood improvement during relaxation training among HIV-seropositive men. J Behav Med. 2000;23:107-122.
39. Regelson W, Loria R, Kalimi M. Hormonal intervention: "buffer hormones" or "state dependency". The role of dehydroepiandrosterone (DHEA), thyroid hormone, estrogen and hypophysectomy in aging. Ann N Y Acad Sci. 1988;521:260-273.
40. Herbert J. Neurosteroids, brain damage, and mental illness. Exp Gerontol. 1998;33:713-727.
41. Malik AS, Narayan RK, Wendling WW, Cole RW, Pashko LL, Schwartz AG, Strauss KI. A novel dehydroepiandrosterone analog improves functional recovery in a rat traumatic brain injury model. J Neurotrauma. 2003;20:463-476.
42. Lee WC, Chen SM, Wu MC, Hou CW, Lai YC, Laio YH, Lin CH, Kuo CH. The role of dehydroepiandrosterone levels on physiologic acclimatization to chronic mountaineering activity. High Alt Med Biol. 2006;7:228-236.
43. Wright DC, Geiger PC, Holloszy JO, Han DH. Contraction- and hypoxia-stimulated glucose transport is mediated by a Ca2+-dependent mechanism in slow-twitch rat soleus muscle. Am J Physiol. 2005;288:E1062-E1066.
44. Dill RP, Chadan SG, Li C, Parkhouse WS. Aging and glucose transporter plasticity in response to hypobaric hypoxia. Mech Ageing Dev. 2001;122:533-545.
45. Ivy JL, Kuo CH. Regulation of GLUT4 protein and glycogen synthase during muscle glycogen synthesis after exercise. Acta Physiol Scand. 1998;162:295-304.
46. Ivy JL, Zderic TW, Fogt DL. Prevention and treatment of non-insulin-dependent diabetes mellitus. Exerc Sport Sci Rev. 1999;27:1-35.
47. Cassin S, Gilbert RD, Bunnell CE, Johnson EM. Capillary development during exposure to chronic hypoxia. Am J Physiol. 1971;220:448-451.
48. Stray-Gundersen J, Chapman RF, Levine BD. "Living high-training low" altitude training improves sea level performance in male and female elite runners. J Appl Physiol. 2001;91:1113-1120.

49. RodriGuez FA, Casas HC, Casas M, Pages T, Rama RN, Ricart A, Ventura JL, Ib Ez J, Viscor GS. Intermittent hypobaric hypoxia stimulates erythropoiesis and improves aerobic capacity. Med Sci Sports Exerc. 1999;31:264-268.
50. Laaidi M, Laaidi K, Besancenot JP. Temperature-related mortality in France, a comparison between regions with different climates from the perspective of global warming. Int J Biometeorol. 2006;51:145-153.
51. Wang HT, Chen SM, Lee SD, Chen KN, Liu YF, Kuo CH. The role of DHEA-S in the mood adjustment against negative competition outcome in golfers. J Sports Sci. 2009;27:291-297.
52. Regelson W, Loria R, Kalimi M. Hormonal intervention: "buffer hormones" or "state dependency". The role of dehydroepiandrosterone (DHEA), thyroid hormone, estrogen and hypophysectomy in aging. Ann N Y Acad Sci. 1988;521:260-273.
53. Grillon C, Pine DS, Baas JM, Lawley M, Ellis V, Charney DS. Cortisol and DHEA-S are associated with startle potentiation during aversive conditioning in humans. Psychopharmacology 2005;29:1-8.
54. Møller N, Beckwith R, Butler PC, Christensen NJ, Orskov H, Alberti KG. Metabolic and hormonal responses to exogenous hyperthermia in man. Clin Endocrinol. 1989;30:651-660.
55. Holmang A, Bjorntorp P. The effects of cortisol on insulin sensitivity in muscle Acta Physiol Scand. 1992;144:425-431.
56. Rosenfeld RG, Wilson DM, Dollar LA, Bennett A, Hintz RL. Both human pituitary growth hormone and recombinant DNA-derived human growth hormone cause insulin resistance at a postreceptor site. J Clin Endocrinol Metab. 1982;54:1033-1038.
57. Kirschbaum C, Bono EG, Rohleder N, Gessner C, Pirke KM, Salvador A, Hellhammer DH. Effects of fasting and glucose load on free cortisol responses to stress and nicotine. J Clin Endocrinol Metab. 1997;82:1101-1105.
58. Rizza RA, Mandarino LJ, Gerich JE. Cortisol-induced insulin resistance in man: impaired suppression of glucose production and stimulation of glucose utilization due to a postreceptor detect of insulin action. J Clin Endocrinol Metab. 1982;54:131-138.
59. Shein NA, Horowitz M, Alexandrovich AG, Tsenter J, Shohami E. Heat acclimation increases hypoxia-inducible factor 1 alpha and erythropoietin receptor expression: implication for neuroprotection after closed head injury in mice. J Cereb Blood Flow Metab. 2005;25:1456-1465.
60. Maloyan A, Eli-Berchoer L, Semenza GL, Gerstenblith G, Stern MD, Horowitz M. HIF-1 alpha-targeted pathways are activated by heat acclimation and contribute to acclimation-ischemic cross-tolerance in the heart. Physiol Genomics. 2005;23:79-88.
61. Katschinski DM, Le L, Heinrich D, Wagner KF, Hofer T, Schindler SG, Wenger RH. Heat induction of the unphosphorylated form of hypoxia-inducible factor-1alpha Is dependent on heat shock protein-90 activity. J Biol Chem. 2002;277:9262-9267.
62. Fan JL, Cotter JD, Lucas RAI, Thomas K, Wilson L, Ainslie PN. Human cardiorespiratory and cerebrovascular function during severe passive hyperthermia: effects of mild hypohydration. J Appl Physiol. 2008;105:433-445.

63. 1Shein NA, Horowitz M, Alexandrovich AG, Tsenter J, Shohami E. Heat acclimation increases hypoxia-inducible factor 1-alpha and erythropoietin receptor expression: implication for neuroprotection after closed head injury in mice. J Cereb Blood Flow Metab. 2005;25:1456-1465.
64. Lucidi P, Laureti S, Santoni S L, M., Busciantella-Ricci N, Angeletti G, Santeusanio F, De Feo P. Administration of recombinant human growth hormone on alternate days is sufficient to increase whole body protein synthesis and lipolysis in growth hormone deficient adults. Clin Endocrinol. 2000;52:173-179.
65. Lucidi P, Laureti S, Santoni S, Lauteri M, Busciantella R, Angeletti G, Santeusanio F, De Feo P. Administration of recombinant human growth hormone on alternate days is sufficient to increase whole body protein synthesis and lipolysis in growth hormone deficient adults. Clin Endocrinol. 2000;52:173-179.
66. Yang SC, Chen CY, Liao YH, Lin FC, Lee WC, Cho YM, Chen MT, Chou CH, Kuo CH. Interactive effect of an acute bout of resistance exercise and dehydroepiandrosterone administration on glucose tolerance and serum lipids in middle-aged women. Chin J Physiol. 2005;48:23-29.
67. Alessandro DG, Massimo S, Claudia S, Simone P, Stefano L, Andrea RG. Oral dehydroepiandrosterone supplementation modulates spontaneous and growth hormone-releasing hormone-induced growth hormone and insulin-like growth factor-1 secretion in early and late postmenopausal women. Fertil Steril. 2001;76:241-248.
68. Brooke A, Kalingag L, Miraki-Moud F, Camacho-Hubner C, Maher K, Walker D, Hinson J, Monson J. Dehydroepiandrosterone (DHEA) replacement reduces growth hormone (GH) dose requirement in female hypopituitary patients on GH replacement. Clin Endocrinol. 2006;65:673-680.
69. Boonstra VH, Mulder PGH, de Jong FH, Hokken-Koelega ACS. Serum dehydroepiandrosterone sulfate levels and pubarche in short children born small for gestational age before and during growth hormone treatment. J Clin Endocrinol Metab. 2004;89:712-717.
70. Nasrallah MP, Arafah BM. The Value of Dehydroepiandrosterone Sulfate Measurements in the Assessment of Adrenal Function. J Clin Endocrinol Metab. 2003;88:5293-5298.
71. Schmale AH. Adaptive Role of Depression in Health and Disease. In: Senay EC, ed. Separation and Depression: Clinical and Research Aspects. Washington: Scott, J.P.; 1973.
72. Saito M, Ikeda M, Seki H. A study on adaptive action development-interference of mood states with perceived performance and perceived efficacy. Int J Ind Ergon. 2007;5:48-51.
73. Keller MC, Nesse RM. Is low mood an adaptation? Evidence for subtypes with symptoms that match precipitants. J Affect Disord. 2005;86:27-35.
74. Lane AM, Terry PC. The nature of mood: Development of a conceptual model with a focus on depression. J Appl Sport Psychol. 2000;12:16-33.
75. Doyle J, Parfitt G. The effect of induced mood states on performance profile areas of perceived need. J Sports Sci. 1999;17:115-127.

76. Fabian TJ, Dew MA, Pollock BG, Reynolds CF, Mulsant BH, Butters MA, Zmuda MD, Linares AM, Trottini M, Kroboth PD. Endogenous concentrations of DHEA and DHEA-S decrease with remission of depression in older adults. Biol Psychiatry. 2001;50:767-774.
77. Wolkowitz OM, Reus VI, Roberts E, Manfredi F, Chan T, Raum WJ, Ormiston S, Johnson R, Canick J, Brizendine L, Weingartner H. Dehydroepiandrosterone (DHEA) treatment of depression. Biol Psychiatry. 1997;41:311-318.
78. Rabkin JG, McElhiney MC, Rabkin R, McGrath PJ, Ferrando SJ. Placebo-controlled trial of dehydroepiandrosterone (DHEA) for treatment of nonmajor depression in patients with HIV/AIDS. Am J Psychiatry. 2006;163:59-66.
79. Southwick SM, Vythilingam M, Charney DS. The psychobiology of depression and resilience to stress: Implications for prevention and treatment. Annu Rev Clin Psychol. 2005;1:255-291.
80. Rasmusson AM, Vasek J, Lipschitz DS, Vojvoda D, Mustone ME, Shi Q, Gudmundsen G, Morgan CA, Wolfe J, Charney DS. An increased capacity for adrenal DHEA release is associated with decreased avoidance and negative mood symptoms in women with PTSD. Neuropsychopharmacology. 2004;29:1546-1557.
81. Lane AM. Mood and Human Performance: Conceptual, Measurement, and Applied Issues. Hauppauge, NY: Nova Science Publishers; 2007.
82. Drossman DA, Leserman J, Li Z, Keefe F, Hu YJB, Toomey TC. Effects of coping on health outcome among women with gastrointestinal disorders. Psychosom Med. 2000;62:309-317.
83. Vaughn AA, Roesch SC. Psychological and physical health correlates of coping in minority adolescents. J Health Psychol. 2003;8:671-683.
84. Panjari M, Davis SR. DHEA therapy for women: effect on sexual function and wellbeing. Hum Reprod Update. 2007;13:239-248.
85. Morgan CA, III, Southwick S, Hazlett G, Rasmusson A, Hoyt G, Zimolo Z, Charney D. Relationships among plasma dehydroepiandrosterone sulfate and cortisol Levels, symptoms of dissociation, and objective performance in humans exposed to acute stress. Arch Gen Psychiatry. 2004;61:819-825.

Adaptation Biology and Medicine (Volume 6: Cell Adaptations and Challenges)
Editors: P. Wang, C.-H. Kuo, N. Takeda and P.K. Singal
Copyright © 2011 Narosa Publishing House Pvt. Ltd., New Delhi, India

29. The Role of Cytoprotective HO-1 and its Induction by Plant Adaptogens and Phytochemicals in the Development of an Enhanced Cellular Adaptive Resistance

Nicola C. Frenkel[1] and Fred A.C. Wiegant[1, 2*]

[1]University College Utrecht, The Netherlands

[2]Faculty of Science, Department of Biology, Utrecht University, The Netherlands

Abstract

Extracts of plant adaptogens (such as *Eleutherococcus senticosus, Schisandra chinensis and Rhodiola rosea*) increase the resistance of a variety of organisms against the negative effects of different stress conditions. Incubation of a variety of cell types with different extracts of plant adaptogens has been shown to invariably result in the induction of heme oxygenase (HO-1), which is one of the key components of the cellular stress response and a sensitive biomarker of oxidative stress. HO-1 is the rate-limiting enzyme in the process of degrading potentially toxic free heme into carbon monoxide, biliverdin and free iron. In this paper, the role of HO-1 as a crucial enzyme in fortifying cellular defense capacities against oxidative stress, is described. The induction mechanism of HO-1 and especially the relevance of the Nrf2/ARE pathway is reviewed. Finally, the molecular mechanism and background of the potential benefit of plant adaptogens and HO-1-inducing phytochemicals is discussed in relation to hormesis.

Keywords: Heme oxygenase, phytochemicals, adaptogens, oxidative stress

Introduction

Plant Adaptogens & Enhanced Adaptive Resistance to Stress Conditions

The term 'adaptogen' was first applied to pharmacologically active compounds that could induce a state of enhanced resistance in mammals which allowed them to counteract and adapt to the negative effects of different stress conditions (Brekhman and Dardymov 1969). Extracts of a limited number of plants (such as *Eleutherococcus senticosus, Schisandra chinensis and Rhodiola rosea*) also known as 'plant adaptogens', have been shown to increase physical performance and to increase the resistance of a variety of organisms against stress conditions (Wagner *et al.* 1994; Boon-Niermeijer *et al.* 2000; Panossian and Wagner 2005; Wiegant *et al.* 2008). Extracts of plant adaptogens were also reported to

*Address for Correspondence: Corresponding Address: Dr. F.A.C. Wiegant Faculty of Science Department of Biology Cellular Dynamics Utrecht University P.O.Box 80056 3508 TB Utrecht, The Netherlands Email: f.a.c.wiegant@uu.nl

increase resistance to oxidative stress in different cell types (Brown *et al.* 2002; DeSanctis *et al.* 2004; Wiegant *et al.* 2008) and to significantly increase the lifespan and stress resistance both in *C. elegans* (Wiegant *et al.* 2006, 2009) and in *Drosophila melanogaster* (Jafari *et al.* 2007), probably by alleviating oxidative stress. In a double blind placebo-controlled pilot experiment we showed that plant adaptogens were able to reduce biophoton emission in a group of volunteers, which reflects a reduction of oxidative free radicals (Schutgens *et al.* 2009). Focusing on the mechanism of action of plant adaptogens that are able to fortify cellular defense capacities, it was shown that heme oxygenase (HO-1) was invariably induced in different cell cultures (Wiegant *et al.* 2008). Since HO-1 is a key component in the cellular anti-oxidative defense, induction of HO-1 by adaptogens is suggested to serve an important function in the development of enhanced cellular adaptive response.

Oxidative Stress, Cellular Defense Systems and HO-1

Oxidative stress is mediated mainly by reactive oxygen species (ROS) and only for a smaller part by reactive nitrogen species (RNS). Reactive species are constantly generated as by-products of oxidative respiration (Surh *et al.* 2008). Moreover, cells are also exposed daily to toxic, mutagenic and carcinogenic chemicals which constitute another source of these harmful species (Shih *et al.* 2007). These reactive species cause the disruption of the physiological function of cells by attacking crucial cellular components such as lipids, DNA and proteins as well as by modulating and interfering with important signal transduction pathways (Valko *et al.* 2007; Shih *et al.* 2007; Surh *et al.* 2008).

Since these reactive species can be detrimental for a cell's functioning, a general antioxidant defense system is in place that protects cells against the daily exposure to low physiological levels of these harmful radicals, thereby maintaining the homeostatic state of the cells. Elements of the general cellular antioxidative defense include for instance glutathione (GSH) and the enzymes superoxide dismutase (SOD) and catalase (Halliwell and Gutteridge 1999). However, when the balance between these reactive species and the basic antioxidant defense system that scavenges them is disturbed, resulting in too many oxidants to be scavenged properly, a state of 'oxidative stress' results (Shih *et al.* 2007; Surh *et al.* 2008). The danger for cells to be in such a state of oxidative stress has been illustrated by its association with several severe diseases, such as cardiovascular diseases and neurodegenerative diseases, such as Alzheimer's and Parkinson's (Shih *et al.* 2007). Moreover, since these oxidants are capable of modulating signal transduction cascades as well as attacking the genetic material directly, it has also been observed that a state of oxidative stress causes a predisposition for carcinogenesis (Prawan *et al.* 2005).

A second system of cellular defense is in place to aid cells at times of oxidative stress. This cellular defense system is comprised of several enzymes which are named the Phase II drug metabolizing enzymes. These enzymes, such as NAD(P)H:quinine oxidoreductase (NQO1), glutathione S-transferase (GST), glutathione synthase, γ-glutamyl transpeptidase (γ-GT), γ-glutamate cysteine ligase (γ-GCL), UDP-glucuronosyltransferase (UGT), aldo-keto-reductase and heme oxygenase-1 (HO-1), are also called detoxifying and antioxidant enzymes, since they offer cytoprotection against the harmful state of oxidative stress (Lee and Surh 2005; Eggler *et al.* 2008). These enzymes are crucial to maintaining the homeostatic balance of the cells' environment as they neutralize and remove excess oxidants, allowing the restoration of redox homeostasis (Li and Kong 2009). In this list of enzymes, HO-1 fulfills a specific role; its induction is considered to be one of the most sensitive indications of cellular oxidative stress and its upregulation has been recognized as a crucial event in the development of oxidative stress resistance (Rytter *et al.* 2006). The observation that plant adaptogens were able to induce HO-1 as well as to increase adaptive resistance against oxidative stress conditions was therefore of interest (Wiegant *et al.* 2008).

In this chapter the role of this crucial enzyme HO-1 in the combat against oxidative stress and in the generation of adaptive resistance will be discussed. First, the enzyme itself will be shortly introduced, after which its beneficial effects and the possibility and current developments of using these properties in clinical setting will be focused upon. Then the actual induction of HO-1 gene expression will be considered. Lastly, this chapter will then be concluded with a discussion elucidating the current questions on the link between adaptogens and this enzyme.

Heme Oxygenase-1, a Crucial Phase II Enzyme

The enzyme HO-1 has been highly conserved in bacteria, algae, plants and animals indicating the importance of its existence as the enzyme that catalyses the breakdown of heme into carbon monoxide (CO), iron and biliverdin (Maines 2000). The biliverdin reductase quickly reduces the formed biliverdin into bilirubin (Li *et al.* 2007).

Heme is crucial to the transport of oxygen in the blood and serves as an essential part of the heme-containing enzymes, such as NADPH-oxidase, nitric oxide synthase and guanylate cyclase, which are indispensable for aerobic cells. Unbound heme, however, is toxic to cells as it possesses strong oxidant potential (Srisook *et al.* 2005). Interestingly, it was observed that ROS promote processes such as structural alteration, destabilization, fragmentation and proteolysis of these heme-containing enzymes, resulting in the release of toxic unbound heme. Fortunately, at these times of oxidative stress, HO-1 is induced and is capable of eliminating the damaging heme released by ROS. Therefore, HO-1 performs the crucial task of ridding the cell of toxic and detrimental heme at times of oxidative stress when even more of this damaging compound is released (Srisook *et al.* 2005). In addition, during the process of heme degradation by HO-1 several antioxidants are formed (Srisook *et al.* 2005). Indeed, its breakdown products CO and bilirubin possess anti-inflammatory, anti-apoptotic and anti-oxidant properties (Immenschuh and Ramadori 2000; Kirkby and Adin 2006). Therefore, the induction of HO-1 gene expression results in both the elimination of toxic unbound heme as well as in the production of cytoprotective CO and bilirubin (Srisook *et al.* 2005).

Furthermore, HO-1 has been observed to confer important protection and beneficial properties regarding numerous serious illnesses, such as cardiovascular diseases, neurodegeneration and carcinogenesis (Immenschuh and Schröder 2006; Abraham and Kappas 2005; Kirkby and Adin 2006; Maines 2000; Lee and Surh 2005; Mattson and Cheng 2006; Syapin 2008).

The Cytoprotective Effects of the Break-Down Products of HO-1

Thus, as stated above, at times of oxidative stress, induced HO-1 gene expression results in the increased breakdown of heme which due to its pro-oxidant properties proves detrimental when present at high levels. Moreover, not only does HO-1 induction leads to increased heme breakdown, the products of this process (CO, Fe, biliverdin and bilirubin) also contribute to the cytoprotective effect of HO-1 because of their anti-inflammatory, anti-apoptotic and anti-oxidant effects (Kirkby and Adin 2006; Rytter *et al.* 2006; Dulak *et al.* 2008).

Carbon monoxide, until recently viewed as nothing more than a toxic air pollutant, has proven that besides its dangerous properties the gas might also have cytoprotective effects as a therapeutic window allows dosages where it possesses anti-apoptotic, vasodilatory and anti-inflammatory effects. The vasodilatory properties of CO are similar to that of the well-known vasodilator nitric oxide (NO), namely by binding to the enzyme guanylyl cyclase which induces the production of cGMP leading to vascular smooth muscle relaxation (Kirkby and Adin 2006; Dulak *et al.* 2008). The cytoprotective anti-inflammatory and anti-apoptotic effects of CO have been observed and suggested to occur via MAPK signaling pathways. Of the three subfamilies of these MAPK signaling cascades, namely ERK, JNK and p38 MAPK pathways, mostly the latter has been observed to be involved in these cytoprotective effects. For instance, in the case of lung injury, the anti-inflammatory and anti-apoptotic effects of CO appear to be mediated via the p38 MAPK pathway (Kirkby and Adin 2006). CO has been observed to modulate the inflammatory pathway by reducing pro-inflammatory cytokine production, such as Il-2, TNF and IL-1β, as well as increase the production of anti-inflammatory cytokines, such as IL-10 (Kirkby and Adin 2006).

Biliverdin, quickly reduced to bilirubin, contributes to the cytoprotective effect of HO-1 action. While biliverdin appears to be non-toxic, bilirubin was thought of as dangerously toxic since high levels of this compound can cause hyperbilirubinemia which can contribute to neonatal jaundice, kernicterus and brain damage in newborns (Kirkby and Adin 2006). However, bilirubin was also shown to possess cytoprotective properties due to its strong anti-oxidant activity (Immenschuh and Ramdori 2000). Indeed, research observed that super-induced HO-1 resulted in bilirubin-mediated cytoprotection of cardiomyocytes and neurons when experiencing oxidative stress (Kirkby and Adin 2006). Besides its powerful anti-oxidant effects, bilirubin also possesses anti-inflammatory capacities. It has been suggested that these effects are mediated by the inhibition of leukocyte adhesion and rolling. However, as of yet the mechanism of the anti-inflammatory effects of bilirubin remains to be fully elucidated (Kirkby and Adin 2006).

The production of iron (Fe^{2+}) is beneficial to an organism as long as it proceeds to be in small amounts as iron is an essential co-factor to numerous cellular enzymes and proteins. When present in large quantities, however, iron is severely cytotoxic due to its strong pro-oxidant nature which leads to the production of detrimental ROS (Immenschuh and Ramadori 2000). The cytotoxic breakdown product iron has been observed to induce the expression of the iron sequestering ferritin (Tsuchihashi *et al.* 2004). Its increased expression reduces the cytotoxicity of iron in the cell. Intriguingly, in several studies it has been observed that the induction of HO-1 is connected to an upregulation of ferritin as well (Kirkby and Adin 2006).

Therapeutic Application of HO-1

By enhancing the adaptive resistance of cells and tissues against (oxidative) stress conditions, HO-1 has been observed to possess cytoprotective effects in the case of many different diseases, such as atherosclerosis, sepsis, diabetes, lung injury, ischemia, hypertension, acute renal failure, cardiac hyperthrophy, ocular vascular disease and neurodegenerative diseases, such as Alzheimer's, Parkinson's and Pick's disease (Rytter *et al.* 2006; Abraham *et al.* 2007). The enzyme has therefore become the focus of research on the possibility of its use in a preventive and therapeutic setting.

Indeed, recent studies suggest that overexpressing HO-1 may result in novel beneficial treatments for the states of damaging oxidative stress that occur with these diseases (Rytter *et al.* 2006; Abraham *et al.* 2007). One of the possibilities is using gene therapy, namely the insertion of specific genes into cells that are either deficient or underexpressing these genes, to superinduce HO-1 gene expression (Abraham *et al.* 2007).

The development of using adenoviral vectors and lentiviral vectors offers great potential. The former possesses the ability to integrate into the genome of targeted cells, to be applied to various cell types, to highly express the specific gene of interest and to transfer this gene independent of active cell division. The use of lentiviral vectors also offers many advantages, such as their capability of transfecting cells at both mitotic and post-mitotic stages and therefore including brain tissues and other non-dividing tissues to its targets. These vectors can also integrate stably into cell genomes resulting in long-term expression of the gene. So far, the delivery of the human HO-1 gene in several animal model studies has shown successful and promising results in the long-term overexpression of this gene (Abraham *et al.* 2007). Therefore, the possibility of using gene therapy in the therapeutical application of the beneficial effects of HO-1 has potential. However, the obstacles that surround every field of study interested in gene therapy remain.

Besides the application of gene therapy, research has also turned its focus to the different breakdown products of HO-1 and their use in clinical setting. Several studies have presented evidence of the cytoprotective effects of CO in lungs, kidneys, small intestines, liver and pancreatic tissues of rats (Rytter *et al.* 2006; Kirkby and Adin, 2006). However, the novelty of these studies make that the exogenous delivery of CO to human patients in clinical settings is still unfeasible as the logistics of dosage, safety and timing of delivery remain to be established (Kirkby and Adin 2006). The use of bilirubin in therapeutical applications has been observed in recent research as a promising treatment for atherosclerosis, coronary heart disease and organ transplantations (Mayer 2000; Kirkby and Adin 2006). As with CO, the effective dose, timing of administration and safety of bilirubin treatment remain to be established for human patients (Kirkby and Adin, 2006). However, the current studies on using these cytoprotective break-down products of heme degradation offer great promise.

Another possibility is to induce HO-1 expression using plant extracts including plant adaptogens (Wiegant *et al.* 2008) and *Ginkgo biloba* (Chen *et al.* 2001) as well as specific phytochemicals such as curcumin, sulforaphane or resveratrol that have all been reported to induce HO-1 gene expression (Lee and Surh 2005; Prawan *et al.* 2005; Li *et al.* 2007).

EXPRESSION AND INDUCTION OF HEME OXYGENASE-1

The expression of HO-1 differs with cell type, species, intensity and duration of the used stimuli as is known for many different stress-responsive genes (Prawan *et al.* 2005). The current research on this enzyme has shown that the induction of HO-1 involves many different signaling cascades (Prawan *et al.* 2005). First, since the research is most advanced on the level of transcription factors that are known to induce HO-1 expression, these factors will be discussed shortly.

Transcription Factors involved in HO-1 Gene Induction

The current consensus is that the transcriptional control of HO-1 gene expression induction is mainly exerted via antioxidant response elements (ARE) which are cis-acting elements present in the promoter regions of Phase II enzymes. It is through these elements that the Phase II enzymes have been observed to be expressed (Kang *et al.* 2005). Thus, ARE present in the 5' flanking region of the HO-1 promoter have been thought to be responsible for the induction of its gene expression in response to oxidative stress (Itoh *et al.* 2007).

At first, the transcription factor AP-1 was suspected to bind to these ARE elements sequences and induce HO-1 gene expression, since the ARE sequences closely resembled the AP-1 binding sequences, named TPA-responsive elements (TRE) (Alam and Cook 2006; Prestera *et al.* 1995). However, currently another transcription factor is seen as the most likely candidate for ARE-binding, namely NF-E2 related factor 2 (Nrf2), as its binding sequences also show high similarity to the ARE sequences (Itoh *et al.* 1997). Moreover, this transcription factor was recently observed to play a crucial role in the ARE-mediated gene expression induction of Phase II enzymes, including HO-1 (Prawan *et al.* 2005; Jaiswal 2004). Furthermore, it was also observed that the ARE sequences resembled MARE sequences, namely the Maf recognition elements. This discovery indicated that small Maf proteins might also play an important role in the activation of ARE-mediated gene expression. Recent studies have shown that indeed a Nrf2-small Maf heterodimer is most likely responsible for the induction of HO-1 gene expression (Itoh *et al.* 1997; Jaiswal 2004).

The induction of HO-1 gene expression in response to oxidative stress is induced via the Nrf2-ARE pathway, but there are several other transcription factors that can induce the expression of this enzyme as well, the previously mentioned AP-1 for instance. This is a dimeric combination of basic leucine zipper proteins of the widely distributed Jun and Fos families of DNA binding proteins which are involved in many cellular processes, such as proliferation, differentiation, apoptosis and the response to stress (Alam and Cook 2007). Another transcription factor that is involved in HO-1 induction is NF-κB which plays an essential role in the regulation of processes such as the immune response, cellular stress response, proliferation, apoptosis and embryonic development (Alam and Cook 2007).

Recently, the transcription factors hypoxia-inducible factor-1 (HIF-1) and 3'5'-cyclic monophosphate (cAMP)-responsive element binding protein (CREB) were also observed to possibly play a role in the induction of HO-1. However, current research is insufficient to fully elucidate the roles of these two transcription factors (Prawan *et al.* 2005). Binding elements to most of these transcription factors have indeed been identified in the promoter of this specific gene (Alam and Cook 2007).

The Nrf2-ARE Mediated Induction of HO-1 Expression

Since the discovery was made that Nrf2 is crucial to the induction of Phase II enzyme gene expression and thus also to that of HO-1, research has focused on elucidating the pathway wherein Nrf2 binds ARE and hereby induces gene expression (Li and Kong 2009) (Figure 1). Under normal conditions Nrf2 is sequestered by Keap1 in the cytoplasm. The binding of Keap1 to the actin cytoskeleton anchors it in the cytoplasm. First it was believed that a reaction with the cysteine thiol moieties present in Keap1 would spur the release of Nrf2 as it would induce a conformational change in Keap1. After its release, Nrf2 would then have been free to translocate to the nucleus where it could bind to ARE elements sequences (Surh *et al.* 2008; Li and Kong 2009).

However, it is now believed that the mechanism by which Nrf2 is allowed to translocate to the nucleus is more complicated than initially suggested. Recent research has focused on two mechanisms that could influence the nuclear translocation of Nrf2, namely (a) the impediment of the Cul3-dependent E3 ubiquitination and proteasomal degradation of Nrf2 which allow for the cytoplasmic levels of Nrf2 to be stabilized and (b) post-translational modifications such as phosphorylations of Nrf2 to occur (Li and Kong 2009; Eggler *et al.* 2008).

Ubiquitination and Degradation of Nrf2 and Keap1

Keap1 is believed to serve as a bridge between Nrf2 and the Cullin3-based E3-ligase ubiquitination complex where the latter marks the transcription factor for proteasomal degradation by the 26S proteasome (Li and Kong 2009; Srisook *et al.* 2005). The continuous ubiquitination and proteasomal degradation of Nrf2 maintains a constant low concentration of Nrf2 in the cytoplasm (Li and Kong 2009). Therefore, HO-1 expression induction is low or non-existent, depending on cell type (Li *et al.* 2007). It has recently been suggested that the impediment of Nrf2 ubiquitination and degradation plays an important role in the induction of HO-1 expression (Eggler *et al.* 2008). Moreover, two cysteine residues of Keap1, C273 and C288, were observed to be required for efficient ubiquitination of Nrf2 and their modification is therefore likely to impair the Keap1-ubiquitination complex from ubiquitinating Nrf2 efficiently. Another cysteine residue, namely C151, has been observed to be important to the stabilization and activation of Nrf2 in response to ARE-mediating HO-1-inducing chemicals (Eggler *et al.* 2008). Thus, it has been suggested that the modification of Keap1 cysteine residues leads to a reduced interaction between Keap1 and the ubiquitination complex leading to a reduction in Nrf2 ubiquitination and a consequent increase in Nrf2 levels. This accumulation of Nrf2 in the cytoplasm would then lead to the expression of HO-1 (Eggler *et al.* 2008). More research has recently focused on the possible importance of Keap1 ubiquitination and degradation to the accumulation of Nrf2 and consequent ARE-controlled gene expression (Eggler *et al.* 2008).

Post-Translational Modifications

Besides the influence of the degradation of Keap1 and Nrf2, post-translational modifications also play an important role in the induction of HO-1 gene expression. One of these most important modifications appears to be the phosphorylation of Nrf2 on Ser40 in the Protein Kinase C (PKC) pathway (see Figure 1). Indeed, it was observed that the involvement of the PKC pathway was necessary to the induction of ARE-controlled genes (Kang *et al.* 2005). Therefore, this phosphorylation in response to a state of oxidative stress has been suggested as a critical step in the nuclear translocation of Nrf2 and the consequent HO-1 gene expression induction (Kang *et al.* 2005). Moreover, ARE-controlled gene expression inducers have been shown to stimulate PKC activity, while PKC inhibitors resulted in down-regulated ARE-induced responses (Eggler *et al.* 2008; Numazawa *et al.* 2003). It has been suggested that this phosphorylation could be required for the disruption of the Nrf2-Keap1 complex. However, it was also shown that this phosphorylation was not necessary for the accumulation of Nrf2 in the nucleus, indicating that other mechanisms will most likely be responsible for this accumulation (Jaiswal 2004). As research has supported that this pathway is important to the Nrf2-ARE-mediated induction of Phase II enzymes such as HO-1, the exact mechanism and influence of this pathway remain to be elucidated.

Another pathway that has been observed to play a role in the ARE-induced gene expression is that of the p38 mitogen-activated protein kinase (p38 MAPK) (Figure 1). The activation of this kinase, capable of phosphorylating Nrf2, has been observed to result in an increased affinity and binding between Nrf2 and Keap1. However, there seems to be a discrepancy between different ARE-inducers. Some have been observed to activate p38 MAPK, while others were shown to inhibit the kinase's activity (Eggler *et al.* 2008). Therefore, the phosphorylation of Nrf2 by this kinase requires further research to clarify the influence of this pathway.

Recently, suggestions have been made on the importance of the rearrangement of the cell's actin filaments to the nuclear localization of Nrf2. This rearrangement of the actin cytoskeleton appears to depend on the activity of the PI3-kinase (PI3K) signaling pathway (Figure 1). Kang *et al.* (2005) observed that not only Keap1, but Nrf2 is also capable of binding to actin. In fact, Nrf2 was reported to co-localize with actin in cells and that the nuclear translocation of Nrf2 appeared to be dependent on actin filament rearrangement under PI3K-control (Kang *et al.* 2005). In states of oxidative stress, this actin rearrangement consists of the depolymerization of actin and appears to facilitate Nrf2's translocation to the nucleus (Kang *et al.* 2005). The role of Keap1 in this pathway, however, remains to be investigated.

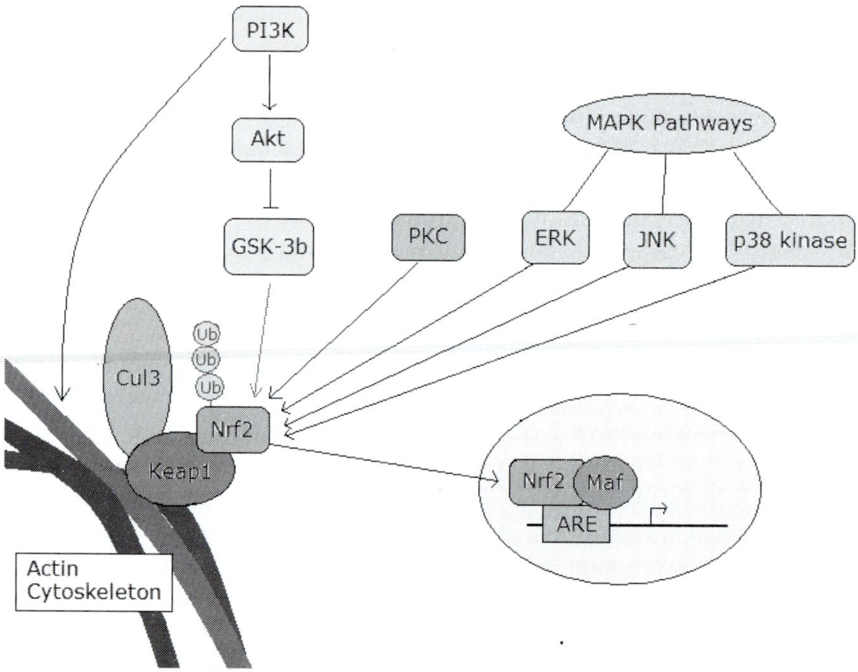

Fig.1. Schematic overview of the different signaling cascades involved in the Nrf2/ARE pathway. The MAP kinases with the sub-families ERK, JNK and especially p38 kinase have been observed to play important roles in the induction of HO-1 via the Nrf2/ARE pathway. Moreover, the PKC and PI3K signaling pathways have also been implicated in the Nrf2/ARE pathway. PKC has been observed to be able to phosphorylate Nrf2 directly at Ser40 which is possibly required for the dissociation of Nrf2 from the Nrf2-Keap1 complex. PI3K affects the actin cytoskeleton rearrangement, thereby also influencing Nrf2 nuclear translocation. Furthermore, the activation of Akt1 kinase via the PI3K pathway is believed to inhibit the GSK-3β kinase, which might negatively influence the nuclear localization of Nrf2.

Before Nrf2 can perform its task as transcription factor, it must be translocated into the nucleus from the cytoplasm. Two phosphorylation reactions have been observed to affect the nuclear import and export of Nrf2. The glycogen synthase kinase-3β (GSK-3β) downstream from the previously mentioned PI3 kinase was reported to directly phosphorylate Nrf2 (Eggler *et al.* 2008). The Akt1 kinase, also downstream of PI3K, was observed to be required for ARE-controlled gene expression induction (Martin *et al.* 2004). Akt1 kinase inactivates GSK-3β, indicating that GSK-3β is a negative regulator of Nrf2 activity (Salazar *et al.* 2006). Indeed, it appears that the phosphorylation of Nrf2 by GSK-3β stimulates its cytoplasmic localization. Thus, the ARE-gene inducers appear to stimulate Akt1 kinase to down-regulate GSK-3β and Nrf2 cytoplasm localization, thereby stimulating ARE-gene induction (Eggler *et al.* 2008) (Figure 1).

Another phosphorylation event was observed to affect the nuclear export of Nrf2. Several hours after ARE-gene induction, Crm1-mediated export of Nrf2 is stimulated. A phosphorylation reaction on Nrf2's Y568 is believed to result in increased interaction of Nrf2 with Crm1 (Eggler *et al.* 2008). The most likely candidate for this phosphorylation is the Fyn kinase, a Src kinase. Thus, ARE-gene inducers also appear to be involved in the Fyn kinase mediated stimulation of Nrf2 export several hours later, probably serving to terminate the signaling response and to shut off the ARE-gene induction (Eggler *et al.* 2008).

Hinge and Latch Model by Li and Kong (2009)

Li and Kong (2009) have recently proposed an alternate model explaining the Nrf2-ARE pathway of HO-1 induction, namely the "Hinge and Latch" model. This model offers an intriguing explanation on the mechanism of achieving a cell's priming or the ability to develop a state of enhanced adaptive resistance. The application of this theory to the notion of adaptive resistance will follow later. First, the theory itself will be explained.

This theory makes use of the observation that two domains of Nrf2, ETGE and DGL motifs, bind with different affinities to Keap1. The ETGE motif possesses high binding affinity for Keap1 and serves as a "hinge", while the DGL motif has a lower affinity, hereby functioning as a "latch" (Li and Kong 2009) (Figure 2). The lysine residues of Nrf2 that are ubiquitinated and mark the molecule for degradation are positioned inbetween these motifs. Li and Kong (2009) propose that when the "latch" is in position, namely the DGL motif is bound to Keap1, these Nrf2 lysines are optimally presented and exposed to the ubiquitination complex (Figure 2).

Li and Kong (2009) suggest that there are two pools of Nrf2 present in a cell. One of the pools is made up of free floating Nrf2 (fNrf2), while the other consists of Keap1-bound Nrf2 (kNrf2). Under basal conditions, the synthesis and constant degradation of Nrf2 are balanced, resulting in a small pool of fNrf2 which leads to low or no levels of activity, depending on cell type. At times of oxidative stress, the ubiquitination and degradation of Nrf2 is hindered leading to a saturation of Keap1's binding capacity. As a result, the fNrf2 pool grows larger. Li and Kong (2009) also suggest that the size of the fNrf2 pool might be a determinate factor in the amplitude of the response, thus the induction of Phase II enzyme gene expression, such as that of HO-1 (Li and Kong, 2009).

This model also suggests an explanation for the fact that the inhibition of Nrf2 degradation is not necessarily linked to the disruption of the Nrf2-Keap1 complex. The non-binding of the DGL motif, while the ETGE motif remains bound results in failure to optimally present Nrf2 to the ubiquitination

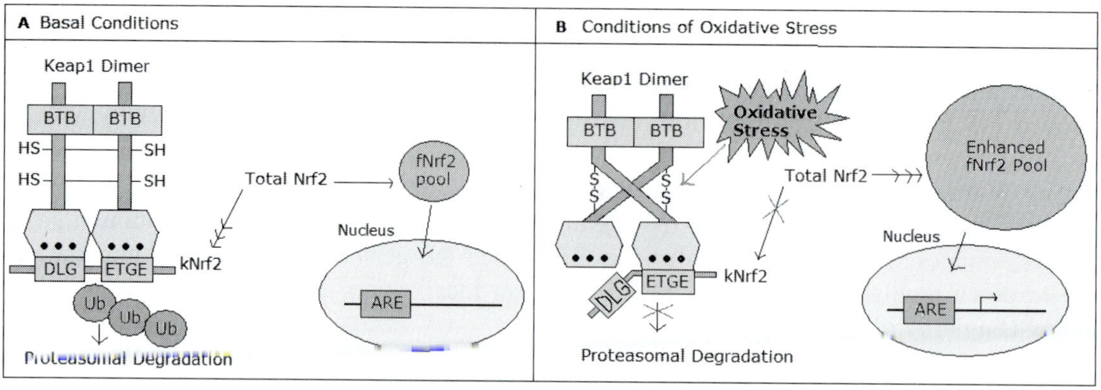

Fig. 2. The Hinge and Latch model of the Nrf2/ARE pathway. The Keap1 dimer functions to present Nrf2 in the right position to the Cul3 E3-ligase Ubiquitination complex. There are two pools of Nrf2 present in the cell, free Nrf2 (fNrf2) and Keap1-bound Nrf2 (kNrf2). Nrf2 binds to Keap1 via its ETGE (hinge) and DLG (latch) motifs. (A) Under basal conditions, the constant Nrf2 ubiquitination (Ub) and consequent proteasomal degradation keep the fNrf2 levels low. (B) Under conditions of oxidative stress, cysteine thiol groups of Keap1 are modified resulting in the changed conformation of Keap1. This conformational change disrupts the two-sites binding and as a result, the ubiquitination of Nrf2 is impeded. Keap1 remains bound to undegraded Nrf2 and becomes saturated. This causes an increase in the fNrf2 pool and ARE-mediated HO-1 gene induction. Adapted from Li and Kong (2009).

complex. This causes the accumulation of fNrf2 and subsequent ARE-controlled gene induction, however, without disrupting the bond between Keap1 and Nrf2 completely (Li and Kong 2009).

As indicated above, this model also provides an interesting angle explaining the formation of adaptive resistance that has been observed at mild oxidative stress conditions. When a cell is treated with ARE-controlled gene expression inducers, Nrf2 ubiquitination is impeded and the fNrf2 pool will grow in size. If the cell is then confronted with oxidative stress while it is in this 'primed' condition, the response that is triggered will be stronger and more effective as the size of the fNrf2 pool appears to determine the amplitude of the response. Therefore, the cell would be more capable of handling oxidative stress due to a built-up adaptive resistance thanks to an increased fNrf2 pool (Li and Kong 2009). Recent research has indeed observed that when treated with adaptogen extracts from *Eleutherococcus* and *Rhodiola*, and subsequently exposed to oxidative stress, cell survival was higher and the oxidative stress in the cells themselves was lower than in the non-treated controls which correlated with the induction of HO-1 (Wiegant *et al.* 2008 and unpublished observations).

Thus, the induction of Phase II enzymes, of which HO-1 plays a crucial role in the cytoprotection of cells, consists of the involvement of several major signaling cascades and while research has made great progress recently on deciphering the mechanisms of HO-1 induction, a full understanding of the complexity of the events leading up to this gene expression induction remains elusive.

Discussion and Conclusion

The induction of HO-1, being a crucial Phase II cytoprotective enzyme, is generally considered to enhance cellular adaptive resistance against a variety of stress conditions. The observation that extracts of plant adaptogens were able to enhance expression of HO-1 led to the speculation that HO-1 induction by plant adaptogens might be an important element of their mechanism of action (Wiegant *et al.* 2008). However, adaptogens do not have the monopoly on inducing HO-1 gene expression and nor on triggering cellular resistance. Indeed, many other plant extracts such as *Ginkgo biloba* (Chen *et al.* 2001) and phytochemicals including compounds such as curcumin, caffeic acid phenethyl ester (CAPE), sulforaphane and resveratrol have all been reported to be able to induce HO-1 gene expression and to activate the Phase II cellular defense system (Lee and Surh 2005; Prawan *et al.* 2005; Li *et al.* 2007).

Recently, other factors and mechanisms have been suggested to explain the specific enhancement of adaptive resistance; including a variety of phytochemicals that have been identified to be responsible for the beneficial effects of eating fruits and vegetables (Mattson and Cheng 2006). These phytochemicals serve a variety of functions in plants and play an important role in their color, smell, flavor and texture (Son *et al.* 2008). In order to explain how these phytochemicals are able to exert beneficial cytoprotective effects, it was initially believed that these compounds were able to counteract oxidative stress by functioning as anti-oxidants. Some phytochemicals indeed do possess anti-oxidant properties which enable them to scavenge oxidants. However, current research observed that phytochemicals can also exert their beneficial effects by other mechanisms.

In this respect, it was discovered that many of these phytochemicals actually evolved as toxins. These toxic compounds were in place to protect the plants against damaging organisms such as insects, bacteria, yeast and the like (Mattson and Cheng 2006). It would seem counterintuitive that these toxic compounds are responsible for the beneficial properties of plants and vegetables. Indeed, phytochemicals are often carcinogens and/or neurotoxins at high concentrations. It is at low dosages, however, that these phytochemicals exert their beneficial effects (Mattson and Cheng 2006). It is therefore believed that phytochemicals participate in the hormesis concept which states that while at high dosages a chemical proves inhibitory and toxic, when taken at lower levels a chemical can possess stimulatory or beneficial effects (Son *et al.* 2008). Focusing on plant adaptogens, these extracts were also shown to exert detrimental effects at high concentrations but beneficial effects at lower concentrations which is in line with the hormesis concept. In addition, it was shown that in cell cultures plant adaptogens can exert mild oxidative stress which initiates the induction of endogenous cellular defense mechanisms including the expression of HO-1, which leads to an enhanced state of adaptive resistance (Wiegant *et al.* 2008). Mild oxidative stress at sub-toxic dosages causes the activation of multiple signaling cascades leading to the expression of the Phase II enzymes such as HO-1 that confer protection and an adaptive resistance against more severe states of oxidative stress (Son *et al.* 2008).

If these adaptogens are not alone in their function of inducing HO-1 and enhancing adaptive resistance, then perhaps they possess other beneficial functions that explain a more exclusive effect thought to be characteristic for these plant extracts. Adaptogens have, for instance, also been suggested to confer their beneficial effects via the hypothalamic-pituitary-adrenal (HPA) axis (Wagner *et al.* 1994; Panossian and Wagner 2005). This HPA axis is part of the organism's stress system that has been suggested to function in the bodily reactions against stress as well as adaptation against stress by balancing the release of certain hormones, such as adrenaline and corticosteroids (Panossian and Wagner 2005). Another important part of the stress system has also been implicated in the functioning of adaptogens, namely the sympatho-adrenal system (SAS) which functions in rapid responses to strain or stress conditions (Panossian and Wagner 2005). Therefore, adaptogens could very well exert their beneficial effects on organismal level as well as on cellular level by inducing Phase II enzyme gene expression. To further establish the effects of adaptogens as well as the previously mentioned phytochemicals on both cellular and on organism level, current research would therefore benefit from studies focusing on the elucidation of the molecular mechanisms underlying cellular adaptive resistance at these different levels.

References

1. Abraham NG and Kappas A. Heme oxygenase and the cardiovascular-renal system. Free Radical Biology & Medicine 2005; 39: 1-25.
2. Abraham NG, Asija A, Drummond G and Peterson S. Heme Oxygenase-1 Gene Therapy: Recent Advances and Therapeutic Applications. Current Gene Therapy 2007; 7: 89-108.
3. Alam J and Cook JL. How Many Transcription Factors Does It Takes to Turn On the Heme Oxygenase-1 Gene? Am J Respir Cell Mol Biol. 2007; 36: 166-174.
4. Boon-Niermeijer EK, van den Berg A, Wikman G and Wiegant FAC. Phyto-adaptogens protect against environmental stress-induced death of embryos from the freshwater snail Lymnaea stagnalis. Phytomedicine 2000; 7: 389-399.
5. Brekhman II, Dardymov IV. New substances of plant origin which increase non-specific resistance. Ann Rev Pharmacol. 1969; 9: 419-430.
6. Brown RP, Gerbang PL and Ramazanov Z. Rhodiola rosea: a phytomedicinal overview. Herbal Gram 2002; 56: 40-52.
7. DeSanctis R, De Bellis R, Scesa C, Mancini U, Cuchiarini L and Dacha M. In vitro protective effect of Rhodiola rosea extract against hypochlorous acid-induced oxidative damage in human erythrocytes. BioFactors 2004; 20: 147-159.
8. Dulak J, Loboda A and Jozkowicz A. Effect of Heme Oxygenase-1 on Vascular Function and Disease. Current Opinion of Lipidolology 2008; 19: 505-512.
9. Eggler AL, Gay KA and Mesecar AD. Molecular mechanisms of natural products in chemoprevention: Induction of cytoprotective enzymes by Nrf2. Mol Nutr Food Res. 2008; 52: 84-94.
10. Halliwell B and Gutteridge JMC. Free radicals in Biology and Medicine (3rd ed.). Oxford University Press, Oxford, 1999.
11. Immenschuh S and Ramadori G. Gene Regulation of Heme Oxygenase-1 as a Therapeutic Target. Biochemical Pharmacology 2000; 60: 1121-1128.
12. Immenschuh S and Schröder H. Heme oxygenase-1 and cardiovascular disease. Histology and Histopathology 2006; 21: 679-685.
13. Itoh K, Chiba T, Takahashi S, Ishii T, Igarashi K, Katoh Y, Oyake T, Hayashi N, Satoh K, Hatayama I, Yamamoto M and Nabeshima Y-I. An Nrf2/Small Maf Heterodimer Mediates the Induction of Phase II Detoxifying Enzyme Genes through Anti-oxidant Response Elements. Biochemical and Biophysical Research Communications 1997; 236: 313-322.

14. Jafari M, Jeffrey S, Felgner JS, Bussel II, Hutchili T, Khodayari B, Rose MR and Mueller LD. Rhodiola: a promising anti-aging Chinese herb. Rejuvenation Research 2007; 10: 587-602.
15. Jaiswal AK. Nrf2 Signaling In Coordinated Activation of Antioxidant Gene Expression. Free Radical Biology and Medicine 2004; 36: 1199-1207.
16. Kang KW, Lee, SJ and Kim SG. Molecular Mechanism of Nrf2 Activation by Oxidative Stress. Antioxidants & Redox Signalling 2005; 7: 1664-1673.
17. Kirkby KA and Adin CA. Products of heme oxygenase and their potential therapeutic applications. Am J Physiol Renal Physiol. 2006; 290: 563-571.
18. Lee J-S and Surh Y-J. Nrf2 as a novel molecular target for chemoprevention. Cancer Letters 2005; 224: 171-184.
19. Li C, Hossieny P, Wu BJ, Qawasmeh A, Beck K and Stocker R. Pharmagologic Induction of Heme Oxygenase-1. Antioxidants & Redox Signalling 2007; 9: 2227-2239.
20. Li W and Kong A-N. Molecular Mechanisms of Nrf2-Mediated Antioxidant Response. Molecular Carcinogenesis 2009; 48: 91-104.
21. Maines MD. The Heme Oxygenase System and Its Functions in The Brain. Cellular and Molecular Biology 2000; 46: 573-585.
22. Martin D, Rojo AI, Salinas M and Diaz R. Regulation of heme oxygenase-1 expression through the phosphatidylinositol 3-kinase/Akt pathway and the Nrf2 transcription factor in response to the antioxidant phytochemical carnosol. J Biol Chem. 2004; 279: 8919-8929.
23. Mattson M and Cheng A. Neurohormetic phytochemicals: low-dose toxins that induce adaptive neuronal stress responses. Trends in Neurosciences 2006; 29: 632-639.
24. Mayer M. Association of serum bilirubin concentration with risk of coronary artery disease. Clin Chem 2000; 46: 1723-1727.
25. Numazawa S, Ishikawa M, Yoshida A and Tanaka S. Atypical protein kinase C mediates activation of NF-E2-related factor 2 in response to oxidative stress. Am J Physiol Cell Physiol. 2003; 285: C334-C342.
26. Panossian A and Wagner H. Stimulating Effect of Adaptogens: An Overview with Particular Reference to their Efficacy following Single Dose Administration. Phytother Res. 2005; 19: 819-838.
27. Prawan A, Kundu JK and Surh Y-J. Molecular Basis of Heme Oxygenase-1 Induction: Implications for Chemoprevention and Chemoprotection. Antioxidants & Redox Signalling 2005; 7: 1688-1703.
28. Rytter SW, Alad J and Choi AMK. Heme Oxygenase-1/Carbon Monoxide: From Basic Science to Therapeutic Applications. Physiol Rev. 2006; 86: 583-650.
29. Salazar M, Rojo AI, Velasco D and DeSagarra RM. Glycogen synthase kinase-3b inhibits the xenobiotic and antioxidant cell response by direct phosphorylation and nuclear exclusion of the transcription factor Nrf2. J Biol Chem. 2006; 281: 14841 -14851.
30. Schutgens FWG, Neogi P, Wijk van EPA, Wijk van R, Wikman G and Wiegant FAC. The Influence of Adaptogens on Ultraweak Biophoton Emission: a Pilot-Experiment. Phytother Res. 2009; 23: 1103-1108.

31. Shih P-H, Yeh C-T and Yen G-C. Anthocyanins induce the Activation of Phase II Enzymes through the Antioxidant Response Elememt Pathway against Oxidative stress-Induced Apoptosis. Journal of Agricultural and Food Chemistry 2007; 55: 9427-9435.
32. Son TG, Camandola S and Mattson MP. Hormetic Dietary Phytochemicals. Neuromolecular Medicine 2008; 10: 236-246.
33. Srisook K, Kim C and Cha Y-N. Molecular Mechanisms Involved in Enhancing HO-1 Expression: De-repression by Heme and Activation by Nrf2, The "One-Two" Punch. Antioxidants & Redox Signalling 2005; 7: 1674-1686.
34. Surh Y-J, Kundu JK and Na H-K. Nrf2 as a Master Redox Switch in Turning on the Cellular Signaling Involved in the Induction of Cytoprotective Genes by Some Chemopreventitive Phytochemicals. Planta Med. 2008; 74: 1526-1539.
35. Surh Y-J, Kundu JK, Li MH, Na H-K and Cha Y-M. Role of Nrf2-mediated heme oxygenase-1 upregulation in adaptive survival response to nitrosative stress. Arch Pharm Res. 2009; 32: 1163-1176.
36. Syapin PJ. Regulation of haeme oxygenase-1 for treatment of neuroinflammation and brain disorders. British Journal of Pharmacology 2008; 155: 623-640.
37. Tsuchihashi S, Fondevila C and Kupiec-Weglinski JW. Heme Oxygenase System in Ischemia and Reperfusion Injury. Annals of Transplantation 2004; 9: 84-87.
38. Valko M, Leibfritz D, Moncol J, Cronin MTD, Mazur M and Telser, J. Free radicals and antioxidants in normal physiological functions and human disease. Int J Biochem & Cell Biol. 2007; 39: 44-84.
39. Wagner H, Nörr H and Winterhoff, H. Plant adaptogens. Phytomedicine 1994; 1: 63-76.
40. Wiegant FAC, Surinova S, Ytsma E, Langelaar-Makkinje M, Post JA and Wikman G. Life-span enhancing effects of plant adaptogens in C.elegans. in: Abstract book of the International Association for Adaptive Medicine. (VIII World Congress) - ISAM-Moscow. 2006: p61.
41. Wiegant FAC, Limandjaja G, de Poot SAH, Bayda LA, Vorontsova ON, Zenina TA, Langelaar-Makkinje M, Post JA and Wikman G. Plant adaptogens activate cellular adaptive mechanisms by causing mild damage. In: 'Adaptation Biology and Medicine: Health Potentials'. Eds. Lukyanova L, Takeda N. and Singal PK, Narosa Publishers, New Delhi, 2008; Vol. 5: 319-332.
42. Wiegant FAC, Surinova S, Ytsma E, Langelaar-Makkinje M, Wikman G and Post JA. Plant adaptogens increase lifespan and stress resistance in C.elegans. Biogerontology 2009; 10: 27-42.

Adaptation Biology and Medicine (Volume 6: Cell Adaptations and Challenges)
Editors: P. Wang, C.-H. Kuo, N. Takeda and P.K. Singal
Copyright © 2011 Narosa Publishing House Pvt. Ltd., New Delhi, India

30. Propylthiouracil and Steroidogenesis

Po-Ling Yu[1,2], Hsiao-Fong Pu[2], and Paulus S. Wang[2,3,4,5]

[1]Department of Surgery, Taipei City Hospital, Taipei 10431; [2]Department of Physiology, School of Medicine, and [3]Laboratory Animal Center, National Yang-Ming University, Taipei 11221; [4]Department of Medical Research and Education, Taipei City Hospital, Taipei 10431; and [5]Department of Medical Research and Education, Taipei Veterans General Hospital, Taipei 11217, Taiwan, Republic of China

Summary

Propylthiouracil (PTU) has been used as one of the effective drugs in treating hyperthyroidism. Traditionally, the side effects of applying PTU in treating patients with hyperthyroidism were thought as consequences of the suppression of thyroid function. These results which have been investigated by our laboratory in past years showed that PTU directly reduced the steroidogenesis in cells of rats, e.g. adrenocortical cells, testicular interstitial cells, Leydig cells and granulosa cells, by inhibition of adenylyl cyclase or steroids synthesis enzymes via mechanisms independent of thyroid hormones and gonadotropin releasing hormone (GnRH). Hopefully, these observations can be applied in clinical use in the future, especially using PTU in treatment of patients with hyperthyroidism.

Keywords: Hyperthyroidism, steroidogenesis, adrenocortical cells, Leydig cells, granulosa cells

Introduction

Propylthiouracil (PTU) is commonly used in treating hyperthyroidism by the mechanism of inhibiting thyroid hormone synthesis and conversion of thyroxine (T4) to triiodothyronine (T3). Clinical studies have reported that the most common side effects of PTU in treating hyperthyroidism included leukocytopenia (Cooper, 1984) and jaundice (Deidiker and deMello, 1996; Levy, 1993). However, it had been seldom investigated that the effects of PTU on other organs, except thyroid, e.g. adrenal cortex and gonadal organs. Adrenocorticotropic hormone (ACTH) delays the response of corticosterone in neonate rat with hypothyroidism treated with PTU (Henning *et al.* 1986). The concentration of corticostetrone, not ACTH decreased after treatment with PTU in rats (Sanchez-Franco *et al.* 1989). Transient neonatal hypothyroidism, induced with the PTU, results in dramatic increases in both testis size and sperm production in the adult rat (Joyce *et al.* 1993; Kirby *et al.* 1997). However, PTU treatment led to a significant reduction in GnRH-stimulated luteinizing hormone (LH) production without reduction of testosterone (Kirby *et al.* 1997). The numbers of Leydig cells increased but declined in steroidogenic function in the adult rat after neonatal hypothyroidism (Hardy

*Address for Correspondence: Dr. Paulus S. Wang Department of Physiology School of Medicine National Yang Ming University Shih-Pai, Taipei 11221, Taiwan Republic of China. Tel: +886-2-2826-7082 Fax: +886-2-2826-4049 E-mail: pswang@ym.edu.tw

et al. 1993). It was unclear whether PTU affected the extra-thyroid organs directly or indirectly by the hypothyroidism induced with PTU. The purpose of this manuscript is to summarize the previous studies regarding of PTU and steroidogenesis in our laboratory in the past ten years.

Chronic Effects of PTU on Adrenal Cortex: Corticosterone Secretion by Rat Zona Fasciculata-reticularis (ZFR) Cells

We investigated the direct effects of PTU on corticosterone secretion by ZFR cells of rats both in vivo and in vitro. Male rats were divided into 4 groups and then injected subcutaneously with saline, PTU, PTU plus T4, or T4 once daily for 2 weeks. After 2 weeks, rats were decapitated or received ACTH, intravenously. ZFR cells from normal, saline-, PTU-, PTU plus T4-, or T4-treated rats were incubated with ACTH, forskolin, 8-bromo-cAMP (8-Br-cAMP), deoxycorticosterone +/- PTU (1, 2, or 5 mg/mL) at 37° C for 2 hours. Corticosterone concentrations in plasma and cell media, and cAMP production in ZFR cells were determined by radioimmunoassay. The chronic effects of PTU on the activities of steroidogenic enzymes in ZFR cells were measured by the amounts of intermediate steroidal products separated by thin-layer chromatography. The results showed that the basal and ACTH-stimulated levels of plasma corticosterone in PTU-treated rats were lower as compared to saline-treated animals. Both basal and ACTH-stimulated corticosterone secretion were inhibited by PTU dose > 2 mg/mL in rat ZFR cells. The cAMP production induced by forskolin was lower in PTU-, PTU plus T4-, or T4-treated rats than in saline-treated animals. Chronic administration of PTU or PTU plus T4 inhibited the 3β-hydroxysteroid dehydrogenase (3β-HSD), 21β-hydroxylase and 11β-hydroxylase activities. Administration of PTU (1, 2, and 5 mg/mL) suppressed the basal, ACTH-, 8-Br-cAMP-, forskolin-, and deoxycorticosterone-stimulated corticosterone secretion in rat ZFR cells. Likewise, PTU dose > 2 mg/mL inhibited the ACTH and 8-Br-cAMP-stimulated levels of intracellular cAMP in rat ZFR cells. These results suggested that PTU counteracted both basal and ACTH-induced adrenal steroidogenesis through their attenuation of the activity of 11β-hydroxylase and cAMP production in rat ZFR cells (Lo *et al.* 1998).

Acute Effects of PTU on Gonads

Testosterone Production in Rat Interstitial Cells and Leydig Cells

To investigate the acute effects of PTU on the androgen production, we examined the testosterone secretion by rat testicular interstitial cells. We found that the plasma testosterone concentration was decreased with intravenous infusion of PTU, but the concentration of plasma T4 was unaffected. Exposure of anterior pituitary tissue to PTU in vitro affect neither basal nor GnRH-stimulated LH release. PTU (3-12 mM) inhibited both the basal and the human chorionic gonadotropin (hCG)-stimulated release of testosterone from rat testicular tissue in vitro; at the highest dose of PTU (12 mM) inhibited the forskolin- or 8-Br-cAMP-stimulated release of testosterone. The 25-OH-cholesterol-stimulated release of pregnenolone and testosterone by the testicular interstitial cells was inhibited by PTU. The results suggested that the inhibitory actions of PTU on testosterone secretion are exerted, at least in part, at the testicular level through a thyroid status independent mechanism involving a reduction of P_{450} side-chain cleavage enzyme (P_{450} scc)activity and, the conversion of cholesterol to pregnenolone (Chiao *et al.* 2000a).

In further study, PTU was also found to inhibit both basal and evoked-testosterone release in rat Leydig cells including challenge of hCG, forskolin, and 8-Br-cAMP respectively. PTU inhibited pregnenolone production by diminishing the function of P450scc in Leydig cells. Besides, PTU also regulated steroidogenesis by diminishing Steroidogenic acute regulatory (StAR) mRNA expression. These results suggested that PTU acted directly on rat Leydig cells to diminish testosterone production by inhibiting P450scc activity and StAR expression (Chiao *et al.* 2002).

PTU on Testosterone Production in Monkeys

We examined the acute effects of PTU on testosterone secretion in monkeys. Male monkeys were infused intravenously with PTU for 30 min. Blood samples were collected at several time intervals. Monkey testicular interstitial cells were cultured with PTU, hCG or forskolin, at 34°C for 1 h. In another study, steroidogenesis in monkey testicular interstitial cells was examined. PTU decreased plasma testosterone but not plasma T4 and LH levels in monkeys. Administration of PTU resulted in a dose-dependent inhibition of basal and hCG-, as well as forskolin-stimulated testosterone release by monkey testicular interstitial cells. PTU also diminished the stimulatory effects induced by androstenedione. These results suggest that PTU inhibits testosterone secretion via a mechanism independent of the secretion of T4 and LH in primates. The inhibitory mechanism of PTU on testosterone production involves a decreased activity of 17β-hydroxysteroid dehydrogenase (17β-HSD) and post-cAMP pathways (Chiao *et al.* 2000b).

PTU on Rat Granulosa Cells

The acute effect of PTU on the production of progesterone in rat granulosa cells was investigated (Chen *et al.*2003). PTU decreased the basal and hCG-stimulated release of progesterone from rat granulosa cells. PTU also attenuated the stimulatory effects of forskolin and 8-Br-cAMP on progesterone release from rat granulosa cells. The activities of both cytochrome P450scc and 3 -HSD were inhibited by PTU in rat granulosa cells. PTU decreased the hCG-increased amount of StAR protein in rat granulosa cells. These results suggest that PTU decreases the progesterone release by granulosa cells via a thyroid-independent mechanism involving the inhibition of cAMP pathway, and the activities of intracellular calcium, steroidogenic enzyme, and StAR protein functions (Chen *et al.* 2003).

Furthermore, our recent results showed that 16-hour treatment of PTU was able to stimulate pregnenolone production in rat granulosa cells although progesterone production was diminished by PTU through 3β-HSD inhibition (Chen *et al.* submitted). In this study, we found that PTU treatment enhanced the conversion of cholesterol into pregnenolone while P450scc was not increased. The levels of StAR in both total cell lysate and mitochondrial fraction were significantly increased by PTU treatments. The binding of steroidogenic factor-1 (SF-1) with StAR promoter region was also enhanced by PTU treatment, which suggested that PTU could upregulate StAR gene expression. In addition to SF-1 regulation, we found that extracellular signal-regulated kinase (ERK)/Mitogen-activated protein kinase (MAPK) activation was also an important regulator to PTU-stimulated StAR protein expression evaluated by the effects of ERK/MAPK inhibitor, PD98059. These findings suggested that PTU possessed opposite roles on the productions of progesterone and its precursor, pregnenolone. The increased conversion of cholesterol into the pregnenolone after 16-hour treatment of PTU might compensate the inhibition of progesterone production by short-term PTU treatment (Chen *et al.* submitted).

PTU on Aldosterone Secretion by Rat Zona Glomerulosa (ZG) Cells

In this experiment, we prepared ZG cells from adrenocortical tissues of male rats, and reacted with angiontension II, ACTH, A23187 (a calcium ionophore), cyclopiazonic acid (CPA, a specific inhibitor of Ca^{2+}-ATPase in the sarcoplasmic and endoplasmic reticulum), forskolin, 8-Br-cAMP and steroidogenic precursors (e.g. 25-OH-cholesterol, pregnenolone, progesterone, deoxycorticostrerone and cortisol). PTU dose-dependently decreased the aldosterone secretion by ZG cells in response to the above chemicals. PTU decreased StAR protein expression, but not the P450scc during steroidogenesis of aldosterone. These results suggested that PTU inhibited aldosterone production via cAMP-, and calcium channel-pathways as well as steroidogenic enzymes including P450scc, 3β-HSD, 11β-hydroxylase and StAR (Pu *et al.* submitted).

References

1. Cooper DS. Antithyroid drugs. N. Engl. J. Med. 1984; 311: 1353-1362.
2. Deidiker R, deMello DE. Propylthiouracil-induced fulminant hepatitis: case report and review of the literature. J. Pediatr. Pathol. Lab. Med. 1996; 16: 845-852.
3. Levy M. Propylthiouracil hepato-toxicity. A review and case presentation. Clin. Pediatr. 1993; 32: 25-29.
4. Henning SJ, Leeper LL, Dieu DN. Circulating corticosterone in the infant rat: the mechanism of age and thyroxine effects. Pediatr. Res. 1986; 20: 87-92.
5. Sanchez-Franco F, Fernandez L, Fernandez G, Cacicedo L. Thyroid hormone action on ACTH secretion. Horm. Metabol. Res. 1989; 21: 550-552.
6. Joyce KL, Porcelli J, Cooke PS. Neonatal goitrogen treatment increases adult testis size and sperm production in the mouse. J. Androl. 1993; 14: 448-455.
7. Kirby JD, Arambepola N, Porkka-Heiskanen T, Kirby YK, Rhoads ML, Nitta H, Jetton AE, Iwamoto G, Jackson GL, Turek FW, Cooke PS. Neonatal hypothyroidism permanently alters follicle-stimulating hormone and luteinizing hormone production in the male rats. Endocrinology 1997; 138: 2713-2721.
8. Hardy MP, Kirby JD, Hess RA, Cooke PS. Leydig cells increase their numbers but decline in steroidogenic function in the adult rat after neonatal hypothyroidism. Endocrinology 1993; 132: 2417 -2420.
9. Lo MJ, Wang SW, Kau MM, Chen JJ, Fang VS, Ho LT, Wang PS. Pharmacological effects of propyl-thiouracil on corticosterone secretion in male rats. J. Invest. Med. 1998; 46: 444-452.
10. Pu HF, Lee HW, Lin CW, Wang PS. Effects of propylthiouracil on the production of aldosterone in adrenal zona glomerulosa cells of male rats. (submitted)
11. Chiao YC, Lin H, Wang SW, Wang PS. Direct effects of propylthiouracil on testosterone secretion in rat testicular interstitial cells. Br. J. Pharmacol. 2000a; 130: 1477-1482.
12. Chiao YC, Cho WL, Wang PS. Inhibition of testosterone production by propylthiouracil in rat Leydig cells. Biol. Reprod. 2002; 67: 416-422.
13. Chen JJ, Wang SW, Chien EJ, Wang PS. Direct effect of propylthiouracil on progesterone release in rat granulosa cells. Br. J. Pharmacol. 2003; 139: 1564-1570.

14. Chiao YC, Chao HL, Huang SW, Wang PS. Direct effects of propylthiouracil on testosterone production in monkeys. Chin. J. Physiol. 2000b; 43: 55-59.
15. Chen MC, Wang SW, Kan SF, Tsai SC, Wu YC , Wang PS. Stimulatory effects of Propylthiouracil on pregnenolone production through upregulation steroidogenic acute regulatory protein expression in rat granulosa cells. Toxicol. Sci. 2010. (In Press)

Adaptation Biology and Medicine (Volume 6: Cell Adaptations and Challenges)
Editors: P. Wang, C.-H. Kuo, N. Takeda and P.K. Singal
Copyright © 2011 Narosa Publishing House Pvt. Ltd., New Delhi, India

31. High-Altitude Diseases: The Implications of Genes, Genetics and Drugs

Tsering Stobdan, Zahara Ali and M.A. Qadar Pasha*

Functional Genomics Unit, Institute of Genomics and Integrative Biology, Delhi, India

Abstract

High-altitude (HA) is characterized as one of the extreme environmental conditions on this planet. The chronic or acute exposure to such environment may lead an individual to various HA diseases. Although the immediate physiological responses constructively shape an individual to adapt to the hypoxic environment, there are people genetically predisposed for better performance to such conditions. Although the idea of comparing such inter-individual variation at the DNA level is classical, only few empirical studies have succeeded in documenting causal links between molecular variation and its therapeutic applications e.g. for pharmacogenetic studies evaluating the influence of sequence variations on drug response. Here we review candidate genes encoding drug target receptors or enzymes, projected to prevent HA disease, simultaneously highlighting high-altitude pulmonary edema as an example. We include comments about interactions between genetic variants and its therapeutic entity, as well as its selective role among native population. By giving adequate consideration on the pharmacogenetic issues, the foremost objective of the present review is to advance our understanding onto individual drug-response to the drugs used, to treat HA diseases and to develop strategies for HA genetic predisposition.

Keywords: Pharmacogenetic, pulmonary edema, hypoxia, hypoleanic, Monge's disease, adaptive medicine?

Introduction

The environment at high-altitude (HA) is governed by simple laws of barometric pressure. With increasing altitude there is less overlying atmospheric mass, so that pressure decreases (Figure 1). Concomitantly as the air becomes thinner, the density of air decreases, and so too does the pressure of air. This means that with increasing altitude the number of oxygen molecules per breath is proportionately reduced, which in turn reduces the amount of oxygen available to the blood and tissues in the body, a condition known as hypoxia. In addition, the environment at HA is also characterized by diurnal temperature variation of hot and scorching days to freezing nights with strong winds and low humidity. These factors further add to the deleterious effect of hypobaric-hypoxia. It is estimated that >140 million people live permanently at an altitude of >2500m and every year an additional 40 million

*Address for Correspondence: M.A. Qadar Pasha, PhD; Functional Genomics Unit, Institute of Genomics and Integrative Biology, Mall Road, Delhi - 110 007, INDIA Ph: +91 11 27666156, Fax: +91 11 27667471, Email: qpasha@igib.res.in

travel to these heights for various reasons. These figures may have doubled in the recent past due to rapid advancement in means for easy and quick access to these altitudes.

HA environment causes in humans many physiological and biochemical changes which enable the body to function optimally in the low oxygen environment (Gleadle and Ratcliffe, 1998). The initial and most important adjustments are an increase in the frequency and depth of breathing. Simultaneously, the heart rate for a given O_2 uptake becomes higher with more severe hypoxia (Reeves *et al.* 1987) as well as evidences about the role of sympathetic nervous system in hypoxic/HA environment (Koller *et al.* 1988, Asano *et al.* 1997). Thus, adaptation is primarily an evolutionary process whereby an organism after several generations becomes better able to live in a particular habitat, especially on traits in the pathway of oxygen uptake and utilization to adapt to hypoxic environment may refrain natives to the stressful environment. Genetically, long term exposure to HA provides a selective pressure towards the beneficial alleles of certain loci. At the same time the exposure to acute or chronic hypoxic environment may lead few individuals to HA disease (Basnyat and Murdoch, 2003) viz, high-altitude pulmonary edema (HAPE), high-altitude cerebral edema (HACE), acute mountain sickness (AMS) and chronic mountain sickness (CMS) (Monge's disease). Based on increasing probability of experiencing health problems the international society of mountain medicine has defined altitude as High (8,000 - 12,000 ft), Very High (12,000 - 18,000 ft), and Extremely High (>18,000 ft) (Figure 1).

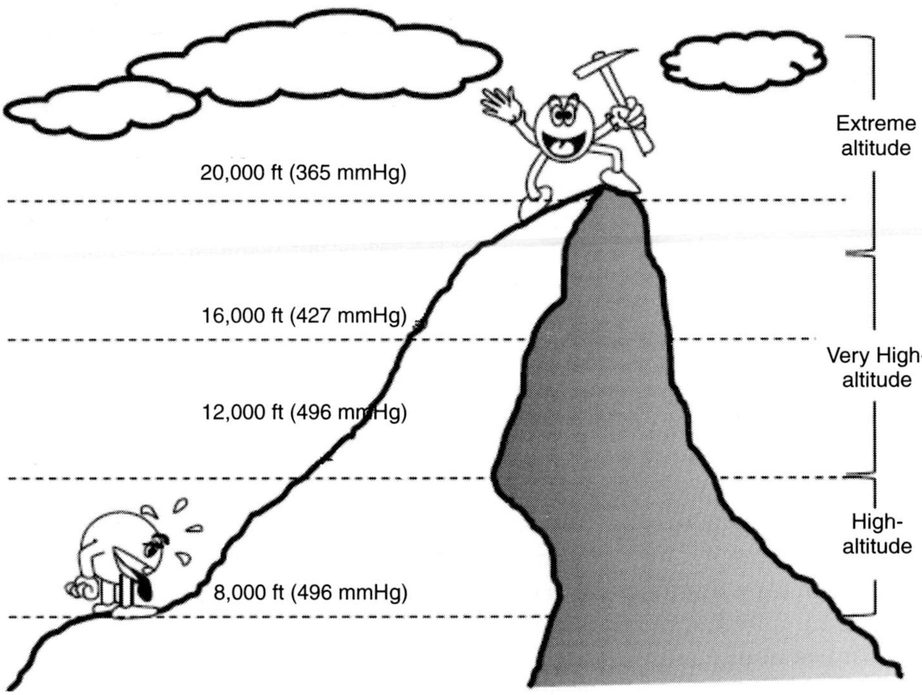

Fig. 1. How high is high-altitude? A marked variability in its definition is observed based on individual response to high-altitude. A person may perform well at extreme altitude, while another may find difficult to breath even at relatively low altitude.

Therefore, among various factors, variations at certain loci may also predispose individuals to HA illness. While, few individuals may perform strenuous activity at extreme altitude while others cannot (Figure 1), a growing knowledge about genetic influence on HA diseases combined with the clinical data hold promise for the identification of genetic markers for altitude predisposition. The recent explosion of knowledge about the human genome, fueled by the tools of molecular biotechnology and data from the human genome projects have enabled researchers around the globe to unravel the genetic underpinning of human diseases more rapidly.

The present review serves to illustrate the genetic predisposition to HA and to highlight the association of genetic variants with HA illness, a concept within the ambit of the discipline called 'Adaptive Medicine'. Emphasizing HAPE as an example the purpose of this review is to provide an overview of the potential contributions of genetic variability to HA disease and the possible impact of this variability on drug response.

High Altitude Pulmonary Edema (HAPE)

HAPE is a non-cardiogenic pulmonary edema that may develop in otherwise healthy individuals upon exposure to hypoxia above 2500m (Hultgren, 1996, Swenson *et al.* 2002, Scherrer *et al.* 1999, Basnyat and Murdoch, 2003, Bartsch *et al.* 2001). This disorder is primarily caused by an increase in pulmonary capillary pressure (Maggiorini *et al.* 2001), which may lead to stress failure (Costello *et al.* 1992, Tsukimoto *et al.* 1991) thus leading to high-permeability edema (Figure 2). Subsequently, a decreased fluid clearance from the alveoli may further contribute to such non-cardiogenic pulmonary edema (Bartsch *et al.* 2003). The high pulmonary artery pressure that leads to a protein-rich and mildly hemorrhagic edema, and no elevations in a number of proinflammatory cytokines, eicosanoid and nitric oxide (NO) metabolites (Swenson *et al.* 2002) suggests that the inflammatory reaction is rather a consequence than a causative factor in this HA disorder (Kleger *et al.* 1996). Since, it is also considered that exaggerated pulmonary pressure, possibly a prerequisite, may not always be sufficient to cause HAPE, a defective alveolar fluid clearance may represent a second important pathogenic mechanism (Sartori *et al.* 2000). While all sojourners do not suffer, those who have had HAPE once, run an unpredictable but significant risk of recurrence. These observations do indicate the involvement of genetic factors in its etiology. Our efforts, in understanding the genetic role in HAPE, indicated association of genetic variants in Angiotensin-I converting enzyme (*ACE*), endothelin 1 (*END1*) (Charu *et al.* 2006) and nitric oxide synthase 3 (*NOS3*) (Ahsan *et al.* 2004, Ahsan *et al.* 2006); nevertheless their inconsistence in other population (Weiss *et al.* 2003) signals towards complex multigenic and multifactorial interaction. Although there are various reports covering different aspects of HAPE e.g. clinical, pathophysiological or pharmacological, but the literature on pharmacogenetics or genetic predisposition to HAPE is still scant. The following paragraph briefly highlights the role of genetics in different therapeutics and prophylactic measures in HAPE (Figure 2).

1. Predisposition to Hypoxia

Since the barometric pressure directly determines the inspired partial pressure of oxygen, any physiological responses induced by HA exposure aim at increasing arterial oxygen saturation (SaO_2). It is therefore unambiguous to say that hypoxia is the major detrimental factor of any HA illness (Figure 2) more so because a prompt descent to a lower altitude is the most effective treatment for any

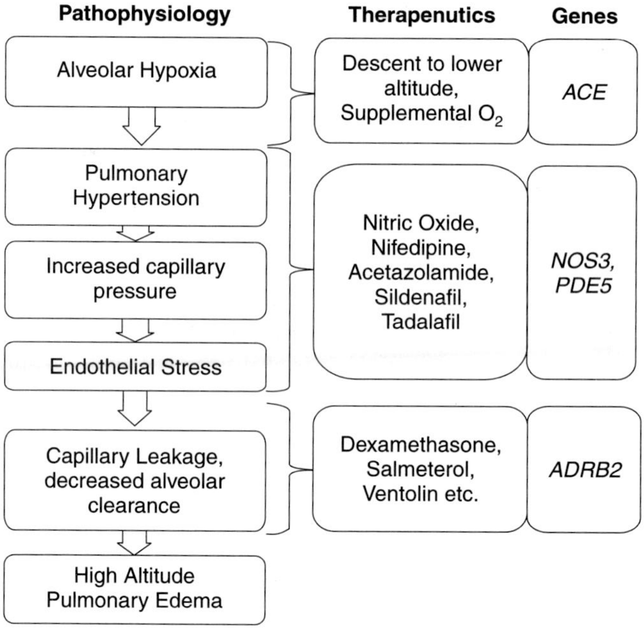

Fig. 2. Schematic presentation of the pathophysiological steps that lead to HAPE, the associated therapies and few of the corresponding candidate genes.

HA illness. However, in case rapid descent is not possible, the supplemental oxygen is the most preferred alternative (Zafren *et al.* 1996). Acute hypoxia leads to an initial increase in ventilation (Powell *et al.* 1998). The ventilatory response to hypoxia also depends on the pattern and intensity of hypoxic exposure and involves several physiological mechanisms. In addition, the inter-individual variability in acclimatization to altitude indicates that the genetic components play important role in individual response to hypoxia. The notion is supported by various studies e.g. in infants born at HA the role of hereditary factors for SaO_2 was predominantly noted. It was observed that the newborns of natives had higher SaO_2 at birth and during the first four months of life than the newborns of the migrants (Niermeyer *et al.* 1995); moreover, this characteristic is retained even after a prolong stay at low altitude, implicating an intrinsic hypoxic adaptation among natives (Zhou *et al.* 2008).

Angiotensin I-Converting Enzyme I/D Polymorphism and SaO_2

Among several candidate genes proposed, the polymorphism in Angiotensin I-converting enzyme (*ACE*) whereby an alu sequence with insertion/deletion (I/D) of a 287-base-pair in intron 16 was observed to be multifunctional, although the role of this alu has yet to be validated. Because of its relevance, the polymorphism has been targeted by several investigators. The variants were associated with SaO_2 (Woods *et al.* 2002), endurance performance (Montgomery *et al.* 1998), HA adaptation (Qadar Pasha *et al.* 2001) as well as HAPE (Hotta *et al.* 2004,Charu *et al.* 2006). The D allele is hypothesized to contribute to the hyperresponsive pulmonary vascular resistance (PVR) and pulmonary vascular resistance index (PVRI) in acute hypoxia (Hotta *et al.* 2004). Of equal significance, the I-allele was found to be associated with higher SaO_2 at HA (Bigham *et al.* 2008, Woods *et al.* 2002) which may be of consequence when performing physical work at altitude or ascending to over 8000m (Thompson *et al.* 2007). An optimal transfer of O_2 although requires a coordinated increase of pulmonary ventilation and efficient pulmonary exchange, nonetheless the role of genetic components in inter-individual variability in SaO_2 seem to play a significant role. Hence the SaO_2 was studied for its genetic influence.

2. Targeting Pulmonary Artery Pressure

The elevated pulmonary artery pressure (PAP) is the most striking feature in HAPE (Dehnert *et al.* 2007) and even other HA diseases like HA pulmonary hypertension (Scherrer *et al.* 2006), CMS (Ge and Helun, 2001) etc. The prophylactic administration of drugs, effective in lowering PAP, support the concept that high PAP has an important role in the development of HAPE (Bartsch *et al.* 1991). Contrary to this, HA natives are better protected (Beall *et al.* 2001). Therefore increased PAP in correspondence to a dramatic fall in SaO_2 may culminate in HAPE.

Endothelial Nitric Oxide Synthase

Recently, it is proposed that exacerbation of hypoxic pulmonary vasoconstriction by exertion would induce stress failure of the distal lung structure, promoting alveolar flooding with a protein rich and hemorrhagic edema (She *et al.* 2009). Therefore, hypoxia and exertion are considered as the two main factors in development of HAPE leading to increased pulmonary capillary pressure (Maggiorini *et al.* 2001).

Besides O_2, inhalation of NO has also been successfully employed to treat HAPE patients. At few instances the combined use of inhaled NO and O_2 were used together for additive effects on pulmonary hemodynamics and even greater effects on gas exchange are also studied (Anand *et al.* 1998). These results indicate that O_2 and NO, albeit, act on separate but interactive mechanisms in the pulmonary vasculature.

NO is associated with 'hypoxia-induced' cytoskeletal rearrangements and membrane leakiness of endothelial cells, a cyclic guanosine monophosphate (cGMP)-dependent phenomenon (Kolluru *et al.* 2008). Alternatively, NO inhalation in HAPE patients also results in decreased PAP and improved ventilation-perfusion mismatch (Scherrer *et al.* 1996). In a case-control study, significantly lower NO levels were found in HAPE patients compared to healthy controls at HA (Ahsan *et al.* 2004) suggesting that NO improves the ventilation/perfusion ratio and lowers the alveolar to arterial oxygen tension difference. Perhaps, that explains the higher NO levels in healthy HA natives (Ahsan *et al.* 2005). Further, the functional advantage of high NO concentration in the lungs of HA native or the lowlander at HA is targeted to offset ambient hypoxia by enhancing the uptake of O_2 from the lungs, which presumably improves delivery of O_2 to peripheral tissues (Beall *et al.* 2001). Therefore, NO is, thus, an important endogenous vasodilator that regulates pulmonary vascular tone and maintains physiological low pulmonary vascular resistance (Hampl and Herget, 2000) and by virtue of it, endothelial nitric oxide synthase (*NOS3*) is a key catalytic enzyme responsible for providing basal pulmonary NO.

In a significant development, a correlation study of genotype to phenotype (*NOS3* vs. NO) reveals that the combination of wild-type genotypes GG and BB of Glu298Asp (G894T) and 4B/4A (the latter a 27-bp variable number of tandem repeat, VNTR) polymorphisms of *NOS3* contribute to higher NO levels (Ahsan *et al.* 2005, Ahsan *et al.* 2006). A consistent finding of positive association of the G894T and 4B/4A polymorphisms of the *NOS3* with HAPE in Japanese (Droma *et al.* 2002), Indian (Ahsan *et al.* 2004) populations and a similar trend in Caucasians (Weiss *et al.* 2003) (Figure 3) indicated to its universal pharmacogenetic appeal. It is thus evident that the G and B alleles at the G894T and 4B/4A loci contribute to higher NO levels (Ahsan *et al.* 2005) that lead to higher NO levels in the lungs. This phenomenon compensates ambient hypoxia with higher pulmonary blood flow and O_2 delivery without the consequences of higher PAP (Hoit *et al.* 2005). Contrary to this any impairment of the NO pathway could contribute to enhanced hypoxic pulmonary vasoconstriction that is central to the pathogenesis of HAPE (Berger *et al.* 2005). The functional significance of G894T variant has yet to be demonstrated at the molecular level. However, reduced promoter activity for the 786C allele, results in diminished endothelial NO synthesis (Nakayama *et al.* 1999).

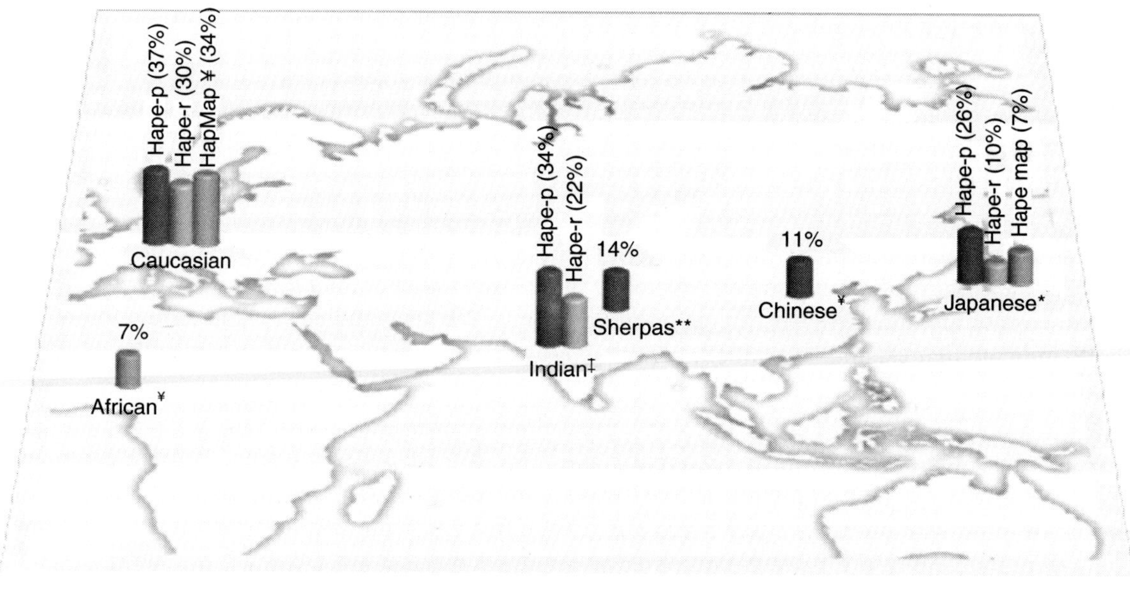

Fig. 3. Global distribution of T (risk) allele of G894T polymorphism of *NOS3*. Comparison of frequency (%) of the T allele of *NOS3* gene from different studies ([‡]Ahsan et al. 2004, **Droma et al. 2002, [†]Weiss et al. 2003, **Droma et al. 2006) including the [¥]HapMap study.

Phosphodiesterase Type 5 (PDE5)

PDE5 is a cGMP binding, cGMP-specific hydrolytic enzyme. Sildenafil and tadalafil are inhibitors of PDE5 and effective pulmonary vasodilators. The *PDE5A* gene in humans has been mapped to chromosome 4q26 containing 23 exons. It has three isoforms i.e. *PDE5A1, PDE5A2* and *PDE5A3* differing only in the N-terminal sequence that are presumably generated through alternative splicing of the first exon of the *PDE5A* gene (Lin *et al.* 2002).

Sildenafil attenuates hypoxia-induced pulmonary hypertension and has beneficial pulmonary hemodynamic effects even when *NOS3* activity is impaired (Zhao *et al.* 2001); whereas, tadalafil decreases systolic PAP and by virtue of that may reduce the incidence of HAPE (Maggiorini *et al.* 2006). Individually, sildenafil protects against the development of altitude-induced pulmonary hypertension (Aldashev *et al.* 2005) and improves gas exchange (Richalet *et al.* 2005). The drug when combined with the non-selective phosphodiesterase-inhibitor theophylline is suggested to impede the two main pathophysiological features of HAPE: the deleteriously high PAP (sildenafil's task) on one hand and the activation of an inflammatory cascade (theophylline's task) on the other hand (Kleinsasser and Loeckinger, 2002). Although it looks promising, further research is necessary to determine the efficacy and safety of PDE5 inhibitors as a drug of choice. It would further be interesting to investigate the *PDE5* genotypes in relation to these drugs.

The association between response to sildenafil and *ACE* I/D and *NOS3* G894T genotypes in patients with erectile dysfunction (ED) was studied (Eisenhardt *et al.* 2003). Analysis of the response to sildenafil revealed that the individuals with the D allele were less likely to respond to sildenafil. The response to sildenafil was also found to be significantly associated with G protein beta 3 subunit (*GNB3*) C825T genotype status in patients with ED (Sperling *et al.* 2003). The study reported that although the *GNB3* genotype distribution of patients with ED exactly matched to that of healthy controls, the analysis of the response to sildenafil revealed a significant association of C825T allele status with the erectile response to sildenafil. In the group with TT genotype, a response of 90.9% was observed while only 50.9% and 48.9% response in patients with CC and TC genotype, respectively. A similar study approach on HAPE would comprehensively elaborate the viability of drug on individuals with a particular genetic setup.

Additional Genes/Targets

The other important drugs used for the treatment of HAPE consists of nifedipine, tested positive for prophylactic measure (Bartsch *et al.* 1991, Oelz *et al.* 1989, Oelz *et al.* 1992), acetazolamide, associate changes in the respiratory exchange ratio due to altered CO_2 production kinetics (Faoro *et al.* 2007) and dexamethasone, stabilizes the capillary endothelium, along with reduction in hypoxic pulmonary vasoconstrictive response and improved maximum oxygen uptake (Maggiorini *et al.* 2006, Fischler *et al.* 2009). No studies are available on the association of genetic variants of the target molecules of these drugs with HAPE.

3. Targeting Alveolar Fluid Clearance

Endothelial disruption due to haemodynamic shear forces is not accepted universally as the only possible explanation for HAPE (Mortimer *et al.* 2004). In addition to the PAP, predisposition to HAPE is also associated with an imbalance between the amount of alveolar fluid secreted and its reabsorption. Therefore, therapeutic targets acting directly or indirectly on alveolar fluid reabsorption need equal attention.

Beta(2)-Adrenoceptors (ADRB2)

ADRB2 are important modulators of vascular tone, particularly in the pulmonary circulation. At HA, hypoxia inhibits alveolar fluid reabsorption and decreases Na, K-ATPase activity (Dada *et al.* 2003,

Litvan *et al.* 2006); whereas, norepinephrine was found to increase the activity and abundance of the Na, K-ATPase thereby increasing alveolar fluid reabsorption via the activation of both alpha (1)- and beta-adrenergic receptors (Azzam *et al.* 2004). It is therefore suggested that Na, K-ATPase endocytosis and alveolar epithelial dysfunction can be prevented with beta-adrenergic agonists (Vadasz *et al.* 2008, Litvan *et al.* 2006, Mutlu and Sznajder, 2005). Moreover, a substantial progress has been made in understanding the role of ADRB2 in regulating lung fluid clearance (Sakuma *et al.* 1997, Frank *et al.* 2000, Licker *et al.* 2008, Mutlu *et al.* 2004, Jayr *et al.* 1994, Dumasius *et al.* 2001). Salmetrol, a beta-adrenergic agonist is found to be a useful prophylactic drug to reduce the risk of HAPE (Sartori *et al.* 2002, Mutlu and Sznajder, 2004) presumably by increasing alveolar fluid clearance.

The *ADRB2*, a 413-amino-acid (G protein-coupled) receptor is encoded by an intronless gene, *ADRB2*, located on chromosome 5q31-32 (Kobilka *et al.* 1987). A member of G protein-coupled receptors (GPCRs), the largest family of plasma membrane receptors (~1,000 members), and the principal catecholamine receptor in the lungs, ADRB2 is one of the most important targets for current therapeutics. Enormous efforts are being made in bringing a better molecular picture of this receptor and recent achievement in presenting crystal structures (Cherezov *et al.* 2007, Rasmussen *et al.* 2007) cleared ways for rapid progress. Of importance to pulmonary function, the receptor is distributed throughout the lungs especially smooth muscles of small airways, in the airway epithelium, the pulmonary vasculature, and the pulmonary lymphatics (Hamid *et al.* 1991, McGraw and Liggett, 2005, Mutlu and Factor, 2008). Among the common polymorphism in this gene, the variations of arginine 16 to glycine (Arg16Gly) and glutamine 27 to glutamic acid (Gln27Glu) are the most frequently reported. There are several studies reporting the association between *ADRB2* polymorphisms and the pathogenesis of various diseases (Bhatnagar *et al.* 2005, Ellsworth *et al.* 2005, Jazdzewski *et al.* 2007, Matheson *et al.* 2006, McLaren *et al.* 2007, Stanzione *et al.* 2007, Zak *et al.* 2008). Further *ADRB2* genotypes have also been associated with therapeutic responses to beta (2) agonists in asthma (Israel *et al.* 2000) but not in chronic obstructive pulmonary disease (Kim *et al.* 2008).

A comparative study between two populations native to different altitudes for its possible role in HA adaptation did not show any significant difference (Rupert *et al.* 2000). Significant elevation of sympathetic activity in AMS (Loeppky *et al.* 2003) further led to the analysis of its common haplotype for association with AMS but without success (Wang *et al.* 2007). A recent study indicated association of genetic variants of *ADRB2* with differences in lung fluid accumulation in humans, subjects homozygous for Arg16 had greater susceptibility for lung fluid accumulation relative to subjects homozygous for Gly at this position (Snyder *et al.* 2007). Further, its role in improved O_2 delivery during altitude acclimatization (Mazzeo and Reeves, 2003) and in endurance performance (Wolfarth *et al.* 2007, Moore *et al.* 2001) marked *ADRB2* as a candidate gene for HAPE. Although, the literature on the influence of *ADRB2* gene polymorphisms on HA illness and pharmacogenetics is in its infancy, an increased understanding of this gene in relation to drugs may lead to improvements in the understanding and treatment of HA maladies.

Conclusion

Hypoxia is the major cause for most of the deleterious effects at HA and descending to a lower altitude is always the first and best choice for treatment. The association of SaO_2 with *ACE* I/D confers a genetic basis to hypoxia tolerance. Furthermore, the physiological adaptation under acute hypoxic exposure is still a subject of discussion, the genetic adaptation among natives e.g. higher SaO_2, higher

NO production etc. are evidently an ideal example to natural selection under extreme environmental condition. What remains to be uncovered is a better understanding of the relationship between these genes (genetic polymorphisms) and environment. This may also answer the genetic basis of HAPE with respect to its therapeutic targets. At present, the viability of the drugs used for treating HAPE either alone or in combination needs detail examination, as the drugs currently prescribed for HAPE are not specifically designed for this disease. Additionally, the drug response studies pertaining to individual with certain genetic background may provide ways for pharmacogenetic studies. For example genetic association study of the genes, targeted directly or indirectly by these drugs viz. *NOS3*, *PDE5*, *ADRB2*, etc provide evidence as well as a lead to inter-individual variability in drug efficacy thus drawing route to individualized medicine.

References

1. Ahsan A, Charu R, Qadar Pasha M A, Norboo T, Charu R, Afrin F, Ahsan A, Baig M A. eNOS allelic variants at the same locus associate with HAPE and adaptation. Thorax 2004; 59(11): 1000-1002..
2. Ahsan A, Mohd G, Norboo T, Baig M A, Qadar Pasha M A. Heterozygotes of NOS3 polymorphisms contribute to reduced nitrogen oxides in high-altitude pulmonary edema. Chest 2006; 130(5): 1511-1519.
3. Ahsan A, Norboo T, Baig M A, Qadar Pasha M A. Simultaneous selection of the wild-type genotypes of the G894T and 4B/ 4A polymorphisms of NOS3 associate with high-altitude adaptation. Ann Hum Genet 2005; 69(Pt 3): 260-267.
4. Aldashev A A, Kojonazarov B K, Amatov T A, Sooronbaev T M, Mirrakhimov M M, Morrell N W, Wharton J, Wilkins M R. Phosphodiesterase type 5 and high altitude pulmonary hypertension. Thorax 2005; 60(8): 683-687.
5. Anand I S, Prasad B A, Chugh S S, Rao K R, Cornfield D N, Milla C E, Singh N, Singh S, Selvamurthy W. Effects of inhaled nitric oxide and oxygen in high-altitude pulmonary edema. Circulation 1998; 98(22): 2441-2445.
6. Asano K, Mazzeo R S, McCullough R E, Wolfel E E, Reeves J T. Relation of sympathetic activation to ventilation in man at 4300 m altitude. Aviat Space Environ Med 1997; 68(2): 104-110.
7. Azzam Z S, Adir Y, Crespo A, Comellas A, Lecuona E, Dada L A, Krivoy N, Rutschman D H, Sznajder J I, Ridge K M. Norepinephrine increases alveolar fluid reabsorption and Na,K-ATPase activity. Am J Respir Crit Care Med 2004; 170(7): 730-736.
8. Bartsch P, Maggiorini M, Ritter M, Noti C, Vock P, Oelz O. Prevention of high-altitude pulmonary edema by nifedipine. N Engl J Med 1991; 325(18): 1284-1289.
9. Bartsch P, Mairbaurl H, Swenson E R, Maggiorini M. High altitude pulmonary oedema. Swiss Med Wkly 2003; 133(27-28): 377-384.
10. Bartsch P, Swenson E R, Maggiorini M. Update: High altitude pulmonary edema. Adv Exp Med Biol 2001; 502: 89-106.
11. Basnyat B, Murdoch D R. High-altitude illness. Lancet 2003; 361(9373): 1967-1974.
12. Beall C M, Laskowski D, Strohl K P, Soria R, Villena M, Vargas E, Alarcon A M, Gonzales C,

Erzurum S C. Pulmonary nitric oxide in mountain dwellers. Nature 2001; 414(6862): 411-412.

13. Berger M M, Hesse C, Dehnert C, Siedler H, Kleinbongard P, Bardenheuer H J, Kelm M, Bartsch P, Haefeli W E. Hypoxia impairs systemic endothelial function in individuals prone to high-altitude pulmonary edema. Am J Respir Crit Care Med 2005; 172(6): 763-767.
14. Bhatnagar P, Gupta S, Guleria R, Kukreti R. beta2-Adrenergic receptor polymorphisms and asthma in the North Indian population. Pharmacogenomics 2005; 6(7): 713-719.
15. Bigham A W, Kiyamu M, Leon-Velarde F, Parra E J, Rivera-Ch M, Shriver M D, Brutsaert T D. Angiotensin-converting enzyme genotype and arterial oxygen saturation at high altitude in Peruvian Quechua. High Alt Med Biol 2008; 9(2): 167-178.
16. Charu R, Stobdan T, Ram R B, Khan A P, Qadar Pasha M A, Norboo T, Afrin F. Susceptibility to high altitude pulmonary oedema: role of ACE and ET-1 polymorphisms. Thorax 2006; 61(11): 1011-1012.
17. Cherezov V, Rosenbaum D M, Hanson M A, Rasmussen S G, Thian F S, Kobilka T S, Choi H J, Kuhn P, Weis W I, Kobilka B K, Stevens R C. High-resolution crystal structure of an engineered human beta2-adrenergic G protein-coupled receptor. Science 2007; 318(5854): 1258-1265.
18. Costello M L, Mathieu-Costello O, West J B. Stress failure of alveolar epithelial cells studied by scanning electron microscopy. Am Rev Respir Dis 1992; 145(6): 1446-1455.
19. Dada L A, Chandel N S, Ridge K M, Pedemonte C, Bertorello A M, Sznajder J I. Hypoxia-induced endocytosis of Na,K-ATPase in alveolar epithelial cells is mediated by mitochondrial reactive oxygen species and PKC-zeta. J Clin Invest 2003; 111(7): 1057-1064.
20. Dehnert C, Berger M M, Mairbaurl H, Bartsch P. High altitude pulmonary edema: a pressure-induced leak. Respir Physiol Neurobiol 2007; 158(2-3): 266-273.
21. Droma Y, Hanaoka M, Ota M, Katsuyama Y, Koizumi T, Fujimoto K, Kobayashi T, Kubo K. Positive association of the endothelial nitric oxide synthase gene polymorphisms with high-altitude pulmonary edema. Circulation 2002; 106(7): 826-830.
22. Dumasius V, Sznajder J I, Azzam Z S, Boja J, Mutlu G M, Maron M B, Factor P. beta(2)-adrenergic receptor overexpression increases alveolar fluid clearance and responsiveness to endogenous catecholamines in rats. Circ Res 2001; 89(10): 907-914.
23. Eisenhardt A, Sperling H, Hauck E, Porst H, Stief C, Rubben H, Muller N, Siffert W. ACE gene I/D and NOS3 G894T polymorphisms and response to sildenafil in men with erectile dysfunction. Urology 2003; 62(1): 152-157.
24. Ellsworth D L, Coady S A, Chen W, Srinivasan S R, Boerwinkle E, Berenson G S. Interactive effects between polymorphisms in the beta-adrenergic receptors and longitudinal changes in obesity. Obes Res 2005; 13(3): 519-526.
25. Faoro V, Huez S, Giltaire S, Pavelescu A, van O A, Moraine J J, Guenard H, Martinot J B, Naeije R. Effects of acetazolamide on aerobic exercise capacity and pulmonary hemodynamics at high altitudes. J Appl Physiol 2007; 103(4): 1161-1165.
26. Fischler M, Maggiorini M, Dorschner L, Debrunner J, Bernheim A, Kiencke S, Mairbaurl H, Bloch K E, Naeije R, Brunner-La Rocca H P. Dexamethasone but not tadalafil improves exercise

capacity in adults prone to high-altitude pulmonary edema. Am J Respir Crit Care Med 2009; 180(4): 346-352.

27. Frank J A, Wang Y, Osorio O, Matthay M A. Beta-adrenergic agonist therapy accelerates the resolution of hydrostatic pulmonary edema in sheep and rats. J Appl Physiol 2000; 89(4): 1255-1265.
28. Ge R L, Helun G. Current concept of chronic mountain sickness: pulmonary hypertension-related high-altitude heart disease. Wilderness Environ Med 2001; 12(3): 190-194.
29. Gleadle J M, Ratcliffe P J. Hypoxia and the regulation of gene expression. Mol Med Today 1998; 4(3): 122-129.
30. Hamid Q A, Mak J C, Sheppard M N, Corrin B, Venter J C, Barnes P J. Localization of beta 2-adrenoceptor messenger RNA in human and rat lung using in situ hybridization: correlation with receptor autoradiography. Eur J Pharmacol 1991; 206(2): 133-138.
31. Hampl V, Herget J. Role of nitric oxide in the pathogenesis of chronic pulmonary hypertension. Physiol Rev 2000; 80(4): 1337-1372.
32. Hoit B D, Dalton N D, Erzurum S C, Laskowski D, Strohl K P, Beall C M. Nitric oxide and cardiopulmonary hemodynamics in Tibetan highlanders. J Appl Physiol 2005; 99(5): 1796-1801.
33. Hotta J, Hanaoka M, Droma Y, Katsuyama Y, Ota M, Kobayashi T. Polymorphisms of renin-angiotensin system genes with high-altitude pulmonary edema in Japanese subjects. Chest 2004; 126(3): 825-830.
34. Hultgren H N. High-altitude pulmonary edema: current concepts. Annu Rev Med 1996; 47: 267-284.
35. Israel E, Drazen J M, Liggett S B, Boushey H A, Cherniack R M, Chinchilli V M, Cooper D M, Fahy J V, Fish J E, Ford J G, Kraft M, Kunselman S, Lazarus S C, Lemanske R F, Martin R J, McLean D E, Peters S P, Silverman E K, Sorkness C A, Szefler S J, Weiss S T, Yandava C N. The effect of polymorphisms of the beta(2)-adrenergic receptor on the response to regular use of albuterol in asthma. Am J Respir Crit Care Med 2000; 162(1): 75-80.
36. Jayr C, Garat C, Meignan M, Pittet J F, Zelter M, Matthay M A. Alveolar liquid and protein clearance in anesthetized ventilated rats. J Appl Physiol 1994; 76(6): 2636-2642.
37. Jazdzewski K, Bednarczuk T, Stepnowska M, Liyanarachchi S, Suchecka-Rachon K, Limon J, Narkiewicz K. beta-2-adrenergic receptor gene polymorphism confers susceptibility to Graves disease. Int J Mol Med 2007; 19(1): 181-186.
38. Kim W J, Oh Y M, Sung J, Kim T H, Huh J W, Jung H, Lee J H, Kim E K, Lee J H, Lee S M, Lee S, Lim S Y, Shin T R, Yoon H I, Kwon S Y, Lee S D. Lung function response to 12-week treatment with combined inhalation of long-acting beta2 agonist and glucocorticoid according to ADRB2 polymorphism in patients with chronic obstructive pulmonary disease. Lung 2008; 186(6): 381-386.
39. Kleger G R, Bartsch P, Vock P, Heilig B, Roberts L J, Ballmer P E. Evidence against an increase in capillary permeability in subjects exposed to high altitude. J Appl Physiol 1996; 81(5): 1917-1923.
40. Kleinsasser A, Loeckinger A. Are sildenafil and theophylline effective in the prevention of high-altitude pulmonary edema?. Med Hypotheses 2002; 59(2): 223-225.

41. Kobilka B K, Dixon R A, Frielle T, Dohlman H G, Bolanowski M A, Sigal I S, Yang-Feng T L, Francke U, Caron M G, Lefkowitz R J. cDNA for the human beta 2-adrenergic receptor: a protein with multiple membrane-spanning domains and encoded by a gene whose chromosomal location is shared with that of the receptor for platelet-derived growth factor. Proc Natl Acad Sci U S A 1987; 84(1): 46-50.
42. Koller E A, Drechsel S, Hess T, Macherel P, Boutellier U. Effects of atropine and propranolol on the respiratory, circulatory, and ECG responses to high altitude in man. Eur J Appl Physiol Occup Physiol 1988; 57(2): 163-172.
43. Kolluru G K, Tamilarasan K P, Rajkumar A S, Geetha P S, Rajaram M, Saleem N K, Majumder S, Jaffar Ali B M, Illavazagan G, Chatterjee S. Nitric oxide/cGMP protects endothelial cells from hypoxia-mediated leakiness. Eur J Cell Biol 2008; 87(3): 147-161.
44. Licker M, Tschopp J M, Robert J, Frey J G, Diaper J, Ellenberger C. Aerosolized salbutamol accelerates the resolution of pulmonary edema after lung resection. Chest 2008; 133(4): 845-852.
45. Lin C S, Chow S, Lau A, Tu R, Lue T F. Human PDE5A gene encodes three PDE5 isoforms from two alternate promoters. Int J Impot Res 2002; 14(1): 15-24.
46. Litvan J, Briva A, Wilson M S, Budinger G R, Sznajder J I, Ridge K M. Beta-adrenergic receptor stimulation and adenoviral overexpression of superoxide dismutase prevent the hypoxia-mediated decrease in Na,K-ATPase and alveolar fluid reabsorption. J Biol Chem 2006; 281(29): 19892-19898.
47. Loeppky J A, Icenogle M V, Maes D, Riboni K, Scotto P, Roach R C. Body temperature, autonomic responses, and acute mountain sickness. High Alt Med Biol 2003; 4(3): 367-373.
48. Maggiorini M, Brunner-La Rocca H P, Peth S, Fischler M, Bohm T, Bernheim A, Kiencke S, Bloch K E, Dehnert C, Naeije R, Lehmann T, Bartsch P, Mairbaurl H. Both tadalafil and dexamethasone may reduce the incidence of high-altitude pulmonary edema: a randomized trial. Ann Intern Med 2006; 145(7): 497-506.
49. Maggiorini M, Melot C, Pierre S, Pfeiffer F, Greve I, Sartori C, Lepori M, Hauser M, Scherrer U, Naeije R. High-altitude pulmonary edema is initially caused by an increase in capillary pressure. Circulation 2001; 103(16): 2078-2083.
50. Matheson M C, Ellis J A, Raven J, Johns D P, Walters E H, Abramson M J. Beta2-adrenergic receptor polymorphisms are associated with asthma and COPD in adults. J Hum Genet 2006; 51(11): 943-951.
51. Mazzeo R S, Reeves J T. Adrenergic contribution during acclimatization to high altitude: perspectives from Pikes Peak. Exerc Sport Sci Rev 2003; 31(1): 13-18.
52. McGraw D W, Liggett S B. Molecular mechanisms of beta2-adrenergic receptor function and regulation. Proc Am Thorac Soc 2005; 2(4): 292-296.
53. McLaren N, Reed D M, Musch D C, Downs C A, Higashi M E, Santiago C, Radenbaugh P A, Allingham R R, Richards J E, Moroi S E. Evaluation of the beta2-adrenergic receptor gene as a candidate glaucoma gene in 2 ancestral populations. Arch Ophthalmol 2007; 125(1): 105-111.
54. Montgomery H E, Marshall R, Hemingway H, Myerson S, Clarkson P, Dollery C, Hayward M, Holliman D E, Jubb M, World M, Thomas E L, Brynes A E, Saeed N, Barnard M, Bell J D,

Prasad K, Rayson M, Talmud P J, Humphries S E. Human gene for physical performance. Nature 1998; 393(6682): 221-222.

55. Moore G E, Shuldiner A R, Zmuda J M, Ferrell R E, McCole S D, Hagberg J M. Obesity gene variant and elite endurance performance. Metabolism 2001; 50(12): 1391-1392.
56. Mortimer H, Patel S, Peacock A J. The genetic basis of high-altitude pulmonary oedema. Pharmacol Ther 2004; 101(2): 183-192.
57. Mutlu G M, Factor P. Alveolar epithelial beta2-adrenergic receptors. Am J Respir Cell Mol Biol 2008; 38(2): 127-134.
58. Mutlu G M, Koch W J, Factor P. Alveolar epithelial beta 2-adrenergic receptors: their role in regulation of alveolar active sodium transport. Am J Respir Crit Care Med 2004; 170(12): 1270-1275.
59. Mutlu G M, Sznajder J I. beta(2)-Agonists for treatment of pulmonary edema: ready for clinical studies? Crit Care Med 2004; 32(7): 1607-1608.
60. Mutlu G M, Sznajder J I. Mechanisms of pulmonary edema clearance. Am J Physiol Lung Cell Mol Physiol 2005; 289(5): L685-L695.
61. Nakayama M, Yasue H, Yoshimura M, Shimasaki Y, Kugiyama K, Ogawa H, Motoyama T, Saito Y, Ogawa Y, Miyamoto Y, Nakao K. T-786-->C mutation in the 5'-flanking region of the endothelial nitric oxide synthase gene is associated with coronary spasm. Circulation 1999; 99(22): 2864-2870.
62. Niermeyer S, Yang P, Shanmina, Drolkar, Zhuang J, Moore L G. Arterial oxygen saturation in Tibetan and Han infants born in Lhasa, Tibet. N Engl J Med 1995; 333(19): 1248-1252.
63. Oelz O, Maggiorini M, Ritter M, Noti C, Waber U, Vock P, Bartsch P. Prevention and treatment of high altitude pulmonary edema by a calcium channel blocker. Int J Sports Med 1992; 13 Suppl 1: S65-S68.
64. Oelz O, Maggiorini M, Ritter M, Waber U, Jenni R, Vock P, Bartsch P. Nifedipine for high altitude pulmonary oedema. Lancet 1989; 2(8674): 1241-1244.
65. Powell F L, Milsom W K, Mitchell G S. Time domains of the hypoxic ventilatory response. Respir Physiol 1998; 112(2): 123-134.
66. Qadar Pasha M A, Khan A P, Kumar R, Grover S K, Ram R B, Norboo T, Srivastava K K, Selvamurthy W, Brahmachari S K. Angiotensin converting enzyme insertion allele in relation to high altitude adaptation. Ann Hum Genet 2001; 65(Pt 6): 531-536.
67. Rasmussen S G, Choi H J, Rosenbaum D M, Kobilka T S, Thian F S, Edwards P C, Burghammer M, Ratnala V R, Sanishvili R, Fischetti R F, Schertler G F, Weis W I, Kobilka B K. Crystal structure of the human beta2 adrenergic G-protein-coupled receptor. Nature 2007; 450(7168): 383-387.
68. Reeves J T, Groves B M, Sutton J R, Wagner P D, Cymerman A, Malconian M K, Rock P B, Young P M, Houston C S. Operation Everest II: preservation of cardiac function at extreme altitude. J Appl Physiol 1987; 63(2): 531-539.
69. Richalet J P, Gratadour P, Robach P, Pham I, Dechaux M, Joncquiert-Latarjet A, Mollard P, Brugniaux J, Cornolo J. Sildenafil inhibits altitude-induced hypoxemia and pulmonary hypertension. Am J Respir Crit Care Med 2005; 171(3): 275-281.

70. Rupert J L, Monsalve M V, Devine D V, Hochachka P W. Beta2-adrenergic receptor allele frequencies in the Quechua, a high altitude native population. Ann Hum Genet 2000; 64(Pt 2): 135-143.
71. Sakuma T, Folkesson H G, Suzuki S, Okaniwa G, Fujimura S, Matthay M A. Beta-adrenergic agonist stimulated alveolar fluid clearance in ex vivo human and rat lungs. Am J Respir Crit Care Med 1997; 155(2): 506-512.
72. Sartori C, Allemann Y, Duplain H, Lepori M, Egli M, Lipp E, Hutter D, Turini P, Hugli O, Cook S, Nicod P, Scherrer U. Salmeterol for the prevention of high-altitude pulmonary edema. N Engl J Med 2002; 346(21): 1631-1636.
73. Sartori C, Allemann Y, Trueb L, Lepori M, Maggiorini M, Nicod P, Scherrer U. Exaggerated pulmonary hypertension is not sufficient to trigger high-altitude pulmonary oedema in humans. Schweiz Med Wochenschr 2000; 130(11): 385-389.
74. Scherrer U, Sartori C, Lepori M, Allemann Y, Duplain H, Trueb L, Nicod P. High-altitude pulmonary edema: from exaggerated pulmonary hypertension to a defect in transepithelial sodium transport. Adv Exp Med Biol 1999; 474: 93-107.
75. Scherrer U, Turini P, Thalmann S, Hutter D, Salmon C S, Stuber T, Shaw S, Jayet P Y, Sartori-Cucchial C, Villena M, Allemann Y, Sartori C. Pulmonary hypertension in high-altitude dwellers: novel mechanisms, unsuspected predisposing factors. Adv Exp Med Biol 2006; 588: 277-291.
76. Scherrer U, Vollenweider L, Delabays A, Savcic M, Eichenberger U, Kleger G R, Fikrle A, Ballmer P E, Nicod P, Bartsch P. Inhaled nitric oxide for high-altitude pulmonary edema. N Engl J Med 1996; 334(10): 624-629.
77. She J, Shen J, Goolaerts A, Song Y, Shen C, Hong Q, Bai C. Stress failure plays an major role in development of High altitude pulmonary oedema in rats. Eur Respir J 2009.
78. Snyder E M, Beck K C, Turner S T, Hoffman E A, Joyner M J, Johnson B D. Genetic variation of the beta2-adrenergic receptor is associated with differences in lung fluid accumulation in humans. J Appl Physiol 2007; 102(6): 2172-2178.
79. Sperling H, Eisenhardt A, Virchow S, Hauck E, Lenk S, Porst H, Stief C, Wetterauer U, Rubben H, Muller N, Siffert W. Sildenafil response is influenced by the G protein beta 3 subunit GNB3 C825T polymorphism: a pilot study. J Urol 2003; 169(3): 1048-1051.
80. Stanzione R, Di A E, Evangelista A, Barbato D, Marchitti S, Zanda B, Pirisi A, Quarta G, Volpe M, Rubattu S. Beta2-adrenergic receptor gene polymorphisms and risk of ischemic stroke. Am J Hypertens 2007; 20(6): 657-662.
81. Swenson E R, Maggiorini M, Mongovin S, Gibbs J S, Greve I, Mairbaurl H, Bartsch P. Pathogenesis of high-altitude pulmonary edema: inflammation is not an etiologic factor. JAMA 2002; 287(17): 2228-2235.
82. Thompson J, Raitt J, Hutchings L, Drenos F, Bjargo E, Loset A, Grocott M, Montgomery H. Angiotensin-converting enzyme genotype and successful ascent to extreme high altitude. High Alt Med Biol 2007; 8(4): 278-285.
83. Tsukimoto K, Mathieu-Costello O, Prediletto R, Elliott A R, West J B. Ultrastructural appearances of pulmonary capillaries at high transmural pressures. J Appl Physiol 1991; 71(2): 573-582.

84. Vadasz I, Dada L A, Briva A, Trejo H E, Welch L C, Chen J, Toth P T, Lecuona E, Witters L A, Schumacker P T, Chandel N S, Seeger W, Sznajder J I. AMP-activated protein kinase regulates CO_2-induced alveolar epithelial dysfunction in rats and human cells by promoting Na,K-ATPase endocytosis. J Clin Invest 2008; 118(2): 752-762.

85. Wang P, Koehle M S, Rupert J L. Common haplotypes in the beta-2 adrenergic receptor gene are not associated with acute mountain sickness susceptibility in Nepalese. High Alt Med Biol 2007; 8(3): 206-212.

86. Weiss J, Haefeli W E, Gasse C, Hoffmann M M, Weyman J, Gibbs S, Mansmann U, Bartsch P. Lack of evidence for association of high altitude pulmonary edema and polymorphisms of the NO pathway. High Alt Med Biol 2003; 4(3): 355-366.

87. Wolfarth B, Rankinen T, Muhlbauer S, Scherr J, Boulay M R, Perusse L, Rauramaa R, Bouchard C. Association between a beta2-adrenergic receptor polymorphism and elite endurance performance. Metabolism 2007; 56(12): 1649-1651.

88. Woods D R, Pollard A J, Collier D J, Jamshidi Y, Vassiliou V, Hawe E, Humphries S E, Montgomery H E. Insertion/deletion polymorphism of the angiotensin I-converting enzyme gene and arterial oxygen saturation at high altitude. Am J Respir Crit Care Med 2002; 166(3): 362-366.

89. Zafren K, Reeves J T, Schoene R. Treatment of high-altitude pulmonary edema by bed rest and supplemental oxygen. Wilderness Environ Med 1996; 7(2): 127-132.

90. Zak I, Sarecka-Hujar B, Krauze J. Cigarette smoking, carrier state of A or G allele of 46A>G and 79C>G polymorphisms of beta2-adrenergic receptor gene, and the risk of coronary artery disease. Kardiol Pol 2008; 66(4): 380-386.

91. Zhao L, Mason N A, Morrell N W, Kojonazarov B, Sadykov A, Maripov A, Mirrakhimov M M, Aldashev A, Wilkins M R. Sildenafil inhibits hypoxia-induced pulmonary hypertension. Circulation 2001; 104(4): 424-428.

92. Zhou Z N, Zhuang J G, Wu X F, Zhang Y, Cherdrungsi P. Tibetans retained innate ability resistance to acute hypoxia after long period of residing at sea level. J Physiol Sci 2008; 58(3): 167-172.

Adaptation Biology and Medicine (Volume 6: Cell Adaptations and Challenges)
Editors: P. Wang, C.-H. Kuo, N. Takeda and P.K. Singal
Copyright © 2011 Narosa Publishing House Pvt. Ltd., New Delhi, India

32. Effect of Ginsenoside Rb1 on Glucose Uptake and Expression of Glucose Transporters in Animal Cells

Su-Fen Huang[1], Yu-Chuan Huang[2], Jheng-Yu Lin[1], Wen-Liang Chang[3], Tsu-Chung Chang[1*]

[1]Department of Biochemistry, [2]Institute of Preventive Medicine and [3]School of Pharmacy, National Defense Medical Center, Taipei, TAIWAN

Abstract

Ginsenosides are major components of ginseng and are attributed for most of its pharmacological activities. Despite extensive studies, the mechanism of action of ginseng remains to be clarified. In an effort to elucidate the mechanism of the prominent gluco-regulatory activity of ginseng, in this study, we measured the effect of ginsenoside Rb1 on glucose uptake in different cell systems including intestinal, adipocytes, and skeletal muscle myotubes. At submicromolar concentrations, Rb1 inhibited the rate of glucose transport across the differentiated Caco-2 cell monolayer to about 67% of the original rate. Consistent with the effect on glucose uptake rate, Rb1 conferred a significant and paralleled alteration on both the protein and mRNA expression levels of the Na^+/glucose cotransporter 1 (SGLT1) gene. In contrast to the intestinal cells, Rb1 exerted a stimulus effect on glucose uptake of the 3T3-L1 adipocytes and skeletal muscle myotubes. Western blot analysis revealed that Rb1 also enhanced GLUT4 expression in these cells. Further studies demonstrated that the stimulus effect of Rb1 on glucose uptake might be associated with the activation of AMP-dependent protein kinase (AMPK), which is one of the key pathways in mediating glucose uptake in insulin-independent manner. In summary, we demonstrated the Rb1 inhibited glucose uptake in intestinal cells while enhance the insulin-independent manner in adipocytes and skeletal muscle myotubes. These findings are important in understanding the hypoglycemic properties and potential applications of ginseng and ginsenosides.

Keywords: ginsenoside, Na^+/glucose cotransporter, GLUT4, Caco-2 cells, 3T3-L1 adipocytes, skeletal muscle myotubes

Introduction

Insulin plays a key regulatory role in stimulating the transport of blood glucose into peripheral tissues through GLUT4 transporter, which is mainly expressed in skeletal muscle and adipose cells. In these

*Address for Correspondence: T.C. Chang Department of Biochemistry National Defense Medical Center P.O. Box 90048, Nei-Hu 114, Taipei, Taiwan, R.O.C. Tel: 886-2-87923100 ext. 18820 Fax: 886-2-87924820
E-mail: tcchang@mail.ndmctsgh.edu.tw

target cells, insulin stimulates the translocation and redistribution of GLUT4 glucose transporter from specific intracellular compartments to the plasma membrane where it facilitates glucose uptake (1). The insulin-mediated glucose uptake is impaired in insulin resistance, which results in the decreased glucose uptake into muscle or adipose cells. Insulin resistance is a major metabolic disorder of diabetes, which is associated with serious conditions such as cardiovascular and kidney disease. Despite effective therapeutic drugs for diabetes treatment, there are undesirable side effects and their precise action mechanism remains to be completely clarified. Natural products provide an important source for searching new drugs to prevent metabolic disorders and as new therapeutic approaches (2).

Ginseng is one of the most widely used herbal medicines with a long history in Asian countries. Ginseng exerts a wide range of beneficial effects such as anti-aging, improving cognitive performance, and enhancing metabolic functions (3). The main active components in ginseng are ginsenosides, which are generally believed to contribute to ginseng's pharmacological actions (3). Historically having been used in treating diabetes, ginseng is known for its prominent hypoglycemic activity (4). Ginseng was shown to decrease postprandial glycemia in both non-diabetic and type 2 diabetes subjects (5). The aqueous extract of ginseng was shown capable of producing hypoglycemia in both glucose-loaded healthy animals and in animals with experimentally induced diabetes (6). The proposed mechanisms for the hypoglycemic effect of ginseng may involve enhanced insulin secretion (7), increased insulin sensitivity (6), and reduced glucose absorption (8) or a combination of these modes. Despite extensive literature documenting the gluco-regulatory activity of ginseng, the mechanisms responsible for ginseng's hypoglycemic effect still need to be elucidated.

In an attempt to clarify the effect of ginseng on glucose uptake, the effect of ginsenoside Rb1 on glucose uptake was studied. Three different cell models were used, including the differentiated intestinal Caco-2 cells and cells derived from insulin target tissue, the differentiated 3T3-L1 adipocytes and multi-nucleated human skeletal muscle myotubes. The Caco-2 cells will spontaneously differentiate into an enterocyte-like phenotype and have been used as an in vitro model system for the study of nutrient and drug transport in the intestinal lumen (9). Caco-2 cells were shown to express proteins involved in absorption of sugar nutrients, including the Na^+/glucose cotransporter (SGLT1) and facilitative transporter GLUT2 (10). The differentiated 3T3-L1 adipocytes and skeletal muscle myotubes were shown to express both GLUT1 and GLUT4.

In this study, we demonstrated that ginsenoside Rb1 effectively inhibited glucose uptake in intestinal Caco-2 cells but stimulated glucose uptake in adipocytes and myotubes. The effect on glucose transport coincides very well with the specific glucose transporter expression levels in these cell systems. Our findings provide the molecular basis for hypoglycemic activity of ginsenoside Rb1, which is a promising candidate in the development of an anti-diabetic drug.

Materials and Methods

Extraction and Isolation of Ginsenoside Rb1

Dried roots of Panax notoginseng (5.05 kg) were extracted repeatedly with EtOH (10 L × 5) at room temperature. The combined EtOH extract was evaporated in vacuo to yield dark-brown syrup (778.19 g). The syrup was partitioned between n-hexane and 80% MeOH to give n-hexane layer (20.01 g), MeOH layer (741.41 g), and insoluble-layer (16.77 g), respectively. The MeOH layer was chromatographed on an Amberlite XAD-4 column by elution with H_2O, 50% MeOH/H_2O, and 100%

MeOH, gradually, to give 3 fractions, H_2O fraction (202.02 g; fraction 1), 50% MeOH/H_2O fraction (68.35g; fraction 2), and 100% MeOH fraction (471.04g; fraction 3). Fraction 2 was chromatographed on Lobar (RP-18, 75% MeOH/H_2O) to give ginsenoside Rb1 (5.53 g) (11). Fraction 3 was subjected to SiO_2 column and eluted with $CHCl_3$-MeOH-H_2O (65: 35: 10) and MPLC (C-8, 75% MeOH/H_2O) to give ginsenoside Rb1 (25.14 g, Figure 1) and ginsenoside Rg1 (29.54 g) (11).

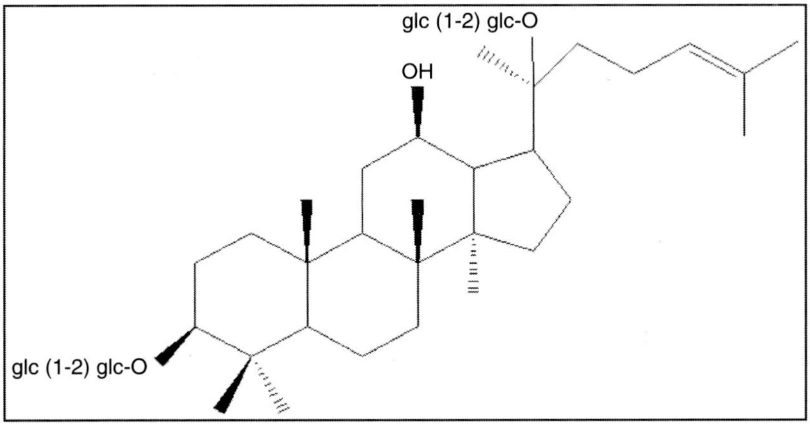

Fig. 1. Chemical structure of ginsenoside Rb1.

Cell Culture and Differentiation

The established human intestinal epithelial Caco-2 cell line and 3T3-L1 pre-adipocytes were obtained from American Type Culture Collection (Rockville, Maryland, USA). Cells were routinely grown in cultured Dulbecco's minimal essential medium (DMEM) supplemented with 100 IU/mL penicillin, 100 μg/mL streptomycin, 1% nonessential amino acid and 10% of fetal bovine serum for Caco-2 cells or 10% calf serum (CS, Hyclone, Logan, UT, USA) for 3T3-L1 cells in 5% CO_2 at 37 °C. The growth and differentiation of Caco-2 cells was carried out as described (12). 3T3-L1 cell differentiation was induced by treating the cells with differentiation inducers (DMEM containing 0.5 mM 3-isobutyl-1-methylxanthane (IBMX), 0.25 μM dexamethasone, 10 μg/mL insulin, and 10% fetal bovine serum (FBS, Gibco, Gaithersburg, MD, USA) for 48 h. The cells were re-fed with DMEM supplemented with 10 μg/mL insulin and 10% FBS for the following 48 h and changed every 2 days. More than 90% of the cells expressed the adipocyte phenotype between 8-10 days after the initiation of differentiation and were used for the experiments. Human primary skeletal muscle cells were from Cambrex Corporation (East Rutherford, NJ, USA). The cells were cultured in DMEM supplemented with 20% FBS and induced to differentiate into multi-nucleated myotubes with DMEM supplemented with 2% horse serum for 5-7 days.

Glucose uptake Assay

Glucose uptake in human CaCo-2 cells was performed as previously described (12). Glucose uptake in 3T3-L1 adipocytes and human skeletal muscle myotubes was performed using [^{14}C] 2-deoxy-D-glucose (2-DG), a nonmetabolizable analog of D-glucose, as the substrate. Cells cultured in 24-well plates and differentiated for 8-10 days were incubated in the serum-free DMEM containing 0.2 % BSA, followed by incubation in the media without serum but with compound for various doses. The cells were preincubated in the presence or absence of various concentrations of Rb1 or 100 nM of insulin for 24 h. Prior to uptake study, cells were incubated for 3 h at 37 °C in transport buffer Krebs-Ringer's phosphate buffer (KRP) (20 mM HEPES, 137 mM NaCl, 4.7 mM KCl, 1.2 mM $MgSO_4$, 1.2 mM KH_2PO_4, 2.5 mM $CaCl_2$, and 2 mM pyruvate, pH 7.4 and 0.2 % BSA). Transport experiments were started by replacing the KRP buffer containing 0.1 mM 2-DG in which 0.2 μCi [14C] 2-DG (200 μCi/mL, Amersham Biosciences, Little Chalfont, Bucks, U.K.) was added for 5 min. Cells were washed twice with ice-cold PBS, solubilized with 0.2% SDS, and the amount of radioactivity measured with a scintillation counter (TopCount, Packard BioSciences, Meriden, CT, USA). Protein concentration of each sample was quantified using BCA (bicinchoninic acid) protein assay kit (Pierce, Rockford, IL, USA)

Western Blot Analysis

Caco-2 cells, 3T3-L1 preadipocytes, and skeletal muscle myotubes were plated on culture dishes (6 cm diameter) at a density of $1x10^6$ cells/dish. The cells were left to undergo differentiation prior to treatment with ginsenosides for 24h. The cells were then washed and lysed in 0.2 mL of lysis buffer (1% NP-40, 50 mM Tris-Cl, pH 7.4, 180 mM NaCl, 1 mM EDTA, 1 mM PMSF, 1 mM NaF, 10 mM Na_3VO_4) for 30 min at 4 °C. For study of GLUT4 expression level in 3T3-L1 adipocytes and human myotubes, the cytoplasmic membrane fraction of these cells was prepared using subcellular proteome extraction kit from EMD Biosciences (Darmstadt, Germany). Protein concentrations were determined and equal amount of protein in cell lysates (20 μg) were separated by 10% SDS-PAGE gel and transblotted onto PVDF membrane. Immunoblotting was performed using anti-human antibodies for GLUT2, GLUT4, SGLT1, AMP-dependent kinase (AMPK) and phosphorylated AMPK (AMPK-P) and tubulin. Signals were visualized with an enhanced chemoluminescence kit (ECL, Amersham, UK) followed by exposure to X-ray films.

Quantitative Analysis of Glucose Transporter Transcripts

Relative levels of gene transcripts expressed in CaCo-2 cells were performed by quantitative real time PCR (Q-PCR). Total RNA was isolated using Trizol (Invitrogen, Irvine, CA, USA). RNA (1 μg) and was reverse-transcribed to cDNA using the High-Capacity cDNA Reverse Transcription Kits (Applied Biosystems, Darmstadt, Germany). The reaction mixtures (20-μL total volume) contained 2 μL of serially diluted cDNA, 10 μL of Taqman Universal PCR Master Mix (Applied Biosystems), and 1 μL of either SGLT1 (assay ID Hs00165793_ml), GLUT2 (assay ID Hs00165775_ml), or GAPDH primer mix (assay ID Hs99999905_m1, Applied Biosystems). Two independent triplicate experiments were performed for the selected genes, and the obtained threshold cycle (Ct) values were averaged. Data are presented as the fold differences in gene expression normalized to the GAPDH as endogenous reference, and relative to the untreated control cells. All of the information about primers sequences and probes used are available from web page of Applied Biosystems.

Statistical Analysis

All results were representative of at least three independent experiments. Statistical analysis was performed using the SPSS 11.0 for Window software. The analysis of variance (ANOVA) was used to determine the statistical significance of differences between the values for the various experimental and control groups. Data were expressed as mean ± S.D. and a p value < 0.05 was considered statistically significant.

Results

Rb1 Modulates Glucose Uptake Activity in Caco-2 Cells

In order to examine the effect of Rb1 of Caco-2 cell survial, the cells were incubated with increasing concentrations of Rb1 and viable cells were analyzed 48 h after treatment. As shown in **Figure 2**, Rb1 exhibited no apparent cytotoxicity as the cell survival rate is similar over the concentrations analyzed. Using confluent and differentiated Caco-2 cell monolayer, the effect of Rb1 on glucose uptake in human Caco-2 cells was studied. A saturated glucose concentration was used in this study to minimize the rapid metabolic effect of glucose in cells. To eliminate nonspecific absorption, the uptake of L-glucose was measured simultaneously to correct for D-glucose taken up passively. As shown in **Figure 3**, the Caco-2 cell monolayer with intact tight junction exhibited steady state rate of glucose transport at 1.54 ± 0.09 nmole/min. With increasing Rb1 concentration, glucose uptake rate decreased from 1.54±0.09 to 1.03±0.02 nmole/min across the differentiated Caco-2 monolayer over the

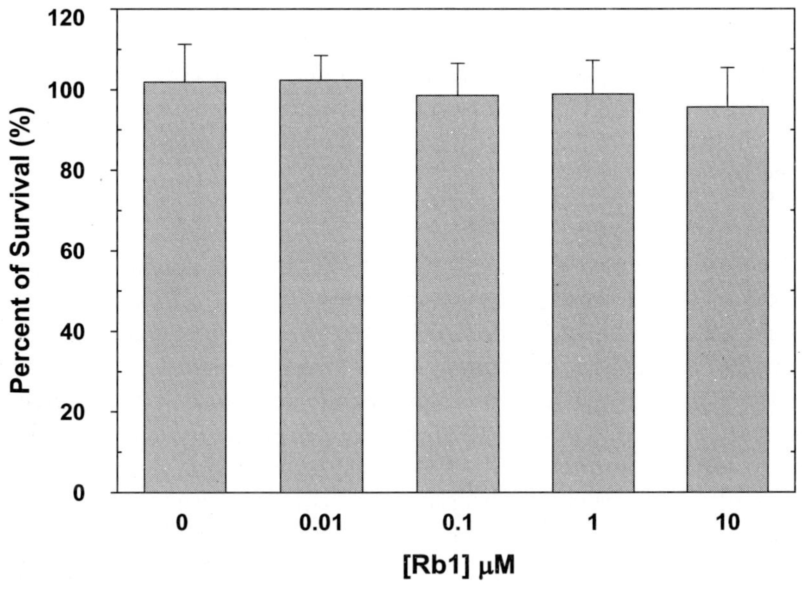

Fig. 2. Cytotoxicity of Rb1 on Caco-2 cells. Growth inhibition of indicated cells after treatment with Rb1 for 48 h. Cell viability was determined. The percentage of growth was calculated, with 100% representing control cells treated with solvent. The results represent mean ± SD (n = 3).

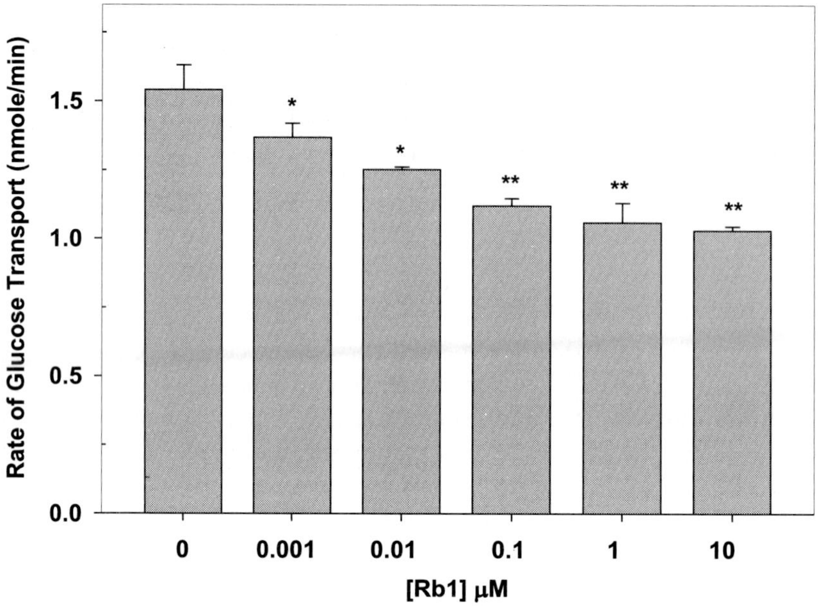

Fig. 3. Rb1 suppresses glucose uptake in human intestinal Caco-2 cells. The cell monolayer was pretreated with Rb1 ginsenoside for 24 h before carrying out the assays. Glucose transport was measured as described in Materials and Methods. Results represent mean ± SD (n = 3), *p < 0.05, **p < 0.01 vs. sample without Rb1.

concentration range of Rb1 assayed, which represent a decrease to 66.9% of the control, untreated cells. The Rb1 reached a maximum inhibition on glucose uptake in these cells with concentration over 1 μM. These results indicated that Rb1 exerted a marked inhibitory effect on glucose uptake in human intestinal Caco-2 cells.

Rb1 Modulates of SGLT1 Gene Expression

The inhibitory effect of ginsenoside Rb1 on the rate of glucose uptake suggested that Rb1 may modulate the expression or activity of glucose transporters. Caco-2 cells were found to express many glucose transporters, including SGLT1 and GLUT2 (10). Western blot analysis was used to exam the effect of Rb1 on the expression of these glucose transporters. Results shown in **Figure 4** demonstrated that the SGLT1 protein level was significantly decreased in Rb1-treated cells. As the Rb1 concentration increased to 1 μM, the SGLT1 expression was gradually reduced to 57% of the original level (**Figure 4A** and **4B**). In contrast to SGLT1, Rb1 displayed no significant induction of the GLUT2 expression. These results suggest that the observed effects on the rate glucose transport is likely due to decreased SGLT1 protein expression levels mediated by Rb1. In addition to protein expression level, Q-PCR was used to analyze the effect of ginsenoside on mRNA levels of SGLT1 and GLUT2. The results shown in **Figure 4C** indicated that, at 0.1 μM, Rb1 down-regulated the SGLT1 mRNA to a significant level by 52% compared with the control. The expression of GLUT2 increased slightly but not significantly in the presence of Rb1. These results demonstrated that Rb1 conferred a parallel alteration on SGLT1 transcript and protein levels. Taken together, our results indicated that Rb1 modulate SGLT1 gene expression at the level of transcription or mRNA stability in Caco-2 cells.

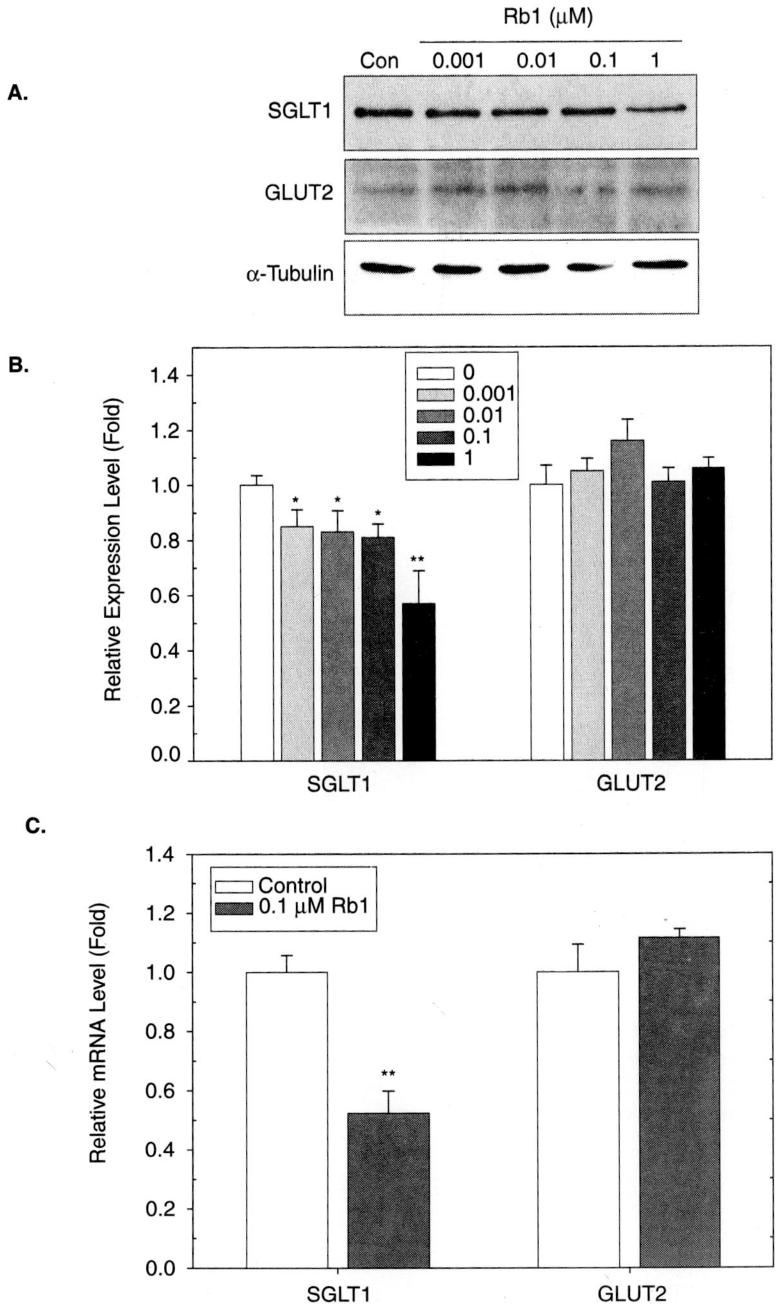

Fig. 4. Effect of Rb1 on SGLT-1 and GLUT2 expression in Caco-2 cells. Cell extracts were prepared from differentiated Caco-2 cells treated with increasing concentration of Rb1 for 24 h. (A) Effect of Rb1 on glucose transporters protein expression level. Immunoblots were analyzed using antibodies against SGLT1 and GLUT2. (B) Quantitative analysis of the results shown in (A). Relative protein expression levels of SGLT1 and GLUT2 were normalized with respect to the level of α-tubulin and expressed as relative fold to the untreated control (Con). (C) Real-time Q-PCR analysis of the relative mRNA levels of SGLT1 and GLUT2 in differentiated Caco-2 cells treated with 0.1 μM Rb1 for 24 h. Results represent mean ±SE (n = 3), *p < 0.05, **p < 0.01 vs. sample without Rb1 (or control).

Rb1 stimulates Glucose Uptake in Differentiated 3T3-L1 Adipocytes and Human Skeletal Muscle Myotubes

Treatment of fully differentiated 3T3-L1 adipocytes or human skeletal muscle myotubes for 24 h with increasing doses of Rb1 resulted in significant enhancement in glucose uptake (**Figure 5**). For 3T3-L1 adipocytes, Rb1 increased glucose uptake maximally to about 164% of the control ($p < 0.05$) at the concentration of 0.01 μM. Higher concentration of Rb1, up to 0.1 μM, is less effective as the lower concentrations. Rb1 also exerted a similar stimulatory effect on glucose uptake in well-differentiated human skeletal muscle myotubes. At 0.001 and 0.01 μM, Rb1 increased glucose uptake to 145 and 152% of the control, respectively ($p<0.05$). Higher dose of Rb1 at 0.1 μM was also less effective. These results indicated that ginsenoside Rb1 displayed a significant effect on glucose uptake in the insulin-sensitive adipocytes and myotubes at submicromolar concentrations.

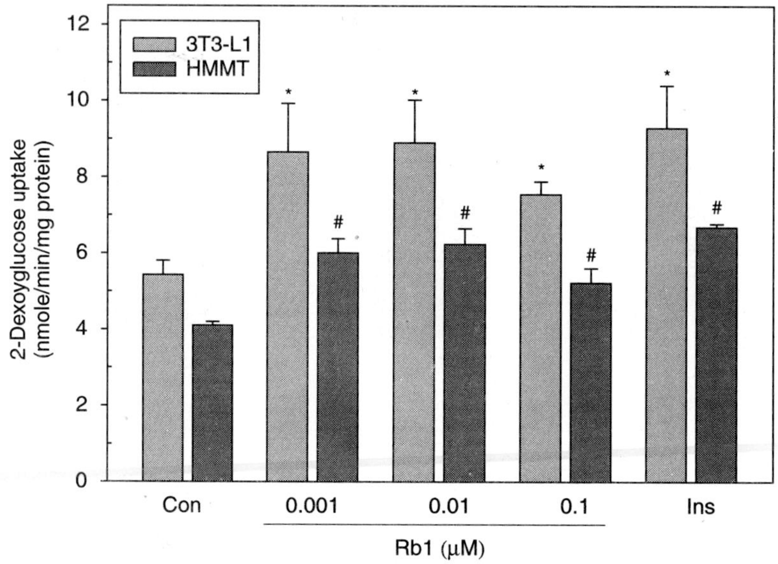

Fig. 5. Rb1 stimulates glucose uptake in 3T3-L1 adipocytes and skeletal muscle myotubes. 3T3-L1 adipocytes and the human skeletal muscle myotubes (HMMT) were pretreated with Rb1 at dose of 0.001 to 0.1 μM, or 100 nM insulin (Ins) for 24 h and the cells were then analyzed for 2-DG uptake. Data represent mean ± SD (n = 3). * p < 0.05 vs. control.

Rb1 Induces GLUT4 Gene Expression in 3T3-L1 Adipocytes and Human Skeletal Muscle Myotubes

Results shown in **Figure 6** indicate that membrane-associated GLUT4 protein level was significantly increased in Rb1-treated 3T3-L1 adipocytes and human skeletal muscle myotubes. Incubation of adipocytes with Rb1 significantly increased GLUT4 protein level with maximum increment at 0.01 μM (143% of the control, $p < 0.05$) over the concentration range studied. For skeletal muscle myotubes, the maximum induction was also at 0.01 μM (134% of the control, $p < 0.05$). These results suggested that induction of GLUT4 expression is essential for Rb1-mediated stimulation in insulin-independent glucose uptake in both adipocytes and myotubes.

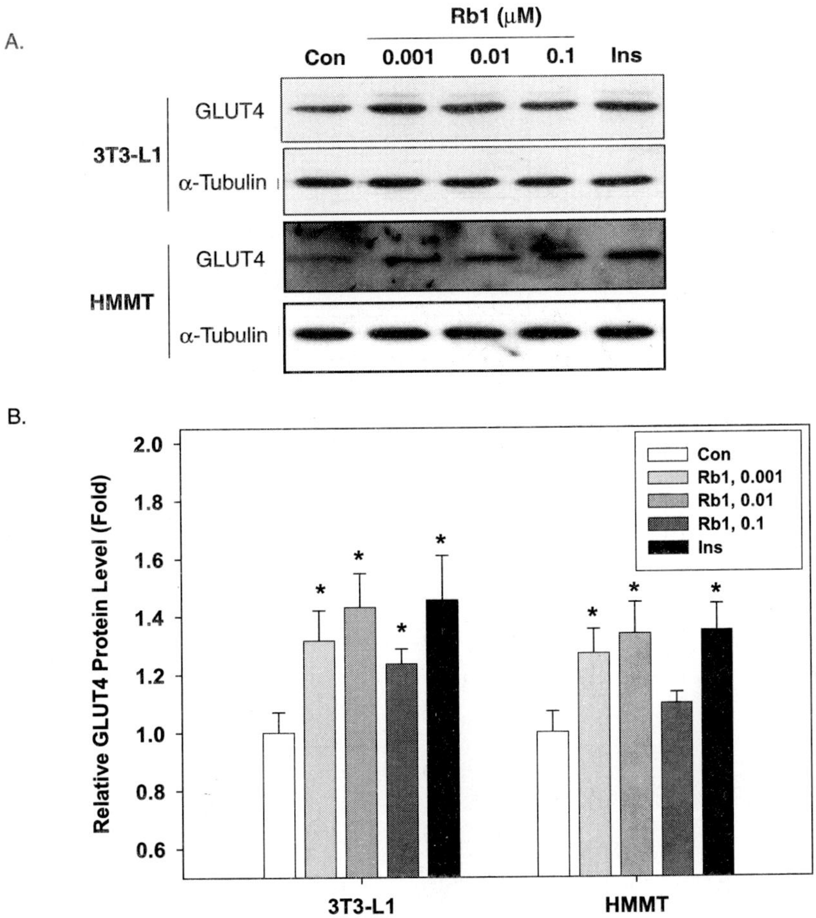

Fig. 6. Rb1 stimulates GLUT4 expression in 3T3-L1 adipocytes and skeletal muscle myotubes. 3T3-L1 adipocytes and human skeletal muscle myotubes (HMMT) were incubated with or without the indicated concentrations of Rb1, or 100 nM of insulin (Ins) for 24 h. (A) Representative immunoblots of GLUT4 expression in 3T3-L1 adipocytes and skeletal muscle myotubes; and (B) quantification of GLUT4 levels in as shown in (A). Data represent mean ± SD (n=3). * $p < 0.05$ vs. the control (Con).

Rb1 Activates AMPK Pathways in 3T3-L1 Adipocytes and Human Skeletal Muscle Myotubes

To elucidate the signaling pathways involved in Rb1-mediated activities, the effect of Rb1 on the activation of AMPK was analyzed. Results shown in **Figure 7** indicate that phosphorylation level of AMPK was significantly increased in Rb1-treated 3T3-L1 adipocytes and human skeletal muscle myotubes. Incubation of adipocytes with Rb1 significantly increased in AMPK phosphorylation with maximum increment at 0.001 μM (162 % of the control, $p<0.05$) over the concentration range studied. For skeletal muscle myotubes, the induction of AMPK phosphorylation at 0.001 μM was 146% of the control, $p < 0.05$. As expected, no activation on AMPK was found in insulin treated cells. These results suggested that activation of AMPK signaling pathway may underlie the insulin-independent mechanism of ginsenoside Rb1 in adipocytes and myotubes.

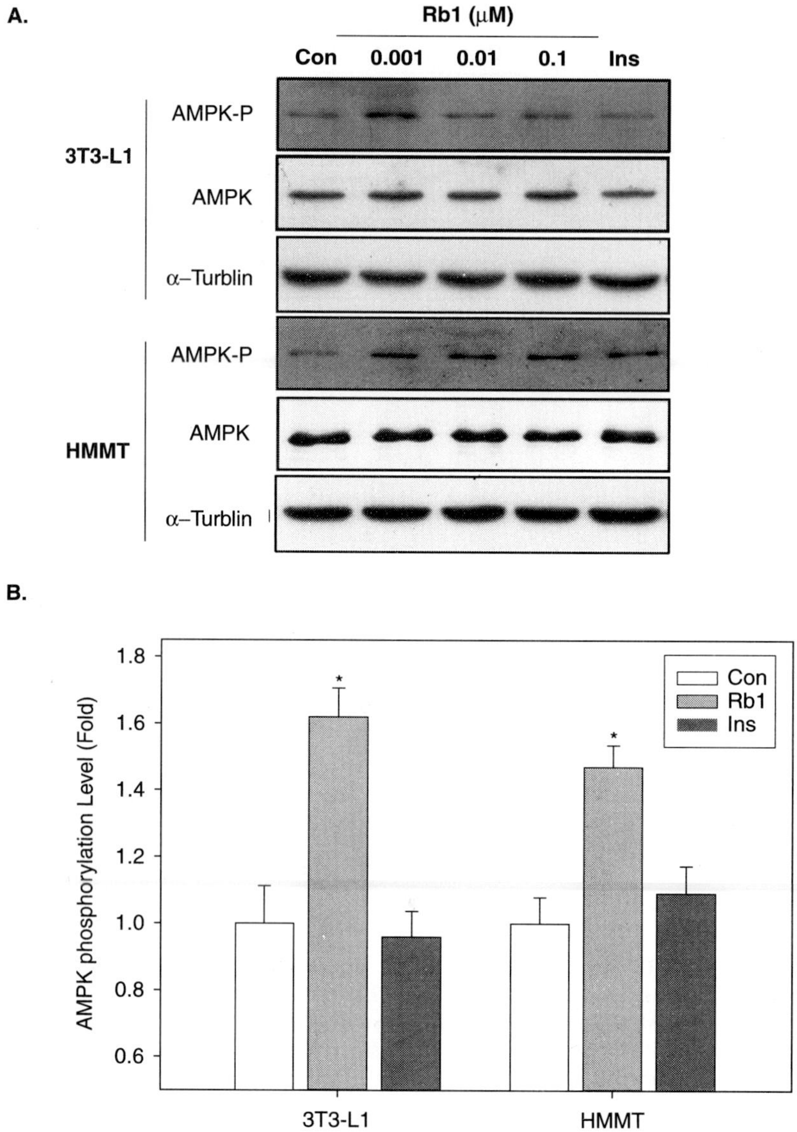

Fig. 7. Effect of Rb1 on AMPK activation in 3T3-L1 adipocytes and skeletal muscle myotubes. 3T3-L1 adipocytes and human skeletal muscle myotubes (HMMT) were incubated with or without the indicated concentrations of Rb1, or 100 nM of insulin (Ins) for 24 h. (A) Representative immunoblots of AMPK phosphorylation; and (B) quantification of control (Con), 0.01 μM Rb1, and insulin results shown in (A). Data represent mean ± SD (n=3). * $p < 0.05$ vs. the control.

Discussion

Metabolic syndrome is a complex and chronic progressive syndrome that leads to adverse function of many organs. Among metabolic syndrome components, hyperglycemia was the main contributor of the associated disorders, which can exacerbate defective glucose disposal by interfering with insulin action in insulin-target tissues. Thus, regulation of glucose uptake is critical for treatment of metabolic syndrome as well as diabetes. Regulation of glucose transporter gene expression is an important mechanism in the regulation of glucose uptake in animal cells. Glucose transport in adipocytes is regulated by facilitative glucose transporters, GLUT1 and GLUT4 (13). GLUT1 is essentially for basal glucose transport in various tissues, whereas the GLUT4 is expressed in insulin-responsive tissues, including adipose and skeletal muscle cells. The study of glucose uptake in adipocytes is important in that there is a strong correlation between abnormal fatty acid metabolism and the development of insulin resistance in muscle cells (14).

The activity of the ginsenoside Rb1 in facilitating glucose uptake into 3T3-L1 adipocytes, human skeletal muscle myotubes, and intestinal cells was investigated in this study. The hypoglycemic effect of ginseng is one of the most studied pharmacological activities of this traditional medicine. Although many in vitro and animal studies have repeatedly showed the hypoglycemic effects of ginseng, the efficacy evidence of ginseng is still inconclusive (15). A likely explanation for the variable hypoglycemic effects of different ginseng preparations is the marked difference in the profile of ginsenosides (16). To avoid any ambiguous effect, purified ginsenoside was used in this study. No apparent cytotoxicity in Rb1 at dose up to 10 μM toward the differentiated 3T3-L1 cells or skeletal muscle myotubes (data not shown) was found in our study. We used submicromolar concentrations (0.001 to 0.1 μM) of ginsenoside Rb1 that are much lower than most of the studies for ginsenosides from different sources (17, 18). The concentrations used in this study are important in that they could decrease the side effects or other unwanted activities.

In this study, we first analyzed the effect of ginsenoside Rb1 on glucose transport in the differentiated Caco-2 cells. Our results showed that Rb1, a protopanaxadiol, significantly inhibited glucose uptake in the Caco-2 intestinal cell system. With increasing Rb1 concentrations, glucose uptake is inhibited to about 67% of the original control. Rb1 reached maximum inhibition at concentrations greater than 0.01 μM. Associate with changes in glucose uptake rate, we found that Rb1 conferred significant and parallel alterations in SGLT1 mRNA and protein levels in Caco-2 cells. The SGLT1 is the primary carrier protein responsible for transport of glucose from the lumen of the intestine across the brush border membrane of intestinal epithelial cells (19). Since there was no apparent and significant change in the expression levels of GLUT2, we conclude that the altered glucose uptake rate across Caco-2 cell monolayer is primarily through modulation of SGLT1 gene expression level. The close correlations in the magnitude of the increases or decreases in SGLT1 mRNA and protein suggest that the exposure of the cells to Rb1 leads to a decrease in abundance of the SGLT1 transcript, and that in turn gives rise to a corresponding increase or decrease in the levels of SGLT1 protein. Thus, in human Caco-2 cells, the Rb1-mediated expression of SGLT1 is primarily controlled at the level of mRNA abundance. This is important, as inhibition of glucose uptake by intestinal glucose transporters would be beneficial for glucose homeostasis and represent an effective way in treating hyperglycemia or metabolic syndrome (20).

In addition to the effect on intestinal cells, in this study, we also found that ginsenoside Rb1 significantly stimulated glucose uptake in both 3T3-L1 adipocytes and human skeletal muscle myotubes with a potency comparable to that of insulin. Overexpression of GLUT4 has been shown to improve insulin sensitivity, suggesting that GLUT4 is a potential therapeutic target for the treatment of insulin resistance (21). It is thus important for us to investigate the effect of Rb1 on GLUT4 gene expression in adipocytes and muscle myotubes. In this study, we demonstrated that, in a manner very similar to insulin, prolonged exposure of Rb1 induced significant increase in the expression GLUT4 protein and enhanced glucose uptake in these cell models. To understand the mechanism in regulation of the effect of Rb1, we also examined the signaling pathways that might be involved in the insulin-independent activities of ginsenoside Rb1. Herbal medicines have been shown to display significant effects in modulating glucose uptake in various experimental systems. It is known that the stimulation of glucose uptake by insulin in adipocytes or myocytes requires the activation of insulin receptor-associated phosphatidylinositol-3 kinase (PI3K) signaling pathway (1). However, for insulin-independent mechanism, AMPK has been shown to play a critical role in regulation of glucose transport in muscle and adipocytes (22). Here, we also demonstrated the involvement and activation of AMPK in Rb1-mediated activities. Apparently, further study is required to elucidate signaling mechanisms involved in the action of Rb1. AMPK is an important therapeutic target for regulating whole-body energy balance. Compounds that activate AMPK showed improved hypoglycemic effect by facilitating glucose uptake and GLUT4 translocations and expression. It has been shown that many compounds or drugs with beneficial activity to metabolic syndrome are known to activate AMPK with metformin as the prominent example (23, 24).

In conclusion, we showed that ginsenoside Rb1 effectively inhibited glucose transport across the Caco-2 cell monolayer, which correlates significantly with SGLT1 gene expression. We also showed that Rb1 enhanced glucose uptake, increased GLUT4 expression, and activated AMPK in 3T3-L1 adipocytes and skeletal muscle myotubes. As one of the most popular traditional medicines used around the world, ginsenosides present as the promising candidate as the insulin-like natural medicine. Despite the detailed mechanism of these compounds required further investigation; our findings provide evidence for the ginsenoside with marked insulin-like activity and potential in the treatment of metabolic syndrome or diabetes. In addition, this study may also provide an opportunity to develop Rb1 as antidiabetic agents or acting as dietary adjuncts to existing therapies.

Acknowledgment

This study is supported by grants from National Science Council (NSC 97-2320-B-016 -003 -MY3 to T-C. C.) and from the Ministry of National Defense (DOD 99C10-04 to T-C. C.). The authors are also grateful to Nuliv Science Co., Taipei, Taiwan, for financial support of this work.

References

1. Zaid H, Antonescu CN, Randhawa VK, Klip A. Insulin action on glucose transporters through molecular switches, tracks and tethers. Biochem J. 2008; 413:201-215.
2. Moller DE. New drug targets for type 2 diabetes and the metabolic syndrome. Nature 2001; 414:821-827.
3. Attele AS, Wu JA, Yuan CS. Ginseng pharmacology: multiple constituents and multiple actions. Biochem Pharmacol. 1999; 58:1685-1693.
4. Sotaniemi EA, Haapakoski E, Rautio A. Ginseng therapy in non-insulin-dependent diabetic patients. Diabetes Care 1995; 18:1373-1375.

5. Vuksan V, Sievenpiper JL, Koo VY, Francis T, Beljan-Zdravkovic U, Xu Z, Vidgen E. American ginseng (Panax quinquefolius L) reduces postprandial glycemia in nondiabetic subjects and subjects with type 2 diabetes mellitus. Arch Intern Med. 2000; 160:1009-1013.
6. Ohnishi Y, Takagi S, Miura T, Usami M, Kako M, Ishihara E, Yano H, Tanigawa K, Seino Y. Effect of ginseng radix on GLUT2 protein content in mouse liver in normal and epinephrine-induced hyperglycemic mice. Biol Pharm Bull. 1996; 19:1238-1240.
7. Waki I, Kyo H, Yasuda M, Kimura M. Effects of a hypoglycemic component of ginseng radix on insulin biosynthesis in normal and diabetic animals. J Pharmacobiodyn. 1982; 5:547-554.
8. Onomura M, Tsukada H, Fukuda K, Hosokawa M, Nakamura H, Kodama M, Ohya M, Seino Y. Effects of ginseng radix on sugar absorption in the small intestine. Am J Chin Med. 1999; 27:347-354.
9. Hidalgo IJ, Raub TJ, Borchardt RT. Characterization of the human colon carcinoma cell line (Caco 2) as a model for intestinal epithelial permeability. Gastroenterology 1989; 96:736-749.
10. Harris DS, Slot JW, Geuze HJ, James DE. Polarized distribution of glucose transporter isoforms in Caco-2 cells. Proc Natl Acad Sci. USA 1992; 89:7556-7560.
11. Sanada S, Kondo N, Shoji Y, Tanaka O, Shibata S. Studies on saponins of ginseng structures of ginsenosides R0, Rb1, Rb2, Rc and Rd. Biol Pharm Bull. 1974; 22:421-428.
12. Chang T-C, Huang S-F, Yang T-C, Chan FN, Lin H-C, Chang W-L. The effect of ginsenosides on glucose uptake in human Caco-2 cells is mediated through altered Na+/coupled glucose transporter 1 expression. J Agri Food Chem. 2007; 55:1993-1998.
13. Scheepers A, Joost HG, Schurmann A. The glucose transporter families SGLT and GLUT: molecular basis of normal and aberrant function. JPEN J Parenter Enteral Nutr. 2004; 28:364-371.
14. Birnbaum MJ, Diabetes: Dialogue between muscle and fat. Nature 2001; 409:672-673.
15. Sievenpiper JL, Arnason JT, Leiter LA, Vuksan V. Decreasing, null and increasing effects of eight popular types of ginseng on acute postprandial glycemic indices in healthy humans: the role of ginsenosides. J Am Coll Nutr. 2004; 23:248-258.
16. Sievenpiper JL, Arnason JT, Vidgen E, Leiter LA, Vuksan V. A systematic quantitative analysis of the literature of the high variability in ginseng (Panax spp.): should ginseng be trusted in diabetes? Diabetes Care 2004; 27:839-840.
17. Shang W, Yang Y, Jiang B, Jin H, Zhou L, Liu S, Chen M. Ginsenoside Rb1 promotes adipogenesis in 3T3-L1 cells by enhancing PPARgamma2 and C/EBPalpha gene expression. Life Sci. 2007; 80:618-625.
18. Park S, Ahn IS, Kwon DY, Ko BS, Jun WK. Ginsenosides Rb1 and Rg1 suppress triglyceride accumulation in 3T3-L1 adipocytes and enhance beta-cell insulin secretion and viability in Min6 cells via PKA-dependent pathways. Biosci Biotechnol Biochem. 2008; 72:2815-2823.
19. Wood IS, Trayhurn P. Glucose transporters (GLUT and SGLT): expanded families of sugar transport proteins. Br J Nutr. 2003; 89:3-9.
20. Wagman AS, Nuss JM. Current therapies and emerging targets for the treatment of diabetes. Current Pharmaceutical Design. 2001; 7:417-450.

21. Leturque A, Loizeau M, Vaulont S, Salminen M, Girard J. Improvement of insulin action in diabetic transgenic mice selectively overexpressing GLUT4 in skeletal muscle. Diabetes 1996; 45:23-27.
22. Huang S, Czech MP. The GLUT4 glucose transporter. Cell Metab. 2007; 5:237-252.
23. Tan MJ, Ye JM, Turner N, Hohnen-Behrens C, Ke CQ, Tang CP, Chen T, Weiss HC, Gesing ER, Rowland A, James DE, Ye Y. Antidiabetic activities of triterpenoids isolated from bitter melon associated with activation of the AMPK pathway. Chem Biol. 2008; 15:263-273.
24. Levri KM, Slaymaker E, Last A, Yeh J, Ference J, D'Amico F, Wilson SA. Metformin as treatment for overweight and obese adults: a systematic review. Ann Fam Med. 2005; 3:457-461.

Adaptation Biology and Medicine (Volume 6: Cell Adaptations and Challenges)
Editors: P. Wang, C.-H. Kuo, N. Takeda and P.K. Singal
Copyright © 2011 Narosa Publishing House Pvt. Ltd., New Delhi, India

33. Household Food Insecurity in Canadian Adolescents: Adaptation of a Lifestyle and Body Mass Index Consequences

Anju Bajaj[1], Dan D. Ramdath[2], Mohammed Moghadasian[3], Pawan K. Singal[1*]

[1]Institute of Cardiovascular Sciences, Department of Physiology, [2]Guelph Food Research Center Agriculture and Agri-Food Canada, Guelph, Ontario, Canada. [3]Department of Human Nutritional Sciences, University of Manitoba, Winnipeg

Summary

We examined the extent to which household food insecurity might influence lifestyle choices and nutritional status in Canadian adolescents. Data from the Canadian Community Health Survey cycle 3.1 (2005) were used. Household Food security status (HFSS) was cross tabulated in three age groups (12-14, 15-17, 18-19 yrs), gender, level of household education, total household income, racial origin, living arrangements, physical activity, fruit and vegetable consumption and Body Mass Index (BMI). A total of 16,306, representing 3,352,761 individuals participated. Of these 92.3% lived in homes that were food secure, but 18-19 year olds experienced the highest HFSS (94.4%). Regardless of age; income and education were positively associated ($p<0.001$) with HFSS. In addition, adolescents living in homes with two parents had a higher HFSS, compared to those having other living arrangements. Racial origin was only significantly ($p<0.001$) related to HFSS for 12 -17 year olds; in this age range caucasians and visible minorities had 92.8% and 86.0% HFSS respectively. Visible minorities experienced severe food insecurity more than three times as often as a Caucasians (1.5% versus 0.4%). Physical activity level was reduced in the 15-17 yrs old with food insecurity. In the 15-17 and 18-19, year olds vegetable and fruit consumption increased with HFSS ($p<0.05$). Food insecurity was associated ($p<0.05$) with a higher BMI regardless of the age group, but the association was strongest for 15 to 17 year olds ($p<0.005$). In this group, the prevalence of obesity among food secure was 4.7% compared to 17.4% among severely food insecure homes. Among Canadian adolescents HFSS is associated with many adaptations with respect to lifestyle behaviours. Important being physical inactivity and increase in BMI with food insecurity. The study suggests interventions in the targeted group to improve household food security.

Key Words: Obesity; Food Insecurity; Lifestyle; Body Mass Index.

Introduction

Food insecurity is recognized as an important determinant of health. According to the Food and Agriculture Organization, "Food insecurity exists when people live with hunger and fear of starvation.

*Address for Correspondence: Dr. Pawan K. Singal R3021- 351 Tache Avenue Winnipeg, MB R2H 2A6
Tel: 204-235-3631 Fax: 204-233-6723 Email: psingal@sbrc.ca

Conversely, food security exists when all people, at all times, have access to sufficient, safe and nutritious food to meet their dietary needs and food preferences for an active and healthy life" (FAO 2006). In 2004, it was estimated that food insecurity affected almost one in ten Canadian households (Kirkpatrick, 2008). Health Canada in 2004 estimated that 412,300 Canadian households with children were food insecure: 317,100 were moderately food insecure and 95,200 were severely food insecure (Health Canada, 2004).

It has been suggested that food insecurity is a marker of consumption of fewer fruit and vegetables, milk products, lower fibre intakes, higher energy density, and heightened risk of inadequate intakes of nutrients among adolescents and adults (Kirkpatrick, 2008). There have been reports suggesting an association between food insecurity and being overweight. Possible explanations include: (a) food insecurity may lead to selection of obesity promoting types of food; (b) food insecurity may lead to overeating when food becomes available; and (c) food insecurity might lead to metabolic changes due to feast or famine eating patterns that may result in more efficient use of food (Gulliford, 2006; Rose, 2006). Thus it is likely that food insecurity creates barriers to practicing healthy lifestyle behaviours.

Many factors influence healthy lifestyle practices; these include age, education, income, living arrangements, geographical location, genetics, health status, nutrition knowledge, familial relationships among others. The purpose of this current study was to: (a) analyze the status of household food security for adolescents aged 12 to 19 years old; (b) identify the impact that sociodemographic factors have on household food security status; and (c) analyze the impact of food security on healthy lifestyle practices. Specifically, fruit and vegetable consumption, physical activity and Body Mass Index (BMI) levels.

Methods

Data was obtained from the Canadian Community Health Survey, Cycle 3.1(CCHS, 2005), conducted by Statistics Canada in 2005. The CCHS is a cross-sectional survey that gathers information regarding health status, health care utilization and health determinants for the Canadian population (Cycle 3.1 Guide). The CCHS is conducted on a two-year collection cycle. The first year of the survey cycle".1", is a general population health survey that is administered to a relatively large sample size of approximately 130,000 participants. The second year of the survey cycle ".2", is focused on specific health topics and has a much smaller sample size of approximately 30,000 participants. CCHS Cycle 3.1 includes responses from persons age 12 or older, living in privately occupied dwellings in 122 health regions covering all provinces and territories. A total of 132,221 subjects participated in this survey. Excluded from the sampling frame were individuals living on Indian Reserves/Crown lands, institutional residents, full-time members of the Canadian Forces, and residents of certain remote regions. The CCHS covered approximately 98% of the Canadian population age 12 and over (Health Canada, 2005; www.hc-sc.gc.ca).

The CCHS Cycle 3.1 survey had 3 sections that required approximately 40 minutes to complete. Section 1 was 25 minutes of common content, which was asked of all respondents; 5 minutes of sub-sample content and 10 minutes of optional content (Health Canada 2005).Household food security status was determined based on a set of 18 questions aimed to estimate whether households were able to afford the food they needed in the previous 12 months. There were 10 questions that were specific to experiences of adults in the household, while the other 8 questions were specific to children under the age of 18 years. The CCHS categorizes food security into the following four distinct household situations and in this study the following definitions from Health Canada (Health Canada, 2005) have been used:

Food secure (FS) – Household members show no or minimal evidence of food insecurity; **Food insecure without hunger (FI)** – Household members feel anxious about running out of food or compromise on the quality of foods they eat by choosing less expensive options. Little or no reduction in the household members' food intake is reported; **Food insecure with moderate hunger (Moderate)** – Food intake for adults in the household has been reduced to an extent that implies that adults have repeatedly experienced the physical sensation of hunger. In most (but not all) food insecure households with children, such reductions are not observed at this stage for children; and, **Food insecure with severe hunger (Severe)** – At this level, all households with children have reduced the children's food intake to an extent indicating that the children have experienced hunger. Adults in households with and without children have repeatedly experienced more extensive reductions in food intake.

To provide reliable estimates and to ensure a sufficient sample size from each health region within the 10 provinces and 3 territories, a three-step sampling strategy was used. The sample was allocated among the provinces according to their respective populations, the number of health regions each province contained, and finally each province's sample was allocated among its health regions proportionally to the square root of the estimated population in each health region (Health Canada, 2005). Three sampling frames were applied: (a) an area frame (49% of respondents); (b) a list frame of telephone numbers (50%); and (c) a random digit-dialling frame (1%). The survey was either administered over the telephone or in person, or in some cases both.

Data Analysis: Household food security status was condensed into three age groups (12 to 14, 15 to 17, and 18 to 19 years) and cross-tabulated with gender, household income, household education level, living arrangements, and cultural/racial origin. The influence of food security on BMI, physical activity level, and fruit and vegetable consumption was also cross-tabulated.

Total household income was classified into five income brackets (Canadian dollars): less than $15,000, $15,000-$29,999, $30,000-$49,999, $50,000-$79,999, and $80,000 and over. Household education level was categorized into 4 groups: less than secondary school graduation, secondary school graduation, some post-secondary education, and post-secondary graduation. Living arrangements were grouped into 3 categories: (i) child living with 1 parent; (ii) child living with 2 parents; and (iii) other or not stated.

The BMI categories included combined underweight and acceptable weight (18.5 – 24.9), overweight (25 – 30), obese (30+). The physical activity index categorizes respondents as being "active", "moderately active", or "inactive" based on the total daily energy expenditure values (kcal/kg/day). Fruit and vegetable consumption classified the respondent based on the total number of times per day he/she ate fruits and/or vegetables. The categories were: less than 5 servings, 5 to 10 servings, or 10 or more servings.

The data was analyzed using SPSS for Windows (SPSS Science. Chicago. USA; version 16.0) and Microsoft Excel. Population estimates were calculated with SPSS using the weights provided by Statistics Canada. Excel was used to merge data into larger tables and calculate proportions. In order to perform statistical analyses on the results, normalized weights needed to be calculated and applied to the sample. Normalized weights are calculated by taking the mean of the master weights, which is the number Canadians represented by each participant in the sample. The mean of the master weights in the Canadian adolescent sample is 206. The master weights of each individual participant were then divided by this value to give the normalized weight for each individual. Significant associations were examined using chi-squared tests in SPSS and the level of significance was set at $p < 0.05$.

Results

The Canadian Community Health Survey covered approximately 98% of the population age 12 and older. A total of 16,306 adolescents, age 12–19 years participated, representing 3,352,761 Canadian individuals. This number was estimated using the following weightage: Completed the survey via telephone = 23%; completed the survey in person= 71%; and completed the survey via both telephone and in person = 6%. Socio demographic characteristics of the sample size are summarized in Table 1.

Table 1. Sociodemographic Characteristics of the study cohort, consisting of 16,306 Canadian Adolescents extracted from the Canadian Community Health Survey Cycle 3.1

Characteristics	Proportion of Sample Size (%)[a]
Sex (n = 16,306)	
Male	50.4%
Female	49.6%
Age (y) (n = 16,306)	
Age 12 – 14	37.9%
Age 15 – 17	37.7%
Age 18 – 19	24.5%
Total Annual Household Income (Canadian $) (n = 12,267) (Missing = 4039)	
> $ 15,000	4.6%
$ 15,000 - $29,999	8.9%
$ 30,000 - $49,999	14.8%
$ 50,000 - $79,999	21.3%
$ 80,000 +	75.2%
Highest Education in the Household (n = 13,345) (Missing = 2961)	
Less Than Secondary School Graduation	5.1%
Secondary School Graduation	11.3%
Some Post-Secondary Education	6.3%
Post-Secondary Graduation	59.2%
Racial Origin (n = 15,919) (Missing = 387)	
Caucasians	80.5%
Visible Minority[b]	17.1%
Living Arrangements (n = 16,145) (Missing = 161)	
Living with one parent	20.1%
Living with two parents	66.3%
Other	12.6%

[a]Percentages may not add up to 100% due to rounding; participants who did not respond to a question were not included in the analysis of that question

[b]In Canada, "visible minority" is defined as non-white and non-First Nations people; it includes Arabs, Chinese, Japanese, Koreans, blacks, Pacific Islanders, Southeast Asians, Latin Americans, and West Asians

Household Food Security Status (HFSS): Age, Gender, Cultural/Racial Origin

Household food security of Canadian adolescents in the sample was: 92.3% FS (Food Secure), 5.2% FI (Food Insecure), 1.9% moderately FI and 0.6% severely FI. These results are outlined in Figure 1. In the 12-14 year old group 91.2% reported FS, 5.9% FI, 2.3% moderate, and 0.6% severe FI. The 15-17 year old group reported a similar pattern of household food security; 92.1% FS, 5.4% FI, 2.0% moderately FI, and 0.6% severely FI. The 18–19 year old group had a slightly higher rate of food security with a proportion of 94.4% FS, 3.9% FI, 1.3% moderately FI, and 0.5% severely FI. Males and females showed similar results regardless of the age range. Interestingly, there were significantly more females among the 12-14 and 18-19 year olds experiencing source food insecurity. (Martin 2007). It was found that ethnicity was significantly associated ($p<0.001$) with household food insecurity in adolescents from the ages of 12 to 17 (caucasians, 92.8% FS and visible minority, 86.0% FS), however among the 18-19 year olds this association was not significant.

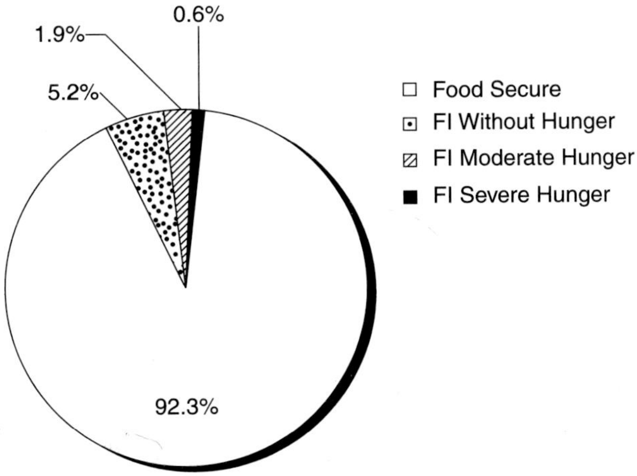

Fig. 1. Household food security in Canadian adolecents, age 12-19 yrs. FI, Food Insecurity

Household Food Security Status (HFSS): Income, Education, Living Arrangement, Fruit and Vegetable Consumption

Total household income was significantly ($p < 0.001$; chi-square test) associated with food security, regardless of age. As household income increased, so did the likelihood of that household becoming food secure. Households with a total income of $80,000 or more experienced the highest prevalence of food security (98.9%) versus households earning less than $15,000 (63%).

There was a significant ($p < 0.001$) positive association between highest education in the household and food security, in all three age groups. Households in which post-secondary graduation was attained, it experienced 94.0% household food security. Household food security prevalence for households attaining less than secondary education, showed a lower (77.4%) prevalence of food security. The lowest proportion of food insecurity with severe hunger (0.4%) came from households with albeit one member completing post-secondary education. Chi-squared tests revealed that there was a significant ($p < 0.001$) association between living arrangement and household food security, regardless of age group. Children living with 2 parents had the highest incidence of food security (95.7%), followed by the "other" category (88.1%), and finally children living with 1 parent (84.1%).

There was a significant association between household food security and fruit and vegetable consumption among the 15 to 17 year olds ($p<0.05$) and 18 to 19 year olds ($p<0.005$). In both age groups, children residing in food secure homes had the lowest incidence of consuming less than 5 servings of fruit and vegetables a day (55.2% and 59.0%). The consumption of fruits and vegetables tend to increase as the household food security level increase. An interesting result was that the highest incidence of consuming more than 10 servings a day came from 18 – 19 year olds in food insecure – without hunger homes.

Household Food Security and Physical Activity Level and BMI

According to Chi-squared tests the relationship between household food security and physical activity level was only significant ($p<0.001$) in the 15 to 17 year olds. Children living in food secure homes had the lowest proportion of physical inactivity levels (26.6%), whereas 36.0% of children in severe food insecure homes were physically inactive. The highest level of physically active children was in homes where hunger was not an issue. There was a significant relationship for all three-age categories, although the strongest association was noted for 15 to 17 year olds. As the household food security situation worsened, so did the frequency of adolescent obesity as reported in Table 2. The proportion of 15 to 17 year old food secure adolescent reporting obesity was a mere 4.7% compared to 17.4% of adolescents living in severely food insecure homes. The present study shows that low-income families have a higher prevalence of food insecurity. It is likely that these families cannot afford to have a well-balanced meal.

Epidemiological evidence suggests that regular fruit and vegetable intake helps promote health and prevent chronic diseases, especially diseases associated with obesity (Kristjansdottir, 2009). According to Canada's Food Guide recommendations adolescents should be consuming 5 to 10 servings of fruits and vegetables on a daily basis. One serving would be, for example, a medium sized banana, one medium celery stalk, or 125 millilitres (1/2 cup) of fruit juice.

Table 2. Frequency of BMI classifications in Canadian adolescent from the Canadian Community Health Survey Cycle 3.1, by Household food security status

Age Group and Number	Food Status	BMI		
		Underweight/Normal	Overweight	Obese
12 to 14 years olds	Food Secure	79.1%	16.7%	4.2%
n = 4588	Food Insecure Without Hunger	73.0%	19.1%	7.9%
	Food Insecure Moderate Hunger	75.5%	19.4%	5.1%
	Food Insecure Severe Hunger	68.0%[##]	28.0%[##]	4.0%
15 to 17 year olds	Food Secure	81.7%	13.6%	4.7%
n= 4447	Food Insecure Without Hunger	77.4%	15.5%	7.1%
	Food Insecure Moderate Hunger	69.6%	19.0%	11.4%
	Food Insecure Severe Hunger	60.9%[##]	21.7%[##]	17.4%[##]
18 to 19 year olds	Food Secure	77.4%	16.2%	6.4%
n= 3016	Food Insecure Without Hunger	66.4%	25.2%	8.4%
	Food Insecure Moderate Hunger	69.4%	19.4%	11.1%
	Food Insecure Severe Hunger	57.1%[##]	35.7%[##]	7.1%

The average consumption rate for Canadian adolescents is 4.5 servings a day; while adults consume an average of 5.2 servings a day (Garriguet, 2004). Based on data from the 2004 Canadian community health survey, cycle 2.2, 59% of Canadian adolescents were reported to consume fruit and vegetables less than 5 times day. Individuals who consume fruits and vegetables more frequently are less likely to be overweight or obese. The present study confirmed that as the household food security situation worsens, fruit and vegetable consumption declines. One important factor is the food price which is an important determinant of food choice; the lowest-cost diets tend to be the least healthy, consisting of high-energy dense foods rather than nutrient dense foods (Darmon, 2008).

Some of the major barriers to physical activity include a preference for indoor entertainments (television, video games, and computer), low energy level, time constraints, and societal factors (such as lack of parental support). Food insecure children may have higher frequencies of physical inactivity because they lack the energy necessary to take part in physical activities. In a study focused on the consequences of food insecurity, subjects reported experiencing hunger pangs, episodes of fatigue or illness, all related to insufficient food (Hamelin, 1999). Physical manifestations of hunger include reduced learning ability in children, due to their lack of concentration, loss of productivity for adults at work, and an overall feeling of energy depletion (Hamelin, 1999). Family dynamics also play a role in healthy lifestyle practices, behaviour modeling, reinforcement, support, family rules and restrictions; these may be important influences on childhood obesity because children's eating, physical activity and leisure time pursuits develop within the context of the family (Timperio, 2008).

Food insecurity has been largely viewed as a low-income related issue; indeed our study found that income and education were positively associated with food security. Disadvantaged segments of the population suffer higher rates of obesity, diabetes, cardiovascular disease, osteoporosis, dental caries and some forms of cancer, all of which have a direct link to nutrition and diet (Darmon, 2008). Health Canada has found that the prevalence of household food insecurity was higher in certain sub-populations, including households with low incomes, households with social assistance as the main source of income, off-reserve aboriginals households, and households with children – in particular, those headed by a female parent, alone. (Health Canada, 2005).

Poor dietary habits are not solely reliant on income; nutrition knowledge also plays a role in food selection. As mentioned above educational attainment is also associated with household food security, the higher the education level of the household the more food secure. Higher levels of education are positively linked to higher levels of income; a healthier lifestyle is more affordable. Income is not the only attribute that is correlated with educational attainment, concerns regarding health maintenance and perceptions of how to achieve health are also relevant. Households with higher education considered healthful eating and the inclusion of fruits and vegetables in daily meals to prevent disease development (Riediger, 2007; Casey, 2006). Parental intake and home availability/accessibility were some of the most consistently supported determinants of fruit and vegetable consumption in young people, clearly suggesting that family environment influences dietary behaviours of young people (Pearson, 2009). If parents are aware of the benefits of healthy eating they are more likely to engage in healthy lifestyle practices.

Living arrangements play a role in the incidence of food security; this effect may be secondary to total household income and household education level. Statistical analysis identified 11.6% difference in household food security occurrence between two parent households. Studies have shown that adolescents report eating the same foods their parents eat (Pearson, 2009). A supportive and cohesive family structure may be more conducive to taking part in healthy lifestyle practices such as fruit and vegetable consumption and physical activity. Mendelson (2007) has argued that the role of parents in the development of healthy eating and physical activity patterns is critical from the early stages of life and warrants further attention (Mendelson, 2007).

This study shows that individuals from the following groups are at higher risk: low household income, low household educational attainment, single parent homes with one or more children, and visible minorities. Food insecurity is a multi-faceted problem which has no one solution. Macro and micro level approaches are necessary to address the dilemmas that many lower socioeconomic families face on a daily basis. Strategies to increase food security should ensure the health of children and adolescents (Kirkpatrick, 2008).

Limitations

The household food security module was not a mandatory part of the Canadian community health survey, cycle 3.1. Only six provinces administered the optional food security module to its participants, this makes it difficult to generalize results across Canada. Another limitation to the present study was that food security is a concept problem that may be difficult for younger participants to understand. Our study focused on the responses of adolescents from the ages of 12 to 19, the younger children included in this survey may not have provided accurate responses to certain questions, such as, household income; this can be considered as a general pitfall of self-reported studies. Also because adults often try to shield their children from the effects of household food insecurity, the children may not be aware that food insecurity exists in their home. The body mass index that was used in this study was self-reported and, therefore, may introduce discrepancies in the data.

Acknowledgments

Dr. Singal is the holder of the Naranjan Dhalla Chair in Cardiovascular Research, supported by the St. Boniface Hospital & Research Foundation.

References

1. CaseyPH , Pippa M. Simpson, Jeffrey M. Gossett, Margaret L. Bogle, Catherine M. Champagne, Carol Connell, David Harsha, Beverly McCabe-Sellers, James M. Robbins, Janice E. Stuff, Judith Weber. “The Association of Child and Household Food Insecurity with Childhood Overweight Status”*Pediatrics;* 118 e1406-e1413, 2006.
2. Darmon, N., Drewnowski, A. “Does social class predict diet quality?” *The American Journal of Clinical Nutrition;* 87: 1107 – 1117, 2008.
3. FAO/FSAU. Integrated Food Security and Humanitarian Phase Classification: Technical Manual Version 1. Nairobi, FAO/ FSAU Technical Series IV.11. 2006.
4. Garriguet, D. 2004. “Overview of Canadians’ Eating Habits”. Ottawa, Ont: Statistics Canada. Also available at http://dspsd.tpsgc.gc.ca/Collection/Statcan/82-620-M/82-620-MIE2006002.pdf (accessed October 13, 2009).

5. Hamelin, A., Habicht, J., Beaudry, M. "Food insecurity: consequences for the household and broader social implications". *The Journal of Nutrition*; 129 (2S Suppl): 525S – 528S, 1999.

6. Kirkpatrick, S., & Tarasuk, V. "Food Insecurity in Canada: Considerations for Monitoring". *Canadian Journal of Public Health*; 99: 324 – 327, 2008.

7. Kristjansdottir, A., De Bourdeaudhuij, I., Klepp, K., Thorsdottir, I. "Children's and parent's perceptions of the determinants of children's fruit and vegetable intake in a low-intake population". *Public Health Nutrition.* 16: 1 – 10. [Epub ahead of print] 2009.

8. Martin, K., Ferris, A. "Food insecurity and gender are risk factors for obesity". *Journal of Nutrition Education and Behaviour*; 39: 31 – 36, 2007.

9. Mendelson, R. "Think tank on school-aged children: nutrition and physical activity to prevent the in obesity". *Applied Physiology, Nutrition, and Metabolism.* ; 32: 495 – 499, 2007.

10. Pearson N, Biddle SJ, Gorely T. "Family correlates of fruit and vegetable consumption in children and adolescents: a systematic review". *Public Health Nutr.* ; 12:267-283, 2009.

11. Riediger, N.,Shooshtari, S., Moghadasian, M "The influence of sociodemographic factors. on patterns of fruit and vegetable consumption in Canadian adolescents". *Journal of the American Dietetic Association;* 107: 1511 –1518, 2007.

12. Rose D, Bodor JN. "Household food insecurity and overweight status in young school children: results from the early childhood longitudinal study". *Paediatrics*; 117:464–473, 2006.

13. Shields, M. Measured Obesity: "Overweight Canadian children and adolescents. Nutrition: Findings from the Canadian Community Health Survey". *Statistic Canada Cat. No. 82-620-MWE. 2004.*

14. Skinner, K., Hanning, R., Tsuji, L. "Barriers and supports for healthy eating and physical activity for First Nation youths in northern Canada". *International Journal of Circumpolar Health;* 65: 148 – 161, 2006.

15. Timperio A, Salmon J, Ball K, Baur LA, Telford A, Jackson M, Salmon L, Crawford D."Family physical activity and sedentary environments and weight change in children". *Int J Pediatr Obes*; 3:160-167, 2008.

Zalesin K., Franklin B., Miller W., Peterson E., McCullough P. "Impact of Obesity on Cardiovascular Disease". *Endocrinology and Metabolism Clinics of North America*; 37: 663 – 687, 2008.

Adaptation Biology and Medicine (Volume 6: Cell Adaptations and Challenges)
Editors: P. Wang, C.-H. Kuo, N. Takeda and P.K. Singal
Copyright © 2011 Narosa Publishing House Pvt. Ltd., New Delhi, India

34. Beyond Fatness and Fitness in Singaporean Youths

Michael Chia, PhD*

Physical Education and Sports Science, National Institute of Education, Nanyang Technological University, SINGAPORE

Summary

This discourse evokes critical thinking and brings to the forefront the loud refrain that fatness is associated with current and future health ailments among youths, and that adequate amounts of physical activity and fitness help keep diseases in check. There are issues associated with physical activity and fitness that are debatable- for example, has physical fitness among youth declined over time? How physically active are youths of today? To what extent does physical fitness and physical activity track over time? No easy answers to these questions and issues are available as both constructs are influenced by many factors and these factors may be different in different countries and cultures. The merits of free play among young children and adolescents are emphasised, the threats to play and allure of cyber gaming are articulated and results of a pilot programme called PRIDE for PLAY in schools in Singapore are discussed. The importance and implications of play in building interest in sport, recreation and physical education are alluded to.

Keywords: Body mass index, tsunami, obesity, metabolic disorders, cyber gaming

Introduction

Concerns about obesity among young people, even in developing and developed countries such as Taiwan and Singapore in the Far East are legitimate as obesity overwhelms western nations such as the USA like a 'tsunami' of epic proportions. In adults, for instance, sedentary lifestyles and physical activity insufficiency are associated with clusters of diseases that threaten lifespan. These include cardiovascular diseases, musculoskeletal disorders, psychological disorders, pulmonary diseases, cancers and metabolic disorders (Kimm and Obarzanek, 2002). The associations between physical activity insufficiency and adult disease afflictions are generally secure and uncontested. In young people, obesity causes the three Ds- Discomfort, Disease and Despair (Ho, 1985). There is physical discomfort because an excess of body mass means that movement or locomotion (e.g. stair-climbing, jogging, jumping) that involves the carriage of body mass requires greater energy expenditure; social

*Address for Correspondence: Dr. Michael Chia, Head Department of Physical Education and Sports Science National Institute of Education Nanyang Technological University 1 Nanyang Walk SINGAPORE 637616
Email: michael.chia@nie.edu.sg

discomfort because they stand out and emotional discomfort as they are often at the receiving end of insensitive remarks from peers about their body size (Chia and Wang, 2003). There is despair especially during the puberty when body awareness is heightened and many suffer from low physical self-esteem. Indeed, some teenagers consider overweight people lazy, dishonest and would rather date someone with a criminal record rather than an overweight person (Sobal, Nicolopoulos and Lee, 1995). In Singapore, juvenile obesity, explained as being above 120% of the ideal weight for height, among school-going youths aged between seven and 18 years, is about 10 % but given the pervasiveness and entrenchment of sedentary lifestyles among Singaporean youths and the popularity and ease of availability of high energy density fast foods, there are fears that the relatively low obesity rates may not be sustainable in the longer term (Chia, 2009).

Physical Fitness in Youths

As obesity and sedentary lifestyle become more pervasive among young people, some argue that the physical fitness of youths has declined over time. Physical fitness in Singapore is measured using specific test items- timed distance runs (e.g. 1.6 km, 2.4 km, 0.8 km or 1.5 km) (maximal cardiovascular function), 1-minute sit-up (abdominal muscular endurance), pull-up/inclined pull-up (upper body muscular endurance), sit-and-reach (trunk and hamstring flexibility), standing broad jump (lower limb power), and 4x10m shuttle run (sprint agility). Singapore Sports Council cross-sectional data on physical fitness of youths aged between 11 and 18 years (where N>200 000) tested by certified personnel, over a 21 year period (1981-2002), showed little or non-significant change in physical fitness for all the test items, for both male and female subjects (Chia, 2008, 2009). Therefore there is no conclusive evidence that the physical fitness of Singaporean youths has declined over the years, even after 20 years (Chia, 2008, 2009). There is some evidence that there are secular changes in physical fitness over time. Sample data reported by Quek et al (1993), which analysed a cross-sectional sample of 3 263 students aged 12-19 years, showed that over a 12-year period (1980-81 versus 1991-92), standing broad jump, 4x10 m shuttle run improved at a rate of + 0.25% and + 0.41% per year, on the average while 2.4 km run performance declined at an average rate of -0.16 % per year.

Analysis of physical fitness data (N = 6622) provided by the Ministry of Education of Singapore, using latent growth curves, garnered from 138 schools and examined retrospectively over a 4-year-period (aged 13-16 years) revealed that the development of physical fitness with age followed a quadratic trend- that there were declining improvements over time and that male subjects showed higher peaks of physical fitness than female subjects where the improvements in physical fitness with age was less discernable (Chia, unpublished data). Comparatively, equivalent data on 102 765 Taiwanese youths, aged between 9 and 18 years, showed that both body mass and physical fitness changes non-linearly with age; the relationship of fitness and age is parabolic- peaks are sharper for boys than in girls and that physical fitness declines with increasing BMI (Huang and Malina, 2007).

Physical fitness test performances tend to be body-mass dependent (e.g. pull-up, timed distance runs, agility runs, standing broad jump) and will therefore be positively biased toward those with a smaller body mass. This may explain why up to 90% of the 10 schools in terms of physical fitness test performance results over the decades have been all-girl schools in Singapore.

Research truisms inform that the stability of physical fitness in the near-term is modest but is moderate-to-low over the years, as physical fitness cannot be stored and will decay through disuse.

Issues raised about the physical fitness of youths include- not over-testing youths for physical fitness since fitness changes in youths can largely be explained by changes in muscle mass in male and female subjects; how to make physical fitness personally meaningful for youths and what sorts of functional physical fitness do youth require and what are the results of physical fitness of youths used for? These issues remain largely unaddressed and require further research and expert discussion.

Benefits of Regular Physical Activity in Youth

Physical activity is the natural inclination of all young people, regardless of race, language or religion in Singapore. Physical activity is the capstone for the growth and development of toddlers, children, adolescents and young adults. Among adolescent and young adults, physical activity through sports and games provide opportunities to build a sport-loving culture, for social inter-course, for stress-release and for the pursuit of healthy lifestyles (Chia, 2007). Importantly, a routine of daily physical activity, whether incidental or deliberate provide a counterbalance to the sedentary activities of modern living. Moreover, there is modest tracking of physically active behaviours from adolescence to young adulthood. Some longitudinal data show that about 47% of physically active adolescents continued with physically active behaviours as young adults in a group of Finns, tracked over 6 years (Raitakan *et al*, 1994). However, these data referred to a different cultural context from Singapore and local research in this area is advised.

An entrenched cycle of hypercaloric consumption and hypocaloric expenditure over time usually results in energy storage leading to overweight and if uncurbed, to obesity. Emergent pilot data suggest that overweight adolescents have poorer verbal memory recall than normal weight adolescents among the top 10% of academic achievers, based on examination results (Wong *et al*, 2007). In the cited study, in all 8 trials using an objective verbal memory recall of random words using a computer-based protocol (Rey Auditory Verbal Learning Test), overweight boys and girls recalled correctly fewer words than their normal-weight peers. Social-economic status was not a significant predictor of cognitive verbal recall performance among the subjects. These data demonstrate a significant relationship between impaired cognitive verbal memory recall and excessive body weight among youths. Further research is advised with other youths of different academic capabilities.

Physical Activity of Youths

Since there are no gold standards for estimating physical activity in youths, both subjective methods (e.g. self-report questionnaire on physical activity recall) and objective methods (e.g. motion sensors like accelerometers, heart-rate monitors and pedometers) are used. Physical activity guidelines recommended by experts range between at least 60 to 90 minutes daily or for most days of the week (Cavill *et al*, 2002; Andersen *et al*, 2006). These activity bouts can be accumulated and are recommended to be of at least a moderate intensity (HR > 140 beats per minute). In terms of daily accumulated step count, some researchers advocate a minimum of daily accumulated steps of 12000 for girls and 15000 for boys (Tudor-Locke *et al*, 2004), mainly based upon current prevalence information, while others recommend daily accumulated step count based upon BMI and body fat (Duncan Schofield and Duncan, 2007). These are up to 13 000 steps for girls and 16 000 for boys to stay within recommended body composition in the normal range.

Physical activity recall by youths using questionnaire almost always shows that young people have adequate levels of habitual physical activity in Singapore (Wang *et al*, 2006) and in Taiwan (Chinese Dietetics Society, 2002) but these are not affirmed by more objective methods like heart-rate monitoring or motion sensors (Chia, 2008, 2009). For instance, a 7-day physical activity recall by 280 youths aged 10 years and 15 years, of normal body weight wore HR monitors for 10 hours to gauge physical activity consumption over two weekdays and a weekend, revealed that 41 % of primary school participants and 38 % of secondary school participants experienced moderate-to-vigorous PA. HR data, over the weekday and weekend, on the contrary, showed that 86-94 % of primary school participants and 94-99 % of secondary school participants did not experience any moderate-to-vigorous PA (HR 140-159; HR >160 bpm).

Survey data from the Health Promotion Board in Singapore (2002) suggest that adolescents spent at least 15 hours per weekday engaged in sedentary activities such as being at school (6 hrs), watching TV/working on computer (up to 4.5 hrs), doing homework (up to 3 hrs) and transport (2 hrs). These data are affirmed by HR monitored data over 10 hrs where 9 and 15 year old youths spent 86-94 % of the monitored time on a weekday at HR<120 bpm. On the weekend, 96-99 % of the monitored time was spent at HR<120 bpm (Chia, 2008, 2009).

A 12-country survey that involved 12 000 participants aged between 8-24 years, from Japan, Taiwan, South Korea, Hong Kong, China, Malaysia, Indonesia, India, Vietnam and Philippines, among others revealed that Singaporean youth spent an average of 13 hours engaged in multi-media activities like watching television, going on-line, listening to music, emailing and playing cyber games. Indeed Singaporean youths are reported as the busiest in Asia by craming 42 hours worth of activities in 24 hours by multi-tasking. Interestingly, the same survey showed that comparatively more affluent countries had youths rating lower on the 'happiness' meter, with Singaporean youth scoring 73 %, ahead of Hong Kong (65 %) and Japan (55 %). In contrast, youths from India, Philippines and Indonesia scored 97, 89 and 84 % respectively on the same 'happiness' meter (AsiaOne, 2009). The relative popularity of cyber-gaming compared to playing the physical sport was articulated by Chia (2009) in several keynote presentations to various audiences. This is summarised in Table 1. Chia (2009) proposes that in order to purposefully engage youths in the physical nature of sport, innovative teaching approaches using game media like Wii Games and Wii Sports, can be integrated in teaching the physical sport or game.

Table 1. Cyber versus physical sport

Cyber sport	Physical sport
• Anytime, anywhere, whenever- easy access to players and computer terminal	• Limited by time, number of players, location and access to facility
• Not weather dependent- played in comfortable setting	• Weather-dependent
• Not dependent on physical fitness	• Dependent on physical fitness
• Does not depend on a high level of physical skill	• Dependent on physical skill
• Game skill level is not important for game enjoyment	• Game skill level is important for game enjoyment
• Relative ease in skill mastery	• Time-consuming to master skill
• Gratification and enjoyment are immediate	• Enjoyment and gratification are delayed
• Low level of physical exertion	• Great deal of physical exertion

Research truisms for physical activity reveal that physical inactivity behaviours track better than physical activity behaviours (Raitakan *et al*, 1994) and that there are different clusters of youth with distinct characteristics that can explain or predict their physical activity or physical inactivity behaviours (Wang *et al*, 2006). Much more research is needed to better understand the different contexts that youths are physically active or not, in order to appropriately engage young people so that they will find physical activity personally meaningful.

Professing Play

Research informs all that physical play is beneficial to the holistic development of young children and that the benefits of physical play out-weigh the disincentives of physical play such as accidents and physical injuries (Chia, 2007). Play is natural and is a simple joy that is a most cherished part of childhood. Play fosters creativity, promotes imagination, practices dexterity and encourages emotional, cognitive and physical strength. Indeed, play is important for the healthy development of the brain (Tamis-LeMonda *et al.* 2004). Through play, children can explore the world, interact and engage with peers and adults, practising mastery and resiliency and acquiring competencies that will help them to face future uncertainty (Chia, 2008). Play has strong potential for the holistic development of young people, encompassing physical, social, emotional, mental and spiritual dimensions.

Undirected play, allows children to learn to share, to care, to negotiate, to get along in groups, to resolve conflicts and also to learn self-advocacy skills (Hurwitz, 2003). When play is child-directed, children practise decision-making, move at their own, discover their own interests and children ultimately engage in the passions that they wish to pursue. When play is adult-directed, which is common, children acquiesce to adult standards and this sometimes compromises creativity, leadership and group skills (MacDonald, 1993). Therefore adult-based interventions should be tampered to allow for a greater licence for children to freely mingle and interact, rather than to inhibit physical activity without intent.

Play builds active and healthy bodies. Encouraging unstructured play every day of the week which is interspersed throughout the day could increase the habitual physical activity of children. This may help ameliorate childhood obesity or delay it, cause stress alleviation and allow for 'good mood' hormones, which are released into the circulation when physical activity thresholds are achieved (Chia and Wang, 2002). Parents who spend time at play with young children get a glimpse of the world through the 'child's eyes'. Apart from opportunities to offer gentle and nurturing guidance, the interactions also promote enduring relationships between the child and parent (Smith, 1995). Interestingly, play is also used to foster academic and social-emotional learning. Play helps children to adjust to the school environment, heightens children's readiness to learn, enhances positive learning behaviours and fine-tunes problem-solving skills (Fisher, 1992).

No other medium apart from physical play allows for so much gain for so little investment. Persuasive anecdotal evidence garnered from the parents of pre-school children who make their transition into primary school and from principals of primary schools, shows that young pupils miss the daily opportunities for physical play in their pre-schooling years that become limited and more restrictive once they enter primary school.

Threats to Play

Young people face many threats to play and often they lose the licence for play because of environmental, social and regulatory prohibitions in and outside of school. In the push for academic excellence in Singapore and Taiwan, it is easy for free play and even physical education classes to be cancelled and used for academic work, and enrichment classes in academics when examinations loom. Interestingly, when PE classes are cancelled, these are often not compensated at some future time while 'missed' academic classes are always 'made-up' for. Survey data on Taiwanese children and adolescents (Chinese Dietetics Society, 2002), revealed that they were not physically active after school because they had no time (49%), they had no friends to play with (21%) and they had no place for play (19%). Survey data on Singaporean youths already show high levels of physical inactivity and increasingly, physical play is being substituted by play in cyberspace where the allure of cyber sports and passive entertainment (e.g. video-watching and listening to music) may be more attractive than the allure of physical sports among young people (National Youth Survey, 2002).

Researchers, education specialists and physical activity protagonists should re-energise their efforts in exploring intervention programmes that facilitate increased daily physical activity engagement through play, either through expanded physical education sessions, in terms of duration and frequency, or co-curricular sport and game activities conducted during and after-school hours and other events that 'regularise' physical activity.

Pride for Play

PRIDE for PLAY is an acronym for Personal Responsibility in Daily Effort for Participation in Lifelong Activity for Youths (Chia, 2006). MOE advocates that schools should emphasize holistic development of pupils- this means paying sufficient and appropriate attention to the mental, physical, emotional, social and environmental development of all pupils and students. Weekly, compulsory physical education classes of 70 minutes in schools and the associated sports activities after classes are inadequate to meet emergent physical activity guidelines of a daily accumulation of at least 90 minutes of physical activity of at least moderate intensity (Andersen *et al.* 2006). Daily play sessions that are exclusive of an active daily recess, physical education classes taught by trained specialists and after-school sport sessions, can provide many developmental and holistic health benefits that may carry over into adulthood. A school environment that is play-encouraging, play-enabling and play-inviting can provide a useful, innovative and natural way of inculcating a love for movement and help redress a serious trend of physical activity insufficiency while youngsters engage in electronic gaming activities.

Pilot initiatives in two primary schools for the PRIDE for PLAY programme were conducted (Chia *et al.* 2007). This involved working closely with the school leaders and all teachers involved to forge a strong partnership for action and subsequent evaluation. The pilot study involved an infusion of daily physical play of between 20 to 45 minutes during curriculum hours, either as stand-alone additional play and physical activity session or as part of an extended recess.

Outcome deliverables like pre-and post school-based physical activity, assessed using pedometers, social-emotional learning (SEL) outcomes among pupils, assessed using age-modified questionnaires and indicators of academic achievement were garnered. SEL components measured included:- self-awareness (recognizing emotions, accurate self-perception, recognizing needs, strengths & values & self-efficacy); social awareness (empathy, respect for others, perspective taking & appreciating diversity); self-management (stress management, self-motivation, discipline, goal-setting, organizational skills); relationship management (communication & social engagement, cooperation, negotiation, conflict management & seeking help) and responsible decision-making (problem-solving, personal, social & ethnic responsibility).

For one pilot school that involved 270 Primary 2 pupils, daily step count in school was significantly increased by 24% from 3742 to 4642. In the other pilot school that involved 225 Primary 2 pupils, daily step count in school increased by 10% from 4520 to 4984 (Chia, 2009). These amounted to achieving 39-42% of the daily recommended step count for girls and 31-33% of the recommended step count for boys (Tudor-Locke et al, 2004).

Social and emotional learning outcomes of the pupils involved with the PRIDE for PLAY programme were also significantly improved with no compromise to academic standards in both schools. Normal-weight pupils demonstrated greater gains than overweight pupils but there were no sex differences in the SEL outcomes. The principals of both schools raised the following points that are instructive for the success of PRIDE for PLAY- the programme had the support and buy-in from the key stakeholders- school management, teachers, parents and the pupils themselves; there was great enthusiasm among pupils for the programme as they were free to be themselves, and had the freedom to decide how they wanted to play and there was increased teacher-pupil understanding; reduced incidences of pupil indiscipline, no increase in the number of physical accidents and no drop in academic achievement among the pupils.

School leadership was instrumental in the initial success of PRIDE for PLAY and will play an even more important part in the continued success and sustainability of the play programme. According to Leithwood *et al* (2008), the success of health-promoting schools pivot on the following factors which were omnipresent in PRIDE for PLAY- shared vision, clear directions, beliefs and values, well-developed staff; well-organised school, structure and procedures and very good management of the teaching and learning programmes.

Other schools have translated, customised and adapted the PRDIE for PLAY programme into daily physical education. Further research directions associated with PRIDE for PLAY include garnering more secure evidence of the social-emotional learning outcomes as a result of programme implementation in a wider spectrum of schools in Singapore. As Singapore endeavours to build a Sporting Culture in the population, it is commonsensical to foster and inculcate a play culture among her youths. After all, sport without play is playing sport devoid of fun and enjoyment.

References

1. Anderson LB, Harro M, Sardinha L, *et al.* Physical activity and clustered cardiovascular risk in children: a cross-sectional study (The European Youth Heart Study). *The Lancet* 2006, 368:299-304.
2. AsiaOne. Singaporean youth busiest in Asia *http://news.asiaone.com/News/Education/Story/A1Story20090325-130910.html.* accessed 1 August 2009.
3. Cavill N, Biddle S, Sallis JF. Health enhancing physical activity for young people: statement of the United Kingdom Expert Consensus Conference. *Pediatric Exercise Science*, 2001, 13:12-25.

4. Chia M, Wang J. (2003). Fat, unfit and dissatisfied: relationships among perceived physical self-worth, body weight satisfaction and physical fitness among primary school children. *ACHPER Healthy Lifestyles Journal,* 2003, 50:14-19.
5. Chia M, Wang J. Physical activity, exercise & health of the mind in young people. *Journal of Physical Education & Recreation*, 2002, 8: 64-67.
6. Chia M. *Physical activity and fitness among youths*. Keynote address to Kokushikan University. Japan. 11 June 2009.
7. Chia M. Physical inactivity among children and adolescents in Singapore- a paradoxical issue. *Acta Kinesiologica*, 2008, 2: 7-15
8. Chia M. Pitfalls of physical inactivity and innovative intervention programmes to reduce physical inactivity and increase physical activity among Singaporean youths. *Sport Science,* 2009, 2, 1: 105-110
9. Chia M. PRIDE for PLAY. Keynote address to1st Singapore Heart Foundation-National Institute of Education National Seminar *Hearty Children-Sturdy Future-managing obesity in schools.* Suntec City, Singapore. 21 January 2006.
10. Chia M. PRIDE for PLAY: personal responsibility in daily effort for participation in lifelong activity for youths: a Singaporean context. *Journal of Sports Science and Medicine,* 2007, 6:374-379.
11. Chia, M. *The young and the restless: how active and fit are they?* International workshop at Chiayi, Taiwan. 8-9 December 2008.
12. Chinese Dietetics Society Survey 2002. Asian Food Information Centre *http://www.afic.org/* accessed 1 August 2009.
13. Duncan JS, Schofield G, Duncan EK, *et al.* Step count recommendations for children based on body fat. *Preventive Medicine,* 2007, 44, 42-44.
14. Fisher EP. The impact of play on development: a meta-analysis. Play Culture, 1992, 5: 159. 15. Health Promotion Board Youth Survey 2002. Health Promotion Board. *http://www.hpb.gov.sg/* accessed 1 August 2009.
15. Ho TF. Eleventh Haridas Memorial Lecture. Childhood obesity in Singapore primary school children: epidemiological review and anthropometric evaluation. *Journal of Singapore Paediatric Society,* 1985, 27: 5–40.
16. Huang Y, Malina RM. BMI and Health-Related Physical Fitness in Taiwanese Youth 9-18 Years. *Medicine and Science in Sports and Exercise,* 2007, 39: 701-708
17. Hurwitz SC. To be successful-let them play. *Child Education*, 2003, 79: 101-102.
18. Kimm SYS, Obarzanek E. Childhood obesity: a new pandemic of the new millennium. *Pediatrics,* 2002, 110:1003-1007.
19. MacDonald KB. *Parent-Child Play: Descriptions and Implications*. Albany, NY: State University Press, 1993.
20. National Youth Survey. Health Promotion Board, Ministry of Education, Singapore, 2002.
21. Raitakan OT, Kimmo VK, Porkka ST, *et al.* Effects of persistent physical activity and inactivity on coronary risk factors in children and young adults: The cardiovascular risk in young Finns study. *American Journal of Epidemiology, 1994, 140* (3), 195-205.

22. Smith D. How play influences children's development at home and in school. *Journal of Physical Education and Dance*, 1995, 66: 19-23.
23. Sobal J, Nicolopoulos V, Lee J. Attitudes about overweight and dating among secondary school students. *International Journal of Obesity, 1995,* 19: 376-381.
24. Tamis-LeMonda CS, Shannon JD, Cabrera NJ *et al.* Fathers and mothers at play with their 2- and 3-year-olds: contributions to language and cognitive development. *Child Development*, 2004, 75: 1806-1820.
25. Tudor-Locke C. Pangrazi RP, Corbin CB, *et al.* BMI-referenced standards for recommended pedometer-determined steps/day in children. *Preventive Medicine*, 2004, 38: 857–864.
26. Wang J, Chia M, Quek JJ, *et al.* Patterns of physical activity, sedentary behaviours and psychological determinants of physical activity among Singaporean school children. *Journal of Sport and Exercise Psychology*, 2006, 4: 227-249.
27. Wong P, Boh G, Wang J, Chia M. Impact of obesity on verbal memory performance among top academic achievers in Singapore. *Asian Journal of Exercise and Sports Science*, 2007, *4*, 47-55.

Index